D0849003

Modern Applied Algebra

Modern Applied Algebra

Garrett Birkhoff

Department of Mathematics
Harvard University

and

Thomas C. Bartee

Division of Engineering
and Applied Physics
Harvard University

McGraw-Hill Book Company

New York, St. Louis, San Francisco, Düsseldorf,
London, Mexico, Panama, Sydney, Toronto

This book was set in Modern by The Maple Press Company, and printed on permanent paper and bound by The Maple Press Company. The designer was Marsha Cohen; the drawings were done by Joseph Buchner. The editors were Donald K. Prentiss and Maureen McMahon. Sally R. Ellyson supervised the production.

Modern Applied Algebra

Library of Congress Catalog Card Number 70-88879

05381

1234567890 MAMM 79876543210

Preface

The name "modern algebra" refers to the study of algebraic systems (groups, rings, Boolean algebras, etc.) whose elements are typically *non-numerical*. By contrast, "classical" algebra is basically concerned with algebraic equations or systems of equations whose symbols stand for real or complex numbers. Over the past 40 years, "modern" algebra has been steadily replacing "classical" algebra in American college curricula.

The past 20 years have seen an enormous expansion in several new areas of technology. These new areas include digital computing, data communication, and radar and sonar systems. Work in each of these areas relies heavily on modern algebra. This fact has made the study of modern algebra important to applied mathematicians, engineers, and scientists who use digital computers or who work in the other areas of technology mentioned above.

This book attempts to present those ideas and techniques of modern algebra which have proved most useful for these areas. Although some separation is inevitable between the discussion of mathematical principles and their applications, we have tried to weave together the underlying ideas and the technical algorithms which are based on them.

The book begins by presenting and illustrating the notions of set, function, mathematical induction, binary relation, and graph. Discussions are also included of Boolean algebras, monoids, morphisms, and other basic algebraic concepts. This material constitutes two chapters.

In the next chapter we introduce the concepts of finite-state machine and Turing machine, along with representation schemes for these types of machines, and systematic techniques for reducing the number of states in finite-state machines. Chapter 4 provides an introduction to the digital computer programming language called ALGOL and to its syntax.

Chapter 5 introduces the axiomatic approach which is so characteristic of modern algebra. Boolean algebras are defined formally by appropriate postulates, and the properties of Boolean are deduced from these postulates. Boolean algebra is then related to logic, gating networks, and ALGOL programming, after which the canonical form for Boolean polynomials is derived. Chapter 6 centers around the concept of *optimization,* beginning with schemes for finding paths of least cost through networks. Then techniques are presented for describing the gating networks used in digital computers, and for simplifying them (and thereby reducing their cost) using Boolean polynomials. Finally, a method is described for realizing an arbitrary finite-state machine by means of gates and flip-flop (memory) elements.

Chapter 7 presents an axiomatic treatment of monoids and groups. It covers much more about monoids and somewhat less about groups than do most books on modern algebra. Chapter 8 applies some of these ideas to data communications systems in which noise may lead to errors in the messages transmitted. Using the standard "binary symmetric model" of communications theory to describe the probability of error, this chapter describes techniques for generating, coding, and decoding group codes so as to optimize their efficiency in detecting and correcting errors. This is followed by a chapter on lattices, which shows how far-reaching generalizations of Boolean algebra can be derived from the study of partial ordering relations.

Chapter 10 deals with rings and fields. This chapter emphasizes the various kinds of rings which can arise in applications, relates morphisms to ideals, and discusses unique factorization and Gaussian elimination. Chapter 11 studies polynomial rings and applies polynomials to the construction and analysis of error-correcting and error-detecting codes.

Finite or Galois fields are studied in Chapter 12 and are used there to derive a special class of codes called Bose-Chaudhuri-Hoquenghem codes. Chapter 13 then introduces difference equations and a particular class of codes which are based on difference equations and are used with autocorrelation measurements in communications and radar systems.

Chapter 14, the final chapter, then passes from the finite systems which have been studied up to this point to infinite systems. After it is shown that the real numbers are uncountable, the concept of *computability* is introduced and related to the concept of a Turing machine. Finally, the notion of machine computability is related to ideas of mathematical linguistics and to the classification of programming languages.

A number of different courses can be developed from the material in this book, which contains many starred sections which can be omitted if time requires. At Harvard, the material is presented in two half-year courses: Applied Mathematics 106, primarily an advanced undergraduate course, and Applied Mathematics 206, a graduate course. No specific prerequisites are listed for Applied Mathematics 106. Because of the basic nature of the material covered, these courses are prerequisites for a number of other courses. Applied Mathematics 206 also includes a short introduction to real and complex matrices, with emphasis on special properties of matrices arising in various applications; about one-third of the term is devoted to this material.

The authors owe a great deal to John Lipson, who carefully proofread two drafts of the book in manuscript form, who cooperated in writing our exposition on ALGOL, and who wrote our ALGOL programs. They also owe much to Aspi Wadia, who also proofread the manuscript, and wrote out carefully worded solutions for many of the exercises. Finally, they were helped by criticisms and suggestions from many friends and colleagues, including especially Donald Anderson, Marshall Hall, Sesumo Kuno, Donald MacLaren, Albert Meyer, Werner Rheinboldt, Hartley Rogers, David Schneider, and Hao Wang.

Cambridge, Massachusetts *Garrett Birkhoff*
April 11, 1969 *Thomas C. Bartee*

Contents

Preface v

Chapter 1: Sets and Functions 1

1-1. Sets and subsets 1
1-2. Boolean algebra 3
1-3. Functions 8
1-4. Inverses 12
1-5. Functions on S to S 13
1-6. Sums, products, and powers 16
1-7. Peano axioms 20
1-8. Finite induction 22
★*1-9. Pigeonhole principle; division algorithm* 26

Chapter 2: Binary Relations and Graphs 31

2-1. Introduction 31
2-2. Relation matrices 33
2-3. Algebra of relations 35
2-4. Partial orderings 37
2-5. Equivalence relations and partitions 42
2-6. Modular numbers; morphisms 45
2-7. Cyclic unary algebras 48
2-8. Directed graphs 52
2-9. Graphs 55
★*2-10. Directed graphs, II* 58

Chapter 3: Finite State Machines 63

3-1. Introduction 63
3-2. Binary devices and states 65
3-3. Finite-state machines 66
3-4. Covering and equivalence 71
3-5. Equivalent states 75
3-6. A minimization procedure 78
★*3-7. Turing machines* 83
3-8. Incompletely specified machines 87
★*3-9. Relations between states—a minimization procedure* 90

Chapter 4: Programming Languages **99**

4-1. Introduction 99
4-2. Arithmetic expressions 103
4-3. Identifiers: assignment statements 106
4-4. Arrays 109
4-5. **For** *statements* 112
4-6. Block structures in ALGOL 115
4-7. The ALGOL grammar 117
★*4-8. Evaluating arithmetic statements* 122
★*4-9. Compiling arithmetic expressions* 124

Chapter 5: Boolean Algebras **129**

5-1. Introduction 129
5-2. Order 133
5-3. Boolean polynomials 136
5-4. Block diagrams for gating networks 141
5-5. Connections with logic 144
5-6. Logical capabilities of ALGOL 146
5-7. Boolean applications 150
5-8. Boolean subalgebras 153
5-9. Disjunctive normal form 155
★*5-10. Direct products; morphisms* 158

Chapter 6: Optimization and Computer Design **161**

6-1. Introduction 161
6-2. Optimization 162
6-3. Computerizing optimization 166
6-4. Logic design 171
6-5. NAND gates and NOR gates 175
6-6. The minimization problem 179
6-7. Procedure for deriving prime implicants 182
★*6-8. Consensus taking* 187
6-9. Flip-flops 189
6-10. Sequential machine design 191

Chapter 7: Monoids and Groups **197**

7-1. Binary algebras 197
7-2. Cyclic monoids; submonoids 200
7-3. Groups 203
7-4. Morphisms; direct products 206
7-5. Examples of groups; postulates 210
7-6. Subgroups 214
7-7. Abelian groups 217
7-8. Groups acting on sets 219

7-9.	*Permutations*	221
7-10.	*Lagrange's theorem*	224
7-11.	*Normal subgroups*	226

Chapter 8: Binary Group Codes **231**

8-1.	*Introduction*	231
8-2.	*Encoding and decoding*	234
8-3.	*Block codes*	238
8-4.	*Matrix encoding techniques*	242
8-5.	*Group codes*	244
8-6.	*Decoding tables*	246
8-7.	*Hamming codes*	252

Chapter 9: Lattices **257**

9-1.	*Lattices and posets*	257
9-2.	*Lattices as posets*	259
9-3.	*Lattices and semilattices*	263
9-4.	*Sublattices; direct products*	265
9-5.	*Distributive lattices*	267
9-6.	*Modular and geometric lattices*	270
★9-7.	*Boolean lattices*	274
★9-8.	*Morphisms and ideals*	275
★9-9.	*Finite Boolean algebras*	276

Chapter 10: Rings and Ideals **281**

10-1.	*Introduction*	281
10-2.	*Integral domains and fields*	284
★10-3.	*Fields of quotients*	288
10-4.	*Subrings*	290
10-5.	*Morphisms of rings*	293
10-6.	*Direct sums*	295
10-7.	*Ideals and quotient rings*	298
10-8.	*Divisibility*	302
10-9.	*Euclidean domains*	303
★10-10.	*Unique factorization theorem*	307
★10-11.	*Prime and maximal ideals*	309
★10-12.	*Gaussian elimination*	310

Chapter 11: Polynomial Rings and Polynomial Codes **315**

11-1.	*The ring* $R[x]$	315
11-2.	*Polynomial rings over fields*	319
11-3.	*Polynomial codes*	321
11-4.	*Advantageous properties*	324
11-5.	*Shift registers*	327

Sets and Functions

1-1. SETS AND SUBSETS

The central concept of modern algebra is that of an algebraic system A. Such a system consists of a *set* of *elements* $a_1,a_2,a_3, \ldots$ combined by *operations* such as addition or multiplication.

The concept of set is used so frequently in everyday speech that it has many familiar synonyms (class, collection, flock, and herd, to name a few), some with special connotations. This concept is especially fundamental to mathematics. Some important mathematical sets are the set **Z** of all integers (positive, negative, or zero), the set **Q** of all rational numbers, the set **R** of all real numbers, and the set **C** of all complex numbers.[1] Each of these familiar sets has an infinite number of elements; thus, each is infinite. Moreover, each constitutes an important algebraic system under addition and multiplication.

Nonmathematical sets are usually finite. Any finite set can be de-

[1] The symbols **Z**, **Q**, **R**, and **C** will be used throughout this book for these sets.

scribed by listing all its elements. The resulting list is often enclosed in braces $\{\ \ \}$ (curly brackets); thus the set of all primes between 20 and 40 can be listed as $\{23,29,31,37\}$, and the set of all positive integral divisors of 8 is $\{1,2,4,8\}$. Rearranging the elements of a list does not change the set of its members; thus the set $\{1,2,4,8\}$ is the same as the set $\{8,4,2,1\}$.

Some infinite sets (technically, "countable" sets) can be described in much the same way by indicating how their elements can be enumerated in an infinite sequence. For example, the set $\mathbf{P}$ of all positive integers can be suggestively indicated by writing

$$\mathbf{P} = \{1,2,3, \ldots\}$$

Likewise, the set $\mathbf{N}$ of all nonnegative integers and the set $\mathbf{Z}$ of all integers can be described as follows:

$$\mathbf{N} = \{0,1,2,3, \ldots\}$$
$$\mathbf{Z} = \{0, \pm 1, \pm 2, \pm 3, \ldots\}$$

The general principle is to list enough terms in the sequence so that it is evident how to continue it indefinitely by "recursion" (see Chap. 14 for further details). However, the elements of $\mathbf{R}$ and $\mathbf{C}$ cannot be listed in such an infinite sequence.

To define a set S, we must explain how to answer the following question: Given an element a, does it belong to S or not? When a does belong to S, we say that "a is a member of S," and we write $a \in S$ to symbolize this fact. When it does not belong to S, we write $a \notin S$.

By a *subset* S of the set T we mean a set whose elements all belong to T, that is, a set such that $a \in S$ implies $a \in T$. This relation between S and T is symbolized by writing $S \subset T$ (or $T \supset S$); we also say that "S is *contained* (or included) in T." The inclusion relation $\subset$ so defined has a number of obvious properties, such as

$$S \subset S \qquad \text{for any set } S \tag{1}$$

and

$$S \subset T \qquad \text{and} \qquad T \subset U \text{ imply } S \subset U \tag{2}$$

Property (1) is called the *reflexive law* of inclusion, and (2) is called the *transitive law* of inclusion.

A set is completely determined by its elements. In other words, two sets S and T are equal when they have exactly the same elements; in symbols we have

$$S = T \text{ means that } x \in S \qquad \text{if and only if } x \in T$$

Since $S \subset T$ holds if and only if $x \in S$ implies $x \in T$, we have

$$S = T \qquad \text{if and only if } S \subset T \text{ and } T \subset S \tag{3}$$

Sets and subsets can be specified in many ways. Thus, a subset S of a subset U is often defined as the set of all elements $x \in U$ having some specified property. If this property (some statement about x) is signified by $P(x)$, the definition of S can be written in the symbolic form

$$S = \{x \in U \mid P(x)\} \qquad \text{or (simply)} \qquad S = \{x \mid P(x)\}$$

This symbolic formula is read "S is the set of all members x of U such that $P(x)$ holds."

For example, the formulas

$$E = \{x \mid x \in \mathbf{Z} \text{ and } x = 2y \text{ for some } y \in \mathbf{Z}\}$$
$$\mathbf{P} = \{x \mid x \in \mathbf{Z} \text{ and } x > 0\}$$

describe the set of all even integers and the set of all positive integers, respectively. Of course, we can also write

$$E = \{0, \pm 2, \pm 4, \ldots\}$$

The axiom that each meaningful property $P(x)$ determines a set S_P of elements is sometimes called the *Axiom of Specification* and is a basic axiom of logic. Conversely, to any set S there corresponds the property $x \in S$ of membership in S, or "belonging to S."

The set of all subsets of a given set U is called the *power set* of U and is written $\mathcal{P}(U)$. This power set contains U itself, the void set $\varnothing$, and (if U has $n > 1$ elements) $2^n - 2$ "proper" subsets S, satisfying $\varnothing \subset S \subset U$, $S \neq \varnothing$, and $S \neq U$.

For instance, the power set $\mathcal{P}(\{a,b,c\})$ of the set $U = \{a,b,c\}$ has the six proper subsets $\{a\}$, $\{b\}$, $\{c\}$, $\{a,b\}$, $\{a,c\}$, and $\{b,c\}$. In general, if the set U has n distinct elements $a_1, \ldots, a_n$ (n a positive integer), then U has 2^n different subsets S in all—including the empty (void, or null) set $\varnothing$ which has no members at all and the set U itself. The reason for this is that for each element p there are two possibilities: $p \in S$ or $p \notin S$.

1-2. BOOLEAN ALGEBRA

In mathematics generally, a *binary operation* on a set S is a rule which assigns to each *ordered pair* (a,b) of elements of S some "value" in S. The operation $+$ of arithmetic is such a binary operation on the set $\mathbf{Z}$ of all integers; the value of the sum of a and b is commonly written as $a + b$, placing the operation symbol $+$ between the two operands. Subtraction and multiplication are other familiar binary operations on $\mathbf{Z}$, since for any two integers $m,n \in \mathbf{Z}$, the values of $m - n$ and $m \times n$ are well-determined elements of $\mathbf{Z}$.

The *binary* operation $-$, as used in $4 - 2$ or $3 - 1$, is logically distinct from the *unary* operation $-$, as used in -1, and -2, because the first operation involves two operands and the second operation involves only a single operand. In general, a unary operation on a set S is a rule f which assigns to each element $a \in S$ some well-determined value $f(a) \in S$, the result of performing this operation on a. (Thus, a unary operation on S is just a function $f: S \to S$ in the usual sense; see Sec. 1-3.)

There are two basic binary operations and one unary operation on the subsets of a given set U. These are the Boolean or set-theoretic operations of intersection, union, and complement. The binary operations of intersection and union in $\mathcal{P}(U)$, the power set of all subsets of U, are analogous to the operations of multiplication and addition in $\mathbf{Z}$, as we shall see.

If R and S are subsets of U, their *intersection* $R \cap S$ is the set of all those elements common to both R and S; thus

$$R \cap S = \{x \mid x \in R \text{ and } x \in S\}$$

whereas their *union* $R \cup S$ is the set of all those elements (of U) which belong either to R or to S (or to both); thus

$$R \cup S = \{x \mid x \in R \text{ or } X \in S\}$$

For example, if $E = \{0, \pm 2, \pm 4, \ldots\}$ is the set of all even integers and $\mathbf{P}$ is the set of all positive integers $1,2,3, \ldots$, then $E \cap \mathbf{P}$ is the set of all positive even integers ($E \cap \mathbf{P} = 2,4,6, \ldots$), and $E \cup \mathbf{P}$ is the set of all integers which are not both negative and odd (i.e., not $-1,-3,-5, \ldots$).

The Venn diagrams of Fig. 1-1 illustrate pictorially the meaning of the operations $\cap$ and $\cup$. In both diagrams, U is the rectangular area, and A and B are disks. In Fig. 1-1*b*, C is also a disk, and the shaded area is $(A \cap C) \cup (B \cap C) = (A \cup B) \cap C$.

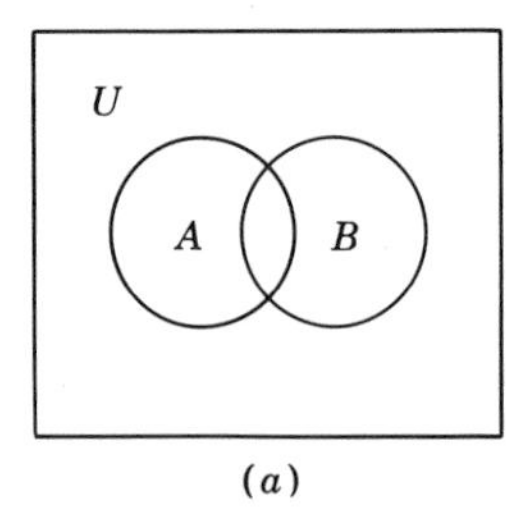

(*a*)

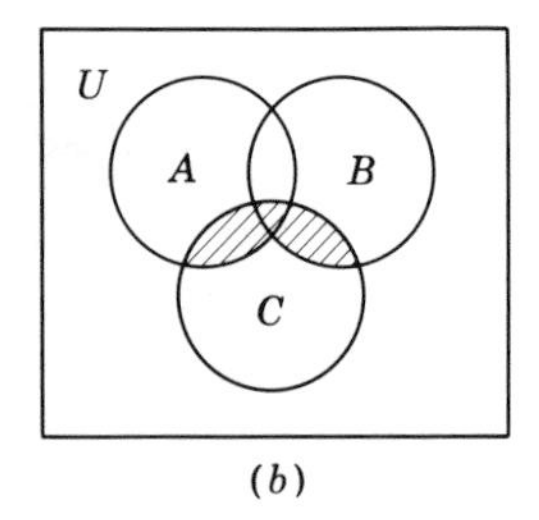

(*b*)

Fig. 1-1. Venn diagrams.

Let $n(S)$ denote the number of elements in S. If R and S are finite sets without a common element (so that $R \cap S = \emptyset$), then the number of elements in $R \cup S$ is the arithmetic sum $n(R \cup S) = n(R) + n(S)$ of the number of elements in R and S, respectively. More generally, we have the equation

$$n(R \cup S) = n(R) + n(S) - n(R \cap S) \qquad \text{for any finite sets } R,S \quad (4)$$

This is true because $n(R \cap S)$ is the number of elements (of U) which are "counted twice" (hence once too often) when we enumerate first R and then S.

As a further example, if $M = \{mk\}$ and $N = \{nk\}$ denote the sets of all integral multiples of fixed integers $m,n \in \mathbf{Z}$ by variable integers $k \in \mathbf{Z}$, then $M \cap N$ is the set of *common multiples* of m and n. [There is an important theorem (due to Euclid) that $M \cap N = \{[m,n]k\}$ is the set of all integral multiples $[m,n]k$ of the least common multiple $[m,n]$ of m and n; see Chap. 10.]

As a final example, if $\mathbf{Q}^+$ denotes the set of all positive rational numbers and $\mathbf{Q}^-$ denotes the set of all negative rational numbers, then $\mathbf{Q}^+ \cup \mathbf{Q}^-$ is the set of all nonzero rational numbers.

When $R \cap S = \emptyset$, R and S are said to be *disjoint*. A division of a set U into a family of disjoint subsets is called a *partition* of U.

For any subset S of U, the set of all elements $x \in U$ which are not in S is called the *complement* of S (in U) and is usually denoted S'; that is, in symbols, $S' = \{x \in U \mid x \notin S\}$. Clearly, if S' is the complement of S, then S is the complement of S'. Two sets S and T are called *complementary* when $T = S'$ or, equivalently, $S = T'$. The complement S' of S (in U) is also characterized by the Boolean equations

$$S \cap S' = \emptyset \qquad \text{and} \qquad S \cup S' = U \qquad (5)$$

Thus the two sets $\{S,S'\}$ define a partition of U into two disjoint subsets which together exhaust U. Thus, in Fig. 1-1*a*, the unshaded area is $(A \cup B)' = A' \cap B'$; in Fig. 1-1*b*, it is

$$[(A \cap C) \cup (B \cap C)]' = (A' \cap B' \cap C) \cup C'$$

The three Boolean operations $\cap$, $\cup$, and $'$ on the subsets of any fixed set U have a number of basic algebraic properties. Some of these are as follows:

L1. $S \cap S = S \qquad S \cup S = S \qquad$ Idempotent laws
L2. $S \cap T = T \cap S \qquad S \cup T = T \cup S \qquad$ Commutative laws
L3. $R \cap (S \cap T) = (R \cap S) \cap T \qquad R \cup (S \cup T) = (R \cup S) \cup T$
Associative laws

L4. $S \cap (S \cup T) = S \cup (S \cap T) = S$ Absorption laws
L5. If $R \subset T$, then $R \cup (S \cap T) = (R \cup S) \cap T$ Modular law
L6. $R \cap (S \cup T) = (R \cap S) \cup (R \cap T)$
$R \cup (S \cap T) = (R \cup S) \cap (R \cup T)$ Distributive laws
L7. $R \cap \varnothing = \varnothing$ $R \cup \varnothing = R$
$R \cap U = R$ $R \cup U = U$ Universal bounds
L8. $R \cap R' = \varnothing$ $R \cup R' = U$ Complementarity
L9. $(S')' = S$ Involution law
L10. $(S \cap T)' = S' \cup T'$ and $(S \cup T)' = S' \cap T'$
de Morgan's laws

For example, let us prove the first distributive law. If $x \in R \cap (S \cup T)$, then $x \in R$, and either $x \in S$ or $x \in T$. In the first case, $x \in R$ and $x \in S$; hence $x \in R \cap S$. Similarly, in the second, $x \in R \cap T$. Hence in both cases, $x \in R \cap S$, or $x \in R \cap T$, so that $x \in (R \cap S) \cup (R \cap T)$. This proves that

$$R \cap (S \cup T) \subset (R \cap S) \cup (R \cap T)$$

Conversely, let $x \in (R \cap S) \cup (R \cap T)$; that is, suppose that $x \in R \cap S$ or $x \in R \cap T$. In either case, $x \in R$, and also in either case, $x \in S$ or $x \in T$; that is, $x \in R$ and $x \in S \cup T$ in any case, whence $x \in R \cap (S \cup T)$. In summary, we have proved that

$$(R \cap S) \cup (R \cap T) \subset R \cap (S \cup T)$$

Combining this result with the final inclusion relation of the preceding paragraph and using Eq. (3), we conclude the first of the two distributive laws L6.

Just as the integers form an *algebraic system* $[\mathbf{Z},+,\times]$ under the two binary arithmetic operations of addition and multiplication, so the power set $\mathcal{P}(U)$ of any set U is an algebraic system $[\mathcal{P}(U),\cap,\cup,']$ under the three set-theoretic operations defined above. To describe the properties of addition and multiplication in $\mathbf{Z}$, the system $[\mathbf{Z},+,\times]$ is usually called a *commutative ring* (see Chap. 10). Likewise, to describe the properties L1 to L10 enumerated above, the algebraic system $[\mathcal{P}(U),\cap,\cup,']$ is called a *Boolean algebra*, the Boolean algebra of all subsets of U. Hence, granted the truth of L1 to L10, we have by definition the following theorem:

Theorem 1. *The power set $\mathcal{P}(U)$ of any set U is a Boolean algebra.*

We shall return to the proofs of L1 to L10 and to their significance in Chap. 5 (and again in Chap. 9). In the meantime, Theorem 1 can be taken simply as a handy summary (with L1 to L10) of some basic properties of the algebra of sets, to be related to simple examples such as the following.

Example 1. Let $U = \{a\}$ be a one-element set. If we write 1 for U and 0 for $\varnothing$, we get the simplest nontrivial Boolean algebra, that is, $[\mathcal{P}(\{a\}),\cap,\cup,']$. Considered as a Boolean algebra, it is defined by the following operation tables:

$\cap$	0	1
0	0	0
1	0	1

$\cup$	0	1
0	0	1
1	1	1

$'$	
0	1
1	0

In such an operation table for a binary operation, the entry in the row beginning with any element a and in the column headed b gives the result of performing the indicated operation on the ordered pair (a,b). Thus, in the tables above, $0 \cup 0 = 0$, $0 \cup 1 = 1$, $1 \cap 0 = 0$, etc.

The next simplest nontrivial Boolean algebra is just an alternative description of $[\mathcal{P}(\{a,b\}),\cap,\cup,']$. In this case, it is convenient to denote $\varnothing$ by 0, $\{a\}$ by S, $\{b\}$ by S' (since it is the complement of $\{a\}$ in $\{a,b\}$), and $\{a,b\}$ by I. In this notation, we get the following Boolean algebra.

Example 2. The four-element Boolean algebra with elements O, S, S', and I is defined by the following operation tables:

$\cap$	O	S	S'	I
O	O	O	O	O
S	O	S	O	S
S'	O	O	S'	S'
I	O	S	S'	I

$\cup$	O	S	S'	I
O	O	S	S'	I
S	S	S	I	I
S'	S'	I	S'	I
I	I	I	I	I

X	X'
O	I
S	S'
S'	S
I	O

EXERCISES A

1. Find a necessary and sufficient condition such that $S + T = S \cup T$, where $S + T = (S \cap T') \cup (S' \cup T)$.
2. (a) Prove that $S \subset S \cap (S \cup T)$ and that $S \supset S \cap (S \cup T)$.
   (b) Prove that $S = S \cup (S \cap T)$.
3. Show that $S \subset T$ if and only if $S' \cup T = U$.
4. (a) Show that each of the three conditions $S \subset T$, $S \cap T = S$, and $S \cup T = T$ on subsets of a given set U implies both of the other conditions (consistency).
   (b) Show that in any Boolean algebra, $S \cap T = S$ holds if and only if $S \cup T = T$ holds.
5. Prove, directly from the universal bounds and consistency rules, that $S \cup U = U$, $S \cap \varnothing = \varnothing$, $S \cap U = S$, $S \cup \varnothing = S$.

6. Check the following against a Venn diagram:
   (a) The distributive laws of L6.
   (b) $(R \cap S) \cup (S \cap T) \cup (T \cap R) = (R \cup S) \cap (S \cup T) \cap (T \cup R)$.
   (c) In (b), show that both regions contain precisely those points which are in a majority (at least two) of the sets R, S, and T.
7. Show that in the Venn diagram of Fig. 1-1 there are exactly $256 = 2^8 = 2^{2^3}$ Boolean combinations of the three disks A, B, and C.

IN EXERCISES 8 AND 9, ASSUME ONLY IDENTITIES L1 TO L10.

8. (a) For $R + S = (R \cap S') \cup (R' \cap S)$, prove the identities $R + S = S + R$, $R + (S + T) = (R + S) + T$, and $R + R = \varnothing$.
   (b) For $S - T = S \cap T'$, prove that $S + T = (S \cup T) - (S \cap T)$.
9. From the definition of a Boolean algebra, prove the following:
   (a) $S + S = \varnothing$, $S + I = S'$, $S + \varnothing = S$, $S + S' = I$
   (b) $S + T = S' + T'$, $(R + S) \cap T = (R \cap T) + (S \cap T)$
   (c) $S - (S \cap T) = S - T = T' - S'$
   (d) $R \cap (S - T) = (R \cap S) - (R \cap T)$

★10. Deduce the identity of Exercise 6*b* by repeated applications of the identities L1–L4 and L6.

★ Starred exercises can be skipped without loss of continuity, throughout the book.

1-3. FUNCTIONS

For given sets S and T, a *function* f with *domain* S and *codomain* T is a rule which assigns to each element $s \in S$ of the domain a single element of the codomain T, called the *value* of f at s and written $f(s)$. We also say that f is a function (or a map or a transformation) on S to T, and we write $f\colon S \to T$ (the arrow is read "to"). The *image* $\operatorname{Im} f$ of $f\colon S \to T$ is the set $f(S)$ of all values $f(s)$ of f for variable $s \in S$; it is a subset of T.

The most straightforward way to define a function is to specify the value which it assigns to each element of its domain. For example, one function f with domain $S = \{a,b,c\}$, codomain $T = \{a,b,c,d\}$, and image $f(S) = \{a,b\}$ is defined by the *assignments*

$$f\colon a \mapsto a,\ b \mapsto a,\ c \mapsto b$$

Another function $g\colon S \to T$ with image $g(S) = \{b,c,d\}$ is defined by the assignments

$$g\colon a \mapsto d,\ b \mapsto c,\ c \mapsto b$$

Note that these image sets can be combined by Boolean operations; thus $f(S) \cup g(S) = \{a,b,c,d\}$ and $f(S) \cap g(S) = \{b\}$ above.

As another example, we define *Peano's successor function* σ to be a function with the set $\mathbf{P}$ of all positive integers for domain and codomain, which assigns to each positive integer n the integer $n + 1$, so that $\sigma(n) = n + 1$; it is a function $\sigma\colon \mathbf{P} \to \mathbf{P}$ from $\mathbf{P}$ to itself. We can also describe the function σ by listing its assignments; thus

$$\sigma\colon 1 \mapsto 2,\ 2 \mapsto 3,\ 3 \mapsto 4,\ \ldots,\ n \mapsto n+1,\ \ldots$$

Clearly, the image Im σ of the function $\sigma\colon \mathbf{P} \to \mathbf{P}$ is the set

$$\sigma(\mathbf{P}) = \{2,3,4,\ \ldots\}$$

which we shall signify by the symbol $\mathbf{P}_2$.

To define a function, we must specify its domain, its codomain, and the value which it assigns to each element of its domain. In symbols, for given $f\colon S \to T$ and $g\colon S_1 \to T_1$, we have $f = g$ if and only if

$$f(x) = g(x) \qquad \text{all } x \in S \tag{6a}$$
$$S = S_1 \qquad \text{and} \qquad T = T_1 \tag{6b}$$

According to this definition, two functions are not strictly identical unless they have the same domain and the same codomain. For example, the assignment $n \mapsto n + 1$ defines a function $\sigma'\colon \mathbf{P} \to \mathbf{P}_2$; we shall not consider σ' as equal to the successor function $\sigma\colon \mathbf{P} \to \mathbf{P}$ described above, because its codomain is not the same: the codomain of σ' does not include the number 1. (Note, however, that $\sigma(\mathbf{P}) = \sigma'(\mathbf{P})$: σ and σ' have the same image.)

For any set S, the *identity* function $1_S\colon S \to S$ maps each element of S onto itself; thus $1_S(s) = s$ for all $s \in S$. Different sets have different identity functions, in accordance with the convention discussed above.

Composition. The *left composite* $g \circ f$ of any two functions is the function obtained by applying them in reverse order. First apply f and then g, provided the domain of g is the codomain of f. More formally, if

$$f\colon S \to T \qquad \text{and} \qquad g\colon T \to U$$

then the left composite is the function $g \circ f\colon S \to U$ defined by

$$(g \circ f)(s) = g(f(s)) \qquad \text{all } s \in S \tag{7}$$

The relation between the three functions f, g, and h can be visualized by drawing the *mapping diagram* shown. This diagram illustrates the idea that one can go from S to U either directly by applying h, or in two steps by applying first f and then g.

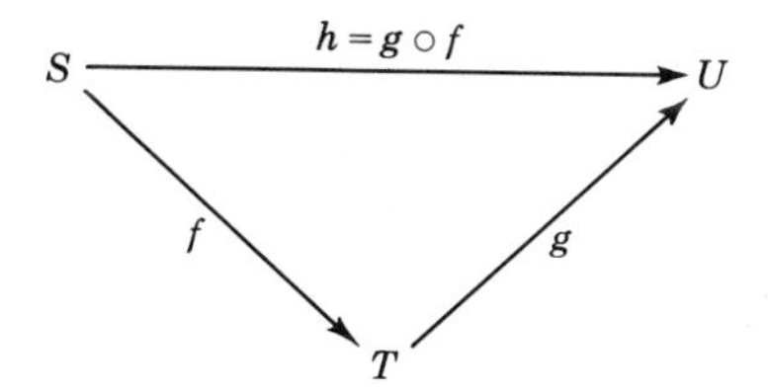

Right Composition. Opposite to the operation of left composition defined above is the operation of *right* composition $f \diamondsuit g = g \circ f$. For example, let $\phi_m: \mathbf{R} \to \mathbf{R}$ be the operation "raise to the mth power": $\phi_m(x) = x^m$. Like the exponent m, the function symbol ϕ_m may be written to the right, as $x^m = x\phi_m$. If we write function symbols $\phi, \psi, \ldots$ to the right, then it is natural to compose them by right composition, since then $(x\phi)\psi = x(\phi \diamondsuit \psi)$. Thus, in the preceding example, we have $x(\phi_m \diamondsuit \phi_n) = (x^m)^n = x^{mn} = x\phi_{mn}$; hence $\phi_m \diamondsuit \phi_n = \phi_{mn}$. In general, right composition has an intuitive advantage in that functions are written in the order in which they act.

Lemma 1. *Whenever the composites involved are defined, composition of functions obeys the double associative law*

$$(h \circ g) \circ f = h \circ (g \circ f) = f \diamondsuit (g \diamondsuit h) = (f \diamondsuit g) \diamondsuit h \tag{8}$$

This is obvious intuitively. Both $(h \circ g) \circ f$ and $h \circ (g \circ f)$ amount to applying first f, then g, and finally h, in that order; the same is true of $f \diamondsuit (g \diamondsuit h)$ and $(f \diamondsuit g) \diamondsuit h$.

Formally, for $f: S \to T$, $g: T \to U$ (as above), and $h: U \to V$ the value assigned to each $m \in V$ is, for all $x \in S$,

$$[(hg)f]x \underset{(hg)f}{=} (hg)(fx) \underset{hg}{=} h(g(fx)) \underset{gf}{=} h((gf)x) \underset{h(gf)}{=} [h(gf)]x$$

where each step can be proved valid by applying (7), the definition of composition, to the particular composite indicated below the equality symbol for that step. By (6*a*) and (6*b*), the definition of equality for functions, this proves the associative law $(hg)f = h(gf): S \to V$. Note that in the proof we have omitted the symbol $\circ$ for brevity, and have signified composition by simple juxtaposition of the symbols for the functions to be composed.

The identity functions 1_S and 1_T on any two given sets S and T obey, for any function $f: S \to T$, the identity laws

$$f \circ 1_S = 1_T \circ f = f \qquad 1_S \diamondsuit f = f \diamondsuit 1_T = f \tag{9}$$

To see this, note that $(f1_S)s = f(1_S s) = f(s)$ for all $s \in S$ by (8), and apply (6*a*) and (6*b*).

A function $f: S \to T$ is *injective*, or an *injection*, when $s \neq s'$ in S implies $f(s) \neq f(s')$ in T, that is, when f carries distinct elements of its domain to distinct elements of its codomain. An injection is often called a *one-one* transformation of S into T. A function $h: S \to T$ is called *surjec-*

tive, or a *surjection* (a fact often indicated by writing $h: S \twoheadrightarrow T$), when its image is the whole codomain, that is, when to each $t \in T$ there exists at least one $s \in S$ with $h(s) = t$. A surjection is thus a map of S *onto* T. Finally, a *bijection* is a function $f: S \to T$ which is both an injection and a surjection; that is, a bijection is a mapping which is both *one-one* and *onto* (to emphasize that a given function $f: S \to T$ is a bijection, we often write $f: S \leftrightarrow T$).

For example, among the functions $\mathbf{Z} \to \mathbf{Z}$, that function which assigns to each integer n its negative $-n$ is a bijection, that which assigns to each n its double $2n$ is an injection but not a surjection, and that which assigns to each n its square n^2 is neither an injection nor a surjection (why not?).

Example 3. For $\mathbf{P}$ the set of all positive integers, let $\sigma: \mathbf{P} \to \mathbf{P}$ be Peano's successor function, defined by the formula $\sigma(n) = n + 1$ for all $n \in \mathbf{P}$. This is an injection, but not a surjection.

Example 4. Let $\mathbf{n} = \{1,2, \ldots ,n\}$ be the set of positive integers $k \leqq n$. Then another basic function is the *cyclic permutation* $\nu_n: \mathbf{n} \to \mathbf{n}$ which maps each $k \in \mathbf{n}$ onto its successor $k + 1$ except for n, which is mapped onto 1. This is a bijection. Thus, for $n = 3$, ν_3 gives the assignments $1 \mapsto 2$, $2 \mapsto 3$, $3 \mapsto 1$; for general values of n, we can write

$$\nu_n: 1 \mapsto 2,\ 2 \mapsto 3,\ 3 \mapsto 4,\ \ldots,\ n - 1 \mapsto n,\ n \mapsto 1$$

Characteristic Functions. Another useful function assigns to each subset S of a given set U its *characteristic function* $e_S: U \to \{0,1\}$, defined by the assignment

$$e_S(x) = \begin{cases} 1 & \text{if } x \in S \\ 0 & \text{if } x \notin S \end{cases} \tag{10}$$

For example, let $U = \{1,2,3\}$, a set also designated by $\mathbf{3}$. Then $e_S: \mathbf{3} \to \{0,1\}$ makes the following assignments:

$$\varnothing \mapsto (0,0,0) \qquad \{1\} \mapsto (1,0,0) \qquad \{2\} \mapsto (0,1,0) \qquad \{3\} \mapsto (0,0,1)$$
$$\{1,2\} \mapsto (1,1,0) \qquad \{1,3\} \mapsto (1,0,1) \qquad \{2,3\} \mapsto (0,1,1)$$
$$\mathbf{3} = \{1,2,3\} \mapsto (1,1,1)$$

The assignment $b: S \to e_S$ just defined is itself a very important bijection, which will be discussed again in Sec. 1-6. Its domain is the power set $\mathcal{P}(U)$ of all subsets $S \subset U$, and its codomain (and image) is the set $\mathbf{2}^U$ of all functions $e_S: U \to \mathbf{2}$. When $U = \mathbf{3}$, for instance, b establishes a bijection between $\mathcal{P}(\mathbf{3})$ and the set of all sequences of three binary digits. Dropping parentheses and commas, we can signify each element of $\mathcal{P}(\mathbf{3})$ by one of the eight *binary triples* 000, 100, 010, 001, 110, 101, 011, 111. This notation will prove very useful below, where we shall usually think of such binary triples in the natural (dictionary, or lexicographic) order:

$$000, 001, 010, 011, 100, 101, 110, 111$$

1-4. INVERSES

If the functions $f: S \to T$ and $g: T \to S$ have as their (left) composite the identity function $g \circ f = 1_S$, then we call g a *left inverse* of f, and f a *right inverse* of g. If in addition to $g \circ f = 1_S$ we also have $f \circ g = 1_T$, then we call g a *two-sided inverse* of f and, symmetrically, f a *two-sided inverse* of g. An element which has a two-sided inverse is called *invertible.*

Theorem 2. *A function is left-invertible if and only if it is one-one (injective); it is right-invertible if and only if it is onto (surjective).*

Proof. First, we show that any left-invertible function must be one-one. Hence, we suppose that $f: S \to T$ has the left inverse $g: T \to S$ and let $f(s) = f(s')$. Then, by definition,

$$s = 1_S(s) = g(f(s)) = g(f(s')) = 1_S(s') = s'$$

Hence $f(s) = f(s')$ implies $s = s'$, which proves that f is one-one.

To prove the first statement of the theorem, it remains to show the converse: why, given that f is one-one, a left inverse g_1 exists with $g_1 f = 1_S$. If f is one-one (injective), choose any fixed $s_1 \in S$, and define $g_1: T \to S$ by

$$g_1(t) = \begin{cases} s & \text{if } t = f(s) \text{ for some } s \in S \\ s_1 & \text{otherwise, that is, if } t \notin \operatorname{Im} S \end{cases}$$

Then $(g_1 \circ f)(s) = g_1(f(s)) = s$ for all $s \in S$. Hence, by definition, $g_1 \circ f = 1_S$, and f is left-invertible with left inverse g_1, as claimed. Similarly, if $t \in T$ and $fh = 1_T$, then $t = 1_T(t) = f(h(t))$; hence every $t \in T$ is the image $f(s)$ of some $s = h(t) \in S$, and f is onto. Conversely, if $f: S \to T$ is onto, then each $t \in T$ is the image $t = f(s)$ of at least one $s \in S$. For each $t \in T$, we select[1] from the set, then, of all s mapped by f into t an arbitrary representative such as $h(t)$. Thus we get a function $h: T \to S$ such that $f(h(t)) = t$ for all $t \in T$; thus $fh = 1_T$ as claimed.

[1] This argument assumes the Axiom of Choice, which will be discussed critically in Chap. 14.

Corollary. *A function $f: S \to T$ is a bijection if and only if it has both a left inverse g and a right inverse h.*

Theorem 3. *The following properties of a function $f: S \to T$ are equivalent: (i) f is a bijection, (ii) f has both a left inverse g and a right inverse h. If they hold then (iii) any two inverses of f (left, right, or two-sided) are equal. Moreover, the unique inverse of f (denoted f^{-1}) is bijective, and satisfies*

$$(f^{-1})^{-1} = f \tag{11}$$

Proof. The equivalence of (*i*) with (*ii*) simply restates the preceding corollary. Moreover, (*ii*) implies

$$g = g1_T = g(fh) = (gf)h = 1_S h = h \tag{12}$$

This shows that any left inverse of a function f must equal any right inverse, proving (*iii*). Finally, the inverse $g = h$ of the bijection f has f for a two-sided inverse since $gf = 1_S$ and $fh = 1_T$; hence the inverse is also bijective as above, with inverse $(f^{-1})^{-1} = f$, proving (11).

Partial Functions. Although we adhere in this book to the convention that a "function" must assign to every element in its domain a unique value, there are also many important "partial functions" in mathematics, such as $1/x$ and $\log x$, which are defined for some but not for all values of x (for example, not for $x = 0$) and so are only partially defined. In this book, a *partial function* $f: X \to Y$ is a rule which assigns to every element x in some *subset* $D \subset X$ (the *domain* of f) a unique value in Y.

1-5. FUNCTIONS ON S TO S

We next consider in more detail the set S^S of all functions $f: S \to S$ from a given set S to *itself*[1] as an algebraic system $[S^S, \circ]$ under the binary operation of (left) composition. Since the codomain of any such function is the domain of any other function, this operation and its opposite, right composition, as defined in Sec. 1-3, are universally defined; given $f \in S^S$ and $g \in S^S$, the composites $f \circ g$ and $f \diamondsuit g = g \circ f$ always exist in S^S. Moreover, as in (8), the operations of left and right composition both satisfy the following *associative law:*

$$f \circ (g \circ h) = (f \circ g) \circ h \qquad \text{and} \qquad f \diamondsuit (g \diamondsuit h) = (f \diamondsuit g) \diamondsuit h \tag{13}$$

[1] Notice that S^S is a *set* of functions, each of which has S for domain and codomain. Similarly, in this notation, S^T is the set of all functions $g: T \to S$.

for any $f, g, h, \in S^S$. Furthermore, in this algebraic system, the identity function 1_S has the characteristic property of (9); thus

$$1_S \circ f = f \circ 1_S = f \text{ (that is, } f1_S = 1_S f = f) \qquad \text{for any } f \in S^S \quad (14)$$

Note that Eqs. (13) and (14) have the same form for either of the two operations $\circ$ or $\diamondsuit$. As stated earlier, it is usual to omit the symbols for these two operations and to signify composition (in either order) by simply juxtaposing the symbols for the functions composed—just as is done with products in high-school algebra. Thus, (13) states that $f(gh) = (fg)h$ for left and for right composite, in other words, that both left and right composition are associative binary operations. Condition (14) states that 1_S is *absorbed* into any element which it multiplies (by composition on either side).

The square of any function $f: S \to S$ is defined as the function $f^2 = f \circ f = f \diamondsuit f$, that is, as the result of performing f twice in succession. When $f^2 = f$, the function f is called *idempotent*, or a *projection*. For example, the orthogonal projection of the (x,y)-plane onto either the x-axis or the y-axis is idempotent. Observe also that these two projections satisfy $f \circ g = g \circ f$ (that is, $g \diamondsuit f = f \diamondsuit g$) since both $f \circ g$ and $g \circ f$ map every point onto the origin (0,0). In general, pairs of functions f and g such that $fg = gf$ are said to *commute* or to be *permutable*.

For either left or right composition, we can define a *left inverse* of a function $f \in S^S$ as a function $g \in S^S$ such that $gf = 1_S$, and a *right inverse* as a function h such that $fh = 1_S$. The existence of a left inverse of f for left composition is equivalent (by Theorem 2) to the condition that f is one-one (an injection); hence the same is true of the existence of a right inverse of f under right composition. Likewise, the existence of a right inverse of f under left composition is equivalent to f being onto (a surjection).

The existence of a two-sided inverse is equivalent to a function being a bijection, whether left or right composition is used; the exponent -1 is used to indicate this condition. Thus, a function $f \in S^S$ is a bijection if and only if it has an inverse $f^{-1} \in S^S$ with

$$f^{-1}f = ff^{-1} = 1_S \qquad \text{(left and right composition)} \qquad (15)$$

Monoids. The above discussion reveals the set S^S of all functions from a set S to itself as an *algebraic system* $[S^S,\circ,1_S]$ under left composition and an identity. Systems which, like $[S^S,\circ,1_S]$ and $[S^S,\diamondsuit,1_S]$, have an associative binary operation and an identity are called *monoids*. We shall study monoids systematically in Chap. 7.

Example 5. Let $S = \{a,b\}$; define $e = 1_S$ and $\alpha: a \mapsto a,\ b \mapsto a$; $\beta: a \mapsto b, b \mapsto b; f: a \mapsto b, b \mapsto a$. Then $[S^S,\circ,1_S]$ and $[S^S,\diamond,1_S]$ have the multiplication tables

$\circ$	e	f	α	β
e	e	f	α	β
f	f	e	β	α
α	α	α	α	α
β	β	β	β	β

$\diamond$	e	f	α	β
e	e	f	α	β
f	f	e	α	β
α	α	β	α	β
β	β	α	α	β

Note that, as stated earlier, the identity $1 = 1_S$ in the monoid S^S is "absorbed" by its neighbor in any product; thus, $1f = f$, and $\alpha\ 1\ \beta = \alpha\beta$. The elements $1,\alpha,\beta$ are idempotent; that is, $1^2 = 1$, $\alpha^2 = \alpha$, and $\beta^2 = \beta$, whereas f is not since $f^2 = e \neq f$. Also, $\alpha \circ \beta \neq \beta \circ \alpha$ so that the operation $\circ$ (and its opposite $\diamond$) are not commutative. Note further that α and β have no inverses, left or right, whereas f is its own inverse (that is, $f^2 = 1$).

Finite Sets. Finite sets play a special role in applied mathematics because they are the only sets which can actually be constructed physically. Informally, a set is finite when it can be "counted," i.e., when for some positive integer n there is a bijection f from the set $\mathbf{n} = \{1, \ldots ,n\}$ of positive integers $k \leqq n$ to S. If we denote $f(k) \in S$ by s_k, this evidently means we can list all the elements of S in sequence as $\{s_1,s_2, \ldots ,s_n\}$.

It is a familiar fact of childhood experience that, no matter in what order we count the elements of a finite set, we end up with the same number. Stated another way, there is no bijection from $\mathbf{m} = \{1, \ldots ,m\}$ to $\mathbf{n} = \{1, \ldots ,n\}$ unless $m = n$. A more vivid statement of the same principle is the following: we cannot put more than n pigeons in fewer than n pigeonholes, without having at least two pigeons in the same pigeonhole. In other words, any injection from a finite set S to itself is necessarily a surjection. The converse is also true: any surjection of a finite set to itself is an injection.

In Sec. 1-9, we shall sketch proofs of these basic facts. For the present we merely observe that the properties just mentioned are *not* true for injections and bijections of infinite sets. Thus, the Peano successor function $\sigma: \mathbf{P} \to \mathbf{P}$ of Example 3 has infinitely many left inverses; namely, for each $m \in \mathbf{P}$, if we define τ_m by

$$\tau_m(n) = \begin{cases} n - 1 & \text{if } n > 1 \\ m & \text{if } n = 1 \end{cases} \tag{15'}$$

clearly $\tau_m \circ \sigma = 1_P$.

EXERCISES B

1. Let $f(n) = 3n$, $g(n) = 3n + 1$, and $h(n) = 3n + 2$, from **Z** to **Z**. Construct a common left inverse for f, g, and h.
2. (a) If fg is defined and both f and g have left inverses, show that fg has a left inverse.
   (b) Show by an example that the converse is not necessarily true: fg may have a left inverse even if f does not.
3. (a) Show that the composite gf of any two injections $f: S \to T$ and $g: T \to U$ is an injection.
   (b) Prove the same result for surjections.
4. (a) How many surjections are there from a set of three elements onto a set of two elements?
   (b) How many injections are there from a set of three elements into a set of four elements?
5. (a) Consider the mapping $x \mapsto x^2$ on each of the following sets to itself:

$$\mathbf{P}, \quad \mathbf{Z}, \quad \mathbf{Q}, \quad \mathbf{R}, \quad \mathbf{C}$$

   and determine its image and whether or not it is one-one.
   (b) Do the same for $x \mapsto x^3$.
6. (a) Which subsets of $\mathbf{4} = \{1,2,3,4\}$ are represented by the following binary strings: 1001, 0110, 1101, and 0010.
   (b) Show that if the string $\mathbf{n} = n_1n_2n_3n_4$ represents the set S, then the string $\mathbf{n}' = (1 - n_1)(1 - n_2)(1 - n_3)(1 - n_4)$ represents its complement. Illustrate by the subsets of (*a*).

★7. Extend the definition of composite gf to the case in which the domain of g contains the range of f. Show that the result of Exercise 3*a* still holds, whereas that of Exercise 3*b* does not.

8. Prove that if $n(X)$ denotes the number of elements in X, the following hold for any three finite sets:
   (a) $n(A) + n(B) = n(A \cap B) + n(A \cup B)$
   (b) $n(A \cup B \cup C) + n(A \cap B) + n(B \cap C) + n(C \cap A)$
   $= n(A \cap B \cap C) + n(A \cup B) + n(B \cup C) + n(C \cup A)$
9. In a hotly fought battle, at least 70% of the combatants lost an eye, at least 75% an ear, at least 80% an arm, and at least 85% a leg. What is the least number of combatants that lost all four members? (Lewis Carroll)
10. (a) Show that a continuous real function $f: [a,b] \to [c,d]$ is one-one if and only if it is monotonic.
    (b) Show that this function is a bijection if and only if it is monotonic and either (*i*) $f(a) = c$, $f(b) = d$ or (*ii*) $f(a) = d$, $f(b) = c$.

★11. Test the function e^x for being one-one on **R** and on **C**, and find its image in each case.

1-6. SUMS, PRODUCTS, AND POWERS

The commutative and associative laws L2 and L3 have familiar arithmetic counterparts; the same is true of the first distributive law of L6. We shall now show that this is no accident. These and many other laws of arithmetic (for nonnegative integers) can be derived from basic properties of sets and functions.

In general, a finite set S is said to have m elements, or *cardinal number* $\mathbf{m}$, when there is a bijection $b: S \leftrightarrow \mathbf{m} = \{1, \ldots, m\}$; the empty set $\varnothing$ is said to have 0 elements. The addition operation of arithmetic refers to the following new concept.

Definition. Given sets S and T, we write $D = S \sqcup T$ and call the set D a *disjoint sum* of S and T when there exist bijections $i: S \leftrightarrow S^* \subset D$ and $j: T \leftrightarrow T^* \subset D$ of S and T, respectively, into *complementary* subsets S^* and T^* of D.

Clearly, any two disjoint sums of S and T are bijective. The simplest case is that $S \subset U$, $T \subset U$, and $S \cap T = \varnothing$. In this case $S \sqcup T = S \cup T$, setting $i = 1_S$ and $j = 1_T$. Also, from Eq. (4) we have[1]

$$n(S \sqcup T) = n(S) + n(T) \tag{16}$$

The commutative and associative laws for addition of nonnegative integers follow from (16) and the existence of obvious bijections; thus

$$\alpha: S \sqcup T \leftrightarrow \mathrm{T} \sqcup S \qquad \beta: S \sqcup (T \sqcup U) \leftrightarrow (S \sqcup T) \sqcup U \tag{16'}$$

for any three sets S, T, and U.

Cartesian Products. Likewise, products of integers are related to Cartesian products of sets. By definition, the *Cartesian product* $S \times T$ of two sets S and T is defined as the set of all ordered pairs (s,t) of elements s from S and t from T, respectively. Thus, if $\mathbf{R}$ is the set of all real numbers, then $\mathbf{R} \times \mathbf{R}$ is the set of all ordered pairs (x,y) of real numbers. Hence $\mathbf{R} \times \mathbf{R}$ is just the set of all Cartesian coordinates of points in the plane (relative to given axes). Likewise, $\mathbf{Z} \times \mathbf{Z}$ is the square grid (or "mesh") of points in the plane having integral coordinates (i,j).

Again, note the existence of the natural bijections

$$S \times T \leftrightarrow T \times S \qquad S \times (T \times U) \leftrightarrow (S \times T) \times U \tag{17}$$

Note also the multiplication rule for the number of elements in the Cartesian product of two finite sets; thus,

$$n(S \times T) = n(S)n(T) \tag{17'}$$

Just as there are natural injections $f: S \to S \cup T$ and $q: T \to S \cup T$, so there are natural surjections

$$p: S \times T \twoheadrightarrow S \qquad \text{defined by } p(s,t) = s \tag{18}$$

$$q: S \times T \twoheadrightarrow T \qquad \text{defined by } q(s,t) = t \tag{18'}$$

[1] As we shall see in Sec. 2-6, Eq. (16) may also be taken as a definition of addition.

These surjections are usually called *projections* onto the factors of $S \times T$ because of their analogy with the projections of the Cartesian plane onto its coordinate axes.

Exponentiation of Sets. Given two sets S and T, the set of all functions from S to T is designated T^S. If S and T are finite sets, the number of whose elements are $n(S)$ and $n(T)$, respectively, then we have the basic formula

$$n(T^S) = n(T)^{n(S)} \tag{19}$$

This is because for each $s_i \in S$, $f(s_i) \in T$ can be chosen in precisely $n(T)$ different ways; the values of f can be chosen in $n(T) \times \cdots \times n(T)$ ways in all.

Of special interest to us is the case of the two-element domain $\{0,1\} = \mathbf{2}$. As in Sec. 1-3, the formula

$$e_S = \begin{cases} 1 & \text{if } x \in S \\ 0 & \text{if } x \in S' \text{ (that is, if } x \notin S) \end{cases} \tag{20}$$

defines a natural bijection $b\colon \mathcal{P}(U) \leftrightarrow \mathbf{2}^U$ which assigns to each subset $S \subset U$ its characteristic function $e_S \in \mathbf{2}^U$. This bijection also relates subsets of $\mathbf{n} = \{1,2, \ldots ,n\}$ to sequences (strings) of n binary digits; it will be used repeatedly in later chapters.

In general, there are natural bijections

$$T^{R \sqcup S} \leftrightarrow T^R \times T^S \qquad (T \times U)^S \leftrightarrow T^S \times U^S \qquad (T^S)^R \leftrightarrow T^{S \times R} \tag{21}$$

By (20), for finite sets these bijections imply the familiar laws of exponents for positive integers. Specifically, if we substitute from (20) into (21) and use (16) and (17′), we get the three laws

$$a^{r+s} = a^r a^s \qquad (ab)^s = a^s b^s \qquad (a^s)^r = a^{sr} \tag{21′}$$

★ **Universality.** Disjoint sums and Cartesian products have interesting "universal" mapping properties. Thus, if $D = S \sqcup T$ as above and functions $f\colon S \to X$ and $g\colon T \to X$ are given with domains S and T and the same codomain X, then we can define $h\colon D \to X$ for elements $t^* = j(t) \in T^*$ by $h(t^*) = h(j(T)) = g(s)$. This makes $f = h \circ i$ and $g = h \circ j$, and it gives the only function $h\colon D \to X$ with these properties. This proves the following result.

Theorem 4. *There is one and only one way to "fill in" the diagram of Fig. 1-2a with D (a disjoint union of S and T) with a function $h\colon D \to X$ so that the resulting diagram commutes.*

★ Starred discussions can be skipped without loss of continuity, throughout the book.

Explanation. A mapping diagram is said to *commute* when any two mapping paths which go from the same initial set to the same final set give the same mapping. Thus, in Fig. 1-2a, this means that $h \circ i = f$ and $h \circ j = g$; in Fig. 1-2b, this means that $p \circ h = f$ and $q \circ h = g$.

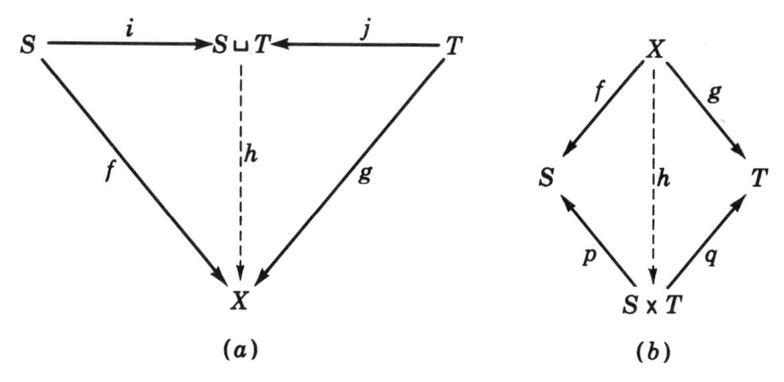

Fig. 1-2. Universal mapping diagrams.

Conversely, one can show that if the functions $i: S \to D$ and $j: T \to D$ have the property that for any set X and functions $f: S \to X$ and $g: T \to X$ such an $h: D \to X$ exists and is unique, then i and j map S and T into complementary subsets S^* and T^* of X; that is, the property of mappings indicated in Fig. 1-2a completely characterizes disjoint unions.[1] If $f: X \to S$ and $g: X \to T$ are any two functions from the same domain X to S and T, respectively, then the equation

$$h(x) = (f(x),g(x)) \qquad \text{each } x \in X \tag{22}$$

defines a unique function $h = f \times g: X \to S \times T$. If p and q are the projections defined by (18) and (18′), respectively, then clearly $p \circ h = f$ and $q \circ h = g$. Conversely, if $h: X \to S \times T$ makes $p \circ h = f$ and $q \circ h = g$, then for any $x \in X$, $h(s) = (s,t)$ implies $s = (p \circ h)(x) = f(x)$ and $t = (q \circ h)(x) = g(x)$.

In summary, we have shown that the function $h = f \times g$ defined by (22) is the only function which can be used to fill in the dotted line in the diagram of Fig. 1-2b so as to make the diagram commute, in the following sense:

$$h \diamondsuit p = p \circ h = f \qquad h \diamondsuit q = q \circ h = q$$

[1] For a proof of the converse stated, see S. Mac Lane & G. Birkhoff, "Algebra," p. 30, Macmillan, 1967; the essential fact is the uniqueness of universals proved in ibid., p. 28.

EXERCISES C

1. (a) Show that a set of n elements has 2^n different subsets.
   (b) Show that if $m < n$, a set of n elements has $(n!)/(m!)(n - m)!$ different subsets of m elements.
2. (a) Prove in detail that there are n^m different function $f: S \to T$ from a set of m elements to a set of n elements.
   (b) How many of the functions of (*a*) are injective?
3. Show that the (capital) letters of the alphabet cannot be represented unambiguously by strings of four symbols, each a dot · or dash —, but that we can do this with strings of dots and dashes allowed to consist of either four, three, or two symbols (Morse code).
4. Let $A = \emptyset$, $B = \{\emptyset\}$. Are A and B equal? Is there a bijection from A to B?
5. For an arbitrary set U, establish a bijection $\mathbf{2}^U \leftrightarrow \mathcal{P}(U)$, where **2** is the set of numbers $\{1,2\}$.
6. Show that in (15′) τ_1 has infinitely many right inverses, whereas τ_2 has only one.
7. Let X, Y, and Z be subsets of U with $Y \cap Z = \emptyset$ and $Y \cup Z = X$. Construct a bijection b: $\mathcal{P}(X) \leftrightarrow \mathcal{P}(Y) \times \mathcal{P}(Z)$.
8. For X, Y, and Z as in Exercise 7 and S an arbitrary set, construct a bijection $S^X \leftrightarrow S^Y \times S^Z$.

★9. Given a function $f: S \to T$, define a new function f^*: $\mathcal{P}(S) \to \mathcal{P}(T)$, which assigns to each $R \subset S$ the set $f^*(R)$ of all $t \in T$ such that $f(r) = t$ for some $r \in R$ (that is, the "image" of R under f). For $g: T \to U$, define g^* similarly. Show that the following mapping diagram is commutative:

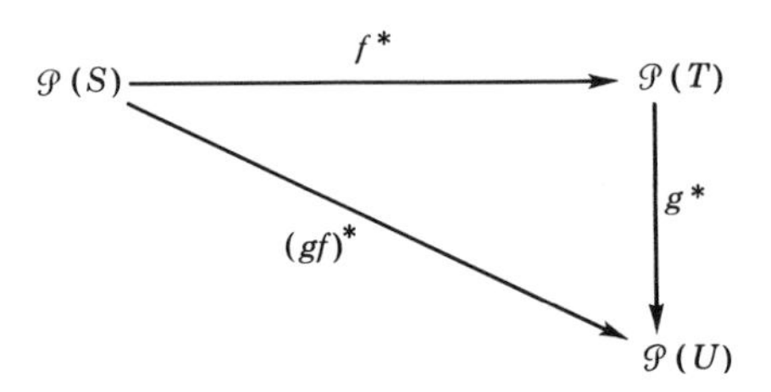

1-7. PEANO AXIOMS

The oldest and most basic mathematical system is the set **P** of all positive integers. Kronecker once said "God made the positive integers; all else is due to man." One commonly thinks of the positive integers as an algebraic system $[\mathbf{P},+,\cdot]$ or $[\mathbf{P},+,\times]$ having two binary operations, addition and multiplication.

However, as was shown by the Italian mathematician G. Peano (1858–1932), this algebraic system can be defined even more simply as an algebra $[\mathbf{P},\sigma]$ with one unary operation σ, that of counting. As a unary algebra $[\mathbf{P},\sigma]$, the positive integers are characterized by a set of three simple conditions on the unary operation $\sigma \in \mathbf{P}^{\mathbf{P}}$, the *successor function*

$\sigma: \mathbf{P} \to \mathbf{P}$. These conditions are the following *Peano axioms* for the successor function:

S1. If $\sigma(m) = \sigma(n)$, then $m = n$ (σ is one-one).
S2. For no $n \in \mathbf{P}$ is $\sigma(n) = 1$.
S3. Let $S \subset \mathbf{P}$ satisfy (*a*) $1 \in S$ and (*b*) $n \in S$ implies $\sigma(n) \in S$. Then $S = \mathbf{P}$.

Axiom S3 is the most powerful of Peano's three axioms S1 to S3; it is called the *induction axiom* because it is the basis of proofs by finite induction (see below).

It is easy to define the binary operations of addition and multiplication on $\mathbf{P}$ in terms of the unary operation σ by induction as follows.

Example 6. For any $m \in \mathbf{P}$, the function $\sigma^m: n \mapsto m + n$ is defined by simple recursion as follows:

$$m + 1 = \sigma^m(1) = \sigma(m) \tag{23}$$
$$m + (n + 1) = \sigma^m(\sigma n) = \sigma(\sigma^m(n)) \tag{24}$$

Consider, for fixed $m \in \mathbf{P}$, the set S_m of all $n \in \mathbf{P}$ such that $\sigma^m(n)$ is defined. By (23), $1 \in S_m$, and by (24), $n \in S$ implies $\sigma(n) \in S_m$. Hence, by axiom S3, $\sigma^m(n) = m + n$ is defined for all $n \in \mathbf{P}$.

Now letting m range over $\mathbf{P}$, we obtain a binary "addition" operation (function) $+: \mathbf{P} \times \mathbf{P} \to \mathbf{P}$, whose properties will be deduced in the next section from Peano's axioms.

Example 7. For any fixed $m \in \mathbf{P}$, the function $p_m: n \mapsto nm$ is defined by simple recursion as follows:

$$p_m(1) = m \tag{25}$$
$$p_m(\sigma(n)) = m + p_m(n) = \sigma^m(p_m(n)) \tag{26}$$

As in Example 6, for any fixed $m \in \mathbf{P}$, the set S of all $n \in \mathbf{P}$ such that $p_m(n)$ is defined comprises all of $\mathbf{P}$. Hence, letting m as well as n vary over $\mathbf{P}$, we get a binary "multiplication" operation on $\mathbf{P}$. Its properties will also be deduced in the next section from Peano's axioms.

Example 8. For any fixed $m \in \mathbf{P}$, the function $e_n: m \mapsto m^n$ is defined as follows:

$$e_n(1) = m \tag{27}$$
$$e_n(\sigma(n)) = m \cdot e_n(m) = p_m(e_n(m)) \tag{28}$$

The familiar laws of exponents can be proved by induction from this definition.

1-8. FINITE INDUCTION

The method of proof by finite induction begins by specifying a sequence of assertions or *propositions* $P(1), P(2), P(3), \ldots$ which associates with each positive integer n an assertion $P(n)$ which is either true or false. In other words, the Principle of Finite Induction asserts that to be sure of the truth of $P(n)$ for all $n \in \mathbf{P}$, it suffices to establish (*i*) the assertion $P(1)$ and (*ii*) the infinite sequence of implications.

$$P(1) \Rightarrow P(2) \Rightarrow P(3) \Rightarrow P(4) \Rightarrow \cdots \Rightarrow P(n) \Rightarrow P(n+1) \Rightarrow \cdots$$

Formally, the Principle of Finite Induction asserts that to prove the truth of every $P(n)$, it suffices to prove that (*a*) $P(1)$, and (*b*) $P(n)$ implies $P(\sigma n)$. (Here, as in Sec. 1-7, $\sigma(n) = \sigma n = n + 1$ denotes the successor of n.) Also, $\sigma(m)$ will sometimes be written in the contracted form σm. Thus (23) might be written $m + 1 = \sigma m(1) = \sigma(m) = \sigma m$.

The Principle of Finite Induction is an almost immediate consequence of the Peano axiom S3 for, if we let $T \subset \mathbf{P}$ be the set of positive integers n for which $P(n)$ is true, then $1 \in T$ by (*i*) above. By (*ii*) above, $n \in T$ implies $\sigma(n) \in T$. Hence, by the Peano Induction Axiom S3, T contains every $n \in \mathbf{P}$. In other words, $P(n)$ is true for all $n \in \mathbf{P}$, thus proving the Principle of Finite Induction.

Finite induction provides a standard technique for proving results in arithmetic and number theory. To illustrate this technique, we next prove that addition is commutative and associative.

Theorem 5. *Define addition in* $[\mathbf{P},\sigma]$ *by* $m + n = \sigma^m(n)$. *Then, for all* $m,n \in \mathbf{P}$

$$m + (n + r) = (m + n) + r \qquad (Associative) \qquad (29)$$
$$m + n = n + m \qquad (Commutative) \qquad (30)$$

To prove (29), we let $P(r)$ denote the proposition that (29) is true for the specified r and all $m,n \in \mathbf{P}$. Then $P(1)$ asserts [setting $r = 1$ in (29)] that $m + \sigma(n) = \sigma(m + n)$, which is true by (24).

We now make the induction hypothesis $P(r)$: that

$$m + (n + r) = (m + n) + r$$

for all m,n. Then, noting that, by (24), $m + \sigma(n) = \sigma(m + n)$, we have

$$m + (n + \sigma(r)) \overset{(24)}{=} m + \sigma(n + r) \overset{(24)}{=} \sigma(m + (n + r)) \overset{P(r)}{=} \sigma((m + n) + r)$$

where we have written above each equality sign its justification (valid for all $m,n \in \mathbf{P}$). Applying (24) again to $m + n$ and r, we obtain

$$\sigma((m + n) + r) \overset{(24)}{=} (m + n) + \sigma(r)$$

Since things equal to the same thing are equal to each other, the above equations imply

$$m + (n + \sigma(r)) = (m + n) + \sigma(r) \qquad \text{for all } m,n \in \mathbf{P}$$

which is $P(\sigma(r))$. This proves (29).

To prove (30), we next set $n = 1$, let $P(m)$ be the proposition that $m + 1 = 1 + m$, and again argue by induction. Since $1 + 1 = 1 + 1$ trivially, $P(1)$ is true. From $P(m)$, there then follow by (23) and (29), respectively, the equations

$$(m + 1) + 1 = (1 + m) + 1 = 1 + (m + 1)$$

which is $P(m + 1)$. This completes the induction, and proves that $P(m)$ is true for every positive integer m.

Now let $Q(n)$ be the proposition that $m + n = n + m$ for all m; we have just proved $Q(1)$. But if $Q(n)$ is granted, then we can prove successively that

$$\begin{aligned} m + (n + 1) = (m + n) + 1 = (n + m) + 1 &= n + (m + 1) \\ &= n + (1 + m) = (n + 1) + m \end{aligned}$$

by (29), $Q(n)$, (29), $P(m)$, and (29) in that order. This proves $Q(n + 1)$, completing the induction. But $Q(n)$ for all $n \in \mathbf{P}$ is obviously equivalent to (30).

Similar proofs by induction can be given for the associative and commutative laws for multiplication for positive integers; thus

$$m(nr) = (mn)r \qquad \text{for all } m,n,r \in \mathbf{P} \tag{31}$$

$$mn = nm \qquad \text{for all } m,n \in \mathbf{P} \tag{32}$$

The three laws of exponents (21′) can also be deduced (for arbitrary $a,r,s \in \mathbf{P}$) from the definitions of Example 8 by finite induction. However, we shall omit these proofs (they are in S. Feferman "The Number Systems," Addison-Wesley, 1964, for example). Instead, we shall prove a related result about functions. For any $f\colon S \to S$, we set

$$f^1 = f \qquad \text{and} \qquad f^{\sigma(n)} = f^n \circ f \tag{33}$$

thus defining $f^n\colon S \to S$ for all $n \in \mathbf{P}$.

Theorem 6. *Let $f: S \to S$ be any function on a set to itself. Then*

$$f^m \circ f^n = f^{m+n} \qquad \text{for all } m,n \in \mathbf{P} \tag{33'}$$

Proof. For $n = 1$, $f^m \circ f^1 = f^m \circ f = f^{\sigma m} = f^{m+1}$ by the definitions of $f^1, f^{\sigma m}$, and $m + 1$ (as σm). Now assume that (33′) holds for some particular $n \in \mathbf{P}$ (induction hypothesis). This implies

$$\begin{aligned} f^m \circ f^{\sigma(n)} &= f^m \circ (f^n \circ f) = (f^m \circ f^n) \circ f \\ &= f^{m+n} \circ f = f^{\sigma(m+n)} = f^{(m+n)+1} \\ &= f^{m+(n+1)} = f^{m+\sigma(n)} \end{aligned}$$

The successive equalities follow by (33), associativity (Sec. 1-3), the induction hypothesis, (33), notational change, (29), and notational rewrite [actually, we verified that $\sigma(m + n) = m + \sigma(n)$ in proving Theorem 5].

Distributive Laws. It is also easy to prove various distributive laws by induction from the Peano axioms. Let $P(r)$ be the proposition that $m(n + r) = mn + mr$ for all $m, n \in P$. Then $P(1)$ follows from (26) and (30). Moreover, granted $P(r)$, we have

$$\begin{aligned} m(n + \sigma r) &= m(n + (r + 1)) = m((n + r) + 1) \\ &= m(n + r) + m = mn + mr + m \qquad (\text{by } P(r)) \\ &= mn + (mr + m) = mn + m\sigma r \end{aligned}$$

Summation Symbol. The summation symbol Σ also has its meaning defined by induction; thus

$$\sum_{j=1}^{1} x_j = x_1 \qquad \text{and} \qquad \sum_{j=1}^{m+1} x_j = \left(\sum_{j=1}^{m} x_j\right) + x_{m+1}$$

One then easily proves by induction on m, the m-term left-distributive law

$$\left(\sum_{j=1}^{m} x_j\right) y = \sum_{j=1}^{m} (x_j y)$$

A similar argument proves the n-term right-distributive law

$$x\left(\sum_{k=1}^{n} z_h\right) = \sum_{k=1}^{n} (xz_k)$$

Combining, we get the *generalized distributive law*

$$\left(\sum_{j=1}^{m} x_j\right)\left(\sum_{k=1}^{n} y_k\right) = \sum_{j=1}^{m} \sum_{k=1}^{n} (x_j y_k)$$

★**Generalized Associative Law.** One learns in high-school algebra the following somewhat imprecisely stated principle: *the result of performing any associative operation on a sequence of n terms depends only on the order in which they are listed and not on the order in which they are combined.*

To actually *prove* this principle, called the *generalized associative law*, from the simple three-term associative law (31), or indeed to even formulate it precisely, requires some rather sophisticated reasoning. Thus, there are actually *five* different ways of combining the four elements x, y, z, and w by a binary operation. If we write $f(x,y) = (xy)$ for brevity, these can be conveniently displayed as follows:

$$(((xy)z)w),\ ((x(yz))w),\ ((xy)(zw)),\ (x((yz)w)),\ (x(y(zw))) \tag{34}$$

The proof that the simple three-term associative law implies the equality of all of these expressions requires *two* chains of equations as follows:

$$(((xy)z)w) = ((x(yz))w) = (x((yz)w)) = (x(y(zw))) \tag{35}$$

$$(((xy)z)w) = ((xy)(zw)) = (x(y(zw))) \tag{35$'$}$$

In (35), the effect of the associative law is just to slide parentheses past letters or strings of (two) letters; the basic parenthesis sequence ((())) is unchanged, just as in (31), which is the three-term associative law assumed, we have the same parenthesis sequence (()) on both sides.

In (35′), however, we treat (xy) as a *single* symbol E and apply (31) to $((Ez)w) = (E(zw))$, sliding) past w and (past E with its parentheses. This requires a more sophisticated use of induction, which will be discussed in Sec. 2-10.

Strings. In algebra, we usually express the generalized associative law symbolically by simply dropping parentheses; thus, in dealing with sums, we write

$$\sum_{k=1}^{4} x_k = x_1 + x_2 + x_3 + x_4 \qquad \sum_{k=1}^{n} x_k = x_1 + \cdots + x_n, \text{ etc.}$$

and add two strings by laying them end to end (and inserting a plus sign).

Generalized Commutative Law. If we do this, we get a much clearer idea of the proof of the generalized commutative and associative law. This law asserts (for addition) that if $\beta\colon \mathbf{n} \to \mathbf{n}$ is any bijection acting on a finite set $\mathbf{n} = (1, \ldots, n)$, then

$$\sum_{k=1}^{n} a_k = \sum_{k=1}^{n} a_{\beta(k)} = \sum_{k=1}^{n} b_k \qquad b_k = a_{\beta(k)} \tag{36}$$

Thus

$$a_1 + a_2 + a_3 + a_4 + a_5 = a_3 + a_5 + a_2 + a_4 + a_1$$

Again, the special two-term commutative law

$$x + y = y + x \qquad \text{for all } x,y$$

implies (36). In outline, the proof consists in repeatedly interchanging pairs of adjacent letters whose subscripts are in reverse order. Thus

$$\begin{aligned} a_3 + a_5 + a_2 + a_4 + a_1 = a_3 + a_5 + a_2 + a_1 + a_4 &= a_3 + a_2 + a_5 + a_1 + a_4 \\ = a_2 + a_3 + a_5 + a_1 + a_4 &= a_2 + a_3 + a_1 + a_5 + a_4 \\ = a_2 + a_1 + a_3 + a_4 + a_5 &= a_1 + a_2 + a_3 + a_4 + a_5 \end{aligned}$$

where the fifth equation includes two changes.

Note also that to prove the generalized commutative law from the two-term identity $x + y = y + x$ in this way (assuming that parentheses can be omitted because of the associative law) clearly involves a more complicated technique of proof by induction than was used to prove Theorem 4. Essentially, one must prove that *any permutation is a product of transpositions*. This will be proved in Sec. 7-9.

★1-9. PIGEONHOLE PRINCIPLE; DIVISION ALGORITHM

We now indicate briefly how to prove two more basic mathematical principles from Peano's axioms. Since the conclusions are familiar facts, we do not include detailed proofs. However, the student is advised to read carefully the definitions and the statements of the main conclusions, which may be described loosely as the *pigeonhole principle* and the *Division Algorithm*. We shall now discuss them in that order.

Inequality. In Secs. 1-1 through 1-6, we referred freely to the set $\mathbf{n} = \{1, \ldots, n\}$ of all $k \leqq n$. We now show one way to define the relation $m \leqq n$ in terms of the successor function, using Peano's axioms.

Namely, call a subset $S \subset \mathbf{P}$ *σ-closed* when $n \in S$ implies $\sigma(n) \in S$. Intuitively, the σ-closed subsets of $\mathbf{P}$ are the void set $\varnothing$ and the "final segments" $[n, \infty] \subset \mathbf{P}$ consisting of some n and all $k \geqq n$. Formally, therefore, we define $m \leqq n$ to mean that any σ-closed subset of $\mathbf{P}$ which contains m must contain n.

Among the properties of inequality in $\mathbf{P}$ are the following:

P1. $n \leqq n$ for all $n \in \mathbf{P}$.
P2. $m \leqq n$ and $n \leqq r$ imply $m \leqq r$.
P3. $m \leqq n$ and $n \leqq m$ imply $m = n$.
P4. For all $m, n \in \mathbf{P}$, either $m \leqq n$ or $n \leqq m$.

Property P1 follows since any subset of $\mathbf{P}$ which contains n trivially contains n. Likewise, if any σ-closed subset $\mathbf{P}$ which contains m must contain n and if any σ-closed subset of $\mathbf{P}$ which contains n must contain r, then any σ-closed subset of $\mathbf{P}$ which contains m is also a σ-closed subset which contains n, and hence r. The proofs of properties P3 and P4 are harder; they require induction and will be omitted. Of these properties, P1 and P2 are obvious from the definition; but the proofs of P3 and P4 (by induction) are harder.[1]

To see the significance of P4, let us define for any $n \in \mathbf{P}$ the initial segment $\mathbf{n} = \{1, \ldots, n\}$ as the list of all $k \leqq n$ in their natural order. Then $m \leqq n$ means the same thing as $\mathbf{m} \subset \mathbf{n}$. This implies that there is an injection from $\mathbf{m}$ to $\mathbf{n}$ and a surjection from $\mathbf{n}$ to $\mathbf{m}$. We now prove the converse of this evident result.

Lemma. *If there exists an injection $h: \mathbf{n} \to \mathbf{m}$, then $n \leqq m$.*

To facilitate the proof, we define $h < k$ to mean the negation of $h \geqq k$, and $h > k$ to mean the negation of $h \leqq k$. By P4, it follows that $h \leqq k$ means precisely that $h < k$ or $h = k$ (exclusive or), whereas $h \geqq k$ asserts that $h > k$ or $h = k$.

Proof. Let $P(n)$ be the proposition that the existence of an injection $f: \mathbf{n} \to \mathbf{m}$ implies $n \leqq m$. Then $P(1)$ is trivial, since $1 > m$ for no $m \in \mathbf{P}$. Now suppose that $P(n)$ is true (induction hypothesis), and let $h: \sigma\mathbf{n} \to \mathbf{m}$ by an injection. Denote $h(\sigma n)$ by s, and observe that there is at most one $r \in \mathbf{n}$ with $h(r) = \sigma m$, since h is an injection. Now define the function $f: \mathbf{n} \to \mathbf{m}$ by

$$f(r) = s \qquad f(k) = h(k) \qquad \text{for } k \neq r \tag{37}$$

This is an injection (why?) hence $n \leqq m$ by $P(n)$; this proves $\sigma n \leqq \sigma m$ and hence $P(\sigma n)$. The proof (by induction) is complete.

We now derive three consequences of the preceding key lemma.

Theorem 7. *If $m < n$, there is no injection $f: \mathbf{n} \to \mathbf{m}$ or surjection $g: \mathbf{m} \twoheadrightarrow \mathbf{n}$.*

Proof. Suppose there were an injection $f: \mathbf{n} \to \mathbf{m}$. Then $n \leqq m$ by the lemma, which excludes $m < n$ by P1–P4. Again, any surjection $g: \mathbf{m} \twoheadrightarrow \mathbf{n}$ has a right inverse $f: \mathbf{n} \to \mathbf{m}$ such that $g \circ f = 1_{\mathbf{m}}$; this f is an injection since it has a left inverse g, which implies that $n \leqq m$ by the lemma.

[1] Proofs may be found in the books by Feferman and Gleason listed at the end of this chapter.

Corollary. *If there is a bijection* $g: m \leftrightarrow n$, *then* $m = n$.

Theorem 8. (*Pigeonhole Principle*) *Let* X *be any finite set. Then a function* $f: X \to X$ *is one-one if and only if it is onto.*

Corollary. *Let* X *be finite. Then, for any function* $f: X \to X$, *the properties of being injective, surjective, and bijective are mutually equivalent.*

The preceding results make the counting of finite sets a very powerful technique of applied mathematics. They do not hold for infinite sets (see Chap. 14).

The second consequence of the lemma refers not only to the set **P** of all positive integers, but also the set **N** of all nonnegative integers, definable as the disjoint sum $\mathbf{P} \sqcup \{0\} = \mathbf{N}$; it is the following.

Theorem 9. (*Division Algorithm*) *Given* $m \in \mathbf{P}$ *and* $n \in \mathbf{Z}$, *there exist* $q \in \mathbf{Z}$ *and* $r \in \mathbf{Z}$ *such that* $n = qm + r$.

Remarks. The letters q and r stand for *quotient* and *remainder* of n under division by m, in the usual sense. The function $\tau: \mathbf{N} \to \mathbf{N}$, defined as σ on **P** and by $\tau(0) = 1$, satisfies axioms T1 to T3, identical with S1 to S3, if 1 is replaced by 0.

Sketch of Proof. We define $q = q(m,n)$ and $r = r(m,n)$ by induction, as functions with domain $\mathbf{P} \times \mathbf{Z}$ and codomain **N**, as follows:

$$r(m,0) = 0 \qquad \text{and} \qquad r(m,\sigma n) = \sigma_m(r(m,n)) \tag{38}$$

where σ_m is a function with domain and codomain $\{0,1, \ldots ,m - 1\}$, satisfying $\sigma_m(m - 1) = 0$ and $\sigma_m(k) = \sigma k$ for $k \neq m - 1$; $q(m,0) = 0$ and

$$q(m,\sigma n) = \begin{cases} q(m,n) & \text{if } r(m,\sigma n) \neq m - 1 \\ \sigma q(m,n) & \text{if } r(m,\sigma n) = m - 1 \end{cases} \tag{39}$$

EXERCISES D

1. Defining m^n suitably, prove by induction the following identities for exponents in **P**: (a) $1^n = 1$, (b) $x^m x^n = x^{m+n}$, (c) $(xy)^n = x^n y^n$, (d) $(r^s)^n = r^{sn}$.
2. Show by induction that if x^n is defined (recursively) by $x^1 = x$ and $x^{\sigma n} = x^n x$, then $a^2 = a$ implies $a^n = a$ for all $n \in P$.
3. (a) Prove by induction that $\sum_{k=1}^{n} k = n(n + 1)/2$.

   (b) Prove similarly that $\sum_{k=1}^{n} = n(n + 1)(2n + 1)/6$.

4. Prove that $\sum_{k=1}^{n} = [n(n+1)/2]^2$.

5. Defining $\binom{r}{s} = (r!)/(s!)/(r-s)!$, prove by induction that

$$\binom{r}{s} + \binom{r}{s-1} = \binom{r+1}{s}$$

for all $r \in \mathbf{P}$ and $s = 0, \ldots, r$. Then use this fact to prove the binomial theorem $(x+y)^n = \sum_{k=0}^{n} \binom{n}{k} x^{n-k}y^k$, also by induction.

★6. In any monoid, define $p_n = \prod_{i=1}^{n} x_i$ by $p_1 = x_1$ and $p_{n+1} = p_n x_{n+1}$. Prove that $p_m p_n = p_{m+n}$.

7. Locate the fallacy in the following "proof" of the proposition $P(n)$: In any set of n elements, all elements are the same.
   Step 1. $P(1)$ is trivially true.
   Step 2. Apply $P(n-1)$ to the subset $a_1, a_2, \ldots, a_{n-1}$ of the set $a_1, \ldots, a_n$, to infer that $a_1 = a_2 = \cdots = a_{n-1}$.
   Step 3. Apply $P(n-1)$ to the subset $a_2, \ldots, a_{n-1}, a_n$ to get

$$a_2 = \cdots = a_{n-1} = a_n$$

   Step 4. From the conclusions of Steps 2 and 3 and the transitivity of equality we can conclude that $P(n-1)$ implies $P(n)$.

8. Prove that $\sum_{k=0}^{n} \binom{n}{k}^2 = \binom{2n}{n}$.

9. Prove by induction that in $\mathbf{P}$ $m + r = m + s$ implies $r = s$.

10. (a) Prove by induction that any product $f_m \circ f_{m-1} \circ \cdots \circ f_1$ (that is, composite) of injections is an injection.
    (b) Obtain the same result for surjections.

11. Prove that in $\mathbf{P}$
    (a) $m < n$ if and only if $m + r = n$ for some $r \in \mathbf{P}$.
    (b) $m \leqq n$ if and only if $m + r = n$ for some $r \in \mathbf{N}$.

12. Prove, by induction, the laws of exponents

$$m^n m^r = m^{n+r} \qquad m^r n^r = (mn)^r \qquad (m^n)^r = m^{nr}$$

GENERAL REFERENCES

BARTEE, T. C., I. L. LEBOW, and I. S. REED: "Theory and Design of Digital Machines," McGraw-Hill, 1962.

BIRKHOFF, G., and S. MAC LANE: "A Survey of Modern Algebra," 3d ed., Macmillan, 1965.

HERSTEIN, I. N.: "Topics in Algebra," Ginn-Blaisdell, 1964.

LIU, C. L.: "Introduction to Combinatorial Mathematics," McGraw-Hill, 1968.
MAC LANE, S., and G. BIRKHOFF: "Algebra," Macmillan, 1967.

THE PRECEDING REFERENCES CONTAIN SUPPLEMENTARY MATERIAL FOR MANY CHAPTERS OF THIS BOOK. THE FOLLOWING REFERENCES ARE SPECIFIC TO CHAPTER 1. SIMILAR LISTS WILL BE GIVEN AT THE END OF OTHER CHAPTERS.

REFERENCES

FEFERMAN, S.: "The Number Systems," Addison-Wesley, 1964.
GLEASON, A. M.: "Fundamentals of Abstract Analysis," Addison-Wesley, 1966.

Binary Relations and Graphs

2-1. INTRODUCTION

By a *binary relation* between two sets X and Y we mean a rule α which decides, for any elements $x \in X$ and $y \in Y$, whether or not x is in the relation α to y. In the former case, we write $x \; \alpha \; y$; in the latter, we write $x \; \alpha' \; y$.

The concept of a binary relation between X and Y generalizes that of a function $f \colon X \to Y$ studied in Chap. 1. For clearly, each function $f \colon X \to Y$ defines a binary relation α_f between X and Y, through the rule

$$x \; \alpha_f \; y \qquad \text{means that} \qquad y = f(x) \tag{1}$$

Conversely, for a given binary relation α between sets X and Y, consider for each $x \in X$ the set of all $y \in Y$ such that $x \; \alpha \; y$. This correspondence defines a function $f \colon X \to Y$ if and only if, for each $x \in X$, there is precisely one $y \in Y$ with $x \; \alpha \; y$. Hence the binary-relation concept includes the function concept as a (very important) special case.

The following example from analytic geometry shows how multiple-valued functions (as well as single-valued functions) define binary relations between their domains and codomains.

Example 1. Let $X = Y = \mathbf{R}$ (the real field), and define $x \alpha y$ to mean that $x^2 + y^2 = 25$. Then $x \alpha y$ if and only if $y = \pm\sqrt{25 - x^2}$; the graph of α is the circle of radius 5 with its center at the origin. In this example, $3 \alpha 4$ and $4 \alpha (-3)$, but $2 \alpha' 3$, for instance.

The word "*graph*," used to describe the circle of points whose coordinates (x,y) are in the relation $x^2 + y^2 = 25$ (that is, in the functional relation $y = \pm\sqrt{25 - x^2}$), is used similarly to describe arbitrary binary relations.

Definition. The *graph* of a binary relation α between sets X and Y is the set $S(\alpha)$ of all pairs $(x,y) \in X \times Y$ such that $x \alpha y$.

In symbols, $S(\alpha) \subset X \times Y$ can be defined by writing

$$S(\alpha) = \{(x,y) \mid x \alpha y\}$$

Example 2. Let $X = \{a,b\}$ and $Y = \{c,d,e\}$. Then a relation α on X and Y is defined by the following list:

$$a \alpha c,\ a \alpha d,\ a \alpha' e,\ b \alpha' c,\ b \alpha' d,\ b \alpha e$$

This relation has the graph $S(\alpha) = \{(a,c)(a,d)(b,e)\}$.

Note that the negation α' of α is also a binary relation between X and Y. If we relabel α' as β, we have the list

$$a \beta' c,\ a \beta' d,\ a \beta e,\ b \beta c,\ b \beta d,\ b \beta e$$

Note that double negation of α gives α back again. These remarks hold for binary relations in general; if we define the *negation* α' of a binary relation α by

$$x \alpha' y \qquad \text{means} \qquad \text{not } x \alpha y \tag{2}$$

then $(\alpha')' = \alpha$.

Any binary relation ρ between finite sets $X = \{x_1, \ldots ,x_m\}$ and $Y = \{y_1, \ldots ,y_n\}$ can be specified by making a table whose rows are headed by the successive elements of X and whose columns are headed by the elements of Y. In the ith row and jth column, we write the entry 1 if $x_i \rho y_j$, and 0 if $x_i \rho' y_j$. Thus, the tables for the α and β of Example 2 are

α	c	d	e
a	1	1	0
b	0	0	1

β	c	d	e
a	0	0	1
b	1	1	0

The preceding tabular representation identifies each relation ρ (for example, α or $\beta = \alpha'$) with the characteristic function [Chap. 1, (10)] of its graph; that is, the entry in the ith row and jth column is

$$e_{S(\rho)}(x_i,y_j) = \begin{cases} 1 & \text{if } x_i \, \rho \, y_j \\ 0 & \text{otherwise} \end{cases} \tag{3}$$

We shall now consider Eq. (3) from another standpoint.

2-2. RELATION MATRICES

For given orderings of the finite sets $X = \{x_1, \ldots ,x_m\}$ and

$$Y = \{y_1, \ldots ,y_n\}$$

the entries in the table of 0's and 1's which define any relation α constitute an $m \times n$ *matrix* (rectangular array) $A = \|a_{ij}\|$ of 0's and 1's; thus

$$a_{ij} = \begin{cases} 1 & \text{when } x_i \, \alpha \, y_j \\ 0 & \text{otherwise} \end{cases} \tag{4}$$

Thus, in Example 2, the matrices A and B so constructed from α and β are

$$A = \begin{bmatrix} 1 & 1 & 0 \\ 0 & 0 & 1 \end{bmatrix} \qquad B = \begin{bmatrix} 0 & 0 & 1 \\ 1 & 1 & 0 \end{bmatrix}$$

Conversely, each $m \times n$ matrix $A = \|a_{ij}\|$ of 0's and 1's defines via (4) a relation $\rho(A)$. We therefore define a (rectangular) matrix consisting of 0's and 1's as a *relation matrix*.

By (3) and (4), the relation matrix $M(\alpha) = A$ of any binary relation α between sets X and Y has as its (i,j) entry a_{ij}, the value of the characteristic function $e = e_{S(\alpha)}\colon X \times Y \to \{0,1\}$ of the graph $S(\alpha) \subset X \times Y$ of the relation α. That is, as in (3) and (4),

$$e(x_i,y_j) = a_{ij} = \begin{cases} 1 & \text{when } (x_i,y_j) \in S(\rho) \\ 0 & \text{otherwise} \end{cases} \tag{5}$$

Thus, for the relation α in Example 2, the characteristic function $e = e_{S(\alpha)}$ is defined by the following assignments:

$$e_{S(\alpha)}\colon \begin{matrix} (a,c) \mapsto 1 & (a,d) \mapsto 1 & (a,e) \mapsto 0 \\ (b,c) \mapsto 0 & (b,d) \mapsto 0 & (b,e) \mapsto 1 \end{matrix}$$

Also, as we have seen, each binary relation α between two sets X and Y is determined by its graph $S(\alpha)$. Conversely, given X and Y, each subset $T \subset X \times Y$ is the graph of one and only one binary relation ρ_T between X and Y. This relation ρ_T is defined by the condition that $x\ \rho_T\ y$ ($x \in X$, $y \in Y$) if and only if $(x,y) \in T$.

Hence, in summary, the correspondences $\alpha \mapsto S(\alpha)$ and $T \mapsto \rho_T$ are inverse bijections which connect the set of *all binary relations* between sets X and Y with the power set $\mathcal{P}(X \times Y)$ of all subsets of $X \times Y$; thus

$$\rho_{S(\alpha)} = \alpha \qquad \text{and} \qquad S(\rho_T) = T \qquad \text{all } \alpha \text{ and } T \tag{6}$$

Moreover $S(\alpha') = [S(\alpha)]'$; the bijection carries negations of relations into complements of sets. For this reason, α' is also called the *complement* of the relation α.

Boolean Operations. The preceding bijections also define from any sets X and Y a Boolean algebra whose elements are the binary relations between X and Y. We define $\wedge$ and $\vee$ by

$$x(\rho \wedge \sigma)y \qquad \text{means} \qquad x\ \rho\ y \text{ and } x\ \sigma\ y \tag{7}$$

$$x(\rho \vee \sigma)y \qquad \text{means} \qquad x\ \rho\ y \text{ or } x\ \sigma\ y \tag{7'}$$

Then we have, from the definitions of $\cap$ and $\cup$ for sets,

$$S(\rho \wedge \sigma) = S(\rho) \cap S(\sigma) \qquad \text{and} \qquad \rho_{T \cap V} = \rho_T \wedge \rho_V \tag{8}$$

$$S(\rho \vee \sigma) = S(\rho) \cup S(\sigma) \qquad \text{and} \qquad \rho_{T \cup V} = \rho_T \vee \rho_V \tag{8'}$$

for any relations between X and Y and subsets $T, V \subset X \times Y$. In particular, for any two complementary relations between sets X and Y, such as α and $\alpha' = \beta$ in Example 2, $S(\alpha \wedge \alpha') = \varnothing$ is the empty set, whereas $S(\alpha \vee \alpha') = X \times Y$. We therefore speak of the intersection and union of any two binary relations α and β between X and Y, referring to $\alpha \wedge \beta$ and $\alpha \vee \beta$, respectively. Finally, we write $\alpha \leqq \beta$ to signify that $S(\alpha) \subset S(\beta)$. In other words,

$$\alpha \leqq \beta \qquad \text{means that} \qquad x\ \alpha\ y \text{ implies } x\ \beta\ y \tag{9}$$

From the properties of Boolean operations on sets listed in Sec. 1-2, we infer the following result as a corollary.

Theorem 1. *The operations of intersection, union, and negation of binary relations satisfy the idempotent, commutative, associative, and absorption laws of Theorem* 1 *in Chap.* 1, *as well as the involution and de Morgan laws.*

Corollary. *The binary relations between any two sets X and Y form a Boolean algebra.*

2-3. ALGEBRA OF RELATIONS

Binary relations between sets have many basic algebraic properties beyond those of Boolean algebras. Thus, if α is any relation on X to Y, we can define a *converse* relation $\breve{\alpha}$ on Y to X by the simple rule

$$y \,\breve{\alpha}\, x \qquad \text{means that} \qquad x \,\alpha\, y \tag{10}$$

Evidently, the relation matrix $\|\breve{a}_{ij}\|$ of $\breve{\alpha}$ is the *transpose* of the relation matrix $\|a_{ij}\|$ of α, and is obtained from it by interchanging rows and columns (when $X = Y$, by reflecting in the main diagonal), so that $\breve{a}_{ij} = a_{ji}$.

It is exceptional that a relation and its converse both correspond to functions. In this case (that $\alpha = \rho_f$ and $\breve{\alpha} = \rho_g$) each $y \in S$ must correspond (since $\breve{\alpha}$ is a function) to one $x \in S$ (f must be onto) and only one $x \in S$ (f must be one-one). Therefore, the functions f and g which correspond to α and $\breve{\alpha}$ must be mutually inverse *bijections*. Their relation matrices (if any) must be (square) *permutation matrices* having a single 1 in each row and each column.

The notion of the composite of two functions can be generalized to relations. Let α and β be relations between X and Y and Y and Z, respectively. Then the composite (or "relative product") $\alpha\beta$ is the relation with

$$x(\alpha\beta)z \qquad \text{if and only if} \quad \text{for some } y \in Y, \quad x \,\alpha\, y \text{ and } y \,\beta\, z \tag{11}$$

When $\alpha = \rho_f$ and $\beta = \rho_g$ are associated with functions, $\alpha\beta = \rho_{f \diamond g}$; the relative product of two relations corresponds to the right composite of the corresponding functions.

Reading (11) from right to left instead of from left to right, we see that $z(\breve{\alpha\beta})x$ if and only if for some $y \in Y$, $z \,\breve{\beta}\, y$ and $y \,\breve{\alpha}\, x$. This proves the interesting identity

$$\breve{\alpha\beta} = \breve{\beta}\breve{\alpha} \tag{12}$$

which generalizes $(fg)^{-1} = g^{-1}f^{-1}$.

Binary Relations on S. A binary relation between a set S and itself (i.e., the special case $X = Y = S$) is called a *relation on* S. A special relation on any set S is the equality relation e on S, defined by the condition that $x \,e\, y$ means $x = y$. Clearly, the equality relation on $\mathbf{n}$ has the identity matrix

$$I = \|\delta_{ij}\| \qquad \text{where } \delta_{ij} = \begin{cases} 1 & \text{if } i = j \\ 0 & \text{if } i \neq j \end{cases}$$

as its relation matrix; this is the square $n \times n$ matrix with 1's on the main diagonal and 0's elsewhere.

Referring to definition (11) of the composite (product) of two relations, we see that this is always defined for relations on the same set S. Moreover, the equality relation e on S satisfies $e\alpha = \alpha e = \alpha$ for all α. Finally, since $x[\alpha(\beta\gamma)]y$ and $x[(\alpha\beta)\gamma]y$ both mean that for some elements $z_1,z_2 \in S$, $x\,\alpha\,z_1$, $z_1\,\beta\,z_2$, and $z_2\,\gamma\,y$, we have the associative law

$$\alpha(\beta\gamma) = (\alpha\beta)\gamma \qquad \text{for any relations } \alpha, \beta, \text{ and } \gamma \text{ on } S \tag{13}$$

This proves the following theorem.

Theorem 2. *Under composition (relation multiplication), the binary relations on a set S form an algebraic system which is (i) associative, and (ii) has the equality relation for identity.*

As was mentioned in Sec. 1-5 where it was observed that the functions $f\colon S \to S$ had the same properties, such a system is called a *monoid* (monoids will be studied in Chap. 7). Hence Theorem 2 can be summarized as follows.

Corollary. *The binary relations on any set form a monoid under relation multiplication.*

Binary relations on S are of many types. When $x\,\alpha\,x$ for all $x \in S$, the relation α is called *reflexive;* when $x\,\alpha\,x$ for no $x \in S$, α is called *irreflexive*. When $x\,\alpha\,y$ implies $y\,\alpha\,x$, the relation is called *symmetric;* if not, it is called *asymmetric*. Trivially, $x\,\alpha\,y$ implies $y\,\alpha\,x$ if $x = y$; conversely, when $x\,\alpha\,y$ and $y\,\alpha\,x$ imply $x = y$, the relation α is called *antisymmetric*.

We have seen that the inclusion relation between sets is reflexive and antisymmetric; hence, so is that between relations.

For example, let $T = \{1,2,3\}$, and let α be the relation on T whose graph is

$$S(\alpha) = \{(1,1)(1,2)(2,1)(2,2)(2,3)(3,2)(3,3)\}$$

This α is reflexive and symmetric but not transitive, because $1\,\alpha\,2$ and $2\,\alpha\,3$, yet $1\,\alpha'\,3$. Note that the graph $S(\alpha^2)$ of α^2 is $T \times T$; α^2 is the universally valid relation on T, with $x\,\alpha^2\,y$ for all $x,y \in T$.

Given a binary relation α on a set S, the relation α is called *transitive* when $x\,\alpha\,y$ and $y\,\alpha\,z$ $(x,y,z \in S)$ together imply $x\,\alpha\,z$. Thus the inclusion relation $\subset$ between sets is transitive. The equality relation e defined above (another notation for $=$) is reflexive, symmetric, and transitive.

In terms of the operations and relations introduced in Sec. 2-2, a binary relation α on S is reflexive if and only if it contains the equality relation e so that $e \le \alpha$, or equivalently $e \wedge \alpha = e$ or $e \vee \alpha = a$. It is symmetric if and only if $\alpha = \breve{\alpha}$. It is antisymmetric if and only if $\alpha \wedge \alpha' \leq e$. It is transitive if and only if $\alpha^2 \leqq \alpha$, where α^2 stands for $\alpha\alpha$. We omit the proofs.

Again, if $S = \mathbf{n}$, a relation on S is evidently reflexive if and only if its relation matrix has all 1's on the diagonal; a relation α is symmetric if and only if its relation matrix $\|a_{ij}\|$ is symmetric (that is, $a_{ij} = a_{ji}$ for all i,j).

Finally, the notion of Cartesian product of functions [Chap. 1, (22)] generalizes to that of the (Cartesian or tensor) product of relations, as follows: let α and β be binary relations on sets A and B to the same set Y. We define the (tensor) *product relation* $\gamma = \alpha \times \beta$ from $A \times B$ to Y as follows:

$$(a,b)\ \gamma\ y \qquad \text{means that} \qquad a\ \alpha\ y \text{ and } b\ \beta\ y \tag{14}$$

Likewise, if ξ and η are binary relations from the same A to sets X and Y, we can define $\zeta = \xi \times \eta$ by the rule

$$a\ \zeta\ (x,y) \qquad \text{means that} \qquad a\ \xi\ x \text{ and } a\ \eta\ y \tag{14$'$}$$

EXERCISES A

1. Test the following relations on **P** for being reflexive, antireflexive, symmetric, antisymmetric, and transitive, respectively:
   (a) $m + n$ is even
   (b) $m + n \leqq 100$
   (c) $m + n$ is odd
   (d) m/n is a power of 2
   (e) m/n is even
   (f) mn is odd
2. Which of the following relations on **Z** is symmetric: (a) $m\ \rho\ n$ means that $m + n$ is a multiple of 3, or (b) $m\ \sigma\ n$ means that $m - n$ is divisible by 3? Which is transitive?
3. Show that if ρ and σ are reflexive and symmetric relations on a set S, then the following conditions are equivalent: (a) $\rho\sigma$ is symmetric, (b) $\rho\sigma = \sigma\rho$, (c) $\rho\sigma = \rho \vee \sigma$.
4. Construct two symmetric relations on the set $\{1,2,3\}$ whose product is not symmetric.
5. Compute the matrices of the relation $\leqq$ and the function "negative of" (opposite) on the sets $\{-1,0,+1\}$, $\{-2,-1,0,1,2\}$.
6. Prove in detail that if ρ is symmetric, then $\rho \vee \rho^2 \vee \cdots \vee \rho^n$ is also a symmetric relation.
7. Show that if $\rho \leqq \sigma$ and τ are all binary relations on X, then $\rho\tau \leqq \sigma\tau$ and $\tau\rho \leqq \tau\sigma$ (monotonicity).
8. Show that a relation r on a set U is a bijection of U if and only if $r\breve{r} = \breve{r}r = 1_U$, the equality relation.
9. (a) Show that there are 256 ternary operations f from $\mathbf{2}^3$ to $\mathbf{2}$.
   (★b) For how many of these f do there exist functions g: $\mathbf{2}^2 \to \mathbf{2}$ and h: $\mathbf{2}^2 \to \mathbf{2}$ such that $f(x,y,z) = g(x,h(y,z))$?

2-4. PARTIAL ORDERINGS

A binary relation on a set S is called a *partial ordering* of S when it is reflexive, antisymmetric, and transitive; such relations are commonly indicated by using the special relation symbol $\leqq$. The hypotheses for a partial ordering may thus be written as follows:

P1. $x \leqq x$ for all $x \in S$.
P2. $x \leqq y$ and $y \leqq x$ imply $x = y$.
P3. $x \leqq y$ and $y \leqq z$ imply $x \leqq z$.

Example 3. The relation $\leqq$ is a partial ordering of the set of all positive integers.

Example 4. The relation $m \mid n$ (meaning that m is a divisor of n) is another partial ordering of the set of all positive integers.

Example 5. For any set U, the relation $S \subset T$ is a partial ordering of the power set $\mathcal{P}(U)$ of all subsets of U.

Definition. A partially ordered set or *poset* is a pair $[S, \leqq]$, where $\leqq$ is a partial ordering of S.

There are many other familiar examples of partial-ordering relations. It is therefore of interest to consider some of the many simple properties of posets in general (others are stated in Exercises B below).

First, the converse of any partial ordering $\leqq$ is again a partial ordering, called the *dual* of $\leqq$ and denoted $\geqq$. Thus, $X \geqq Y$ if and only if $Y \leqq X$. Secondly, posets having only a few elements can be conveniently visualized in terms of their *diagrams*. These diagrams can be drawn by using small circles to signify elements and drawing a rising line from each element to each next larger element.

Thus, Fig. 2-1*a* portrays the power set $\mathcal{P}(\mathbf{3})$ of all subsets of

$$\mathbf{3} = \{1,2,3\}$$

partially ordered by the inclusion relation. Note that the diagram of the dual of a partially ordered set $[S, \leqq]$ is obtained from that of S by turning the latter upside down.

Again, Fig. 2-1 are diagrams of the poset $P = [\{2,3,5,7,14,15,21\}, |]$.

Duality Principle. From the easily verified fact that the converse α of any partial ordering $\breve{\alpha}$ of a set S is itself a partial ordering, it follows that we can replace the relation $\leqq$ in any theorem about posets (in general) by the relation $\geqq$ throughout, without affecting its truth. This metamathematical "theorem about theorems" is called the *Duality Principle* of the theory of posets.

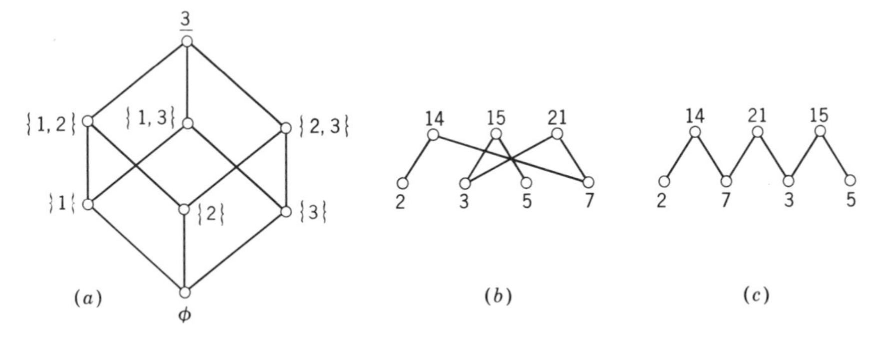

Fig. 2-1. Diagrams of posets.

In the poset $[\mathcal{P}(\mathbf{3}), \subset]$ of Fig. 2-1a, the elements $\varnothing$ and $\mathbf{3}$ are *universal bounds,* in the sense that $\varnothing \leqq x \leqq \mathbf{3}$ for any elements x of $\mathcal{P}(\mathbf{3})$. This concept can be defined in any poset $P = [S, \leqq]$; the elements O and I of S are universal bounds of P (in detail, O is the *least* and I the *greatest* element of P) when

$$O \leqq x \qquad \text{and} \qquad x \leqq I \qquad \text{for any element } x \in S \tag{15}$$

Lemma. *A given poset* $[S, \leqq]$ *can have at most one least element and at most one greatest element.*

Proof. Let O and O^* both be universal lower bounds of $[S, \leqq]$. Then $O \leqq O^*$ (since O is a universal lower bound), and $O^* \leqq O$ (since O^* is also a universal lower bound). Hence, by P2, $O = O^*$. The proof for I is similar.

Posets need not have any universal bounds. Thus, under the usual relation of inequality, the real numbers form a poset $[R, \leqq]$ which has no universal bounds (unless $-\infty$ and $+\infty$ are adjoined to form the "extended reals").

On the other hand, $[R, \leqq]$ is a *simply ordered system* or *chain,* by which we mean that it satisfies P1–P3 (is a poset) and also P4. Given x and y, either $x \leqq y$, or $y \leqq x$.

Evidently, any subset of a chain is a chain. More generally, if $P = [S, \leqq]$ is any poset and $T \subset S$, then $Q = [T, \leqq]$ is also a poset; any subset of a poset is again a poset. This is because P1, P2, P3, and P4 is a "hereditary" property of relations, preserved under restriction of its domain to any subset.

Covering. Associated with any given partial ordering $\leqq$ are many other binary relations. Thus, there is the relation $x < y$, meaning that $x \leqq y$ but $x \neq y$; the relation $x > y$, meaning that $y < x$; and the relation $x \asymp y$ (read "x and y are incomparable"), meaning that neither $x \leqq y$ nor $y \leqq x$ holds. For given $x,y \in [S,\leqq]$, clearly one and only one of the following four alternatives holds: $x = y$, $x > y$, $x < y$, or $x \asymp y$. Another, less obvious relation defined by $\leqq$ is that of covering.

Definition. In a poset $P = [S,\leqq]$, we say that a *covers* b when $a > b$, yet $a > x > b$ for no $x \in S$.

It is this covering relation that is shown in poset diagrams such as those of Fig. 2-1; a rising segment joins each element a with all elements $b,c,$ . . . which cover it. In any *finite* poset, one can conversely determine $\leqq$ from its associated covering relation, as we now show.

Theorem 3. *Let $a < b$ in a finite poset P. Then P contains at least one chain $x_0 = a < x_1 < \cdots < x_l = b$, in which each x_i $(i = 1, \ldots ,l)$ covers its immediate predecessor.*

The proof is by induction on the number n of elements y such that $a < y < b$. If $n = 0$, then b covers a, and thus $P(0)$ is immediate. If $n > 0$, the numbers of y with $a < y < c$ and of z with $c < z < b$ will be at most $n - 1$ since c is excluded. Hence, by induction, there will be (finite) chains of elements, joining a to c and c to b, of the type described in Theorem 3. Putting them together, we get the desired conclusion.

It is suggestive to think of the relation a *covers* b as meaning that a is the immediate superior of b in some hierarchy, such as might be depicted by Fig. 2-1*b*. To get this interpretation, read $x \leqq y$ as "x is y's subordinate."

Definition. An element m of a poset $[S,\leqq]$ is *minimal* when $x < m$ for no $x \in S$; it is *maximal* when $x > m$ for no $x \in S$.

Explanation. By $x < m$, we mean that $x \leqq m$ but $x \neq m$; by $x > m$, we mean that $m < x$.

In any poset with a least element O, clearly O is the only minimal element. However, in an unordered set (where by definition $x \leqq y$ means that $x = y$), every element is minimal. In a chain, a minimal element is least and is therefore unique. In a *finite* poset, the following principle is basic; it asserts *the possibility of consistent enumeration.*

Theorem 4. *Let $[S,\leqq]$ with $S = \{s_1, \ldots ,s_n\}$ be a finite poset. Then the elements of S can be listed as $S = \{x_1, \ldots ,x_n\}$ so that $x_i < x_j$ implies $i < j$.*

Proof. Let $X_m = \{s_1, \ldots ,s_m\}$, where the elements of

$$S = \{s_1, \ldots ,s_n\}$$

are assumed given in any order. We shall construct a sequence of bijections β_m of the sets $\mathbf{m} = 1, \ldots, m$, such that each subset $X_m \subset S$, reordered by β_m as

$$X_m = \{x_1^m, \ldots, x_m^m\} \qquad x_1^m = s_{\beta_m(i)}$$

has the desired property; $x_i^m < x_j^m$ implies $i < j$.

For $m = 1$, β_1 is trivial. Using induction, suppose that $\beta_{n-1}: \mathbf{n-1} \to \mathbf{n-1}$ has been constructed, and let k be the first integer 1 such that $s_n < x_i^{n-1} = x_k^{n-1}$. Form the bijection $\beta_n: \mathbf{n} \to \mathbf{n}$ defined by

$$\beta_n(i) = \begin{cases} i & \text{if } i < k \\ n & \text{if } i = k \\ i + 1 & \text{if } i > k \end{cases}$$

In other words, insert s_n between x_{k-1}^{n-1} and x_k^{n-1}. Then $x_i^n < x_j^n$ with $\{x_i^n, x_j^n\} \subset X_{n-1}$ implies $i < j$ by induction. Also, $s_n = x_k^n < x_j^n$ implies $k < j$ by construction. Finally, $x_i^n < x_k^n = s_n$ implies $x_i^n < x_k^n < x_{k+1}^n$, whence $x_i^n < x_{k+1}^n$ by property P3 (transitivity), whence $i < k + 1$ since $\{x_i^n, x_{k-1}^n\} \subset X_{n-1}$ (induction). Therefore, $i < k$ since $i = k$ is impossible, completing the proof.

Corollary. *Every finite poset P has a minimal element m, such that $x < m$ for no $x \in P$.*

Greatest Lower and Least Upper Bounds. Given a subset S of a poset P, we define $a \in P$ to be a *lower bound* of S when $a \leqq x$ for all $x \in S$, and we define it to be an *upper bound* of S when $a \geqq x$ for all $x \in S$. We define $b \in P$ to be the *greatest lower bound* of S when (*i*) b is a lower bound of S and (*ii*) $b \geqq \tilde{b}$ for any other lower bound of S; in this event, we write $b = \text{glb } S$. Dually, we define $c \in P$ to be the *least upper bound* of S when (*i'*) $c \geqq x$ for all $x \in S$ and (*ii'*) $c \leqq \tilde{c}$ for any other upper bound $\tilde{c}$ of S; in this event, we write $c = \text{lub } S$.

Lemma. *A subset S of a poset P can have at most one glb and at most one lub.*

Proof. If b_1 and b_2 are both glb's of S, then $b_1 \leq b_2$ because b_1 is a lower bound and b_2 is a *greatest* lower bound. Likewise, $b_2 \leqq b_1$ because b_2 is a lower bound and b_1 is a glb of S; that is, $b_1 \leqq b_2$, and $b_2 \leqq b_1$, whence by property P2 $b_1 = b_2$. The proof that lub S is unique is dual.

Posets will be studied further in Sec. 5-2 and Chap. 9.

EXERCISES B

1. Show that the matrix of the relation $i \leqq j$ on the set $\mathbf{n} = \{1,2, \ldots ,n\}$ is triangular.
2. (a) Show that every finite poset has a longest chain.
   (b) Show that every finite poset has a minimal element.
3. Show that a finite poset has a least element if and only if it has exactly one minimal element.
4. (a) Find counterexamples to the statements of Exercise 2 for infinite posets.
   (b) Show that the "if" part of Exercise 3 fails to hold for infinite posets.
5. Show that in a finite poset, $a \leqq b$ if and only if $a = b$ or there exists a finite chain $a = x_0 < x_1 < \cdots < x_l = b$ such that x_i covers x_{i-1} for $i = 1, \ldots ,l$.
6. Let $<$ be any relation on a set P, which is irreflexive and transitive. Define $x \leqq y$ to mean that either $x = y$ or $x < y$. Prove that $\leqq$ is a partial ordering.
7. Let $P = (S, \leqq]$ and $Q = [T, \leqq]$ be two posets. Define $[S \times T, \leqq] = P \times Q$ by having $(s,t) \leqq (s',t')$ mean that $s \leqq s'$ in P and $t \leqq t'$ in Q.
   (a) Show that $P \times Q$ is a poset.
   (b) Show that $P \times Q$ is never a chain unless either P or Q is trivial.
8. In $\mathbf{N} \times \mathbf{N}$, define $(m,n)\ \alpha\ (m_1,n_1)$ to mean that $m \leqq m_1$ and $n \leqq n_1$ in $\mathbf{N}$.
   (a) Show that the relation α is a partial ordering.
   (b) Show that in the poset $[\mathbf{N} \times \mathbf{N}, \alpha]$, every nonvoid subset has a minimal element.

★9. Show that for any two sets A and B, $\mathcal{P}(A \sqcup B) \cong \mathcal{P}(A) \times \mathcal{P}(B)$.

★10. Let P and Q be as in Exercise 7. Define the lexicographic product $P \otimes Q = [S \times T, \leqq]$ by making $(s,t) \leqq (s',t')$ mean that either $s < s'$ in P or $s = s'$ and $t \leqq t'$ in Q.
   (a) Prove that $P \otimes Q$ is always a poset.
   (b) Prove that if P and Q are finite chains, then so is $P \otimes Q$.

★11. Let $P(m,n)$ be a doubly indexed set of propositions, one for each $(m,n) \in \mathbf{N} \times \mathbf{N}$. Let $P(0,0)$ be true, and let $P(m,n)$ imply that $P(m + 1, n)$ and $P(m, n + 1)$ are true. Prove that every $P(m,n)$ is true.

2-5. EQUIVALENCE RELATIONS AND PARTITIONS

A relation E on a set S is called an *equivalence relation* on S when (i.e., if and only if) it is reflexive, symmetric, and transitive. Clearly, the equality relation e is the smallest (least) equivalence relation on S.

Likewise, a *partition* Π of a set S is a collection of subsets S_k of S with the properties (*i*) $S_i \cap S_j = \phi$ if $i \neq j$ (the S_k are disjoint) and (*ii*) $\bigcup S_k = S$ (they exhaust S); that is, each element $x \in S$ is in one and only one S_k, and thus Π decomposes S into various "parts."

Given a partition Π of S, a binary relation $E(\Pi)$ is defined on S by having $xE(\Pi)y$ mean that x and y are in the same subset S_k of Π. Evidently $E(\Pi)$ defined in this way is an equivalence relation.

Conversely, given an equivalence relation E on S, let $p_E\colon S \to \mathcal{P}(S)$ be the function which maps each element $x \in S$ into the set

$$p_E(x) = \{z \in S \mid z\,E\,x\} \subset S \tag{16}$$

By the reflexivity of E, $x \in p_E(x)$, and thus $\cup\, p_E(x) = S$. Now suppose that two such subsets of S ("equivalence classes for E"), $p_E(x)$ and $p_E(y)$, are not disjoint. Then some $z \in S$ must be in both, so that $x\ E\ z$ and $y\ E\ z$. Since E is symmetric, this implies $x\ E\ z$ and $z\ E\ y$; since E is transitive, this in turn implies $x\ E\ y$. Using this result, we see that $y\ E\ t$ implies $x\ E\ y$ and $y\ E\ t$, and thus (again by the transitivity of E) $y\ E\ t$ implies $x\ E\ t$. The conclusion is $p_E(y) \subset p_E(x)$. Reversing the roles of x and y, we see that a repetition of the above argument gives $p_E(x) \subset p_E(y)$. Combining these conclusions, we obtain $p_E(x) = p_E(y)$.

Hence the $p_E(x)$ are disjoint; moreover since $x \in p_E(x)$ for every $x \in S$, $\bigcup p_E(x) = S$. This proves the following result.

Lemma. *Each equivalence relation on a set S defines by* (16) *a partition Π_E of S into disjoint subsets $p_E(x)$.*

The function $p_E\colon S \to \mathcal{P}(S)$ defined by (16) is called the *projection* of S onto the *quotient set S/E*.

Theorem 5. *There is a natural bijection $b\colon E \mapsto \Pi_E$ with inverse $b^{-1}\colon \Pi \mapsto E(\Pi)$ between the set of all equivalence relations E on any given S and the set of partitions Π of S.*

Proof. Given E, Lemma 1 defines Π_E. Conversely, given any partition Π of S, we define $x\ E\ (\Pi) y$ to mean that x and y are in the same subset S_i of S. As in Theorem 3 in Chap. 1, it remains to show that the two functions just defined are inverses; i.e., that $\Pi_{E(\Pi)} = \Pi$ and $E(\Pi_E) = E$. But the proof of the lemma makes it obvious that $\Pi_{E(\Pi)} = \Pi$ and that $E(\Pi_E) = E$ is obvious directly.

Corollary. *If E is any equivalence relation on S, then there is a function $p_E\colon S \to S/E$ from S onto a quotient set $S/E \subset \mathcal{P}(E)$, whose elements are the equivalence classes of E.*

Example 6. Consider the relation E_m defined on $\mathbf{Z}$ by the condition that $x\ E_m\ y$ means $m \mid (x - y)$. This relation is usually written $x \equiv y$ (mod m) and is read "x is congruent to y modulo m." Since $x - x = m0$, E_m is reflexive; since $x - y = mk$ implies $y - x = m(-k)$, E_m is symmetric; and since $x - y = mk$ and $y - z = mk'$ imply $x - z = m(k + k')$, E_m is transitive. Hence, E_m is an equivalence relation. Its equivalence classes are the m arithmetic progressions $\ldots, k - m, k, k + m, k + 2m, \ldots$ for $k = 0,1, \ldots, m - 1$.

Example 7. Let $S = \{a,b,c,d,e\}$, and let E be the equivalence relation whose matrix is shown in Table 2-2a. Then the assignments of the function $p_E\colon S \to \mathcal{P}(S)$ are as in Table 2-2b, and $S/E = \{A,C,E\}$ as indicated.

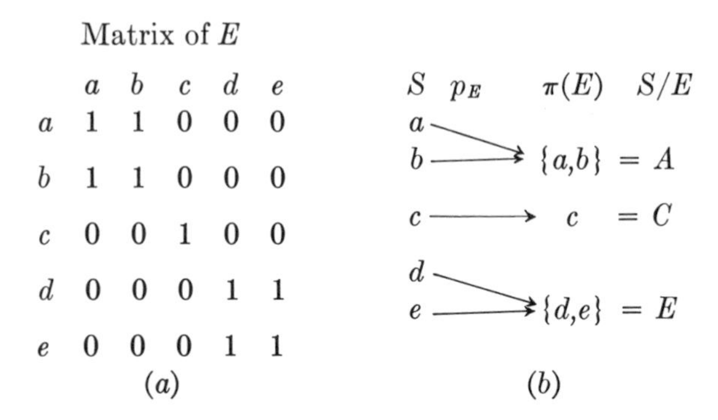

Matrix of E

	a	b	c	d	e
a	1	1	0	0	0
b	1	1	0	0	0
c	0	0	1	0	0
d	0	0	0	1	1
e	0	0	0	1	1

(a)

(b)

Factorization of Functions. The projection $p_E: S \to S/E$ just constructed is the first of three factors into which any function $f: S \to T$ can be factored. The last factor is an "insertion" in the following sense.

Definition. Let $f: S \to T$ be given, together with any subsets $X \subset S$ and $Y \subset T$. We define $f(X)$ as the image of X under f, that is, as the set of all $f(x)$ with $x \in X$. [The set $f(S)$ (the image of the domain) is often called the *range* of $f: S \to T$.] If $f(X) \subset Y$, the function $g: X \to Y$ which is defined by the identity $g(x) = f(x)$ on X is called the *restriction* of f to X and Y. In particular, for $S \subset T$, the restriction of $1_T: T \to T$ to S as domain is called the *insertion* $i: S \to T$ of S into T. (By our convention on the equality of functions, this insertion is not equal to either 1_S or to 1_T, unless $S = T$.)

If $f: S \to T$ is any function with domain S, codomain T, and image $f(S) = Y \subset T$, then clearly f is the right *composite*

$$f = p \diamond b \diamond i \tag{17}$$

of the projection $p: S \to S/E_f$ which maps S onto the quotient set of the equivalence relation E_f defined by f, followed by the bijection $b: S/E_f \leftrightarrow f(S)$ which associates each $y \in f(S)$ with the set $S_i = f^{-1}(y)$ of its antecedents in S, and finally by the (one-one) insertion i of $Y = f(s) \subset T$ into T. Equation (17) is called the *canonical factorization* of f. It shows that any function $f: S \to T$ is determined up to a bijection b by its range $f(S) \subset T$ and the equivalence relation E_f which it induces on S.

EXERCISES C

1. Show that if ρ is reflexive and transitive, $\rho \wedge \breve{\rho}$ is an equivalence relation on S.
2. Show that in Exercise A3, if ρ and σ are equivalence relations, then each of conditions (b) and (c) is necessary and sufficient for $\rho\sigma$ to be an equivalence relation.
3. Two sets ("figures") F and G in the plane are called *isometric* (in symbols, $F \cong G$) when there is a bijection $\mu: F \leftrightarrow G$ which preserves distances. Show that the relation $F \cong G$ is an equivalence relation.

4. Prove that a set of n elements has exactly $2^{n-1} - 1$ partitions into two equivalence classes.
5. (a) Show that a relation α on a set S is both symmetric and antisymmetric if and only if its relation matrix $\|a_{ij}\|$ is "diagonal" (i.e., when $i \neq j$ implies $a_{ij} = 0$).
   (b) Show that this condition is equivalent to $\alpha \leq e$.

★6. Let $\pi(n,k)$ be the number of partitions of a set of n elements into k nonvoid subsets. Prove that

$$\sum_{k=1}^{l} (k!)\pi(n,k) = l^n$$

2-6. MODULAR NUMBERS; MORPHISMS

For each integer $m > 1$, the equivalence relation E_m: $x \equiv y \pmod{m}$ of Example 6 defines a remarkable number system having as elements the equivalence classes of $\mathbf{Z}/E_m$. This number system is called the *modular number system* $\mathbf{Z}_m$; we shall now describe its construction.

First, observe that the equivalence class modulo m containing any given $n \in \mathbf{Z}$ is, by definition, the set of all $n + km$ with $k \in \mathbf{Z}$. Moreover, this set contains a (unique) *least nonnegative element* $r_m(n)$ which, for $n \in \mathbf{N}$, is just the $r(m,n)$ of Chap. 1, (38). The integers $0,1, \ldots, m - 1$ form, therefore, a *complete set of representatives* of the equivalence classes $\{n + km\}$; just one from each equivalence class.

For fixed $m > 1$, we now define $\mathbf{m} = \{0,1, \ldots, m\}$ as this set of representatives, and we define $\mathbf{Z}_m = [\mathbf{m},+,\times]$ as an *algebraic system with two binary operations,* addition and multiplication. These operations are defined by the formulas

$$a \overset{m}{+} b = r_m(a + b) \qquad a \overset{m}{\times} b = r_m(a \times b) \tag{18}$$

where $+$ and $\times$ denote addition and multiplication in $\mathbf{Z}$.

Lemma. *For all n, $n' \in \mathbf{Z}$, we have*

$$r_m(n + n') = r_m(n) + r_m(n') \quad \textit{and} \quad r_m(n \times n') = r_m(n) \times r_m(n') \tag{19}$$

Proof. Evidently, both sides of each equation of (19) are members of $\mathbf{m} = \{0,1, \ldots, m - 1\}$. On the other hand, they differ from $n + n'$ (and so from each other) by integral multiples of m (see Example 6). Hence they must be equal.

We used the symbols $\overset{m}{+}$ and $\overset{m}{\times}$ above to signify addition and multiplication in $\mathbf{m}$ to avoid confusion with addition and multiplication in $\mathbf{Z}$ (which contains $\mathbf{m}$). From now on, however, we shall adopt the standard notation $[\mathbf{Z}_m,+,\cdot]$ in place of $[\mathbf{m},+,\times]$. This notation also brings out more vividly the significance of (19), which simplifies to

$$r_m(n + n') = r_m(n) + r_m(n') \qquad r_m(nn') = r_m(n)r_m(n')$$

That is, r_m is a "morphism" for the operations of multiplication and addition in the following general sense.

Definition. Let $[A,+,\cdot]$ and $[B,+,\cdot]$ be two algebraic systems, each with binary addition and multiplication operations signified by $+$ and $\cdot$, respectively. A function $\theta: A \rightarrow B$ is a *morphism* (for addition and multiplication) when for all $a,a' \in A$,

$$\theta(a + a') = \theta(a) + \theta(a') \qquad \text{and} \qquad \theta(aa') = \theta(a)\theta(a') \tag{20}$$

Equations (20) can also be expressed diagrammatically in terms of the mapping diagram of Fig. 2-2 for addition and in terms of a similar diagram for multiplication.

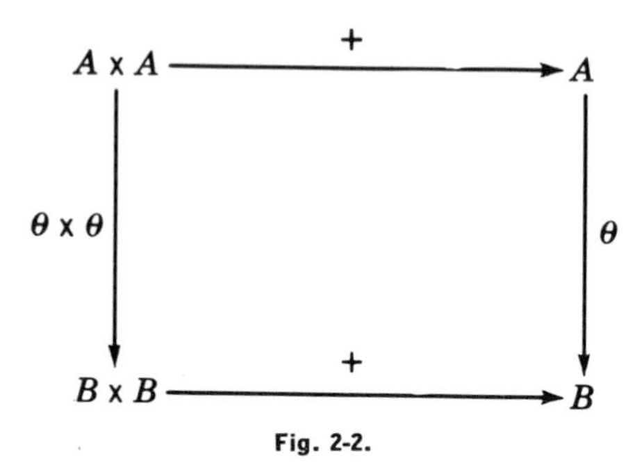

Fig. 2-2.

In general, a morphism (a mapping which "preserves" one or more operations) is called an *epimorphism* when it is onto (surjective), a *monomorphism* when it is one-one (injective), and an *isomorphism* when it is both (a bijection).

Using our new terminology, we can restate our lemma as follows.

Theorem 6. *The mapping which carries each nonnegative integer into its remainder under division by a fixed integer $m > 1$ is an epimorphism for addition and multiplication from $\mathbf{Z}$ to $\mathbf{Z}_m$.*

The image $\mathbf{Z}_4$ of the morphism $r_4: [N,+,\cdot] \rightarrow [\mathbf{Z}_4,+,\cdot]$ referred to above has the addition and multiplication tables displayed in Fig. 2-3.

+	0	1	2	3
0	0	1	2	3
1	1	2	3	0
2	2	3	0	1
3	3	0	1	2

·	0	1	2	3
0	0	0	0	0
1	0	1	2	3
2	0	2	0	2
3	0	3	2	1

Fig. 2-3.

The Substitution Property. In Sec. 2-5, we noted that each function $\phi: S \to T$ defined an equivalence relation E_ϕ on S and that there was a natural bijection $b: S/E_\phi \leftrightarrow \phi(S)$. Conversely, we showed that for a given equivalence relation E on a set S, there was a surjection $\phi_E: S \to S/E$; moreover $E_{\phi_E} = E$. We next ask what better condition is necessary on E_ϕ (respectively, E) in order for ϕ (respectively, ϕ_E) to be a morphism? We shall now show that this condition is expressed in terms of a simple Substitution Property.

Consider first Example 6. In this example the relation E_m is equivalent to the statement that the binary relation E_m, defined by

$$xE_my \qquad \text{means that} \qquad m \mid (y - x) \tag{21}$$

[or equivalently, that $r_m(x) = r_m(y)$] has the following substitution property:

$$xE_my \qquad \text{implies} \qquad \varphi(x)E_m\varphi(y) \tag{21'}$$

The relation E_m defined by (21) also has the following two substitution properties for the binary operations + and · on $\mathbf{Z} \times \mathbf{Z}$ to $\mathbf{Z}$:

$$xE_my \text{ and } x'E_my' \qquad \text{together imply} \qquad (x + x')\ E_m\ (y + y') \text{ and } (xx')\ E_m\ (yy') \tag{22}$$

The properties just described are all special cases of the following concept.

Definition. Let E be an equivalence relation on S, and let f be an $n$ary operation on S. Then the equivalence relation E is said to have the *substitution property with respect to* f if and only if, for all $x_i,y_j \in S$,

$$x_1\ E\ y_1, x_2\ E\ y_2, \ldots, x_n\ E\ y_n \qquad \text{imply} \qquad f(x_1,x_2, \ldots, x_n)\ E\ f(y_1,y_2, \ldots, y_n) \tag{23}$$

Cardinal Numbers. The most fundamental application of the concepts of equivalence relation and Substitution Property is to the class Γ of all sets. We define the relation E on Γ by (*). SET means that there is a bijection $b: S \leftrightarrow T$.

It will be shown in Chap. 14 that E is an equivalence relation which has the Substitution Property for the binary operations of disjoint sum, Cartesian product, and power set. From this fact, the laws of arithmetic for finite and infinite cardinal numbers will be derived there in a few lines.

We can define an $n$ary operation $\tilde{f}$ on the quotient set S/E in a natural way by $\tilde{f}(p_E(x_1), \ldots ,p_E(x_n) \cdot p_E(f(x_1, \ldots ,x_n))$. [The reader should check that the substitution property (23) of E guarantees that f is single-valued.] The connections between morphisms and equivalence relations having the substitution property (23) is now evident; that is, if an equivalence relation E on S has the substitution property with respect to an operation f on S, then the assignment $x \mapsto p_E(x)$ is a morphism from $[S,f]$ to $[S/E,\tilde{f}]$.

We shall return to these ideas in later chapters. They are among the most fundamental concepts of algebra.

Morphisms of Relations. The notion of morphism defined above for binary operations has analogs for relations. Thus, if P and Q are any two posets, a function f: $P \to Q$ is sometimes called an *order morphism* when $x \leqq y$ in P implies $f(x) \leqq f(y)$ in Q. However, morphisms of binary operations are in closer analogy with morphisms of ternary relations. Define a *ternary relation* on a set S as a function α: $S^3 \to \{0,1\}$; we write $(x,y,z)\alpha$ when $\alpha(x,y,z) = 1$, and $(x,y,z)\alpha'$ when $\alpha(x,y,z) = 0$. Then each binary "addition" $+: S \times S \to S$ on a set S defines a ternary relation on S, with $(x,y,z)\alpha$ meaning[1] that $x + y = z$. A function θ: $S \to T$ is evidently a morphism for $+$ (on S and T) if and only if it is a morphism of relations for the ternary relation $x + y = z$, in the sense that $a + a' = b$ in S implies $\theta(a) + \theta(a') = \theta(b)$ in T [see the first equation of (20)].

2-7. CYCLIC UNARY ALGEBRAS

The simplest class of algebraic systems consists of algebras with *one unary* operation. For simplicity, in this chapter[2] we shall call such systems *unary algebras*. We first recall two very simple unary algebras.

Example 8. *The extended Peano algebra* is $[\mathbf{N},\tau]$, where $\mathbf{N}$ is the set of nonnegative integers and $\tau(n) = n + 1$.

[1] In other words, $\alpha(x,y,z)$ takes on the "truth value" of $x + y = z$, which is 1 if the equation is true and 0 if it is false.

[2] Systems with more than one unary operation, but no binary, ternary, . . . relations, are also called unary algebras.

Example 9. The *clock algebra* of the order m is $[\mathbf{m},\sigma_m]$, where $\mathbf{m} = \{1, \ldots, m\}$ and

$$\sigma_m(k) = \begin{cases} k+1 & \text{unless } k = m \\ 1 & \text{if } k = m \end{cases} \tag{24}$$

We can signify the action of any unary operation by an arrow, thus indicating $[\mathbf{N},\tau]$ and $[\mathbf{8},\sigma_8]$ by Fig. 2-4a and b, respectively.

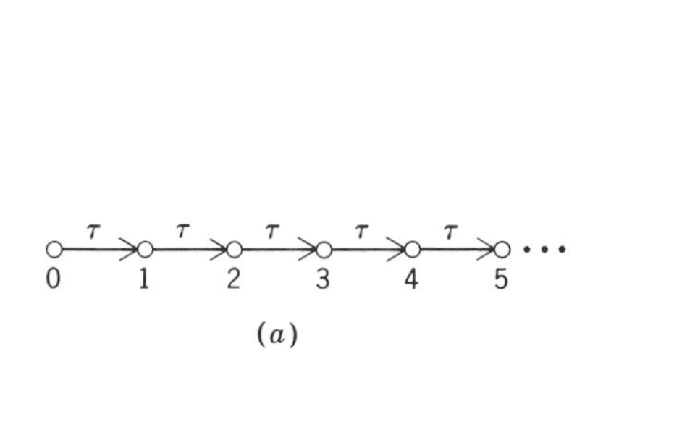

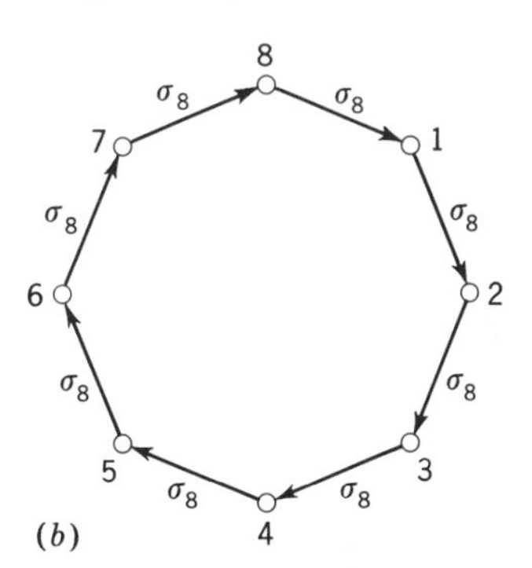

Fig. 2-4.

In any unary algebra $[S,f]$, we define a subset $T \subset S$ to be *f-closed* when $t \in T$ implies $f(t) \in T$. The void set and S are (trivially) always f-closed. In $[\mathbf{m},\sigma_m]$, the only σ_m – closed subsets are the empty set and $\mathbf{m}$; in $[\mathbf{N},\tau]$, for any n, the set of all $k \geqq n$ is τ-closed. Evidently also, the element $0 \in \mathbf{N}$ *generates* $\mathbf{N}$, in the sense that the only τ-closed subset X of $\mathbf{N}$ which contains 0 is $\mathbf{N}$ itself.

The ultimate goal of this section will be to determine the most general unary algebra with one generator. But first, we establish an important preliminary result. Let $[S,f]$ be any unary algebra, and let $a \in S$ be chosen arbitrarily. Consider the mapping

$$\theta: 0 \mapsto a,\ 1 \mapsto f(a),\ 2 \mapsto f^2(a),\ \ldots,\ n \mapsto f^n(a),\ \ldots \tag{25}$$

as a function $\theta: \mathbf{N} \to S$. Since $\tau(n) = n + 1$,

$$\theta(\tau(n)) = \theta(n+1) = f^{n+1}(a) = f(f^n(a)) = f(\theta(n)) \tag{25'}$$

for all $n \in N$. Hence θ is a morphism of unary algebras, in the following sense.

Definition. If $[S,f]$ and $[T,g]$ are two unary algebras, the mapping $\theta: S \to T$ is called a *morphism* when

$$\theta(f(s)) = g(\theta(s)) \qquad \text{for all } s \in S \tag{26}$$

that is, when $\theta \circ f = g \circ \theta$, as in the commutative mapping diagram in Fig. 2-5. (Here and elsewhere, a mapping diagram is called *commutative* when any two paths joining the same initial and final points correspond to sequences of maps having the same right composite.)

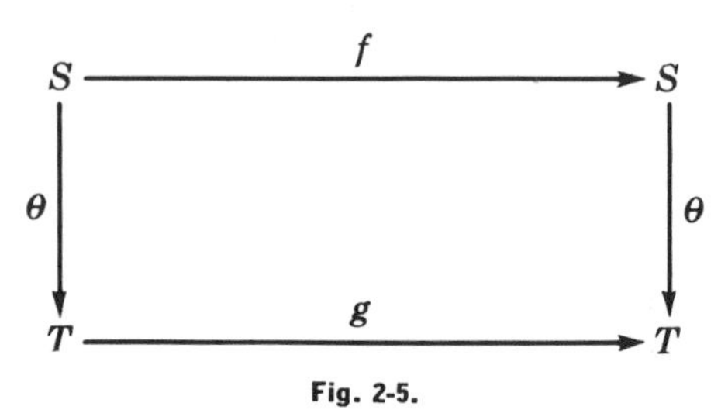

Fig. 2-5.

The result proved in (18) and (19) can be restated in terms of the morphism concept as follows.

Theorem 7. *Let a be any element of any unary algebra* $[S,f]$. *Then there exists a (unique) morphism* θ: $[\mathbf{N},t] \mapsto [S,f]$ *which maps* $0 \in \mathbf{N}$ *into* a.

In other words, $[\mathbf{N},\tau]$ has the *universal mapping property* that any mapping of its generator 0 can be extended to a morphism. This property defines $[\mathbf{N},\tau]$ as the *free unary algebra* with one generator. Although we shall describe other free algebras in the sequel, we shall not discuss this concept further here.

Corollary. *Any unary algebra with one generator is an epimorphic image of* $[\mathbf{N},\tau]$.

Finally, we shall determine (up to isomorphism) all possible unary algebras $[S,f]$ with one generator $a \in S$. Since S is generated by a, $S = \{f^n(a)\}$, the set of all $f^n(a)$, which is closed under f.

Case 1. The $f^n(a)$ are all different. In this case, the correspondence $n \mapsto f^n(a)$ is an isomorphism from $[N,\tau]$ to $[S,f]$.

Case 2. For some first p, there exists a $k < p$ with $f^p(s) = f^k(a)$. Let $p = k + m$. By induction on j, we then have

$$f^{p+j}(a) = f^j(f^p(a)) = f^j(f^k(a)) = f^{k+j}(a) \qquad \text{all } j \in \mathbf{N} \qquad (27)$$

In particular, the diagram of $[S,f]$ has the schematic form indicated in Fig. 2-6. It consists of a loop (cycle) of period m preceded by a "tail" of length k.

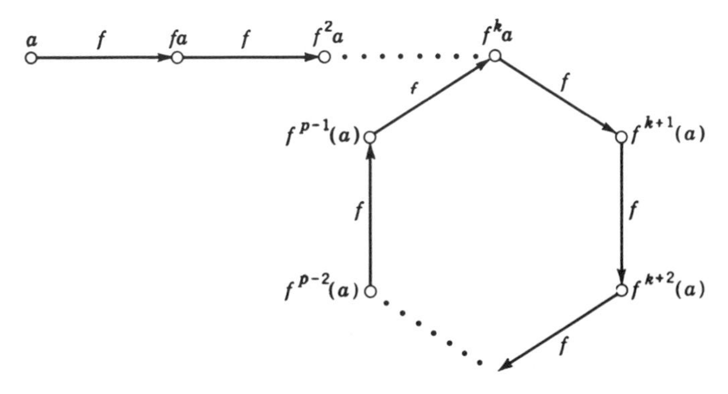

Fig. 2-6.

We therefore define the *unary algebra* U_m^k to have as elements the integers $0,1, \ldots, m + k - 1$ and unary operation $\tau_{m,k}$, with

$$\tau_{m,k}(j) = \begin{cases} j + 1 & j \neq m + k - 1 \\ k & j = m + k - 1 \end{cases}$$

We have proved the following theorem.

Theorem 8. (*Dedekind*) *Every unary algebra with one generator is isomorphic to* $\mathbf{P}$ *or to some* U_m^k $(m \in \mathbf{P},\ k \in \mathbf{N})$.

EXERCISES D

1. Prove that the poset $[\mathbf{n}, \leqq]$ has no nontrivial order automorphism (i.e., isomorphism with itself).
2. Show that any bijection between two finite posets which preserves the covering relation is an isomorphism of the order relation. (*Hint:* Use Exercise B5.)
3. Show that the function $f\colon 0 \mapsto 0,\ 1 \mapsto 3,\ 2 \mapsto 2,\ 3 \mapsto 1$ is an isomorphism for addition in $\mathbf{Z}_4$, but not for multiplication.
4. Define $f\colon \mathbf{4} \to \mathbf{4}$ by $f(1) = 2, f(2) = 3, f(3) = 2,\ f(4) = 1$.
   (a) Show that the powers $f^0 = 1_3, f^1 = f,\ f^2 = f \circ f, f^3 = f^2 \circ f$ are all different functions, but that $f^4 = f^2, f^5 = f^3$.
   (b) Construct a multiplication table for the monoid of the powers of f.
5. Show that any chain of n elements is isomorphic with $[\mathbf{n}, \leqq]$.
6. Show that the assignment $\rho \to \breve{\rho}$ is an isomorphism from $[\mathcal{P}(S \times S), \diamond]$ to $[\mathcal{P}(S \times S), \circ]$ with respect to $\cap$, $\cup$, $'$, as well as an isomorphism of monoids.
7. For given $m,n \in \mathbf{N}$, define the relation E on $\mathbf{N}^m$ by the following statement: $(x_1, \ldots, x_m)E(y_1, \ldots, y_m)$ means $n \mid (x_i - y_i)$ for all $i \in \mathbf{m}$.
   (a) Show that E is an equivalence relation.
   (b) Define addition and multiplication (componentwise) by

$$(x_1, \ldots, x_m) + (y_1, \ldots, y_m) = (x_1 + y_1, \ldots, x_m + y_m)$$
$$(x_1, \ldots, x_m)(y_1, \ldots, y_m) = (x_1y_1, \ldots, x_my_m)$$

Show that E has the substitution property with respect to these operations.

8. Let E be any equivalence relation on $\mathbf{Z}_n$.
   (a) Show that if E has the Substitution Property for addition in $\mathbf{Z}_n$, then it has the substitution property for multiplication.
   (b) Find an equivalence relation on Z_3 which has the Substitution Property for multiplication, but not for addition.
9. Let $A = [S,f]$ be any finite unary algebra with k elements. Define aRb in A to mean that for some $n \in \mathbf{N}$, $f^n(a) = b$.
   (a) Show that this relation is reflexive and transitive.
   (b) Show that A is cyclic if and only if for some $a \in A$, aRb for all $b \in A$.
10. Let P be any poset. Consider the function $\phi: P \to \mathcal{P}(P)$ which assigns to each $a \in P$ the set $A = \phi(a)$ of all $x \leqq a$ in P. Show that this is injective and carries $\leqq$ into $\subset$ (is a monomorphism).

2-8. DIRECTED GRAPHS

A *directed* graph (or *digraph*) $\vec{G} = [N,A,\phi]$ consists of (1) a set N of *nodes* (or *vertices*), (2) a set A of *arcs*, and (3) a function $\phi: A \to N \times N$ which maps each arc $a \in A$ into an ordered pair (p,q) of nodes called its *ends*. An arc (p,p) from a node to itself is called a *loop;* a graph with no loops is called a *loop-free.*

Figure 2-7 shows four directed graphs, each having four nodes. These directed graphs are all loop-free and simple,[1] by which we mean that at most one arc goes from a given node p to any node q. In simple directed graphs, we can signify the unique arc (when it exists) from p to q by $\overrightarrow{pq}$.

Two important families of simple directed graphs, each depending on a parameter $n \in \mathbf{P}$, are the following: A *simple path* $\vec{\Pi}_n$ of length n has $n + 1$ nodes $P_0, \ldots ,P_n$ and n arcs $\overrightarrow{P_kP_{k+1}}$ $(k = 0, \ldots , n - 1)$ joining successive nodes; a *simple circuit* Γ_n of length n has n distinct nodes $p_1, \ldots ,p_n$, and for each $k \neq n$, it has an arc $\overrightarrow{p_kp_{\sigma k}}$ from each p_k $(k < n)$ to p_{k+1} and an arc p_n to p_1. (Thus, a simple circuit of length 1 is a loop from p_1 to itself.) All simple paths of length n are isomorphic, and all simple circuits of length n are isomorphic.

Figure 2-7*a* depicts a simple path of length 3, and Fig. 2-7*b* depicts a simple circuit of length 4 (the graph of the unary algebra U_4^0 of Theorem

[1] Some authors require a simple directed graph to be loop-free, thus restricting Theorem 9 to antireflexive relations.

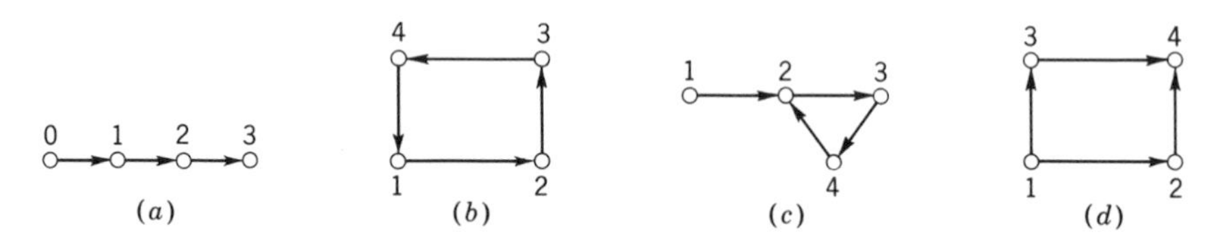

Fig. 2-7. Examples of directed graphs.

8). Figure 2-7*c* shows the graph of the unary algebra U_3^1 of Theorem 8, while Fig. 2-7*d* shows a directed graph which occurs very commonly in commutative diagrams.

Definition. An *isomorphism* between directed graphs $\vec{G} = [N,A,\phi]$ and $\vec{G}^* = (N^*,A^*,\phi^*]$ is a pair of bijections $\beta: N \to N^*$ and $\gamma: A \to A^*$ such that arc **a** goes from node p to node q in $\vec{G}$ if and only if $\phi(\mathbf{a})$ goes from $\beta(p)$ to $\beta(q)$ in $\vec{G}^*$; that is, the condition is that $\phi(a) = (p,q)$ in $\vec{G}$ if and only if $\phi^*(\gamma(a)) = (\beta(p),\beta(p))$ in $\vec{G}^*$, or that the diagram of Fig. 2-8 is commutative.

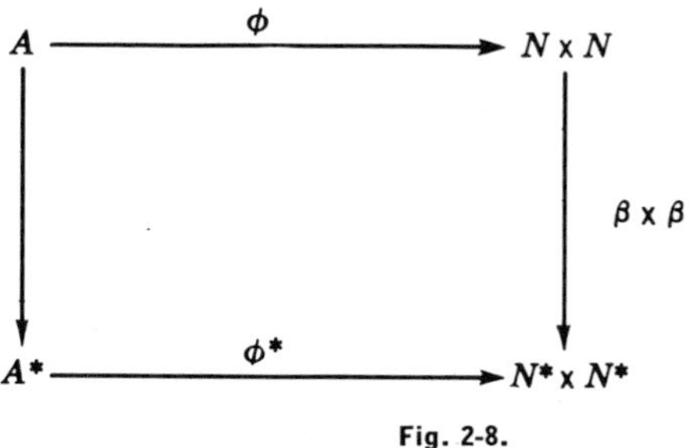

Fig. 2-8.

If $\vec{G}$ and $\vec{G}^*$ are simple, as in the graphs of Fig. 2-7, we have an easier test. A bijection $\beta: N \to N'$ is part of an isomorphism if and only if $\beta \times \beta: \overrightarrow{pq} \mapsto \overrightarrow{p^*q^*}$ $(p^* = \beta(p),\ q^* = \beta(q))$ is a bijection of the set of all arcs $\overrightarrow{pq}$ of G onto the set of all arcs $\overrightarrow{p^*q^*}$ of G, so that

$$\phi^* \circ \gamma = (\beta \times \beta) \circ \phi$$

Two directed graphs are called *isomorphic* when there is an isomorphism between them. Isomorphic directed graphs are usually considered to be the "same," and we shall consider here only properties which are preserved under isomorphism, such as the property of being simple, a simple path, or a simple circuit.

Directed Graphs and Relations. Each directed graph $G = [N,A,\phi]$ determine a binary *successor relation* $\sigma_{\vec{G}}$ on the set N of its nodes. By definition

$$p\sigma_{\vec{G}}q \qquad \text{means that} \qquad \phi(\mathbf{a}) = (p,q) \text{ for some } \mathbf{a} \in A \tag{28}$$

Conversely, each binary relation ρ on a set N defines a simple directed graph $\vec{G} = \vec{G}(\rho) = [N,A(\rho),\psi]$, whose arcs are the $\overrightarrow{pq}$ with $p\rho q$; thus

$$A(\rho) = \{\overrightarrow{pq} \mid p \in N,\ q \in N,\ p\rho q\},\ \psi(\overrightarrow{pq}) = (p,q) \tag{29}$$

It is easy to verify that *if* $\vec{G}$ is a simple directed graph to begin with, then $\vec{G}(\sigma_{\vec{G}}) = \vec{G}$, whereas $\sigma_{\vec{G}(\rho)} = \rho$ for *any* binary relation ρ; that is, we have the following theorem.

Theorem 9. *The class of relational systems* $[U,\rho]$ *is interchangeable with that of simple directed graphs, through the bijections defined by* (28) *and* (29).

As we have already observed, the graphs of Fig. 2-7 are all simple. Moreover, their nodes are ordered numerically; hence, Theorem 9 applies to them. Figure 2-9 shows the 4×4 relation matrices $R = \|\rho_{ij}\|$ of the relations defined from these four graphs by the construction of Theorem 9.

$$\begin{bmatrix} 0 & 1 & 0 & 0 \\ 0 & 0 & 1 & 0 \\ 0 & 0 & 0 & 1 \\ 0 & 0 & 0 & 0 \end{bmatrix} \qquad \begin{bmatrix} 0 & 1 & 0 & 0 \\ 0 & 0 & 1 & 0 \\ 0 & 0 & 0 & 1 \\ 1 & 0 & 0 & 0 \end{bmatrix} \qquad \begin{bmatrix} 0 & 1 & 0 & 0 \\ 0 & 0 & 1 & 0 \\ 0 & 0 & 0 & 1 \\ 0 & 1 & 0 & 0 \end{bmatrix} \qquad \begin{bmatrix} 0 & 1 & 1 & 0 \\ 0 & 0 & 0 & 1 \\ 0 & 0 & 0 & 1 \\ 0 & 0 & 0 & 0 \end{bmatrix}$$

(*a*) (*b*) (*c*) (*d*)

Fig. 2-9. Relation matrices of the graphs of Fig. 2-7.

More generally, if $\vec{G}$ is *any* directed graph, then $\vec{G}(\sigma_{\vec{G}})$ is the simplest directed graph obtained from $\vec{G}$ by "merging" multiple arcs having the same initial and final endpoints.

Matrices of Directed Graphs. If $\vec{G}$ is any directed graph, then any orderings of the sets N,A of its nodes and arcs determines an $|N| \times |A|$ *incidence matrix* $B = \|b_{ij}\|$, by the rule[1]

$$b_{ij} = \begin{cases} -1 & \text{if } a_j \in A \text{ issues from } n_i \in N \\ +1 & \text{if } a_j \in A \text{ goes to } n_i \in N \\ 0 & \text{otherwise} \end{cases} \tag{30}$$

If $\vec{G}$ is *simple,* then one can also describe $\vec{G}$ (relative to any ordering of N) by its $|N| \times |N|$ *successor relation matrix* $\|\sigma_{ij}\|$, where

$$\sigma_{ij} = \begin{cases} 1 & \text{if } n_i \, \sigma_G \, n_j \\ 0 & \text{otherwise} \end{cases} \tag{31}$$

[1] Our convention is that the arrow goes from -1 (tail) to $+1$ (head); Busacker-Saaty (p. 103) and Berge (p. 141) use the opposite convention (names refer to authors of references at end of chapter).

Graphs of Covering Relations. Figure 2-9*a* and *d* above depict *covering* relations in two particular posets, a chain of four elements and the Boolean algebra $\mathbf{2}^2 = \mathcal{P}(\mathbf{2})$, respectively. Likewise, the directed graph of Fig. 2-10 depicts the covering relation in the poset $\mathcal{P}(\mathbf{3})$; it is evidently suggestive of a cube. This example can be generalized as follows.

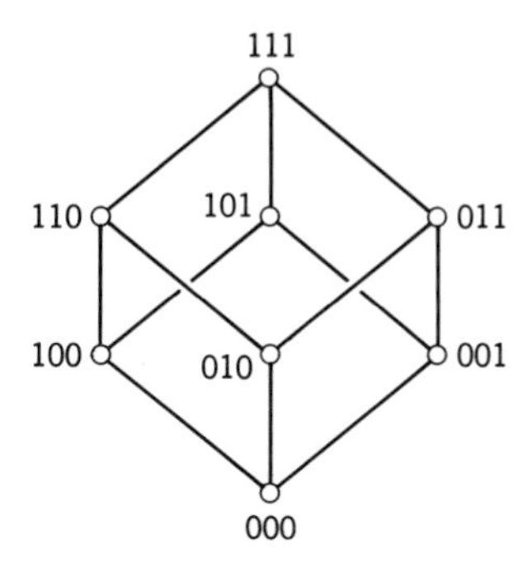

Fig. 2-10.

Example 10. The directed graph of an *n-cube* has nodes corresponding to the 2^n n-digit binary numbers $\mathbf{x} = x_1 \cdots x_n$, $\mathbf{y} = y_1 \cdots y_n$, etc. It is the simple directed graph of the following covering relation: $\mathbf{x}$ *covers* $\mathbf{y}$ if and only if $\mathbf{y}$ can be obtained from $\mathbf{x}$ by changing one digit $x_i = 1$ of $\mathbf{x}$ to $0 = y_i$. Note that this is the covering relation in the poset $\mathcal{P}(\mathbf{n})$, partially ordered by the inclusion relation.

2-9. GRAPHS

Though labeled directed graphs are extremely useful for describing engineering processes, the following closely related concept is more natural from a purely mathematical standpoint and has been more thoroughly studied by mathematicians.

Definition. A *graph* $G = [V,E,\theta]$ consists of (1) a set V of *nodes* (or vertices), (2) a set E of *edges* or *links*, and (3) a function θ which maps each edge $a \in E$ onto a *unordered* pair $(p,q) = (q,p)$ of nodes called its *ends*. An edge (p,p) is called a *loop*.

Example 11. A *complete* n-graph has an edge joining every pair of distinct points. Thus Fig. 2-11*a* and *b* depict a complete 4-graph and a complete 5-graph, respectively.

Example 12. An *n-cube* has 2^n nodes, bijective to the set of all binary n-tuples (sequences of n binary digits). Two nodes are joined by an edge (link) precisely when the corresponding binary numbers differ in just one digit. Figure 2-11c depicts the graph of a 3-cube. The graphs of Fig. 2-11c and d are isomorphic (prove it).

The graphs in Fig. 2-11 are all without loops and also without "parallel" edges a and b having the same two endpoints. Such graphs are called *simple.*

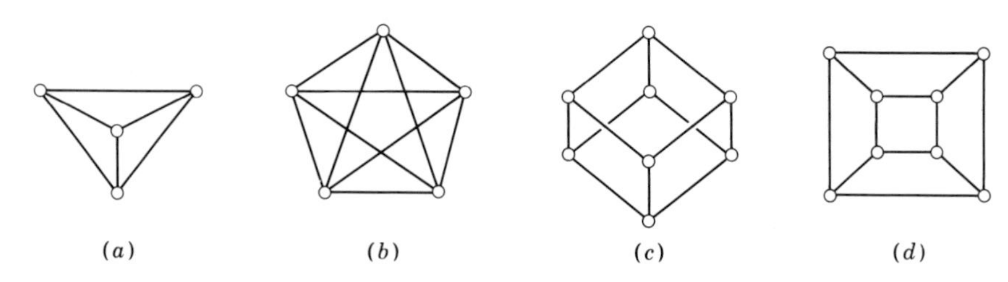

Fig. 2-11. Examples of graphs.

Any graph $G = [N,L]$ is clearly determined by its *incidence* relation ρ_G on E to V, defined by the condition that $l \; \rho_G \; p$ means that $p \in V$ is one end of the link $l \in E$. A simple graph is also determined by the "adjacency" relation α on V, defined by the condition that $p \; \alpha \; q$ means that p and q are joined by a (unique) edge. This relation α is irreflexive and symmetric.

Relative to any listings of the elements of E and V, the incidence relation ρ_G is defined by the *incidence matrix* R_G; relative to any listing of the nodes and edges, the adjacency relation of a simple graph is determined by the (symmetric and irreflexive) *adjacency matrix* $A = \|\alpha_{ij}\|$.

The notions defined for directed graphs in Sec. 2-8 can be extended to undirected or symmetric graphs by associating with each edge the *two* arcs $\overrightarrow{ab}$ and $\overrightarrow{ba}$. We then have the following obvious result.

Theorem 10. *The following correspondences are mutually inverse bijections from the class of all simple graphs to the class of loop-free simple directed graphs with symmetric linking relations:*

$$(i) \; [N,E] \mapsto [N,L], \text{ where } \overrightarrow{pq} \in L \quad \text{if and only if } \overline{pq} \in E \tag{32}$$

$$(ii) \; [N,L] \mapsto [N,E], \text{ where } \overline{pq} \in E \quad \text{if and only if } \overrightarrow{pq} \in L \tag{32'}$$

The bijection of Theorem 10 converts the definitions of simple path and simple circuit, given in Sec. 2-8, into the following definitions. A *chain* of length n in a graph G is a list of nodes $p_i \in V$ with $i = 0, \ldots, n$ and every $\overline{p_{i-1}p_i} \in E$. If also $\overline{p_n p_0} \in E$, then the list of edges $\overline{p_0 p_1}, \ldots, \overline{p_n p_0}$ specified forms a *cycle* of length $n + 1$. If no node occurs twice, the chain (or cycle) is called *simple*. Evidently, if any path connects p to q, there is a simple chain which connects p to q, and any simple chains having the same endpoints define a cycle.

Bipartite Graphs. A graph G is called *bipartite* when there is a partition of its nodes into two nonempty complementary subsets R and B ("red" and "black" nodes, as on a checkerboard), such that no two nodes of R are adjacent and no two nodes of B are adjacent. When G is simple, this means that its adjacency matrix can be put in the form $\begin{pmatrix} 0 & K \\ L & 0 \end{pmatrix}$, where the two diagonal square blocks consist of zeros.

EXERCISES E

1. For a given relation ρ on a graph with n nodes, show that $(e \vee \rho)^n = \bigvee_{k=0}^{n} \rho^k$ is the least transitive relation which contains ρ.
2. Prove rigorously that any path of least length from x to y in a (directed) graph G is a simple path.
3. A node of a symmetric graph is called $\begin{cases}\text{even}\\ \text{odd}\end{cases}$ when its degree is $\begin{cases}\text{even}\\ \text{odd}\end{cases}$. Show that any finite symmetric graph has an even number of odd nodes.
4. Prove that if the relations ρ and σ are symmetric, then the graphs $\vec{G}_\rho + \vec{G}_\sigma$ and $\vec{G}_\rho \times \vec{G}_\sigma$ have symmetric adjacency relations.
5. Prove that the directed graphs of Fig. 2-7*b* and *d* are not isomorphic (that no isomorphism exists between them).

IN EXERCISES 6 AND 7, A EULER GRAPH IS A SYMMETRIC GRAPH ALL OF WHOSE LINKS CAN BE TRAVERSED WITHOUT REPETITION BY A SINGLE PATH.

6. Show that a finite connected graph in which every node is even is a Euler graph.
7. Show that a finite connected graph is a Euler graph if and only if the number of its odd nodes is 0 or 2.
8. A *tree* is a connected (symmetric) graph with no simple cycle.
   (a) Show that a tree with n nodes has $n - 1$ edges.
   (b) Show that a connected graph with n nodes and $n - 1$ edges is a tree.
9. (a) Show that all trees with three nodes are isomorphic.
   (b) Find two nonisomorphic trees with four nodes (draw them), and find three with five nodes.
   (c) Show that every graph with four or five nodes is isomorphic with one of those in your list.

10. Prove that if B is the incidence matrix of a simple directed graph without loops, then its adjacency matrix is obtained from BB^T by setting its diagonal entries equal to zero.
11. A plane *polygonal map* is a finite connected symmetric graph, in which the degree of every node is two or more. Show that a plane polygonal map with n_0 nodes and n_1 edges encloses $n_1 - n_0 + 1 = n_2$ disjoint polygons.
12. A *complete n-graph* is one whose adjacency matrix has all (off-diagonal) elements 1. Show that the complete 4-graph cannot be drawn in a plane without two lines crossing.
13. Let G be a directed graph in which each node has as many entering as departing arcs. Show that there is a directed cycle in G which traverses every arc precisely once.
14. Prove that the product of two bipartite graphs is itself bipartite.

★ 2-10. DIRECTED GRAPHS, II

In any given directed graph $\vec{G} = [N,A,\phi]$ we may consider the *subgraphs* defined by taking the different subsets $N_1 \subset N$ and $A_1 \subset A$ and restricting ϕ to those assignments $\mathbf{a}_1 \mapsto \phi(\mathbf{a}_1)$ for which $\mathbf{a}_1 \in A_1$ and the two endpoints of $\mathbf{a}_1$ all lie in N_1. Corresponding to morphisms of relations, one can also consider *morphisms* of graphs. For example, the images of simple paths and simple circuits of length n under morphisms to a directed graph $\vec{G}$ correspond to the paths and circuits of $\vec{G}$ in the following sense.

Definition. In any directed graph $\vec{G} = [N,A,\phi]$, a *path* of length n is a sequence of $n + 1$ nodes $p_0, \ldots ,p_n$ (not necessarily distinct) and n connecting arcs $\overrightarrow{p_{i-1}p_i} \in A$ $(i = 1, \ldots , n)$. If $p_0 = p_n$, then the path is called a *circuit* of length n. Such a path is called *simple* when its nodes are distinct; such a circuit is called *simple* when $p_0, \ldots ,p_{n-1}$ are all distinct.

Thus, simple paths and simple circuits in a graph $\vec{G}$ amount to monomorphic embeddings of $\vec{\Pi}_n$ and $\vec{\Gamma}_n$ in $\vec{G}$, where $\vec{\Pi}_n$ and $\vec{\Gamma}_n$ are defined as in Sec. 2-8.

Evidently, the directed graph of the covering relation of a poset contains no nontrivial circuits (it is *circuit-free*); moreover, it cannot contain an arc $\overrightarrow{pq}$ and a path of length $n + 1$ with the same endpoints, i.e., a subgraph of the kind shown in the figure. The converse is also true.

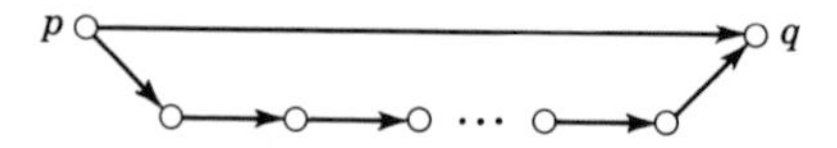

Sums and Products. To build up or "synthesize" various large graphs from small components, we can extend the concepts of disjoint union and Cartesian product of sets to graphs and directed graphs. For instance, if $\vec{G} = [S,A,\phi]$ and $\vec{H} = [T,B,\psi]$ are any two directed graphs, then their *sum* or *disjoint union* $\vec{G} + \vec{H}$ is defined by

$$\vec{G} + \vec{H} = [S \sqcup T, A \sqcup B, \phi \sqcup \psi] \tag{33}$$

For $\vec{G} = \vec{G}(\rho)$ and $\vec{H} = \vec{G}(\tau)$, this amounts to the condition that $\vec{G} \sqcup \vec{H} = \vec{G}(\theta)$, where

$$\begin{array}{llllll} s\,\theta\,s_1 & \text{means} & s\,\rho\,s_1, & t\,\theta\,t_1 & \text{means} & t\,\tau\,t_1 \\ \text{For} & \text{no} & s \in S, & t \in T & \text{is } s\,\theta\,t & \text{or} \quad t\,\theta\,s \end{array} \tag{33'}$$

Hence, if we order the nodes of $\vec{G} + \vec{H}$ so that those of $\vec{G}$ come first, the matrix of the successor relation on $\vec{G} + \vec{H}$ is the *direct sum* of the matrices M and N of the successor relations for $\vec{G}$ and $\vec{H}$ (with nodes in the same order). It has the *reducible* form $\begin{pmatrix} M & 0 \\ 0 & N \end{pmatrix}$, where the diagonal blocks (submatrices) M and N are square.

Again, the *Cartesian product* of the above directed graphs is defined, somewhat analogously, by

$$\vec{G} \times \vec{H} = [S \times T, C, \gamma] \tag{34}$$

where C is the disjoint union $(S \times B) \sqcup (T \times A)$, and

$$\gamma(s,b) = ((s,\psi_1(b)),(s,\psi_2(b))) \qquad \text{for } (s,b) \in S \times B \tag{35}$$

$$\gamma(t,a) = ((\phi_1(a),t),(\phi_2(a),t)) \qquad \text{for } (t,a) \in T \times A \tag{35'}$$

For example, the directed graph of Fig. 2-7d is the Cartesian product of two simple paths of length 1.

One can define the disjoint sum and Cartesian product of any two (symmetric = undirected) graphs similarly. This definition and the derivation of various properties of directed and undirected graphs are left to the reader as exercises (see Exercises E).

Accessibility; Connectedness. In a directed graph $\vec{G}$, node q is said to be *accessible* from node p when there is a path (hence a simple path) from p to q; we shall write this relation $p\,\alpha\,q$.

The algebra of relations gives a simple formula for computing the accessibility relation α on a given graph $\vec{G}(\rho)$ from the relation ρ defining the graph; namely, $p\,\rho^n\,q$ means that there is a path of length n from p

to q, where ρ^n is defined inductively for $k \in \mathbf{N}$ by $\rho^1 = \rho$ and $\rho^{k+1} = (\rho^k)\rho$. Now define the symbol $\bigvee_{k=1}^{r}$ for $r \in \mathbf{N}$ similarly, by

$$\bigvee_{k=0}^{1} \rho^k = e \vee \rho \qquad \text{and} \qquad \bigvee_{k=0}^{r+1} \rho^k = \left(\bigvee_{k=0}^{r} \rho^k\right) \vee \rho^{r+1} \tag{36}$$

where $e = \rho^0$ is the equality relation. For a graph with n nodes, we have

$$\alpha = \bigvee_{k=0}^{n} \rho^k \tag{37}$$

the proof will be left to the reader. That is, the accessibility relation α on $\vec{G}(\rho)$ is the least reflexive and transitive relation which contains ρ; it is the least α such that

$$\alpha \geqq e \qquad \alpha^2 \leqq \alpha \qquad \alpha \geqq \rho \tag{38}$$

If the nodes of $G(\rho)$ are ordered as $p_1, \ldots, p_n$ and $R = \|\rho_{ij}\|$ is the linking relation matrix, then the accessibility relation matrix is

$$A = \bigvee_{k=0}^{n} R^k$$

where R^k denotes the right composite $R \diamond \cdots \diamond R$ (k terms).

A directed graph $\vec{G}$ is said to be *strongly connected* when every node is accessible from every other node. If $\vec{G}$ has n nodes, this is equivalent to the equation $\bigvee_{k=0}^{n} R^k = J$, where J is the *universal relation* whose matrix consists entirely of 1's ($J_{ij} = 1$ for all i,j).

Likewise, in an undirected graph G we can define "q is accessible from p" to mean that "$p = q$ or there is a chain from p to q" (see Theorem 10). This is an equivalence relation on the nodes of G which decomposes G into *connected components* $G_1, \ldots, G_r$ of which G is the sum (disjoint union).

Shortest Paths. In a graph or directed graph, we are often interested in the length of the *shortest* path from node p to node q; this is called the *distance* from p to q and is denoted by $d(p,q)$. In a strongly connected directed graph with successor relation σ, it can be computed by the following algorithm. First, set $Q_0 = \{p\}$. Then, list the nodes $q \neq p$ with $p \,\sigma\, q$, and call the set of such nodes Q_1. Next list all nodes $r \notin Q_0 \cup Q_1$, such that $q \,\sigma\, r$ for some $q \in Q_1$; call this set of nodes Q_2. Recursively, define Q_{n+1} as the set of all nodes $t \notin Q_0 \cup \cdots \cup Q_n$, such that $s \,\sigma\, t$ for some $s \in Q_n$. Then we have $d(p,t) = m$ for all $t \in Q_m$.

Trees. A directed graph which has no circuits (hence no loops) is called *acyclic* (a graph which has no cycles is also called *acyclic*). A connected acyclic graph is called a *tree,* and there is a very amusing theory of trees.

More important for us are a related class of directed graphs called *rooted trees.* By definition, a rooted tree is a tree which has a unique *root* node p_0 from which each other node is accessible by one and only one path. Since every directed path contains a simple subpath, this path must be simple. Of special interest are *binary* rooted trees, in which each nonterminal node is the origin of precisely two arcs. Evidently, the different ways of combining n elements in a binary algebra can be described by the different binary rooted trees with n terminal nodes (listed in a prescribed order), and every such tree has just $2n - 1$ nonterminal nodes. For example, Fig. 2-11 shows the binary rooted trees corresponding to the five different ways of multiplying out a, b, c, and d, if associativity is not assumed; compare Chap. 1, (34).

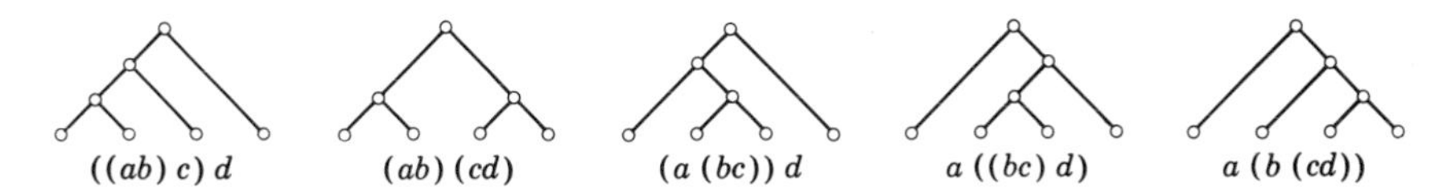

Fig. 2-12. Different binary products.

Labeled Directed Graphs. The suggestive concept of a directed graph has many useful generalizations. For example, one can label the nodes or links of a directed graph or both (as in commutative diagrams). What results is called a *labeled* directed graph. In the next chapter, we shall consider finite-state machines as labeled directed graphs. In Chap. 6, we shall consider the problem of finding "shortest paths" in labeled directed graphs whose arcs are assigned *lengths.*

Graphs of games. We shall conclude this chapter by showing how winning strategies in many two-person games of skill, such as checkers, chess, and tic-tac-toe, can be analyzed using labeled directed graphs.

With each such game, we can associate a graph whose *nodes* are the different possible positions $\left[\text{that is, } \binom{64}{32} = (64!)/(32!)^2 \text{ of 32 chessmen}\right]$ and whose *arcs* are the permitted moves, *labeled* by a symbol designating the player (e.g., white or black) who can make them, going from each position ("state" of the game) to that resulting from the move. Let the rules specify W_0 and B_0 as the sets of positions in which white and black, respectively, have won. Then the sets W and B of *all* winning positions for the two players can be computed recursively, as follows:

(*i*) W_{n+1} consists of all positions *not* in $\bigsqcup_{k=1}^{n} (W_k \cup B_k)$ from which either

(α) if it is white's move, some white-move arc leads to W_n, or

(β) if it is black's move, every black-move arc goes to W_n.

(*ii*) B_{n+1} is defined in complementary fashion.

Then we easily show that W_m and B_m are the positions from which white resp. black surely win in m moves, *if* he plays perfectly.

Example 12. A good example is provided by the game of Nim, played with any number of heaps of matches, from *one* of which each player takes, in each turn (move), as many matches as he pleases. The player who takes the last match wins the game.

If we express the number of matches in each heap in the binary scale and write down the resulting binary numbers in a vertical column, we easily prove that the *winning positions* are those for which the sum of the digits in each column is an even number. This is evident because, the potential loser, confronted by such a position, must change the parity of at least one number, ending with a nonwinning position. To restore the position to a winning one, the potential winner can then take a number of matches from the biggest heap so as to make the sums of digits in every column even. Thus, from four heaps having, respectively, 13, 11, 5, and 10 matches, he could take 9 matches from the first heap.[1]

$$\begin{aligned} 1101 &= 13 \\ 1011 &= 11 \\ 0101 &= 5 \\ 1010 &= 10 \end{aligned}$$

REFERENCES

BECKENBACH, E. (ed.): "Applied Combinatorial Mathematics," Wiley, 1964.
BERGE, C.: "The Theory of Graphs," Wiley-Methuen, 1962.
BUSACKER, R. G., and T. L. SAATY: "Finite Graphs and Networks," McGraw-Hill, 1965.
HALL, M.: "Combinatorial Mathematics," Ginn-Blaisdell, 1967.
HARARY, F.: "Graph Theory and Theoretical Physics," Academic Press, 1967.
ORE, O.: "Theory of Graphs," Am. Math. Society, 1962.

[1] See Hardy and Wright, The Theory of Numbers, Sec. 9-8, for more details.

Finite State Machines

3-1. INTRODUCTION

In the preceding chapters, the notions of set, function, product set, relation, and graph have been defined. In this chapter, these notions will be used to characterize mathematically the functioning of *digital computers*. Thus, *analog computers*, which have continuously varying states, and *hybrid* computers, which are part digital and part analog, will be ignored.

Although there are many kinds of digital computers, they are all *finite-state machines* (or finite *automata*) in the sense of having the following properties.

First, every digital computer consists of a finite set of *elements*, each of which can be in only one of a finite number of stable *states* at any given time. As a result, the computer itself has a finite set of stable states.

Second, every digital computer is *sequential;* that is, the operations of a computer are synchronized by means of a carefully timed sequence of electronic clock signals (currently clocks run from 10^6 to 10^9 signals per second). Accordingly, a computer's states change in an orderly sequence.

Third, a digital computer is basically *deterministic;* that is, given complete information concerning the internal states of the devices (or elements) in the computer and all inputs to the computer, the next action taken by the computer is uniquely determined. This action includes the "next state" which will be taken by each element of the computer, as well as by any output.

Digital computers are broadly classified into *general-purpose* and *special-purpose* digital computers. Since general-purpose computers can be adapted to special purposes, they are of greater interest, and will be our primary concern.

From a functional standpoint, today's general-purpose digital computers are composed of five sections: (1) the *input devices,* (2) the *memory,* (3) the *arithmetic section,* (4) the *control section,* and (5) the *output devices.*

Input devices convey instructions and data to the computer. They may be perforated tapes, punched cards, magnetic tapes, or electrical signals. They may also measure physical quantities and convert them into digital form, which can be accepted by the computer. Output devices convey information from the computer to its users. They include high-speed printers, typewriters with various "alphabets," oscilloscope displays, tapes to control billing machines, and so on. We shall not discuss them further in this book.[1]

To make a computer execute a desired sequence of instructions, a *program* which prescribes the sequence of instructions to be performed by the computer must first be read into the memory section of the computer. Generally, the program will previously have been key-punched into cards or put on tape, and these cards will be read by the input devices into memory.

The computer executes this stored program one instruction at a time, starting with the first instruction and proceeding sequentially. The program itself may also prescribe skipping, or repeating, of steps (see Chap. 4), and this skipping or repeating is controlled. That is, whether or not the program skips from one instruction to another generally depends on the values of data stored in the system.

The data to be used and the program are stored in the same memory elements, this being a feature originated by the late John von Neumann. It should be noted that memory elements are scattered throughout the machine in small groups (that is, the input-output, control, and arithmetic sections also contain limited amounts of memory). Therefore, the

[1] They are discussed, for example, in T. C. Bartee, "Digital Computer Fundamentals," 2d ed., McGraw-Hill, 1966.

part of the computer which has been referred to above as a "memory" is really the mass memory or bulk memory for the computer, consisting of from some thousands to perhaps 10^{10} bistable memory elements, each storing one *bit* of information. The number of these elements depends on the size of the machine.

Arithmetic and factual data stored in the memory of the computer can be processed extremely cheaply and rapidly on modern high-speed computers because these machines have subassemblies which perform millions of logical and arithmetic operations (such as addition or multiplication) each second, at a total cost of a few cents.

The arithmetic section of the computer performs both arithmetic and logical operations. Given two operands, for instance, the arithmetic section is able to add, multiply, divide, or subtract them. It can also perform various Boolean operations (see Chap. 4).

Finally, the control section interprets the instructions which are read from the memory, fetching them in order, or sometimes, as was noted above, transferring instructions in the program depending upon the value of some data stored or computed previously.

3-2. BINARY DEVICES AND STATES

Almost all digital computers, or indeed, digital machines of any sort, are composed of circuits and other devices which are *bistable* in their operation. The reasons for this are largely technical. It is possible to make circuits "go faster" if they are operated at the two extremes of their operating points; it is also possible to make circuits more reliable if they are operated only at extreme points where minor changes in characteristics will not affect the circuit performance. Consequently, the circuits and signals in computers are almost invariably two-valued. As a result, if we consider a specific electrical signal $x(t)$, it will have one of two values (except during transitions).

Figure 3-1 illustrates the representation technique for electrical signals which is most widely used today. A 1 is used to represent a positive electrical signal, and a 0 is used to represent the zero (or ground) signal

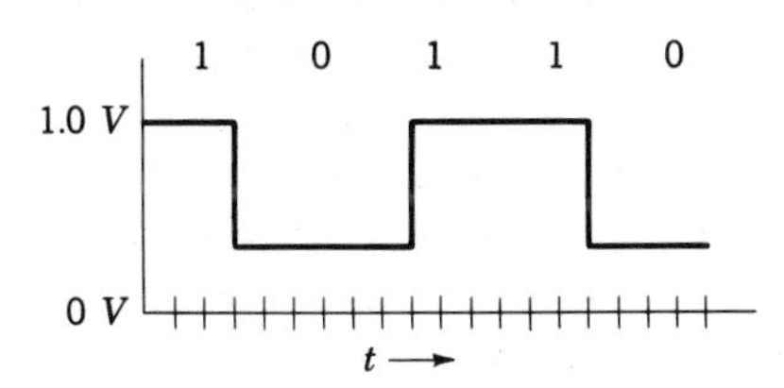

Fig. 3-1. Direct current levels and associated binary values.

level. (Some threshold is set; signals above that threshold are assigned the value 1, and signals below that threshold are assigned the value 0. For example, in resistor-transistor logic the threshold is on the order of 0.75 volt whereas for diode-transistor logic it is about 1.5 volts.)

We shall not discuss further the actual details of the logic levels or the circuitry; for us, the essential property of signals in a computer is that they are binary, and thus the variables in our mathematics will be two-valued also.

Just as the electrical signals used in digital computers are binary, the circuits are also designed so that they are stable in only two states. The ferrite cores used in memories and holes punched or not punched at a given position in a punched card are other examples of *bistable* elements.

From a mathematical standpoint, the state of any machine (automaton) consisting of a finite number r of bistable devices can be specified as follows. Enumerate (in any order) the devices of which it is composed, as $\delta_1, \ldots, \delta_r$. Assign to each stable state of the machine (i.e., of its components) a *state vector* $x = (x_1, \ldots, x_r)$ by giving each x_i the value 1 or 0 according as the device δ_i is in the state labeled 1 or the state labeled 0.

3-3. FINITE-STATE MACHINES

We shall now give a precise mathematical interpretation of the idea of a finite-state machine described above.

Definition. A *finite-state automaton* or *finite-state machine* consists of a 5-tuple $[A,S,Z,\nu,\zeta]$, where A is a finite list of *input symbols*

$$A = \{a_0, a_1, \ldots, a_n\}$$

Z is a list of *output symbols:* $Z = \{z_0, z_1, \ldots, z_m\}$, S is a set of *internal states* $S = \{s_0, s_1, \ldots, s_r\}$, ν is a *next-state function* from $S \times A$ into S, and ζ is an *output function* from $S \times A$ into Z.

A finite-state machine is therefore defined mathematically by three sets and two functions. It acts by "reading" or "accepting" a sequence or "string" of input symbols (the *program*) and then "printing" or delivering, output symbols. A finite-state machine starts in some internal state s_j; it then reads the first input symbol a_k. The combination of s_j and a_k causes an output symbol z_l to be produced through the function ζ and the machine to go to a new state s_r as directed by the function ν. After this, the machine reads the next input symbol, and so on until it finishes the string. As shown in Fig. 3-2, it is suggestive to think of the sequence or

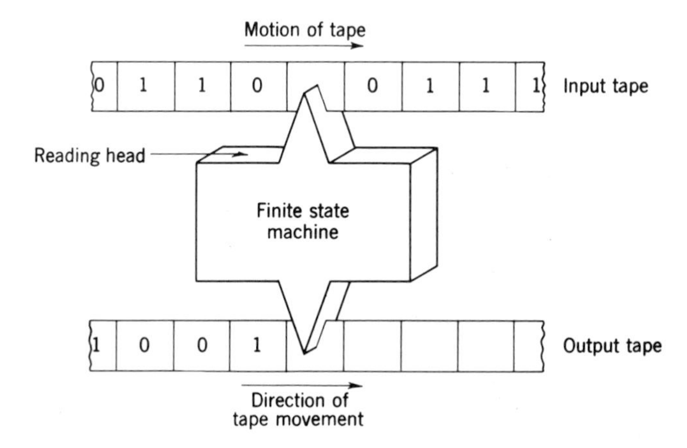

Fig. 3-2. Finite-state machine.

string of input symbols a_k as printed on some *input tape.* The finite-state machine then reads the input symbols successively from the tape, one after another. As each input symbol is read, the machine prints the output symbol z_l determined by ζ on an *output tape* and then assumes the new state determined by ν before reading the next symbol.[1] This will be made clear when state graphs and state tables are explained.

In this definition, the functions ν and ζ for a machine M are assumed to be completely defined; that is, ν and ζ are assumed to have specified values for each element of $S \times A$. Such a machine is said to be *completely specified.* A machine of this type in a given initial, or "starting," state s_i will read any input tape and deliver a uniquely determined output tape. Specifically, there is a function which assigns to each possible initial starting state s_i and any input string for machine M some output string with symbols from Z.

Example 1. A particular three-state machine $M = [A,S,Z,\nu,\zeta]$ can be defined as follows. The machine[2] M has a two-symbol (or two-character) input alphabet $A = \{0,1\}$, a two-character output alphabet $Z = \{0,1\}$,

[1] Electrical engineers and computer designers think of the inputs to their finite-state machines as a sequence of electronic signals or pulses. Each input symbol represents some combination of electronic signals, and these input signals are "clocked" into the machine at regular intervals of time, say, every 10 nanoseconds (10^{-8} seconds). Thus, the electronic engineer thinks of a sequence of input signals $a(1)$, $a(2)$, $a(3)$, . . . , and the symbols from A as labels for representing the values of these signals.

[2] From this point on, the word "machine" will often be used as an abbreviation for "finite-state machine." A finite-state automaton, is, of course, the same thing.

three internal states $S = \{s_0,s_1,s_2\}$, and next-state and output functions defined as

$$\begin{array}{lll} \nu:(0,s_0) \mapsto s_1 & \qquad & \zeta:(0,s_0) \mapsto 0 \\ \quad (1,s_0) \mapsto s_0 & & \quad (1,s_0) \mapsto 1 \\ \quad (0,s_1) \mapsto s_2 & & \quad (0,s_1) \mapsto 1 \\ \quad (1,s_1) \mapsto s_1 & & \quad (1,s_1) \mapsto 0 \\ \quad (0,s_2) \mapsto s_0 & & \quad (0,s_2) \mapsto 1 \\ \quad (1,s_2) \mapsto s_2 & & \quad (1,s_2) \mapsto 0 \end{array}$$

Suppose an input tape is read with the string 0,1,0,1 on it. If the machine is started in state s_0, when it reads the first input symbol 0, it goes to state s_1 and writes 0 on the tape. When 1 is then read, the automaton remains in state s_1 and writes 0 on the tape. When the second 0 is read, the state s_2 is assumed, and 1 is written on the tape.

When the input tape has been read completely, the machine will be in state s_2 and will have written 0,0,1,0 on the (output) tape. Thus, the machine converts the input string 0,1,0,1 (which can be written more compactly as 0101) into the output string 0,0,1,0 (or 0010 in abbreviated notation).

There are two convenient ways to represent this three-state machine. The first is by a labeled directed graph called a *state diagram*. The state diagram for the machine of Example 1 is shown in Fig. 3-3. The nodes of the state diagram graph are labeled with the symbols for internal states. On each arrow is written a pair of symbols a,z, where a is the input symbol which causes the transition indicated by the arrow and z is the output symbol which the machine writes on the tape.

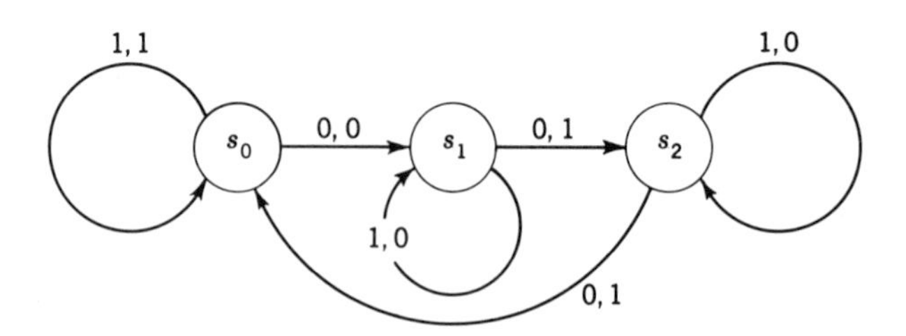

Fig. 3-3. A state diagram.

The second representation is by means of a *state table*. The state table for the machine of Example 1 is shown in Fig. 3-4. Again this is just a tabular representation of the functions ν and ζ.

Both the state diagram and state table have advantages and disadvantages. For computation, the state table is generally superior. On the other hand, it is often easier to see from a state diagram which states

Present state	Next state ν Input 0	Next state ν Input 1	Output ζ Input 0	Output ζ Input 1
s_0	s_1	s_0	0	1
s_1	s_2	s_1	1	0
s_2	s_0	s_2	1	0

Fig. 3-4. A state table.

of the machine are not accessible from other states. Thus, consider Fig. 3-5. This machine has a state s_1 which is not accessible (can never be reached) if the machine starts in s_0 or s_2.

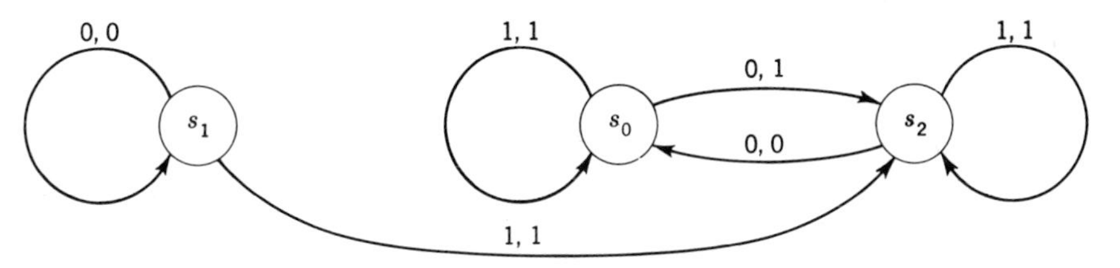

Fig. 3-5. A state diagram for a machine with inaccessible states.

Example 2. The two-state machine depicted in Fig. 3-6 is a *parity-check machine.* This machine reads a sequence of 0's and 1's from an input tape, and the internal state of the machine at any given time depends on

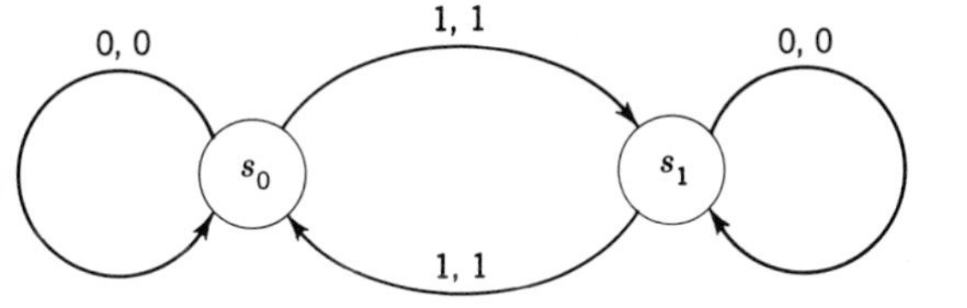

Next state 0	Next state 1	Output 0	Output 1
s_0	s_1	0	1
s_1	s_0	0	1

Fig. 3-6. Parity check machine.

whether the number of 1's read is even (in which case the internal state is s_0) or odd (in which case the internal state is s_1). Its output tape is an exact copy of the input tape.

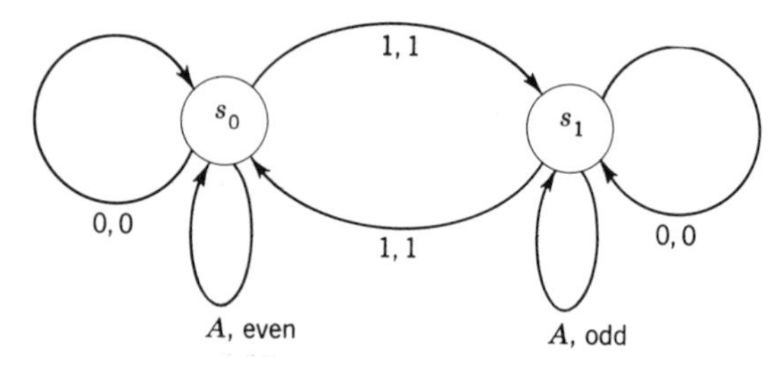

Fig. 3-7. Parity-check machine with print feature.

Example 3. The finite-state machine of Fig. 3-7 is a parity-check machine with the ability to print EVEN or ODD on request. The machine overprints the parity when it reads Q, which is taken to be the symbol asking for information. Thus the input tape $0110Q1110Q$ will be processed to 0110EVEN1110ODD.

EXERCISES A

1. Construct a machine M with alphabet $Z = A = \{0,1\}$, which produces from a given starting state s_0 and any input string $a^0a^2 \cdots$ the output $ooa^0a^1a^2a^3 \cdots$ (i.e., the output is the same as the input delayed two units of time).
2. Draw the state diagram for a machine with input and output alphabets $A = \{0,1\} = Z$, which puts out a 1 whenever it receives four consecutive 1's and puts out a 0 otherwise until it receives three consecutive 0's, after which it puts out 0's only. Thus we have

Input tape	0	0	1	0	1	1	1	1	1	0	1	0	0	0	1	0	1	1	1	1	0
Output tape	0	0	0	0	0	0	0	1	1	0	0	0	0	0	0	0	0	0	0	0	0

For the machine $M = [X,Z,S,\nu,\zeta]$, consider the function $\phi\colon K \to S^S\colon \alpha \to v_\alpha$, where for $s \in S$, $v_\alpha(s) = v(s,\alpha)$. Prove that ϕ is a morphism of semigroups from $[K$, concatenation to $[S^S,\diamond]$.

3. Let $M = [A,S,Z,\nu,\zeta]$ be a finite-state machine, and let the input tape consist of the infinite repetition of the same character $a \in A$.
   (a) Show that the output tape must ultimately be periodic.
   (b) Give numerical bounds on the period and on the length of any preperiodic output as functions of the number of internal states.
4. Let $[T, \cdot\,]$ be a finite semigroup with n elements $t_1, \ldots, t_n$ and associative multiplication $t_it_j = t_{\phi(i,j)}$.
   (a) Construct a "multiplying machine" with $n + 1$ states and $n + 1$ characters ($A = Z = \{0,1, \ldots, n\}$) which will produce from any input string $0,i(1), \ldots, i(r)$ the printout $i(1),i(1)i(2), \ldots, \prod_{k=1}^{r} i(k)$.
   (b) Show that if $[T, \cdot\,]$ has an identity 1, then n states and n output characters are sufficient.

5. Describe a machine with two states which will convert input decimal digits 0,1, . . . ,9 into output characters 0000, 0001, . . . ,1001, respectively, *and* input binary numbers 0000, 0001, . . . ,1111 into the decimal numbers 0,1, . . . ,15, respectively.
6. Let M be a state machine (with 2^{n+1} internal states) which will produce from the starting state s_0 and any two n-bit binary numbers **i** and **j** the state $s_{\mathbf{i}+\mathbf{j}}$ and the output **i** + **j**. Let M, with 2^{2k} internal states, produce from any two k-bit words w_h and w (and starting state b) the state s_{i-}.
7. Let $n = rk$. Describe a machine M^* containing M and $\overline{M}$ as "submachines," which produces from the input sequence $i(1),i(2), \ldots ,i(n),j(1),j(2), \ldots ,j(n)$ a sequence of r $2k$-bit output "words" which is a binary representation $(\tilde{\mathbf{w}}) = \tilde{w}_{k(1)} \cdots \tilde{w}_{k(r)}$ of the product $\mathbf{i} \cdot \mathbf{j}$ of the two n-place binary numbers $\mathbf{i} = i(1)i(2) \cdots i(n)$ and **j**.
8. For the incompletely specified finite-state machine with the state table given below, find
   (a) All equivalent states.
   (b) The minimal state machine.
   (c) Give the surjection from machine to machine for equivalent states.

	0	1	0	1
s_0	s_1	s_0	0	1
s_1	s_0	s_1	1	0
s_2	s_1	s_3	0	1
s_3	s_2	s_3	0	1
s_4	s_3	s_2	1	0
s_5	s_4	s_5	1	0

9. (a) Is the cover relation for states a partial ordering?
   (b) Is the cover relation for machines a partial ordering?

3-4. COVERING AND EQUIVALENCE

A given input tape to a finite-state machine contains a string or an ordered set of symbols from the input alphabet A. Let us denote such an ordered set as **a**. Thus, if $A = \{0,1\}$, then an input tape might contain the symbols 01101, and we would have **a** = 01101. We denote the first symbol on the tape as a^0, the second as a^1, the third as a^2, the fourth as a^3, and the fifth and last as a^4. Then, for **a** = 01101, $a^0 = 0$, $a^1 = 1$, $a^2 = 1$, $a^3 = 0$, and $a^4 = 1$. Hence, in the usual boldface notation for vectors, we may also write $\mathbf{a} = a^0a^1a^2a^3a^4$, or $\mathbf{a} = a^0,a^1,a^2,a^3,a^4$.

Ordered sets of symbols from some set can be given several different names. One finds such terms as *vectors*, (finite) *sequences*, *n-tuples*, *lists*, or *arrays* used to denote ordered sets. In this chapter, we shall usually refer to such ordered sets as *strings,* and we shall reserve the word "vector" primarily to denote an element in a vector space. We shall write the word "sequence" primarily for an infinite ordered set and use "array" as a specific term in our programming languages. Sometimes "n-tuple" will be used for an ordered set of n elements, but for the most part this chapter will use the word "string."

A given $\mathbf{a}$ will then be called a *string* of input symbols, whereas a given output tape will contain a string $\mathbf{z} = z^0, z^1, \ldots, z^{r-1}$ of symbols z^i from Z, and a string $\mathbf{s}$ of internal states s^i will consist of states from S.

Given an arbitrary machine $M = [A,S,Z,\nu,\zeta]$, clearly any input string $\mathbf{a} = a^0,a^1, \ldots ,a^{r-1}$ of length r and initial state $s^0 \in S$ determines a unique string of r states $\mathbf{s} = s^0,s^1, \ldots ,s^{r-1}$ through the repeated action of ν. Recursively,

$$s^{j+1} = \nu(s^j,a^j) \qquad j = 0, \ldots ,r-1 \tag{1}$$

Likewise, a unique output string is determined through the repeated action of ζ, by the formula

$$z^j = \zeta(s^j,a^j) \qquad j = 0, \ldots ,r-1 \tag{2}$$

Hence, if we consider the machine in the starting state s^0 as yielding r-tuples $\mathbf{s} = s^0, \ldots ,s^{r-1}$ and $\mathbf{z} = z^0, \ldots ,z^{r-1}$ from the r-tuple $\mathbf{a} = a^0, \ldots ,a^{r-1}$ in the usual boldface notation, Eqs. (1) and (2) define functions

$$\nu_r\colon S \times A^r \to S^r \qquad \text{and} \qquad \zeta_r\colon S \times A^r \to Z^r \tag{3}$$

defined recursively by M frcm ν and ζ, where A^r is the set of all r-tuples with symbols from A, Z^r is the set of all r-tuples with elements from Z, and S^r the set of all r-tuples from S.

Example 4. For the machine M of Fig. 3-8, let $\mathbf{a} = 0110$ and $s^0 = s_0$. Then

$$\nu_4(s_0,0110) = s_0s_1s_0s_0 \qquad \text{and} \qquad \zeta_4(s_0,0110) = 1001 \tag{4}$$

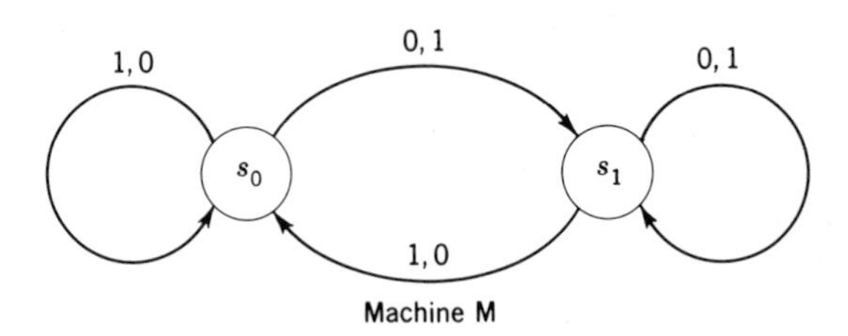

Fig. 3-8. Two-state machine.

One must pay a price for internal states. In general, the more internal states in a machine, the more electronic circuitry is required, the less will be the machine reliability, the more difficult the testing and repairing, etc. We therefore try to reduce the number of states in any given machine as much as possible without reducing its capability.

Therefore, for given input and output alphabets, it is very important to know when a given machine $M = [A,S,Z,\nu,\zeta]$ can be replaced by a smaller machine $\bar{M} = [A,\bar{S},Z,\bar{\nu},\bar{\zeta}]$ which can do everything that M can do in the same number of steps and with fewer internal states (hence presumably more cheaply). The following definition gives a sufficient condition for such a smaller machine $\bar{M}$ to have this capability.

Definition. A machine $\bar{M}$ *covers* a machine M when it has the same input and output alphabets A, Z and when for each internal state $s \in S$ in M, there is a state $\bar{s} = \phi(s)$, ϕ being a function from the internal states of M to the internal states of $\bar{M}$, such that for every positive integer r,

$$\zeta_r(s,\mathbf{a}) = \bar{\zeta}_r(\phi(s),\mathbf{a}) \qquad \text{for all } \mathbf{a} \in A^r \tag{5}$$

A machine which is covered by no smaller machine is called a *minimal state machine.*

We write $\bar{M} \geq M$ to signify that the machine $\bar{M}$ covers the machine M in this sense. The following result is evident, and we omit the proof.

Lemma 1. *The relation of covering is reflexive and transitive.*[1]

Definition. If $\bar{M}$ covers M and M covers $\bar{M}$, then machines M and $\bar{M}$ are *equivalent*, and we write $M \equiv \bar{M}$.

By definition, this means that there is not only a function $\phi\colon S \to \bar{S}$, where S is the set of internal states of M and $\bar{S}$ consists of the internal states of $\bar{M}$ such that (5) holds, but also a function $\psi\colon \bar{S} \to S$ such that

$$\bar{\zeta}_r(\bar{s},\mathbf{a}) = \zeta_r(\psi(\bar{s}),\mathbf{a}) \qquad \text{for all } \bar{s} \in \bar{S} \text{ and } \mathbf{a} \in A^r \tag{6}$$

Corollary. *The relation of equivalence on machines is reflexive, symmetric, and transitive.*

Covering and Morphisms. The relations of covering and equivalence on machines are closely related to the general notion of morphism introduced in Sec. 2-6. This notion can be applied to the same machines M and $\bar{M}$ having the same alphabets A and Z, in the following way.

[1] That is, a "quasi-ordering" in the sense of Sec. 9-1.

Definition. A *morphism* is a mapping $\theta: S \to \bar{S}$ such that

$$\bar{\nu}(\theta(s),a) = \theta(\nu(s,a)) \qquad \text{and} \qquad \bar{\zeta}(\theta(s),a) = \zeta(s,a) \tag{7}$$

for all $s \in S$ and $a \in A$. A morphism which is onto is called an *epimorphism*, and a morphism which is bijective (one-one and onto) is called an *isomorphism* (of machines).

Lemma 1. *If θ is an epimorphism of M onto $\overline{M}$, then given any input string $\mathbf{a} = a^0, a^1, \ldots, a^{n-1}$ and starting state $s^0 \in S$, the output string*

$$\mathbf{z} = z^0, z^1, \ldots, z^{n-1}$$

would also be produced by $\overline{M}$ if the starting state $\bar{s}^0 = \theta(s^0)$.

The proof is by induction on r. Recursively,

$$\begin{aligned} \bar{s}^{n+1} &= \bar{\nu}(\bar{s}^n,a^n) = \bar{\nu}(\theta(s^n),a^n) = \theta(\nu(s^n,a^n)) = \theta(s^{n+1}) \\ \bar{z}^{n+1} &= \bar{\zeta}(\bar{s}^n,a^n) = \bar{\zeta}(\theta(s^n,a^n)) = \theta(\zeta(s^n,a^n)) = \theta(z^{n+1}) \end{aligned}$$

Corollary. *Every epimorphism $\mu: M \to \overline{M}$ defines a covering of $\overline{M}$ by M.*

Likewise, an isomorphism between two machines M and $\overline{M}$ having the same input and output alphabets is simply a bijection $\beta: S \leftrightarrow \bar{S}$ such that for any initial state $s^0 \in S$ and input tape $\mathbf{a} = a^1a^2 \cdots a^n$, M and $\overline{M}$ produce the same output tape and pass through corresponding intermediate states. By this definition,[1] the two machines whose labeled

[1] A more general definition of a morphism of machine $M = (A,S,Z,\nu,\zeta)$ to machine $\overline{M} = (\bar{A},\bar{S},\bar{Z},\bar{\nu},\bar{\zeta})$ is given by Hartmanis and Stearns. They postulate mappings $\alpha: A \to \bar{A}$, $\beta: S \to \bar{S}$ and $\gamma: Z \to \bar{Z}$ such that $\bar{\nu}(\beta(s),\alpha(a)) = \beta(\nu(s,a))$ and $\bar{\zeta}(\beta(s), \alpha(a)) = \gamma(\zeta(s,a))$ for all $s \in S$ and $a \in A$.

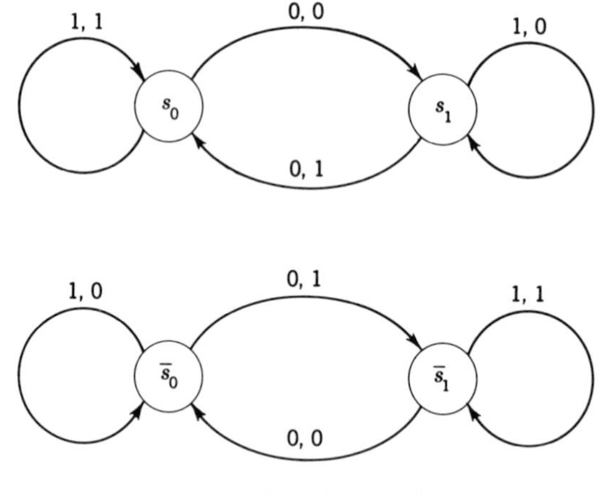

Fig. 3-9.

directed graphs are shown in Fig. 3-9 are isomorphic. The required function is simply

$$\beta: s_0 \mapsto \bar{s}_1 \qquad s_1 \mapsto \bar{s}_0 \tag{8}$$

The above definition of isomorphism for machines says, in effect, that two machines $M = [A,S,Z,\nu,\zeta]$ and $\bar{M} = [A,\bar{S},Z,\bar{\nu},\bar{\zeta}]$, having the same alphabets A, Z, are isomorphic when they have the same number of internal states and there is a bijection β from S to $\bar{S}$ such that for each s_i in S, if we write $\beta(s_i) = \bar{s}_j$ in $\bar{S}$, then for any input tape (if M is started in s_i and $\bar{M}$ in $\bar{s}_j$ and the same input tape is read by both machines), the output tapes will be the same. Then the converse situation also holds: for any $\bar{s}_j \in \bar{M}$, a state $\beta^{-1}(\bar{s}_j)$ can be found in M which has the same effect as s_j. The machine $\bar{M}$ is therefore, in effect, simply M with its states differently labeled, as always with isomorphism in algebra.

3-5. EQUIVALENT STATES

We shall solve below the following design problem. Given a description of a machine M, design a new machine $\bar{M}$ which (*i*) covers (or is equivalent to) M and (*ii*) has the fewest states of all machines which cover M. If the functions ν and ζ are completely specified as we have assumed above (but see Sec. 3-8), this can be done systematically and efficiently. The procedure is to first determine all states in M which are equivalent (according to the following definition) and to then combine or merge equivalent states, thus forming a new machine.

Definition. A state s_i is *r-equivalent* under every input tape $\mathbf{a} \in A^r$ of length r to a state s_j when[1]

$$\zeta_r(s_i,\mathbf{a}) = \zeta_r(s_j,\mathbf{a})$$

We then write $s_i\, E_r\, s_j$, or equivalently, $(s_i,s_j) \in G(E_r)$. If $s_i\, E_r\, s_j$ for all r, we say s_i is equivalent to s_j and write $s_i\, E\, s_j$ [or $(s_i,s_j) \in G(E)$].

Notice first that E_r and E are equivalence relations. Notice also that the equivalence classes E_1 consist of all pairs of states that have the same outputs for single input symbols, that is, $s_i\, E_1\, s_j$ means that

$$\zeta(s_i,a) = \zeta(s_j,a)$$

for all a. This fact can be determined simply by examining the output table. For the machine M of Fig. 3-10, $s_0\, E_1\, s_2$.

Now consider the graph (Sec. 2-1) of the equivalence relation E_r. If $(s_i,s_j) \notin G(E_r)$, then we write $s_i\, E_r'\, s_j$. [As always $G(E_r') \cup G(E_r)$ consists

[1] Recall that an $\mathbf{a} \in A^r$ consists of r input symbols $a^0a^1 \cdots a^{r-1}$.

of the set $S \times S$ of all ordered pairs of states.] Thus, for Fig. 3-10,

$$G(E_1) = \{(s_0,s_2),(s_2,s_0),(s_0,s_0),(s_1,s_1),(s_2,s_2)\}$$

and

$$G(E_1') = \{(s_0,s_1),(s_1,s_0),(s_1,s_2),(s_2,s_1)\}$$

M

Present state	ν		ζ	
	0	1	0	1
s_0	s_2	s_1	0	1
s_1	s_0	s_2	1	0
s_2	s_0	s_1	0	1

$\overline{M}$

Present state	$\bar{\nu}$		$\bar{\zeta}$	
	0	1	0	1
t_0	$\bar{s}_0$	$\bar{s}_1$	0	1
t_1	$\bar{s}_0$	$\bar{s}_0$	1	0

Fig. 3-10. State tables for equivalent machines.

Therefore, the problem of minimizing the number of states in a completely specified machine is to determine those states which are equivalent to each other and then to merge these states into a machine with fewer states. It turns out to be most efficient to compute states which are *not* equivalent first, and then to compute those which *are* equivalent.

In order to investigate this more efficiently, we define two new functions ν^* and ζ^*.

Definition. The function $\nu^*: S \times A^r \to S$ is defined to have value $\nu^*(s^0,\mathbf{a}) = s^{r-1} = \nu(\cdot\ \cdot\ \cdot\ (\nu(\nu(s^0,a^0),a^1), \ldots),a^{r-2})$. Thus $\nu^*(s_i,\mathbf{a})$ is the state s^{r-1} taken by the machine when it started in $s_i = s^0$ and when $\mathbf{a}$ is read from the input tape. The function $\zeta^*: S \times A^r \to Z$ is defined to have value $\zeta^*(s^0,\mathbf{a}) = z^{r-1} = \zeta(\cdot\ \cdot\ \cdot\ \nu(\nu(s^0,a^0),a^1), \ldots ,a^{r-1})$. Thus $\zeta^*(s_i,\mathbf{a})$ is the *last* output z^{r-1} where the machine reads the tape $\mathbf{a}$ and was started in s_i.†

Thus, for the first machine M of Fig. 3-9, with $r = 3$, $\nu^*(s_0,101) = s_1$ and $\nu^*(s_1,101) = s_0$. Also, $\zeta^*(s_0,101) = 0$ and $\zeta^*(s_1,101) = 1$.

Theorem 1. *If* $s_i\ E'\ s_j$, *then either* $s_i\ E_1'\ s_j$ *or for some* $\mathbf{a} = (a^0, \ldots ,a^{r-1})$, $(\nu^*(s_i,\mathbf{a})\ E_1' \nu^*(s_j,\mathbf{a}))$.

Proof. If $\zeta^*(s_i,\mathbf{a}) \neq \zeta^*(s_j,\mathbf{a})$ for some $\mathbf{a} = (a^0, \ldots ,a^{r-1})$, as must be the case if $s_i\ E'\ s_j$, then we can shorten the input sequence $\mathbf{a}$ until the two output sequences associated with s_i and s_j differ only in their last symbols. This makes r as small as is possible for $\zeta^*(s_i,\mathbf{a}) \neq \zeta^*(s_j,\mathbf{a})\mathbf{a} \in A^r$. Now, if $r = 1$, clearly $s_i\ E_1'\ s_j$; if $r > 1$, then $s_i\ E_r'\ s_j$, but $s_i\ E_k\ s_j$ for $k < r$. This

† Notice that here the initial state s^0 is state s_i. Also, note that elements in A^r are of the form $a^0,a^1, \ldots ,a^{r-1}$, an element of Z^{r-1} is of the form $z^0,z, \ldots ,z^{r-1}$.

means that when $\mathbf{a}$ is applied, the last output symbols generated by $\mathbf{a}$ are different if the machine is started in s_i than if it is initially in state s_j.

In order for the two outputs to differ, that is for $\zeta^*(s_i,\mathbf{a}) \neq \zeta^*(s_j,\mathbf{a})$, we must have $\nu^*(s_i,\mathbf{a})\ E_1'\ \nu^*(s_j,\mathbf{a})$. For if $\nu^*(s_i,\mathbf{a}) E_1 \nu^*(s_j,\mathbf{a})$, the final outputs would be the same when a^{r-1} was applied.

This leads us to our strongest theorem.

Theorem 2. *If $s_i\ E_r'\ s_j$ but $s_i\ E_k\ s_j$ for all $k < r$, then $\nu(s_i,a_l)\ E_{r-1}'\ \nu(s_j,a_l)$ for some particular $a_l \in A$.*

Another statement of this is as follows: if $(s_i,s_j) \in G(E_r') - G(E_{r-1}')$, then for some $a_k \in A$, $(\nu(s_i,a_k),\nu(s_j,a_k)) \in G(E_{r-1}') - G(E_{r-2}')$.

This theorem says that two states s_i and s_j which are equivalent under all sequences of length $r - 1$ can become inequivalent for some sequence of length r only if there is some a_k which takes s_i and s_j into some s_l,s_m, where s_l and s_m are not equivalent under some input sequence of length $r - 1$. Thus, at the rth step, we need only examine the states in $G(E_{r-1})$ to see if any pair (s_i,s_j) is taken into some pair (s_l,s_m) where $s_l\ E_{r-1}'\ s_m$. In this case the pair (s_i,s_j) is in the relation $s_i\ E_r'\ s_j$.

If we have determined $G(E_1')$, then $G(E_2')$ consists of $G(E_1')$ and those ordered pairs (s_i,s_j) such that for some a_p, $(\nu(s_i,a_p),\nu(s_j,a_p))$ is a member of $G(E_1')$. In general, we need only examine the pairs generated in $G(E_{r-1}') - G(E_{r-2}')$ [that is, those members of $G(E')$ from the "last pass"] each time. In this way we recursively define the set $G(E')$ and hence $G(E)$, which is simply the complement $S \times S - G(E')$ of $G(E')$ in the Boolean algebra of binary relations on S.

Proof. The proof of the theorem is straightforward. If a pair (s_k,s_l) is in $G(E_{r-1}')$, then it is not in $G(E_r') - G(E_{r-1}')$. Therefore, only those pairs (s_k,s_l) such that for some $\mathbf{a} \in A^r$, $\zeta(s_k,\mathbf{a}) \neq \zeta(s_l,\mathbf{a})$ and for which $\zeta(s_k,\mathbf{a}) = \zeta(s_l,\mathbf{a})$ for all $\mathbf{a} \in A^{r-1}$ need be considered, and these are precisely the pairs which are taken into $G(E_1')$ by the $(r - 1)$st input a^{r-2} and therefore into some pair of $G(E_{r-1}') - G(E_{r-2}')$ by $a^0 \in A$.

Lemma. *If $G(E_r') - G(E_{r-1}') = \varnothing$, then $G(E_r') = G(E_{r+k}')$ for all $k \geq 0$.*

No further ordered pairs of states can be added, since Theorem 2 asserts that only pairs taken by some single input $a_i \in A$ into some pair of $G(E_r') - G(E_{r-1}')$ comprise the change $G(E_{r+1}') - G(E_r')$.

Example 5. Figure 3-11 shows the state table for a completely specified machine. For this machine, $s_1\ E_1'\ s_5$, $s_2\ E_1'\ s_5$, $s_3\ E_1'\ s_5$, and $s_4\ E_1'\ s_5$;

that is, $G(E_1') = \{(s_1,s_5),(s_2,s_5),(s_3,s_5),(s_4,s_5),(s_5,s_1),(s_5,s_2),(s_5,s_3),(s_5,s_4)\}$. A partition could be

$$C_1^1 = \{s_1,s_2,s_3,s_4\} \qquad C_2^1 = \{s_5\}$$

Now, a 0 input takes s_1 into s_1 and s_3 into s_5; that is, $\nu(s_1,0) = s_1$, and $\nu(s_3,0) = s_5$, so the input 01 gives $\zeta^*(s_1,01) = 0$ and $\zeta^*(s_3,01) = 1$, whence $s_1 \, E_2' \, s_3$ and $(s_1,s_3) \in G(E_2')$. Similarly, $s_2 \, E_2' \, s_3$ and $s_4 \, E_2' \, s_3$, so $G(E_2') = G(E_1') \cup \{(s_1,s_3),(s_3,s_1),(s_2,s_3),(s_3,s_2),(s_3,s_4),(s_4,s_3)\}$.

Machine M

Present state	Next state		Output	
	Input		Input	
	0	1	0	1
s_1	s_1	s_2	1	0
s_2	s_1	s_3	1	0
s_3	s_5	s_1	1	0
s_4	s_4	s_2	1	0
s_5	s_4	s_3	1	1

Fig. 3-11. Example of state table.

Further, the input 1 takes s_1 into s_2 and s_2 into s_3; that is, $\nu(s_1,1) = s_2$, and $\nu(s_2,1) = s_3$, so $s_1 \, E_3' \, s_2$ for $\zeta^*(s_1,101) = 0$ and $\zeta^*(s_2,101) = 1$. Similarly, $s_2 \, E_3' \, s_4$, which implies

$$G(E_3') = G(E_2') \cup \{(s_1,s_2)(s_2,s_1)(s_2,s_4)(s_4,s_2)\}$$

Therefore, E_3 partitions S into the equivalence classes

$$C_1^3 = \{s_1,s_4\} \qquad C_2^3 = \{s_2\} \qquad C_3^3 = \{s_3\} \qquad C_4^3 = \{s_5\}$$

An examination of cases shows that $G(E_3') = G(E_4')$. Therefore, $G(E_4) = G(E_3)$, $E = E_3$, and thus $s_1 \, E \, s_4$, but no other pair of states are equivalent.

3-6. A MINIMIZATION PROCEDURE

The minimization procedure to be explained is based on a consideration of equivalence relations on ordered pairs. Consider the state table of Fig. 3-12. First, we determine the members of $G(E_1)$ and $G(E_1')$. These define a partition of the table into two equivalence classes of states; let this partition be π_1, and let its two equivalence classes be

$$C_1^1 = \{1,3,5,7,8\} \qquad C_2^1 = \{2,4,6,9\} \tag{9}$$

Here we have designated s_1 simply 1, s_2 simply 2, etc., which will greatly simplify calculation. Associated with the partition π_1 is the graph $G(E_1)$ of the corresponding equivalence relation. Since E_1 is reflexive and symmetric, the set of pairs in $G(E_1)$ is easily found from the set of (s_i,s_j) for which $\zeta(s_i,a_i) = \zeta(s_j,a_i)$ for all $a_i \in A$. Let G_1 be the $(s_i,s_j) \in G(E_1)$ for which $i < j$, and in general let G_l be the ordered pairs $(s_j,s_j) \in G(E_i)$ for which $i < j$. For the partition π_1 of (9), we have

$$G_1 = \{(1,3)(1,5)(1,7)(1,8)(3,5)(3,7)(3,8)(5,7)(5,8)(7,8)(2,4)(2,6)(2,9)(4,6)(4,9)(6,9)\}$$

$$G_1' = \{(1,2)(1,4)(1,6)(1,9)(2,3)(3,4)(3,6)(3,9)(5,2)(5,4)(5,6)(5,9)(2,7)(4,7)(6,7)(7,9)(2,8)(4,8)(6,8)(8,9)\}$$

	Next state			*Output symbol*		
	a_1	a_2	a_3	a_1	a_2	a_3
s_1	s_2	s_2	s_5	1	0	0
s_2	s_1	s_4	s_4	0	1	1
s_3	s_2	s_2	s_5	1	0	0
s_4	s_3	s_2	s_2	0	1	1
s_5	s_6	s_4	s_3	1	0	0
s_6	s_8	s_9	s_6	0	1	1
s_7	s_6	s_2	s_8	1	0	0
s_8	s_4	s_4	s_7	1	0	0
s_9	s_7	s_9	s_7	0	1	1

Fig. 3-12. State table to be minimized.

Since c_1^1 and c_2^1 are the equivalence classes in our first partition, for inputs of length 1, therefore, $s_1\ E_1\ s_3$, $s_1\ E_1\ s_5$, and $s_1\ E_1'\ s_2$ or $s_1\ E_1'\ s_4$, etc.

Now the members of G_2' are those of G_1' plus the pairs (2,9), (4,9), and (6,9). For instance, a_3 takes (2,9) into (4,7), which is a member of $G(E_1')$. The addition of these pairs to $G(E_2')$ defines a new partition π_2 with

$$C_1^2 = \{1,3,5,7,8\} \qquad C_2^2 = \{2,4,6\} \qquad C_3^2 = \{9\} \tag{10}$$

We now calculate G_3', which consists of all of G_2' plus the pairs (2,6) and (4,6). This is because, for instance, x_2 takes (2,6) into (4,9), which is a member of G_2'. We thus construct another partition π_3 with equivalence classes

$$C_1^3 = \{1,3,5,7,8\} \qquad C_2^3 = \{2,4\} \qquad C_3^3 = \{6\} \qquad C_4^3 = \{9\} \tag{11}$$

Now G'_4 contains G'_3 plus (1,5), (1,7), (3,5), (3,7), (8,5), (8,7) and their transposes so π_4 has the following equivalence classes:

$$C_1^4 = (1,3,8) \qquad C_2^4 = \{2,4\} \qquad C_3^4 = \{5,7\} \qquad C_4^4 = \{6\} \qquad C_5^4 = \{9\}$$

An examination now indicates that no further refinement of this partition results when inputs are applied; therefore $\pi_5 = \pi_4$, and $E_4 = E$.

The design of the new machine is straightforward. Each equivalence class in the final partition becomes a state in the new machine. For instance, we make C_1^4 into $\bar{s}_1$ and C_2^4 into $\bar{s}_2$, getting a five-state machine which covers our original nine-state machine. Since the outputs from each state in each equivalence class must be the same equivalence class of E for single input symbols, those in the original table can be used for the output section of the state table for the new machine. To construct the next-state function, we simply pick a state s_j in each C_i^k, and if $a \in A$ takes this s_j into some state in C_m^k, we set $\mu(s_j,a) = \bar{s}_m$. Notice that no ambiguity can occur since all s_j in C_i^k must go into states in the same C_m^k under the given input $a \in A$.

For the (completely specified) machine in Fig. 3-12, the new machine formed using these rules is shown in Fig. 3-13. This machine is a *minimal state machine* formed from the machine of Fig. 3-12.

In practice, it is not necessary to list every pair in $G(E_i)$ and in $G(E'_i)$. At each step of the refinement of the partition, we simply ask whether or not some input $a_l \in A$ takes a pair (s_i,s_j) in some C_l^k into two different equivalence classes C_m^k and C_n^k. If it does, s_i and s_j must be distinguished in the next partition.

	Next state			*Output symbol*		
	a_1	a_2	a_3	a_1	a_2	a_3
$\bar{s}_1$	$\bar{s}_2$	$\bar{s}_2$	$\bar{s}_3$	1	0	0
$\bar{s}_2$	$\bar{s}_1$	$\bar{s}_2$	$\bar{s}_2$	0	1	1
$\bar{s}_3$	$\bar{s}_4$	$\bar{s}_2$	$\bar{s}_1$	1	0	0
$\bar{s}_4$	$\bar{s}_1$	$\bar{s}_5$	$\bar{s}_4$	0	1	1
$\bar{s}_5$	$\bar{s}_3$	$\bar{s}_5$	$\bar{s}_3$	0	1	1

Fig. 3-13. Minimal state table for machine in Fig. 3-12.

Example 7. Consider the five-state machine in Fig. 3-14. We have $s_0\,E_1\,s_2$, $s_1\,E_1\,s_3$, $s_1\,E_1\,s_4$, and $s_3\,E_1\,s_4$, defining a partition π_1: $C_1^1 = \{s_0,s_2\}$, $C_2^1 = \{s_1,s_3,s_4\}$. Now, a 1 input takes s_3 into the next state s_0; that is, $\nu(s_3,1) = s_0$, and also $\nu(s_4,1) = s_4$. However, $s_0\,E'_1\,s_4$, so $s_3\,E'_2\,s_4$ [for $\zeta_2(s_3,10) = 1$ and $\zeta_2(s_4,10) = 0$].

Present state	Next state		Output	
	0	1	0	1
s_0	s_1	s_2	1	0
s_1	s_4	s_2	0	0
s_2	s_3	s_0	1	0
s_3	s_4	s_0	0	0
s_4	s_4	s_4	0	0

Fig. 3-14. Completely specified machine.

The next partition is π_2, with the equivalence classes

$$C_1^2 = \{s_0,s_2\} \qquad C_2^2 = \{s_1,s_3\} \qquad C_3^2 = \{s_4\} \tag{12}$$

This partition is the final partition, for ν takes each element of a given equivalence class into the same equivalence class under each input. The states s_0 and s_2 can therefore be merged into a single state $\bar{s}_0$, and s_1 and s_3 can be merged into the state $\bar{s}_1$, whereas s_4 is renamed $\bar{s}_2$. A new minimal machine equivalent to the machine in Fig. 3-14 is shown in Fig. 3-15.

	Next state		Output	
	0	1	0	1
$\bar{s}_0$	$\bar{s}_1$	$\bar{s}_0$	1	0
$\bar{s}_1$	$\bar{s}_2$	$\bar{s}_0$	0	0
$\bar{s}_2$	$\bar{s}_2$	$\bar{s}_2$	0	0

Fig. 3-15. Minimal state machine.

EXERCISES B

1. Minimize the number of states in the following machine:

	Next state		Output	
	a_0	a_1	a_0	a_1
1	2	2	1	0
2	3	3	1	0
3	4	4	1	0
4	4	4	0	1
5	5	6	1	1
6	6	5	1	1

2. Minimize the number of states in the following machine:

	Next state		Output	
	a_0	a_1	a_0	a_1
1	9	1	1	1
2	2	2	1	0
3	7	5	0	1
4	2	2	1	0
5	2	2	1	0
6	3	9	1	0
7	6	8	1	0
8	9	9	1	0
9	4	6	0	0

3. Let M and $\overline{M}$ be two (completely specified) machines having p and q states, respectively. If M and $\overline{M}$ have the same input alphabet A, show that a state s_i in M is equivalent to a state $\bar{s}_j$ in $\overline{M}$, if the relation $\zeta_r(s_i,\mathbf{a}) = \zeta_r(\bar{s}_j,\mathbf{a})$ holds for all input tapes of length $r \leq p + q - 1$.
4. Exhibit state tables for two machines M and $\overline{M}$ having the same number of states, which are equivalent but not isomorphic.
5. Draw the state diagrams of the following automata:

	ν		ζ	
	0	1	0	1
1	1	2	0	0
2	2	3	0	0
3	3	4	0	0
4	4	1	0	1

	ν		ζ	
	0	1	0	1
a	b	c	0	0
b	b	c	0	0
c	b	c	0	1

6. Let M and $\overline{M}$, with one input letter a, be given by the state diagram below. Find all the pairs (p,q) of equivalent states s in M and t in $\overline{M}$. Are M and $\overline{M}$ equivalent?

M	ν	ζ
1	1	0
2	1	1
3	3	1
4	3	1

and

$\overline{M}$	ν	ζ
a	b	0
b	b	0
c	a	1
d	d	1

7. Establish a morphism from M to $\overline{M}$ where:

	ν		ζ	
M	0	1	0	1
a	b	c	0	1
b	a	c	0	1
c	c	a	1	0

	ν		ζ	
$\overline{M}$	0	1	0	1
1	1	2	0	1
2	2	1	1	0

8. Consider the following machine: $M = [A,S,\nu,Z,\zeta]$, and

M	ν 0	ν 1	ζ 0	ζ 1
1	1	2	0	1
2	1	3	0	1
3	5	1	0	1
4	4	2	0	1
5	4	3	1	1

where $A = \{0,1\}$
$Z = \{0,1\}$
$S = \{1,2,3,4,5\}$

(a) Find a machine $\overline{M}$ with the minimal number of states which is output-indistinguishable from M for all input tapes of length 2.
(b) Is your machine $\overline{M}$ equivalent to M?

9. (a) Prove that if M_1 is an epimorphic image of M_2 and M_2 is an epimorphic image of M_3, then M_1 is an epimorphic image of M_3.
(b) If $\overline{M}$ is an *epi*morphic image of M, prove that M covers $\zeta(\mathrm{a})M$. [*Hint:* Prove that for each $s \in S$, $\zeta(s,a) = \zeta(f(s),a)\bar{\zeta}(f(s),a)$ by induction on the length of the tape a.]
(★c) Construct machines M and $\overline{M}$ such that M covers $\overline{M}$, yet $\overline{M}$ is not an epimorphic image of M.

10. Find the minimal machine equivalent to the following machine:

	ν a	ν b	ζ a	ζ b
1	1	6	0	0
2	1	4	0	0
3	2	5	1	0
4	5	8	1	1
5	1	3	1	0
6	8	5	1	1
7	6	3	1	1
8	2	5	1	0

★3-7. TURING MACHINES

Historically, the definition of a finite-state machine evolved from a closely related concept introduced in 1936 by the logician Turing. Turing considered a hypothetical "machine" having a finite set S of internal *states* and a single infinitely long tape divided into squares, which the machine could move in either direction, one step at a time. On each square, one could mark any symbol from a finite alphabet A (its vocabulary), but the tape had to be initially blank, except for a finite number of premarked squares. (Loosely speaking, these premarked squares can be considered as the "program" for a given run of the machine.)

The essential difference between Turing machines and the machines defined in Sec. 3-3 is provided by (*i*) the fact that the tape is infinite and (*ii*) the Turing machine's ability to move the tape in either direction. This ability gives the machine a potentially infinite memory which can be called on during its calculations. In addition (actually as a result), input symbols can be looked at more than once. A Turing machine can be defined in several different ways. One definition is the following.

Definition. A *Turing machine* consists of a 5-tuple $[A,S,\nu,\zeta,\delta]$, where A is a finite set of tape symbols $A = \{a_0,a_1, \ldots ,a_n\}$ used for both input and output, S is a finite set of internal states $S = \{s_0,s_1, \ldots ,s_r\}$, ν is a function from $S \times A$ into S, ζ is a function from $S \times A$ into A, and δ is a function from $S \times A$ into the set {L,R,HALT} which will be explained below.

A Turing machine is thus assumed to operate as follows. The machine is started in some initial starting state s^0. When the first symbol on the tape is read, the function ν causes the machine to take a new internal state, the function ζ selects a value a_j which is written on the tape, and δ causes the tape to move left or right one square (L for left and R for right) or causes it to halt.

Figure 3-16 shows a drawing of the Turing machine tape and read-write head. The machine operates by repeating indefinitely the following

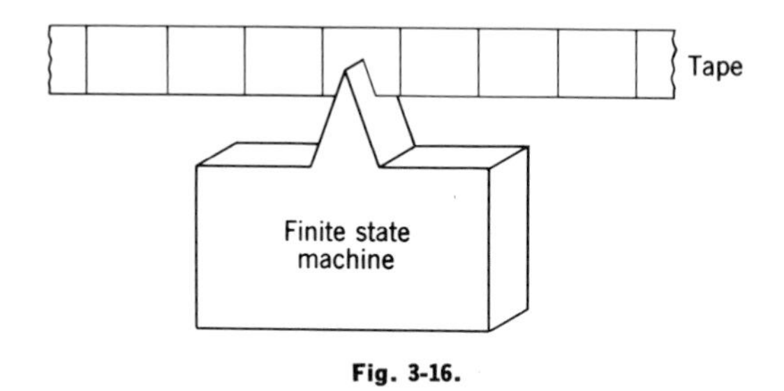

Fig. 3-16.

sequence of operations: it first "reads" a symbol from the tape, then overwrites the symbol read with the symbol selected by ζ (perhaps the same symbol), and finally moves the tape right or left depending on the value of ζ. The tape is assumed to extend without limit in both directions; therefore, it is infinite. However, the string of input symbols written on the tape before the machine starts (and hence at any later stage) must be finite in length.

Example 8. The Turing machine with the two functions listed below is a parity-check machine which examines a specified string of 0's and 1's on a tape and prints an E if the number of 1's is even and a D if the number of 1's is odd. The string of 0's and 1's is preceded and followed by blanks which are indicated by #'s, and then E or D is printed in the first blank square following the string of 0's and 1's. The tape alphabet is

$$A = \{\#,0,1,E,D\}$$

the internal states are $S = \{s_0,s_1,s_2\}$, and s_0 is the initial state. When the machine is to stop, a HALT signal is given to the tape-moving mechanism; thus

$$
\begin{array}{lll}
\nu\colon (s_0,0) \mapsto s_1 & \zeta\colon (s_0,0) \mapsto 0 & \delta\colon (s_0,1) \mapsto \mathrm{L} \\
\quad (s_0,1) \mapsto s_2 & \quad (s_0,1) \mapsto 1 & \quad (s_0,1) \mapsto \mathrm{L} \\
\quad (s_1,0) \mapsto s_1 & \quad (s_1,0) \mapsto 0 & \quad (s_1,0) \mapsto \mathrm{L} \\
\quad (s_1,1) \mapsto s_1 & \quad (s_1,1) \mapsto 1 & \quad (s_1,1) \mapsto \mathrm{L} \\
\quad (s_2,0) \mapsto s_2 & \quad (s_2,0) \mapsto 0 & \quad (s_2,0) \mapsto \mathrm{L} \\
\quad (s_2,1) \mapsto s_1 & \quad (s_2,1) \mapsto 1 & \quad (s_2,1) \mapsto \mathrm{L} \\
\quad (s_0,\#) \mapsto s_0 & \quad (s_0,\#) \mapsto \# & \quad (s_0,\#) \mapsto \mathrm{L} \\
\quad (s_1,\#) \mapsto s_1 & \quad (s_1,\#) \mapsto \mathrm{E} & \quad (s_1,\#) \mapsto \mathrm{HALT} \\
\quad (s_2,\#) \mapsto s_2 & \quad (s_2,\#) \mapsto \mathrm{D} & \quad (s_2,\#) \mapsto \mathrm{HALT}
\end{array}
$$

These functions can be more conveniently listed using a notation used by Turing. In this notation, a Turing machine is described by a finite set of 5-tuples $[s_i,a_j,s_r,z_l,t_n]$,

where s_i = the present state of the machine
a_j = the tape symbol being read on the tape
s_r = the next state of the machine, $s_r = \nu(s_i,a_j)$
z_l = the symbol printed on the tape, $z_l = \zeta(s_i,a_j)$
t_n = move tape left (L), move tape right (R), or HALT

In this notation, the above machine can be written as

$$
\begin{array}{lllll}
s_0 & \# & s_0 & \# & \mathrm{L} \\
s_0 & 0 & s_1 & 0 & \mathrm{L} \\
s_0 & 1 & s_2 & 1 & \mathrm{L} \\
s_1 & 0 & s_1 & 0 & \mathrm{L} \\
s_1 & 1 & s_1 & 1 & \mathrm{L} \\
s_2 & 0 & s_2 & 0 & \mathrm{L} \\
s_2 & 1 & s_1 & 1 & \mathrm{L} \\
s_1 & \# & s_1 & \mathrm{E} & \mathrm{HALT} \\
s_2 & \# & s_2 & \mathrm{D} & \mathrm{HALT}
\end{array}
$$

The following easy theorem shows that Turing machines are at least as powerful as finite-state machines.

Theorem 3. *Let* $M = [\bar{A},S,Z,\nu,\zeta]$ *be a finite-state machine. Set*

$$\bar{A} = (A \cup Z) \sqcup \Lambda$$

(Λ a blank) and

$$\begin{aligned} \bar{\nu}(s_i,a_k) &= \nu(s_i,a_k) & \bar{\nu}(s_i,\Lambda) &= s_i \\ \bar{\zeta}(s_i,a_k) &= \zeta(s_i,a_k) & \bar{\zeta}(s_i,\Lambda) &= \Lambda \\ \delta(s_i,a_k) &= \mathrm{R} & \delta(s_i,\delta) &= \mathrm{HALT} \end{aligned}$$

for all $(s_i,a_k) \in S \times A$. *Then the Turing machine* $T = [\bar{A},S,\bar{\nu},\bar{\zeta},\delta]$ *can do everything that M can do.*

Proof. Given any input tape $\mathbf{a} = a^0a^1 \cdots a^r$ for M, we can mark a^j on square j of the tape for T, leaving all other squares blank. Then T as constructed above will (for $j = 0,1, \ldots ,r$) overprint $z^j = \zeta(s^j,a^j)$ on square j, change to state $s^{j+1} = \nu(s^j,a^j)$, and move to square $j + 1$. When it gets to square $r + 1$, it will halt.

Remarks. If the input alphabet A of M contains a blank, we must use another symbol to designate it in T to avoid ambiguity. We have not defined $\nu(s_i,z_l)$ or $\zeta(s_i,z_l)$, because we don't care what they are. "Incompletely specified" machines with "don't-care" entries will be discussed in the next section.

Example 9. A simple Turing machine which can perform a calculation which cannot be performed by a finite-state machine is a machine which determines whether a tape of the form $\cdots$ ##000 $\cdots$ 0011 $\cdots$ 11## $\cdots$, where the #'s indicate blank squares on the tape, has the same number of 0's and 1's. The tape alphabet is $A = \{0,1,\mathrm{E},\mathrm{D},\#\}$; thus the machine can write blanks, the internal states will be

$$S = \{s_0,s_1,s_2,s_3,s_4,s_5,s_6\}$$

and the 5-tuples describing the machine are

s_0	#	s_0	#	L
s_0	0	s_1	#	L
s_1	0	s_1	0	L
s_1	1	s_1	1	L
s_1	#	s_2	#	R
s_2	1	s_3	#	R
s_3	1	s_3	1	R
s_3	0	s_4	0	R
s_4	#	s_0	#	L
s_3	#	s_5	E	HALT
s_2	0	s_5	0	L
s_5	#	s_6	E	HALT
s_2	#	s_6	D	HALT

The machine prints a D if the number of 0's is equal to the number of 1's, and an E if not. This symbol is printed to the right of the string of 0's and 1's, and the Turing machine then stops.

It is not hard to construct Turing machines which compute functions of numbers appearing on the input tape. It is a common practice with Turing machines to represent the nonnegative integer n by writing $n + 1$ consecutive 1's, and to separate successive integers using a single 0. Thus the tape $\cdots$ ##111011## $\cdots$ would represent the integer 2 followed by the integer 1. The tape $\cdots$ ##111101101011## $\cdots$ represents the integer 3 followed by the integer 1 followed by 0 followed by 1 (the sequence 3,1,0,1).

Example 10. The following Turing machine adds two nonnegative integers which are printed on a tape using the above notation:

s_0	#	s_0	#	L
s_0	1	s_1	1	L
s_1	1	s_1	1	L
s_1	0	s_2	1	L
s_2	1	s_2	1	L
s_2	#	s_3	#	R
s_3	1	s_4	#	R
s_4	1	s_5	#	HALT

This machine converts the two nonnegative integer representations into a single string of 1's equal in number to the sum of the numbers (of 1's) previously on the tape. It is possible to design Turing machines which will add three integers, four integers or even a machine which will add as many integers as may appear on any properly encoded tape, although this is somewhat more complicated.

3-8. INCOMPLETELY SPECIFIED MACHINES

Preceding sections have dealt only with completely specified machines. In actual practice, for most machines the functions ν and ζ are only partially specified. Generally, systems are designed in parts, and certain input combinations either do not occur or do occur at certain places in the input string where we do not care about outputs. This leads to unspecified entries in state tables (or missing links in state diagrams); these are called *don't-care* entries to indicate that what is placed in this position is of no interest. In our tables, a dash will be used to indicate a don't-care position. Figure 3-17 shows a table where the next state when the machine is in state s_1 and a 0 input symbol is read is of no importance; when the machine is in state s_2 and a 0 input symbol is received, the output is a don't-care situation.

Present state	Inputs			
	Next state		Output	
	0	1	0	1
s_0	s_1	s_2	0	1
s_1	—	s_1	1	0
s_2	s_0	s_1	—	1

Fig. 3.17. An incompletely specified machine.

The history of attempts to devise a procedure for minimizing the number of states in incompletely specified machines is interesting. It was once thought that the problem was simply to fill in each don't-care position in the table and the resulting completely specified machine could be best minimized using the procedure previously given for completely specified machines. It was thought that minimizing each resulting machine would yield the minimal state machine. In principle, this would have resolved the problem into that of shortening or making more efficient the above substitution process, and considerable work was done on this.

However, for many machines, no substitution of this sort leads to a minimal machine. In effect, limiting each don't-care state to a particular value unnecessarily limits our ability to minimize. The fact is that if a don't-care state is really "don't care," the values in this don't-care position can change as inputs are received, and this freedom enables further reduction. For example, if (see Fig. 3-12) with the machine in state s_2 and a 0 input, we allow the output symbol to be sometimes a 0 and sometimes a 1 then the machine can be reduced to two states; however, if we insist that the output always be the same, the machine cannot be reduced.

To describe the basic technique for minimizing incompletely specified machines, we shall need the following definitions.

Definition. An input sequence $\mathbf{a} = a^0, \ldots, a^{r-1}$ is *applicable* to a machine started in state s_j when the next state function ν is specified for each input, except possibly the last.

There is therefore a uniquely specified string of internal states associated with starting state s_j and an applicable input string $\mathbf{a}$, except that the state taken by the machine after the last input may be indeterminate (don't-care).

Definition. An output string $\mathbf{z}$ *covers* an output string $\bar{\mathbf{z}}$ if every specified symbol $\bar{z}^j$ in $\bar{z}$ is equal to the corresponding term z^j in $\mathbf{z}$. For instance, $\mathbf{z} = z^0,z^1,z^2 = 0,1,1$ covers $\bar{z} = \bar{z}^0,\bar{z}^1,\bar{z}^2 = -,1,1$; 0,1,0 covers $-$,1,0; while 1,1,0 covers 1,1,0 and $-$,1,0 but does not cover 0,1,0. If $\mathbf{z}$ covers $\bar{\mathbf{z}}$, we then write $\mathbf{z} \geq \bar{\mathbf{z}}$.

Definition. If $\zeta_r(s_k,\mathbf{a}) \geq \bar{\zeta}_r(\bar{s}_j,\mathbf{a})$ for all $\mathbf{a}$ applicable to $\bar{s}_j$ then we write $s_k \geq \bar{s}_j$ and write s_k *covers* $\bar{s}_j$.

Definition. A machine M *covers* a machine $\overline{M}$ when for each state $\bar{s}_j$ in $\overline{M}$ there is a state s_k in M such that $\zeta_r(s_k,\mathbf{a}) \geq \bar{\zeta}_r(\bar{s}_j,\mathbf{a})$ for all $\mathbf{a}$ applicable to $\bar{s}_j$. That is, each state in $\overline{M}$ is covered by some state in M. We then write $M \geq \overline{M}$. Clearly, if s_k and s_j are in the same machine, i.e., if $M = \overline{M}$, then s_k covers s_j if $\zeta_r(s_k,\mathbf{a}) \geq \zeta_r(s_j,\mathbf{a})$ for all $\mathbf{a}$ applicable to s_j, and we write $s_k \geq s_j$.

Now consider the machines in Fig. 3-18. Machine $\overline{M}$ covers machine M, and state $\bar{s}_0$ covers s_0; state $\bar{s}_0$ also covers s_1, and state $\bar{s}_1$ covers s_2.

It is instructive to consider the output strings for M and $\overline{M}$. When the input string 01010 is applied for M initially in state s_1, we get the output sequence $-111-$, and for $\overline{M}$ initially in state s_0 we get 01111. Notice that the first don't-care position in the output for M corresponds to a 0 in $\overline{M}$ and the second to a 1 in $\overline{M}$. In effect, the dash in the table for M has been filled in both by a 0 and by a 1.

Present state	*Next state*		*Output*	
	0	1	0	1
s_0	s_0	s_1	0	1
s_1	s_1	s_2	—	1
s_2	s_0	s_2	1	1

Machine M

Present state	*Next state*		*Output*	
	0	1	1	1
$\bar{s}_0$	$\bar{s}_0$	$\bar{s}_1$	0	1
$\bar{s}_1$	$\bar{s}_0$	$\bar{s}_1$	1	1

Machine $\overline{M}$

Fig. 3-18. An incompletely specified machine.

Because certain of the symbols in a given output sequence may be don't-care symbols, we need still a further relation on output sequences, which we will call compatibility.

Definition. An output string $\mathbf{z}$ is *compatible* with an output string $\bar{\mathbf{z}}$ when $\mathbf{z}$ and $\bar{\mathbf{z}}$ do not differ in any specified position. (If neither $\mathbf{z}$ nor $\bar{\mathbf{z}}$ contain don't-care symbols, then $\mathbf{z}$ must be the same as $\bar{\mathbf{z}}$.) We write $\mathbf{z} \,\gamma\, \bar{\mathbf{z}}$ if $\mathbf{z}$ is compatible with $\bar{\mathbf{z}}$.

Examples are 0,1,—,1 is compatible with 0,1,1,1; 0,—,1,— is compatible with —,0,—,1; but 0,1,0,— is not compatible with 0,1,1,—.

Note that γ is not an equivalence relation. It is reflexive and symmetric but *not* transitive; for example, 0,1,— is compatible with 0,—1, and —,1,0 is also compatible with 0,1,— but 0,—,1 is not compatible with —,1,0.

★3-9. RELATIONS BETWEEN STATES—A MINIMIZATION PROCEDURE

The above considerations lead us to examine in more detail several relations which can be defined on internal states in partially specified machines.

Definition. A state s_i is *output-compatible* with a state s_j when $\zeta(s_i,a) \, \gamma \, \zeta(s_j,a)$ for all $a \in A$. We then write $s_i \overset{1}{\sigma} s_j$. Pairs of states which are not compatible are called *output-incompatible,* and we write $s_k \overset{1}{\sigma'} s_l$.

Two states are therefore *output-compatible* if their outputs are the same (when specified) for all input strings of length 1. Input strings of length 1 are simply single symbols, so we can easily determine this from the state table for the machine.

	0	1	0	1
s_0	s_0	s_2	0	1
s_1	s_1	s_2	—	1
s_2	s_2	s_0	1	1

Fig. 3-19. Example of an incompletely specified machine.

For example, consider Fig. 3-19. In Fig. 3-19, $s_0 \overset{1}{\sigma} s_1$, and $s_1 \overset{1}{\sigma} s_2$, but $s_0 \overset{1}{\sigma'} s_2$ (so σ is not transitive, although it is reflexive and symmetric).

Definition. States s_i and s_j are said to be *compatible* when for any $\mathbf{a} \in A^r$ and for all $\mathbf{a}$ applicable to both s_i and s_j, $\zeta_r(s_i,\mathbf{a}) \, \gamma \, \zeta_r(s_j,\mathbf{a})$. We then write $s_i \, \sigma \, s_j$. If s_k and s_l are not compatible for all $\mathbf{a}$, we write $s_k \, \sigma' \, s_l$. If two states s_i and s_j are compatible for all strings of a fixed length k, that is, if $\zeta_k(s_i,\mathbf{a}) \, \gamma \, \zeta_k(s_j,\mathbf{a})$, $\mathbf{a}$ of length k, then we say s_i and s_j are *k-compatible,* and write $s_i \overset{k}{\sigma} s_j$.

If the output strings generated by a machine M, given any input tape $\mathbf{a}$, are the same in all "specified" entries when M is started in either s_i or s_j, the states s_i and s_j are then compatible ($s_i \, \sigma \, s_j$). In this case, we can combine or *merge* the two states into a single state. Moreover, this new state will generate an output string which is the same as the specified positions in the output string of s_i and s_j and is don't-care only when both outputs are don't-care. This new string will then cover each of the old sequences. If we designate the new state as s, then $\zeta_r(s,\mathbf{a}) \geq \zeta_r(s_i,\mathbf{a})$ for all $\mathbf{a} \in A^r$ applicable to s_i, and $\zeta_r(s,\mathbf{a}) \geq \zeta_r(s_j,\mathbf{a})$ for all $\mathbf{a} \in A^r$ applicable to s_j.

A good test is needed for compatibility, as this relation indicates all possible ways to combine states. However, the relation is not an equivalence relation, so it does *not* tell exactly how to combine states.

Consider Fig. 3-20. We are to determine the members of the compatibility relation σ for the partially specified table of Fig. 3-20*a*. As a first step, list $\overset{1}{G}(\sigma')$, for it is clear that $\overset{1}{G}(\sigma')$ is a subset of $\overset{2}{G}(\sigma')$. $\overset{1}{G}(\sigma')$ contains one element (2,5) as indicated.

Figure 3-20*c* lists the elements of $\overset{1}{G}(\sigma)$ as the headings of the rows of a table containing pairs of next states as entries. The rule for forming this table is as follows: at the intersection of each input symbol a_i and element of $\overset{1}{G}(\sigma)$, (s_l,s_m), list the pair of states into which a_i takes (s_l,s_m); since a_1 takes s_1 into s_2 and s_2 into s_3, we write (2,3) at the intersection of (1,2) and a_1. If one or both of the next states are not specified or if a given pair goes into a single state, we place a — at the intersection.

Now, if some pair [element of $\overset{1}{G}(\sigma)$] is taken into a pair in $\overset{1}{G}(\sigma')$ by some input string, that pair is clearly a member of $G(\sigma')$, because application of that input followed by the input(s) which causes the pair in $\overset{1}{G}(\sigma')$ to have differing outputs will lead to a differing output. For instance, (1,3) is taken into (2,5) by a_4, and (2,5) is in $\overset{1}{G}(\sigma')$ because of differing outputs for input a_3. So a_4,a_3 will lead to a difference in outputs for (1,3). Schematically, we have

$$\text{States}\begin{cases} a_4 & a_3 \\ 1 & 2 \\ 3 & 5 \end{cases}$$

$$\text{Outputs}\begin{cases} — & 0 \\ 0 & 1 \end{cases}$$

We therefore form a list of the members of $\overset{1}{G}(\sigma')$, starting with the elements of $\overset{1}{G}(\sigma')$.† Then we add any pairs taken by some input a_i into

† Notice that only the pairs (i,j) for which $i < j$ are listed. Since the relation is reflexive and symmetric we need not list (i,i) nor (j,i) with $j > i$ if (i,j) is present.

Present state	Next states				Output			
	a_1	a_2	a_3	a_4	a_1	a_2	a_3	a_4
s_1	s_2	—	s_3	s_2	0	—	—	—
s_2	s_3	s_5	s_2	—	0	1	0	—
s_3	s_3	s_4	—	s_5	0	1	—	0
s_4	—	s_1	s_2	—	—	1	—	—
s_5	—	—	s_1	—	—	—	1	—

(*a*) State table.

$$\overset{1}{G}(\sigma) = (1,2)(1,3)(1,4)(1,5)(2,3)(2,4)(3,4)(3,5)(4,5)$$

$$\overset{1}{G}(\sigma') = (2,5)$$

$$\overset{2}{G}(\sigma') = (2,5)(1,3)$$

$$\overset{3}{G}(\sigma') = (1,3)(1,5)(2,5)$$

$$\overset{4}{G}(\sigma') = (2,5)(1,3)(1,5)(2,4)$$

$$G(\sigma) = (1,2)(1,4)(2,3)(3,4)(3,5)(4,5)$$

(*b*) Graphs of relations.

	a_1	a_2	a_3	a_4
(1,2)	(2,3)	—	(2,3)	—
(1,3)	(2,3)	—	—	(2,5)
(1,4)	—	—	(2,3)	—
(1,5)	—	—	(1,3)	—
(2,3)	—	(4,5)	—	—
(2,4)	—	(1,5)	—	—
(3,4)	—	(1,4)	—	—
(3,5)	—	—	—	—
(4,5)	—	—	(1,2)	—

(*c*) Compatibility table.

Fig. 3-20. Technique for determining compatible states.

some pair in $\overset{1}{G}(\sigma')$. Next we add any pairs taken by any input a_j into these pairs because an input string of length 3 will cause a differing output for these pairs. For our table, this yields (1,5) as a member of $\overset{3}{G}(\sigma')$. [Notice that a_3, a_4, a_3 generates differing outputs for (1,5).] We continue this process, bringing more entries in at each step. For our table, (2,4) is next because a_2 takes (2,4) into (1,5).

This process is continued until no further element of σ' can be generated. For a machine with n states no more than $n - 1$ steps need be taken because by that time all state combinations will have been generated. In fact, the process of adding to σ' stops as soon as no further pairs are found.

Notice that there is no reason to list don't-care next states or entries where pairs are taken into a single state by some input; subsequent inputs cannot cause a "splitting" of states and nonmatching outputs.

Also note that we considered less than one-half of all possible ordered pairs, because $\overset{i}{\sigma}'$ is antireflexive and symmetric, so (s_i,s_i) need not be tested; if $s_i \overset{i}{\sigma} s_j$ is a member of one of these relations, we need not test for $s_j \overset{i}{\sigma} s_i$ because it is a member of the relation and similarly for σ'.

In the example of Fig. 3-20, $G(\sigma')$ contains (2,5), (1,3), (1,5), and (2,4), while $G(\sigma)$ contains the rest of the pairs.

We shall now extend our binary relation, since it may be possible to merge more than pairs of states.

Definition. A *compatible class* C_k is a set of internal states $s_i \cdots s_m$ such that for all s_i, $s_j \in C_k$, $s_i \sigma s_j$. A *maximal compatible* class C_l is a compatible class which is properly contained in no other compatible class.

Each maximal compatible class is therefore a collection of states which contains as many states as are mutually compatible since a maximal compatible is not properly contained in any other compatible class. In terms of pairs, if $s_i \sigma s_j$, $s_j \sigma s_k$, and $s_i \sigma s_k$, then we can form (s_i,s_j,s_k). If, however, we have $s_i \sigma s_k$, $s_j \sigma s_k$, but $s_j \sigma' s_i$, then we would have two maximal compatibles, (s_i,s_k) and (s_k,s_j).

A complete set of maximal compatible classes is a listing of the largest sets of states that have the possibility of being merged into a single state. For our machine in Fig. 3-20 we find that the maximum compatibles are (1,2), (1,4), (2,3), and (3,4,5).

Definition. A set of compatible classes is *consistent* if for every class C_k of the set and for any pair s_i and s_j in C_k, the internal states $\nu(s_i,a_{\hat{k}})$, and $\nu(s_j,a_{\hat{k}})$ are elements of some single compatible class C_l, for each $a_{\hat{k}}$.

Definition. A set of compatible classes is *closed* when every internal state of the machine is in at least one compatible class.

Theorem 4. *For a machine M, the states in each compatible class in a set of compatible classes which is consistent and closed can be merged or combined to form a state table for a machine $\overline{M}$ which covers M.*

Sets or pairs of states which are compatible can be merged or combined to form a single state. If a pair of states s_i and s_j is to be merged, simply replace each occurrence of s_j with s_i, fill in any values for $\zeta(s_j,a)$

which are specified for s_j and not for s_i in the output section for s_i, and then strike s_j from the table.

If a set of compatible classes which is consistent and closed has been found, the set of states in a given compatible class C_i can be merged by forming a new machine, where state $\bar{s}_j$ corresponds to compatible class C_j and where each output function $\bar{\zeta}(\bar{s}_j,a_k)$ is specified as $\zeta(s_l,a_k)$ if this is specified for any s_l in C_i, and the next state function $\nu(\bar{s}_j,a_k)$ is specified as $\bar{s}_m$ if for each $s_l \in C_i$, $\nu(s_l,a_k) \in C_m$. The new machine will contain as many states s_i as there were compatible classes in the consistent closed set, and each compatible class will have been merged.

If all the states in a given compatibility class are merged, the new state will cover every state in the compatibility class. The machine which results from merging the (compatible) states in each compatibility class (such that each state in the original machine is in some compatibility class), will clearly yield a new state table with a state to cover each of the original states.

Consistency is necessary for merging; if for some a and for some s_i and s_j in the same C_k $\nu(s_i,a)$ and $\nu(s_j,a)$ are in differing classes C_l and C_m, then we could not merge s_i and s_j.

Theorem 5. *A minimal state machine $\overline{M}$ which covers a machine M results from merging any set of consistent closed compatible classes with the property that no set of consistent closed compatible classes can be found having fewer classes.*

If a machine $\overline{M}$ is to cover a machine M, then there must be a state s_k in $\overline{M}$ which covers each state s_j in M. The compatibility classes of possible mergers which will yield such coverings are listed. The set of compatibility classes chosen must be consistent and closed so the problem is simply one of finding a consistent closed set of compatibility classes with a minimal number of elements. There may be many such sets of compatibility classes, the merging of each of which will yield a minimal state machine, and minimal state machines thereby derived will not necessarily be isomorphic.

Consider the machine in Fig. 3-20. A possible solution is $C_1 = \{1,2\}$ and $C_2 = \{3,4,5\}$, in which case the resulting machine has two states. However, $\nu(s_1,a_1) = s_2$, and $\nu(s_2,a_1) = s_3$, so s_2 and s_3 must both be in the same class C_i, which they are not, and this possible solution is not consistent.

Since no other set of two compatible classes can cover the machine, we must use at least three classes to achieve both covering and consistency. The following is a consistent cover:

$$C_1 = \{s_1,s_2\} \qquad C_2 = \{s_2,s_3\} \qquad C = \{s_4,s_5\}$$

Figure 3-21 shows the resulting minimal state machine. At present, there is no systematic way to perform the step of forming a minimal consistent set of compatible classes which cover a machine. One simply starts

Present state	*Next state*				*Output*			
	a_1	a_2	a_3	a_4	a_1	a_2	a_3	a_4
$\hat{s}_1$	s_2	s_3	s_2	s_1	0	1	0	—
$\hat{s}_2$	s_2	s_3	s_1	s_3	0	1	0	0
$\hat{s}_3$	—	s_1	s_1	—	—	1	1	—

Fig. 3-21. Minimal state machine.

with as few classes as possible and enlarges the number of classes until a consistent machine is discovered.

EXERCISES C

1. Minimize the number of states in the following incompletely specified machine:

	ν		ζ	
	a	b	a	b
1	2	—	0	—
2	1	6	0	0
3	4	—	1	—
4	5	3	0	0
5	—	6	—	1
6	4	—	0	—

$$A = \{a,b\}$$
$$Z = \{0,1\}$$
$$S = \{1,2,3,4,5,6\}$$

2. Find all pairs of states s_i, $\hat{s}_j$ such that $s_i \leqq \hat{s}_j$ for the following two incompletely specified machines:

	Next state		*Output*	
	a_0	a_1	a_0	a_1
s_1	s_3	—	1	—
s_2	s_4	s_3	1	0
s_3	s_2	s_4	—	—
s_4	—	—	1	—
s_5	s_1	—	—	1

	Next state		*Output*	
	a_0	a_1	a_0	a_1
$\hat{s}_1$	$\hat{s}_2$	—	1	0
$\hat{s}_2$	—	$\hat{s}_3$	1	—
$\hat{s}_3$	—	—	1	—
$\hat{s}_4$	$\hat{s}_1$	$\hat{s}_2$	—	1

3. Minimize the number of states in the following machine:

Present state	Next state				Output			
	a_0	a_1	a_2	a_3	a_0	a_1	a_2	a_3
1	2	1	—	—	0	—	—	1
2	2	1	—	2	—	0	—	—
3	1	4	3	—	1	0	0	1
4	1	4	2	2	0	—	0	—
5	2	—	2	—	—	0	—	1

4. Minimize the number of states in the following machine:

States	Inputs	
	a_0	a_1
s_1	s_2, 0	—
s_2	s_1, 0	s_6, 0
s_3	s_4, 1	—
s_4	s_5, 0	s_3, 0
s_5	—	s_6, 1
s_6	s_4, 0	—

5. Minimize the number of states in the following machine:

	Next state			Output		
	a_0	a_1	a_2	a_0	a_1	a_2
1	—	1	2	—	z_0	z_1
2	—	2	1	—	z_0	z_0
3	3	4	—	z_0	z_0	—
4	4	5	—	z_1	z_0	—
5	5	3	—	z_2	z_0	—
6	1	—	1	z_2	—	z_2
7	2	—	2	z_1	—	z_0

6. Minimize the number of states in the following machine:

	Next state			Output		
	a_0	a_1	a_2	a_0	a_1	a_2
s_1	s_3	—	s_2	0	—	—
s_2	s_2	s_5	s_3	0	0	0
s_3	s_2	s_4	—	0	—	—
s_4	s_5	s_1	—	0	—	—
s_5	s_1	s_4	—	—	1	1

7. How many different completely specified state tables can be drawn for a Turing machine with two input symbols X_1 and X_2, two output symbols Z_1 and Z_2, and two internal states Q_1 and Q_2? How many partially specified state tables can be drawn? Do any of the completely specified tables correspond to equivalent machines? Do any of the partially specified tables correspond to equivalent machines? Discuss the relationship between the state table for a Turing machine and the Turing machine. How many different Turing machines with two input symbols, two output symbols, and two internal states are there?
8. A Turing machine has three input and output symbols, S_1, S_2, and S_3. The input tape to the Turing machine is recorded as follows:

blank	X_1	X_2	X_3	X_4	X_5	*blank*	X_1	X_2	. . .

where each variable X_i is either S_1, S_2, or S_3. The Turing machine is to rearrange the sextuples as follows:

blank	X_5	X_4	X_3	X_2	X_1	*blank*	X_5	X_4	X_3	. . .

Draw a state table and flow table for the machine, which is to read and print one symbol X_i at a time. This machine realizes the function $f(X_1,X_2, \ldots ,X_5) = (X_5,X_4, \ldots ,X_1)$.

REFERENCES

ARBIB, M. A. (ed.): "Algebraic Theory of Machines, Languages, and Semigroups," Academic Press, 1968.

BARTEE, T. C.: "Digital Computer Fundamentals," 2d ed., McGraw-Hill, 1966.

HARTMANIS, J., and R. E. STEARNS: "Algebraic Structure Theory of Sequential Machines," Prentice-Hall, 1966.

MC CLUSKEY, E. J.: "Introduction to the Theory of Switching Circuits," McGraw-Hill, 1965.

MC CLUSKEY, E. J., and T. C. BARTEE (eds.): "A Survey of Switching Circuit Theory," McGraw-Hill, 1962.

MC NAUGHTON, R.: Theory of Automata: A Survey, *Adv. in Computers*, **2**:379–421 (1961).

MINSKY, M.: "Computation: Finite and Infinite Machines," Prentice-Hall, 1967.

Programming Languages

4-1. INTRODUCTION

Digital computers have the ability to perform both arithmetic and logical operations with great speed and precision. The computer must be provided with a list of instructions to be followed, however, and this list of instructions must be prepared in a form acceptable to the computer. The preparation of this list of instructions, which is called a *program,* is a major problem, for the computer user generally thinks of the procedure for solving any problem in some "natural" language such as English or German, or in terms of some block diagram outlining the various steps to be taken or, in terms of some set of mathematical relationships. The steps which must be taken to translate this user's formulation into a list of instructions acceptable to the computer represents a very real barrier to the use of the computer. The process of making this translation is called *programming,* and the actual writing of instructions to the computer is called *coding.* Programming therefore includes coding, the final part of preparing a procedure so that it can be carried out on a computer. Naturally, the user would like to "tell" the computer what to do in a manner which appears natural to the problem. If the user were attempting to process numerical data from a space vehicle, for instance, he might like to simply request "Take the mean of these data values, remove any value more than two standard deviations from this mean, list these values, and then recalculate the mean with these values removed."

However, such statements are a far cry from the list of binary words in the final program, which is said to be in *machine language.*

In order to facilitate the programming of the computer, early computer users developed special "artificial" languages, which the computer itself could translate into the ultimate binary computer language. These artificial languages were found to be more convenient and natural for the computer user, and the "translation" processes from the artificial language to the computer language were gradually implemented by special routines prepared for the computer.

The first artificial languages which were developed and the programs for translating these languages into machine language are quite simple and are called *assembler languages.* Let us consider the service performed for the user by assembler languages by means of a simple example.

In the simplest form, the data and instructions in a computer are divided into what are called *words,* so that each word contains either an instruction to the computer or some value such as a number or a name. The memory of the computer is then broken into a set of such words of fixed length, to each of which is assigned a numerical address. Further, in a given word of memory, either an instruction or some data to be used may be stored. For instance, we might find that the first four words of the memory contain the (integer) values 29, 364, 48, and 200, respectively, or four instructions to the computer. In the first case we would say that memory address 1 contains the value 29, memory address 2 contains the value 364, etc.

Here are four typical instructions in assembler language which might be in the instruction repertoire of a typical computer.

1. CLA X: This instruction *clears* a special register called the *accumulator* and then *adds* the value stored at location X in the memory into the accumulator where it remains to be used later.
2. ADD X: This takes the variable stored in location X in the memory and *adds* it to the value currently stored in the accumulator, then storing the sum in the accumulator.
3. MUL X: This takes the value of the number stored in location X in the memory and *multiplies* it by the value currently in the accumulator, then storing this product in the accumulator.

4. STO X: This instruction takes the number in the accumulator and places or stores it in the memory at location X.

In the above, X is a number referring to an address or location in the memory. As was mentioned, the memory in digital machines is divided into a number of such addresses, and in each address it is possible to store a piece of information. In this case the information stored consists of numbers, which are usually binary.

Now, if the number 6 is stored in location X in the memory, the number 7 stored in location Y in the memory, and the number 2 stored in the location Z in the memory, we could get the machine to evaluate $(6 + 7)2$ by issuing to the computer the following sequence of instructions: "Clear the accumulator, add the number at location X to it, add the number at location Y, and multiply this sum by the number at location Z." After this sequence the value "6 plus 7 times 2" would be stored in the accumulator. A "store W" instruction would then place this number in the memory in location W.

In this case, in assembler language, the programmer would simply write

```
CLA  X
ADD  Y
MUL  Z
STO  W
```

If the computer had words of 12 binary digits each (so-called "12-bit words"), an actual input string acceptable to the computer would consist of a sequence of four 12-bit words, that is, 48 binary digits. The translations from the above statements to the 48 binary digits would be difficult and treacherous ones for a human to perform, with many chances for mistakes. An assembler does several very helpful and simple things. It associates with each instruction which can be performed by the computer a mnemonic code, such as CLA for clear and add, and it also enables the computer user to type out letters, or letters and numbers (alphanumeric variables), instead of addresses when he writes his program. The computer itself then translates the mnemonic code into the desired actual binary operation code for the machine and assigns consistent values of addresses to the binary variables written by the programmer.

If the computer uses the first four digits of each instruction word to specify the instruction to be performed, and if the CLA instruction has the binary code name 0100, the ADD instruction has the code name 0110, the MUL instruction has the code name 1110, and the STO instruction has the code name 0110, a program might appear as in the following table, where the output of the assembler program appears on the right, this being the desired machine code:

$$\begin{array}{lll} \text{CLA} & \text{X} & \mapsto 010001000101 \\ \text{ADD} & \text{Y} & \mapsto 011010000110 \\ \text{MUL} & \text{Z} & \mapsto 111010000111 \\ \text{STO} & \text{W} & \mapsto 011011000000 \end{array}$$

where the first four bits indicate the operation to be performed and the last eight bits indicate the addresses for X, Y, Z, and W. This little program stores the values $C(A + B)$ in the location W in the memory, where A is the value stored in location X, B the value in location Y, and C the value in location Z (in our previous example A was 6, B was 7, and C was 2).

The above is a simple example taken from an early assembler program. The use of such assembler languages so reduced the work of the programmer and made the use of the computer available to such an enlarged group of people that assemblers became more and more complicated, providing continually increasing services to computer users. Nonetheless, the use of a language such as the above in writing a program to evaluate even simple mathematical expressions can be quite complicated, particularly when transfers of control based on decisions must be written.

The essential feature of assemblers is that translations must be done on a one-to-one basis; that is, the list of binary instructions to the computer in the translated program is bijective with the list of instructions in the written program and thus not too different from the binary program. For this reason assembler languages are often referred to as *machine languages* although they are in fact quite different.

Once the idea that translation by the computer itself would facilitate the solving of problems, work was begun on *procedure-oriented* languages and translators for these languages, which are called *compilers*. These languages allow for a flexible one-to-many translation of written instructions to the computer into machine coding. The first of these programs was called FORTRAN, an acronym for *FOR*mula *TRAN*slation, which was constructed about 1955. The enormous success of FORTRAN led to the development of quite a large number of computer languages for various types of procedures. These include COBOL for business procedures and ALGOL for scientific work. In FORTRAN, we can simply write the line $D = (A + B) * C$, and after this line has been translated and executed by the computer, the variable D in the system will have the value $(A + B) \times C$. FORTRAN includes not only the facility for writing algebraic expressions directly in a natural-appearing language, but also a number of logical options, by means of a shortened set of pidgin-English-like instructions such as *go to, if, read,* and *do.*

In 1958 a number of scientists met in Zürich, Switzerland, with the intention of formulating a programming language for international scientific use. This language was given the acronym ALGOL, standing for *ALGO*rithmic *L*anguage, whose initial version was called ALGOL 58. By 1960 it was apparent that the initial language needed substantial revision, and a second attempt led to ALGOL 60. In subsequent revisions the language has been called simply ALGOL, and the sections of ALGOL which we shall discuss differ in only minor details from either the original 1960 ALGOL or any of the ALGOLs now in use. ALGOL's popularity as a programming language does not equal that of FORTRAN by any means; however, at this time it is internationally used as a language for writing algorithms or procedures in scientific literature.

One of the principal intentions of the ALGOL language's originators was to make the language *machine-independent,* that is, to so devise the language so that a compiler could be written for any machine of moderate size. This was done to enable programmers to write programs which could be run on any machine having an ALGOL compiler. This universal programming language concept also applies to FORTRAN and COBOL and is not unique to ALGOL. The ALGOL language has come to be widely used to describe algorithms and is a standard language in international journals, as well as being the official language for the Association for Computing Machinery and its publications (FORTRAN is also accepted).[1]

4-2. ARITHMETIC EXPRESSIONS

The ALGOL programming language is built up from basic symbols which include the following:

1. The 26 lowercase and 26 uppercase letters of the English alphabet. a,b, . . . ,z and A,B, . . . ,Z
2. Three groups of operators:

Arithmetic:	$+, -, \times, /, \div, \uparrow$
Relational:	$=, \neq, <, \leq, >, \geq$
Logical:	$\wedge, \vee, \neg, \equiv$

3. Separators and brackets: , . ; : :=) ([ ]
4. Word symbols (denoted throughout by boldface) such as **real, integer, true, go to, for, step, until, begin, end**

In this section, we shall describe ways of forming *arithmetic expressions* in ALGOL. These may refer to combinations of constants, as in elementary arithmetic, or to combinations of variables, as in algebra,

[1] PL-1, which resembles ALGOL but incorporates many FORTRAN features as well, is also coming into increasing use.

or mixtures of the two. We begin with a few simple but essential remarks about the notation for *constants* in ALGOL.

Three *types* of constants are allowed in ALGOL programs: *real* constants, *integer* constants, and *Boolean* constants. There are just two Boolean constants, always designated by **true** and **false,** respectively. However, there are various formats for designating numerical constants.

In ALGOL an *integer constant* designates a whole number $n \in \mathbf{Z}$ with or without a sign; it must be written without a decimal point (or it is interpreted as a real constant) and without a comma. The following are examples of integer constants in acceptable ALGOL format:

$$3 \qquad 0 \qquad +16 \qquad -1764 \qquad 12346$$

As stated, 12,346 or 1,394 are not acceptable because of the embedded commas, and 3.0 will be interpreted as a real constant.

A *real constant* may be written as an integer (if it is one) or in either of the following alternative ways:

(*i*) As a decimal number, such as 3.0, +3.0, 6.47, −367.4325, or 0.004328.

(*ii*) As an integer or ordinary decimal number of the form just described, followed by the depressed symbol $_{\mathbf{10}}$ and then by an integer "exponent" which indicates the power of 10 by which the number before the $_{\mathbf{10}}$ is to be multiplied. Examples are:

$$5.34_{\mathbf{10}}4 \text{ for } 53400 = 5.34 \times 10^4$$
$$-.0687_{\mathbf{10}}{-5} \text{ for } -.0687 \times 10^{-5}$$
$$2_{\mathbf{10}}{+15} \text{ for } 2 \times 10^{15}$$
$$\text{and } 365_{\mathbf{10}}{-12} \text{ for } 365 \times 10^{-12}$$

Clearly, the format (*ii*) just described corresponds to the usual scientific notation and is especially useful for representing numbers of very small or very large magnitude, such as the diameter of an electron or our galaxy, in centimeters.

One must be extremely precise in writing computer programs, and so it is important to note the following additional ALGOL restrictions. A decimal number must not terminate with a decimal point under any circumstances. Thus 10. and $73._{\mathbf{10}}{-6}$ are not acceptable, although 10.0 and $73.00_{\mathbf{10}}{-6}$ are perfectly good ALGOL numbers. Likewise, since only integers are allowed as exponents, $5.34_{\mathbf{10}}2.5$ is not meaningful in ALGOL even though $5.34 \times 10^{2.5}$ is a perfectly well-defined real number.

However, we *can* write $5.34 \times 10^{2.5}$ in ALGOL in the form $5.34 \times 10 \uparrow 2.5$. In general, we can translate any arithmetic or algebraic expression into ALGOL quite easily by adhering rigidly to the following specific conventions.

First, we define an *arithmetic expression* in ALGOL to consist of any set of ALGOL numerical (**integer** or **real**) constants or variables, combined by the five binary operations of addition (+), subtraction (−), multiplication (×), division (/), and exponentiation (↑), with parentheses used to indicate groupings when needed.

It is convenient to write arithmetic expressions with a minimum of parentheses. Thus, in ordinary algebra, we write $a + bc$ and not $a + (bc)$, and we learn in school that $a + bc$ does *not* mean $(a + b)c$; in expressions involving both addition and multiplication, multiplication takes precedence (i.e., is to be performed first). Similar precedence rules must be formulated for ALGOL; they are slightly more elaborate because exponents are not "raised." *In the absence of parentheses*, the following *precedence* conventions are in effect: exponentiation takes precedence over all other arithmetic operations, multiplication and division take precedence over addition and subtraction. Multiplication and division have equal precedence with each other, as do addition and subtraction. In the case of such operations of equal precedence, the order of association is from left to right. In summary, the order of precedence is

1. Exponentiation
2. Multiplication and division
3. Addition and subtraction

In ambiguous cases, associate from left to right.

The above rules should become clear with some examples; for simplicity, we have mixed in some symbols for variables.

Mathematical expression	*ALGOL expression*
$3 + \frac{5}{7}$	$3 + 5/7$
$\frac{3+5}{7}$	$(3 + 5)/7$
$2 + 3^7$	$2 + 3 \uparrow 7$
$\frac{ab}{c}$	$a \times b/c$
$(a/b)c$	$a/b \times c$
$\frac{a}{bc}$	$a/(b \times c)$
$\pi(r - p)^2$	$3.14159 \times (r - p) \uparrow 2$
$\frac{6^{2.437}}{.034^{3.26}}$	$6 \uparrow 2.437/.034 \uparrow 3.26$
$\frac{a^b + c^d}{4q}$	$(a \uparrow b + c \uparrow d)/(4 \times q)$

The above arithmetic expressions were written with a minimum of parentheses. However, extra parentheses are also acceptable in ALGOL. For example, it would have been acceptable to write $a + (b/c)$ instead of $a + b/c$ as an ALGOL expression for $a + \frac{b}{c}$. But we must be careful; it would not be correct to translate a^{bc} into ALGOL as $a \uparrow b \times c$. The latter would be interpreted by a compiler as $a^b c$ because exponentiation takes precedence over multiplication. The correct translation is $a \uparrow (b \times c)$. One must also remember to write $3 \times A$ and not $3A$ for "3 times A" in ALGOL programs.

The symbol ÷. The ALGOL language includes a sixth arithmetic operation, ÷. This can be used only with integer expressions. For these, $a \div b$ has as value the **integer** part of the quotient of a divided by b. Thus $6 \div 2$ has the value 3, $5 \div 2$ has the value 2, and $(3 + 6) \div 4$ has the value 2.

4-3. IDENTIFIERS: ASSIGNMENT STATEMENTS

Just like the variables $x,y,z, \ldots$ of mathematics, ALGOL variables take on *values* in given *sets* which must be specified; and they are also given *names*.

The name of an ALGOL variable is called an *identifier*. An identifier may consist of any number of letters or digits or spaces, but the first character must be a letter. Examples of acceptable identifiers are

x
maximum
TIME
sum
r37BQ2
Largest Diagonal Element
rootl

The following are *not* acceptable identifiers:

3ab (The first character is not alphabetic.)
X,YZ (A comma is not a letter or a digit.)
step (**step** is a special ALGOL word symbol, and compilers insist that such words be *reserved*, i.e., not used as identifiers.)

In mathematics, variables can be of many kinds (real, complex, rational, group element, and so on). In ALGOL just three basic types of variables are admitted: *integer, real,* and *Boolean. Integer variables* take on integer values, *real variables* take on (decimal approximations to) real values, and *Boolean variables* (to be discussed in Chap. 5) take on the values **true** or **false,** as the names suggest.

One must specify the type of each variable by a *type declaration.* Thus the declarations

real $A1$, SIGMA;
integer SUM, alpha, beta;
Boolean X;

announce $A1$ and SIGMA to be real-valued variables; SUM, alpha, and beta can assume only integral values; and X can have only the values **true** or **false.**

We now come to the most important construction of the ALGOL programming language, the *assignment statement.* The general form of the assignment statement is

$$\mathcal{V} := \mathcal{E};$$

where $\mathcal{V}$ is the name of an ALGOL variable and $\mathcal{E}$ is an ALGOL expression. This statement is an instruction or command to the computer to compute the value of the right-hand-side expression $\mathcal{E}$ using the current value of all variables appearing in this expression and to *assign* this value to the left-hand-side variable $\mathcal{V}$. This value then becomes the current value of $\mathcal{V}$, and its previous value is lost.

Examples are

$A := 5;$	gives A the value 5
$B := 5 \times A;$	gives B the value 5 times the current value of A
$C := 5 \times A \times B;$	gives C the value 5 times A times B, or $5AB$

Notice that each statement ends with a semicolon. The sequence of statements

$A := 2;$
$B := 3;$
$C := A \times B;$

gives the variable C the value 6.

We shall now consider a sample ALGOL *block* which illustrates the above concepts. An ALGOL block is just a collection of ALGOL statements and declarations enclosed by the word symbols **begin** and **end.** We shall investigate the block structure of ALGOL in more detail later, but for the time being we can give a simple rule: begin with **begin** and end with **end** as in

```
begin real a,b,c;
      c := 5;
      a := 4.1;
      b := 2 × a + 7;
      c := 3 × a − b;
end
```

Again notice the semicolons. All ALGOL statements not followed by **end** must terminate in semicolons to indicate that the statement is complete. Remember also that all variables which appear within an ALGOL block should be declared by means of a type declaration. In the above a,b,c have been declared as real variables.

The type declaration statement is not executed by the computer but serves only to give the necessary information at compile time (i.e., when the ALGOL program is being translated from ALGOL to machine language) concerning the type of all variables appearing in the block. As such, the type declaration statement is said to be *nonexecutable*, as contrasted with the assignment statement, which is *executable*. We now examine in detail the effect of executing the assignment statements of the above block.

$c := 5;$	c is assigned the value 5.
$a := 4.1;$	a is assigned the value 4.1.
$b := 2 \times a + 7;$	The right-hand expression is evaluated using the current value of a, namely 4.1, and b is assigned the resulting value 15.2.
$c := 3 \times a - b;$	c is assigned the value $3 \times 4.1 - 15.2 = -2.9$, and the previous value of c, namely 5, is lost; i.e., the current value of c is now -2.9.

The next example forcefully illustrates the meaning of the assignment operation := in an assignment statement. The statement

$$N := N + 1;$$

is a perfectly valid ALGOL statement, and the effect of executing this instruction is to increase N by 1. Thus, if the value of N before the execution of the above statement is 5, then the value of N immediately after execution is 6. Clearly the above statement is *not* an equality between the left- and right-hand sides; we strongly emphasize that := denotes a replacement operation and not an equality relation. The reader should read "assign the value of $\mathcal{E}$ to $\mathcal{V}$" or, equivalently, as "replace the value of $\mathcal{V}$ with the value of $\mathcal{E}$" but never as "$\mathcal{V}$ equals $\mathcal{E}$."

Mixed Expressions. In ALGOL, it is acceptable to mix types within a single expression, i.e., to write an expression in which both integer and real-type variables and constants appear.[1] The type of the result of the various arithmetic operations in terms of the types of the operands is determined by the following rules:

1. The result of the operations $+$, $-$, $\times$ is integer when both operands are integers and is real otherwise.
2. The result of $a \uparrow b$ is an integer when a is an integer and b is a nonnegative integer, and is real otherwise.
3. The result of a/b is always real for all four combinations of operands.
4. The operation $a \div b$ is defined only when a and b are of type **integer;** the result is the integer part of the quotient a/b and is of type **integer.**

As examples, we consider execution of the following statements:

$$a := 3.6 + \tfrac{1}{4}; \qquad b := 3 + 2.0; \qquad c := 5 \uparrow (-2);$$
$$i := 5 \uparrow 3; \qquad j := (4 \times 3) \div 2; \qquad k := 4 \times (3 \div 2);$$

In ALGOL, the first three define $a = 3.85$, $b = 5.0$, and $c = 0.04$ of type **real,** and the second three define $i = 125$, $j = 6$, and $k = 4$ of type **integer.**

4-4. ARRAYS

In ALGOL, vectors are called *one-dimensional arrays,* matrices are called *two-dimensional arrays,* and their subscripts are placed in square brackets. Thus, the components of the vector $x = (x_1,x_2,x_3)$ are called elements of the one-dimensional array $x[1:3]$ and written $x[1]$, $x[2]$, $x[3]$. Before an array can be used, the number of its entries must be announced in a type declaration. Thus

real array $X[1:20]$

declares that the array X contains 20 variables $X[1],X[2], \ldots ,X[20]$, each of which is a real variable. The **real** can be dropped; **real array** $X[1:20]$ is compiled in the same way as **array** $X[1:20]$. A **Boolean array** or **integer array** must be declared, however. The description

integer array $X[-5:15]$

[1] Such mixed expressions are not acceptable in most versions of FORTRAN, however.

declares $X[-5], X[-4], \ldots, X[15]$ to be a set of 21 integer-valued variables, with indices running from -5 to 15.

Likewise, the description

$$\textbf{array } A[1:10,\ 1:15]$$

identifies a 10×15 real matrix $A[i,j]$, where i runs from 1 through 10 and j from 1 through 15. Similarly, the declaration

$$\textbf{integer array } C[-1:8,\ 0:9,\ 2:11]$$

establishes a three-way matrix (three-dimensional array with 1,000 integer-valued variables $C[i,j,k]$).

The main effect of such an array declaration on a compiler or other processor is to cause the appropriate number of storage spaces to be reserved for the elements of the array.

Functions. ALGOL processors provide for the use of several of the elementary real functions of analysis; among those usually included are

abs(E)	Absolute value of the expression E
sqrt(E)	Square root of the value of E
sin(E)	Sine of the value of E
cos(E)	Cosine of the value of E
ln(E)	Natural logarithm of the value of E
exp(E)	Exponential function of the value of E, (e^E)

The above functions operate on either real- or integer-type arguments, and yield a value of type **real** in either case. The following examples illustrate the use of ALGOL expressions for functions:

Mathematical expression	*ALGOL expression*
$e^x + y$	exp (x) + y
e^{x+y}	exp ($x + y$)
$\sin^3\left(\frac{x}{y}\right)$	sin (x/y) ↑ 3
$\sin\frac{x^3}{y}$	sin (x ↑ 3/y)
$\frac{-b + \sqrt{b^2 - 4ac}}{2a}$	($-b$ + sqrt (b ↑ 2 − 4 × a × c))/(2 × a)
$\sqrt{1 + \frac{1}{\sin \lvert x \rvert}}$	sqrt (1 + 1/sin (abs (x)))

Caution. The ALGOL language does not recognize complex numbers as such, and the preceding functions are only processed in ALGOL for real values. Hence we would get into trouble with a negative radicand by blindly using the following plausible ALGOL program to compute the roots of the quadratic equation

$$3x^2 + \sqrt{e^\pi}\, x + \ln 8.7 = 0$$

Knowing from high-school algebra that the roots of the equation $ax^2 + bx + c = 0$ are given by

$$\frac{-b \pm \sqrt{b^2 - 4ac}}{2a}$$

we write

```
begin real a,b,c,d, root1, root2;
        a := 3;
        b := sqrt (exp (3.14159));
        c := ln (8.7);
        d := sqrt (b ↑ 2 − 4 × a × c);
        root1 := (−b + d)/(2 × a);
        root2 := (−b − d)/(2 × a)
end
```

EXERCISES A

1. Convert the following ALGOL expressions to conventional mathematical expressions:
   (a) $A \times B - C$
   (b) $A \uparrow (B - C)$
   (c) $A \uparrow B - C$
   (d) $A/B - C \uparrow D$
   (e) $A \div B - C \times D$
2. Evaluate the expressions in Exercise 1 for $A = 2$, $B = 3$, $C = 4$, and $D = 5$.
3. Write the following expressions in ALGOL form without using parentheses:
   (a) $A(B + C) - D$
   (b) $\dfrac{A}{B + C} - D$
   (c) $\dfrac{A^2}{B + C}$
   (d) $(A^N - B^N)(C - D)$
4. Write ALGOL expressions for the following:
   (a) $\sin (A + B^2)$
   (b) $e^B - C$
   (c) Absolute value of (natural log of (A^B))
   (d) $\sqrt{e^A - \sin A}$
   (e) $\sin 4(A + 3^2)$
5. Convert the following ALGOL expressions to conventional mathematical expressions:
   (a) $A - B \times C \times D$
   (b) $A - B + C \uparrow D \times E$
   (c) $A \uparrow D - E \uparrow F$
   (d) $A \times D - E + F \uparrow G$
   (e) $A - B \times C + D$
6. Evaluate the expressions in Exercise 5 for $A = 2$, $B = 3$, $C = -2$, $D = 4$, $E = 5$, $F = 6$, and $G = -2$.

7. Write the following expressions in ALGOL form without parentheses:
   (a) $A^2 + B^C \cdot D$
   (b) $A \cdot B + C^{D^E}$
   (c) $3 \cdot A + BC + (D - E)$
   (d) $4(A + B) + (6(A - C))^2$
8. Write ALGOL expressions for the following:
   (a) $A^2 + B^2 + (C^3 - D)^3$
   (b) $\sin A + \cos^2 B$
   (c) $e^{|x|}$
   (d) $e^{x/y}$
   (e) $\sin^2 e^{x/y}$
9. Write ALGOL expressions for the following:
   (a) $\sin(x) + e^{\cos x}$
   (b) $\sin(x + e^{\cos x})$
   (c) $\sin x + e^{\cos(x+y)}$
   (d) $e^x + e^y$
   (e) $e^x \cdot e^y$

4-5. *FOR* STATEMENTS

In computation, it is often desired to execute a number of statements over and over again with a varying argument which changes its value by uniform steps. For example, consider the computation of the sum

$$S = \sum_{i=1}^{n} i^2 \tag{1}$$

This is achieved by first making $S := 0$ and then executing repeatedly the statement $S := S + i^2$ for $i = 1,2, \ldots ,n$. After the kth iteration, S equals the kth partial sum of squares $1^2 + 2^2 + \cdots + k^2$, and finally, after the last (nth) iteration, S has the required value $\sum_{i=1}^{n} i^2$.

The ALGOL **for** statement provides this iteration capability and represents one of the most powerful features of the language. One form (not the most general but adequate for our purposes) is

for $v := m$ **step** h **until** n **do** T;

where v can be a real or integer variable; m, h, and n are usually constants; and T is a statement. Statement T is executed for each value taken on by v for v ranging from m to n in steps of h. Control passes to the next statement when $v > n$.

Using the **for** statement, we see that the ALGOL statements for the sum of squares example considered earlier become simply

$S := 0$;
for $i := 1$ **step** 1 **until** n **do** $S := S + (i \uparrow 2)$;

Some other illustrative examples of the **for** $\cdots$ **do** ALGOL format are as follows:

1. **for** $i := 1$ **step** 1 **until** 10 **do** S;
   Effect: S is executed for $i = 1, 2, \ldots ,10$.

2. **for** $i :=\ -4$ **step** 2 **until** 7 **do** S;
   Effect: S is executed for $i = -4, -2, 0, 2, 4, 6$. Note that i does not take on the value 7.
3. **for** $x :=$ 0 **step** .1 **until** 1 **do** S;
   Effect: S is executed for $x :=$ 0,0.1,0.2, . . . ,0.9,1.0.
4. **for** $x :=$ 1 **step** $-$ .1 **until** $-$ .5 **do** S;
   Effect: S is executed for $x :=$ 1,0.9,0.8, . . . ,−0.4,−0.5.
5. **for** $x :=$ 5 **step** 1 **until** 4 **do** S;
   Effect: S is not executed at all.

We now present four examples of ALGOL programs.

Example 1. The following ALGOL block generates an array K with $K[i] = i!$ for $i = 1, 2, \ldots, 10$:

```
begin real array K[1:10];
      integer i;
      K[1] := 1.0;
      for i := 2 step 1 until 10 do K[i] := K[i − 1] × i;
end
```

Example 2. Let there be given a sequence of N observations $x_1, \ldots, x_N$ ($N \leqq 500$), and suppose it is desired to compute (1) the mean $m = \left(\sum x_i\right)/N$, and (2) the standard deviation $s = \left[\sum_{i=1}^{N} (x_i - m)^2\right]^{\frac{1}{2}}/(N - 1)$. This can be done by the following ALGOL statements. Assume that the values for the variables m, s, i, N, and the array $X[1]$ through $x[500]$ are stored in memory. Also, assume that the $x[i]$ are real, as are m and s and that i and N are integers. Then we have

```
m := 0;
for i := 1 step 1 until N do
    m := m + x[i];
m := m/N; s := 0;
for i := 1 step 1 until N do
    s := s + (x[i] − m) ↑ 2;
s := sqrt (s)/(N − 1);
```

Example 3. Suppose we wish to evaluate a given polynomial of degree 50 (say), with known coefficients

$$f(y) = \sum_{j=0}^{N} a_j y^j = a_0 + a_1 y + \cdots + a_{50} y^{50} \tag{2}$$

Assuming the values of the $a[i]$ are stored in memory, a straightforward procedure for evaluating this is

```
POLY := 0;
for j := 0 step 1 until 50 do
    POLY := POLY + (a[j] × y ↑ j);
```

For any values of the coefficients $a_j = a[j]$ and y stored in the machine, this program will evaluate (2). This straightforward approach of computing all the powers of x, multiplying by the coefficients, and adding is the simplest to program but requires an unnecessarily large number of arithmetic operations to perform. Indeed, if $y \uparrow j$ is computed recursively by the formulas $y \uparrow 0 = 1$ and $y \uparrow (j + 1) = y \times y \uparrow j$, it would require $N(N - 1)/2$ multiplications and N additions, where N is the degree of the polynomial [which is 50 in (2)]. A more economical procedure is the following.

Example 4. To evaluate the polynomial

$$f(x) = a_N x^N + a_{N-1} x^{N-1} + \cdots + a_0$$

use the following expression with nested parentheses:

$$f(x) = x(\cdots x(x(a_N x + a_{N-1}) + a_{N-2}) + \cdots + a_1) + a_0$$

This requires only N multiplications and N additions, a considerable saving.

The following ALGOL program computes the value F of the polynomial $\sum_{j=0}^{N} a_j x^j$ according to the above formula. It is assumed that the coefficients $a_j = A(j)$ have been stored as real constants, and that the degree $N (1 \leq N \leq 20)$ and the argument x are specified by the integer variable N, and the real variable X, respectively. All these values are assumed to be stored in memory when these statements are performed

```
F := A[N];
for j := N − 1 step −1 until 0 do
    F := X × F + A[j]
```

Note that the controlled variable j of the **for** statement takes on the values $N - 1, N - 2, \ldots, 1, 0$ because of the -1 step increment. The reader should verify that after execution of the **for** statement, F does indeed have the desired value. This can be done by tracing through the program step by step (i.e., "executing" the program) with a low-degree polynomial, say $4x^3 + 2x^2 + 3x + 1$.

4-6. BLOCK STRUCTURES IN ALGOL

ALGOL is designed to be written in block form. A **block** consists of a number of ALGOL statements starting with the word **begin** and terminating with the word **end,** with declarations immediately following the **begin.** Every ALGOL program must be in the form of a block, as illustrated by the following example.

Example 5. Two one-dimensional arrays X and Y each contain 50 elements. The following ALGOL program computes

$$LX = \sqrt{\sum_{j=1}^{50} X_j^2} \qquad \text{(Length of the vector } X\text{)}$$

$$LY = \sqrt{\sum_{j=1}^{50} Y_j^2} \qquad \text{(Length of the vector } Y\text{)}$$

$$\text{INPROD} = \sum_{j=1}^{50} X_j Y_j \qquad \text{(Inner product of } X \text{ and } Y\text{)}$$

```
begin real array X[1:50], Y[1:50];
      real LX, LY, INPROD;
      integer j;
LX := LY := INPROD := 0;
for j := 1 step 1 until 50 do
   begin LX := LX + X[j] ↑ 2;
         LY := LY + Y[j] ↑ 2;
         INPROD := INPROD + X[j] × Y[j]
   end
LX := sqrt (LX); LY := sqrt (LY)
end
```

A *compound statement* consists of a number of ALGOL statements preceded by **begin** and followed by **end**; it differs from a block in that there are no declarations immediately following the **begin.** The statements so enclosed within the **begin** and **end** are treated syntactically as a unit (a single statement). Referring to Example 5, we see that the controlled statement of the **for** statement is a compound statement consisting of three "simple" statements. Note that without the statement parentheses **begin** and **end,** only the statement $LX := LX + X[j] \uparrow 2$ would be executed as the controlled statement, and of course the desired result would not be obtained.

It should be mentioned that the block and compound statement scheme is completely recursive. Thus a statement of a compound statement may itself be a compound statement; this can be illustrated schematically as follows:

```
begin S1;
      S2;
      begin S3;
            S4;
      end
      S5
end
```

The S_i's above stand for statements. Note that ALGOL statements end with semicolons, with the exception that the semicolon may be deleted immediately preceding an **end.**

Also, a program (which is a block) may contain several subblocks which in turn may contain subblocks, etc. The details of block structure need not concern us. We just point out in passing that for large programs requiring large amounts of storage, the block structure allows for efficient utilization of available space.

This presentation will not be concerned with input-output statements, the reading of data into the computer, and the printing of displaying results. This is not to minimize the importance of communication between the computer and the outside world. However, input-output statements depend in general on the nature of the input-output devices available (tapes, card readers, printers, typewriters, oscilloscopes, etc.). For this reason, input-output procedures were not specified in the ALGOL 60 report.[1]

EXERCISES B

1. Write programs which multiply the matrix

$$A = \begin{bmatrix} a_{11} & a_{12} & a_{13} \\ a_{21} & a_{22} & a_{23} \\ a_{31} & a_{32} & a_{33} \end{bmatrix}$$

(a) by a row vector $B = [b_{11}, b_{12}, b_{13}]$
(b) by a column vector

$$B = \begin{bmatrix} b_{11} \\ b_{21} \\ b_{31} \end{bmatrix}$$

[1] Some standardization of ALGOL input-output procedures was agreed upon in 1964, a description of which is given in H. Rutishauser, "Description of ALGOL 60," Chap. 8, Springer, 1967.

2. Write an ALGOL program which sums the a_{ij} in Exercise 1.
3. Write a program with a variable b which has the value of the polynomial $a_0 + a_1x + a_2x^2 + a_3x^3$ after the program has been run.
4. Derive the expression $A := B + C \uparrow D$ from the rules for constructing ALGOL expressions.
5. Write a program that multiplies each a_{ij} in a matrix A by 6.
6. Write a program that multiplies each a_{ij} in a matrix A by 5 if $i = j$ and otherwise leaves the element as is.
7. Write an ALGOL program that forms the inner product $a_{11}b_{11} + a_{12}b_{12} + a_{13}b_{13}$ of two row vectors and gives this value to c.
8. Write an ALGOL program which forms the matrix product AB
   (a) For $A = \|a_{ij}\|$ and $B = \|b_{ij}\|$, both 3×3 real matrices
   (★b) For A and B $m \times n$ and $n \times r$ matrices, $m + n + r \leqq 100$
9. The current value of a real variable A is 3.7 and of a real variable B is -6.3. Explain concisely the effect of executing the statement
   (a) $A := B;$
   (b) $B := A;$
   (c) $C := A;$ (where C is an integer variable)
   (d) $C := B;$ (where C is an integer variable)
10. Translate the following into ALGOL: ($\leftarrow$ means "is assigned the value")
   (a) $z \leftarrow (e^{|x \cdot y|, 1/b})$
   (b) $F \leftarrow \sinh x^2 + \cosh^2 y$ (assume that the functions sinh and cosh are not available; this is the case for most compilers)
   (c) $Y \leftarrow \sum_{k=1}^{N} \cos^k (x^k)/N$ (assume that the integer variable N has been previously assigned a positive value $N \in \mathbf{P}$)
   (d) $h \leftarrow \begin{cases} 1 & \text{if } a,b \neq 0 \text{ and } a,b \text{ have the same sign} \\ 0 & \text{otherwise} \end{cases}$

★11. Suppose that you are using an ALGOL processor that has not compiled any of the elementary functions sin, cos, abs, $\uparrow$, exp, etc.; only the arithmetic operations $+$, $-$, $\times$, $/$ are available. Your problem is to write an ALGOL program to compute $\cos x$ correct to five decimal places for $0 \leqq |x| \leqq 1$. Your solution should consist of
   (a) a mathematical analysis of the problem
   (b) an ALGOL program implementing the results of (a)

4-7. THE ALGOL GRAMMAR

In constructing the programming language ALGOL, its creators gave a complete formal specification of the ways in which acceptable ALGOL programs could be written. Their aim was similar to that of mathematical logicians, who attempted to formalize mathematical reasoning in such a way that a properly programmed computer would be able to decide mechanically whether or not any given mathematical statement was meaningful and if so, whether its assertion was true or false (see Sec. 14.5).

Necessarily, discussions of the laws of any language (its "grammar" and "syntax") are carried on in "words which describe words," that is, in a "metalanguage." By *metalanguage* we mean here a language L_0 which is used to describe some (generally lower-level) language L_1. We now define some properties of ALGOL in terms of the metalanguage of mathematical linguistics.

Definition. A *language* L is a set of strings, each of finite length, in some finite set of symbols called the *vocabulary* of L and designated V.

Given any vocabulary (or alphabet) V, the set of all strings from V is denoted V^*. Thus a language with vocabulary V is defined, in mathematical linguistics, to be a subset $L \subset V^*$. Evidently, any two strings of symbols from V, say $\mathbf{a} = a_1a_2 \cdots a_m$ and $\mathbf{b} = b_1b_2 \cdots b_n$, can be joined together into a single string; thus

$$\mathbf{s} = \mathbf{a} * \mathbf{b} = a_1a_2 \cdots a_mb_1b_2 \cdots b_n$$

by laying them end to end. Here $*$ is called the operation of concatenation.

Clearly, the binary operation of concatenation is *associative;* thus

$$(\mathbf{a} * \mathbf{b}) * \mathbf{c} = \mathbf{a} * (\mathbf{b} * \mathbf{c}) \qquad \text{for all } \mathbf{a}, \mathbf{b}, \mathbf{c} \in V^*$$

Moreover, the algebraic system $[V^*,*]$ is closed under this binary associative operation of concatenation. Such systems are called *semigroups*[1] and will be studied in Chap. 7.

The concept of a language with vocabulary V as an arbitrary subset of V^* is too general to be useful. In particular, statements or sentences in a programming language must be precisely defined. As a result, rules are needed to be used to generate statements and to analyze statements. Mathematical linguistics is concerned with languages which are *produced* recursively by specified "production rules," or *rules of syntax.*

Example 6. Let $V = \{A,B,a,b\}$. Define L by the production rules (*i*) $A \in L$ and $B \in L$; and (*ii*) if $q \in L$ and $r \in \{a,b\}$, then $q * r \in L$. Then L consists of all words beginning with a capital letter A or B, and continuing with a finite number of small letters a or b.

Definition. A *grammar* for a language L is a finite set of rules which enable one to (recursively) *produce* the set $L \subset V^*$ of all acceptable (or meaningful) strings of symbols.

In a programming language, L consists of those strings which constitute meaningful programs. One of the functions of a compiler is to reject a program when it detects an incorrectly formed string and call

[1] Specifically, the semigroup $[V^*,*]$ is called *the free semigroup generated by* V.

this fact and the type of error which has been made to the coder's attention, thus helping him to debug his program.[1]

The metalanguage used to describe the ALGOL grammar on which the ALGOL language is based is itself largely symbolic; it is based on the following conventions:

$\langle X \rangle$ means "the class of things in X"
$::=$ means "is comprised of"
$|$ means "or"

As an example we write

$$\langle\text{digit}\rangle ::= 0 \mid 1 \mid 2 \mid 3 \mid 4 \mid 5 \mid 6 \mid 7 \mid 8 \mid 9$$

which says "a digit in ALGOL is 0 or 1 or 2 or · · · or 9." Likewise, we define the set of letters by the following list:

$$\langle\text{letter}\rangle ::= a \mid b \mid c \mid d \mid e \mid f \mid g \mid h \mid i \mid j \mid k \mid l \mid m \mid n \mid o \mid p \mid q \mid r \mid s \mid t \mid u \mid v \mid w \mid x \mid y \mid z \mid A \mid B \mid C \mid D \mid E \mid F \mid G \mid H \mid I \mid J \mid K \mid L \mid M \mid N \mid O \mid P \mid Q \mid R \mid S \mid T \mid U \mid V \mid W \mid X \mid Y \mid Z$$

This list gives us all 52 lowercase and uppercase letters.

Similarly,

$$\langle\text{arithmetic operator}\rangle ::= + \mid - \mid \times \mid / \mid \div \mid \uparrow$$

lists the arithmetic operators, and

$$\langle\text{logical constant}\rangle ::= \textbf{true} \mid \textbf{false}$$

lists the two logical constants permitted.

The rules of syntax given above are all in what is called *Backus normal form.*[2] This form presents a name for the class of objects being defined to the left of the symbol $::=$, which is followed by either a list of the objects themselves, such as 0, 1, . . . or A, B, . . . , or by the names of other classes of objects and rules for combining them to recursively produce new classes from previously defined classes.

As an example of a recursive definition in Backus normal form, we have the ALGOL definition

$$\langle\text{identifier}\rangle ::= \langle\text{letter}\rangle \mid \langle\text{identifier}\rangle\langle\text{letter}\rangle \mid \langle\text{identifier}\rangle\langle\text{digit}\rangle$$

This defines the class of objects called *identifiers,* first stating that any element of the previously defined class "letter" can be an identifier.

[1] Sometimes compilers attempt to fix errors in construction and then notify a programmer as to the nature of the "fix."

[2] The phrase "Backus-Naur form" or BNF is also used.

The production rule then states that placing a letter or digit to the right of any identifier gives an identifier.

For example, A is an identifier; placing B to its right gives AB, and adding C to the right gives ABC. Likewise, since digits, another previously defined class, can be added to the right, $A1$, $AB1$, $A221$, and $A1R1$ are all valid identifiers. The substance of this production rule is that identifiers must have a letter in the first position to the left and that any combination of letters and digits can follow this letter. Notice also that the class of identifiers is defined recursively; indefinite application of the rules will lead to an infinite set of elements, all of which are possible identifiers.

What is a basic symbol in this language?

⟨basic symbol⟩ ::= ⟨letter⟩ | ⟨digit⟩ | ⟨logical value⟩ | ⟨delimiter⟩

The first three of these have been listed; for delimiter we have

⟨delimiter⟩ ::= ⟨operator⟩ | ⟨separator⟩ | ⟨declarator⟩
⟨operator⟩ ::= ⟨arithmetic operator⟩ | ⟨sequential operator⟩
⟨sequential operator⟩ ::= **go to** | **if** | **then** | **else** | **for** | **do**
⟨separator⟩ ::= ⏨ | : | ; | , | · | := | **step** | **until** | **while**
⟨declarator⟩ ::= **integer** | **real** | **array**

Some of the above have to do with logic and will be discussed in Chap. 5.

The informal description of how to write numbers in ALGOL, given in Sec. 4-2, can now be formalized. Table 4-1 gives the rules for forming numbers in ALGOL. Notice the rules do not tell what "numbers" are, but simply provide instructions for forming legitimate strings of symbols which are called *numbers*.

Using the production rules of Table 4-1 recursively, we could produce an infinite list which included all integers and all decimal fractions with a

Table 4-1. Numbers—syntax

<table>
<tr><td>⟨unsigned integer⟩</td><td>::=</td><td>⟨digit⟩ | ⟨unsigned integer⟩⟨digit⟩</td></tr>
<tr><td>⟨integer⟩</td><td>::=</td><td>⟨unsigned integer⟩ | +⟨unsigned integer⟩ |
−⟨unsigned integer⟩</td></tr>
<tr><td>⟨decimal fraction⟩</td><td>::=</td><td>⟨unsigned integer⟩</td></tr>
<tr><td>⟨exponent part⟩</td><td>::=</td><td>⏨⟨integer⟩</td></tr>
<tr><td>⟨decimal number⟩</td><td>::=</td><td>⟨unsigned integer⟩ | ⟨decimal fraction⟩ |
⟨unsigned integer⟩⟨decimal fraction⟩</td></tr>
<tr><td>⟨unsigned number⟩</td><td>::=</td><td>⟨decimal number⟩ | ⟨exponent part⟩ |
⟨decimal number⟩⟨exponent part⟩</td></tr>
<tr><td>⟨number⟩</td><td>::=</td><td>⟨unsigned number⟩ | +⟨unsigned number⟩ |
−⟨unsigned number⟩</td></tr>
</table>

finite number of digits. But a digital computer, having a finite memory, cannot represent exactly irrational numbers like $\sqrt{2}$ or π. It represents such numbers only *approximately*, "rounding" them off after a finite number of digits. Typically,[1] the binary word length used to store numbers ranges from 32 to 64; hence numbers are given to at most 15 significant decimal digits, in "single precision" arithmetic. In extensive calculations where greater accuracy is required to prevent an excessive build-up of roundoff errors, compilers have been written which permit double precision arithmetic, in which the binary word length is doubled. Multiple precision arithmetic, in which even more significant figures are retained, is also sometimes used.

Tables 4-2, 4-3, and 4-4 give the syntactic rules associated with arithmetic expressions, **for** statements, and assignment statements, respectively. The Backus normal forms used here make wide use of recursion in the definitions. These rules of syntax include all of the expressions and statements which have been used so far.

[1] For example, the IBM 360-65 assigns 32 binary digits and the CDC 6500 48 binary digits to a single precision number. The IBM 85, however, can perform calculations with 112-bit operands and 8-bit scale factors.

Table 4-2. Arithmetic expressions—syntax

⟨adding operator⟩ ::= + | −
⟨multiplying operator⟩ ::= | × | / | ÷
⟨primary⟩ ::= ⟨unsigned number⟩ | ⟨variable⟩ | (⟨arithmetic expression⟩)
⟨factor⟩ ::= ⟨primary⟩ | ⟨factor⟩ ↑ ⟨primary⟩
⟨term⟩ ::= ⟨factor⟩ | ⟨term⟩⟨multiplying operator⟩⟨factor⟩
⟨simple arithmetic expression⟩ ::= ⟨term⟩ | ⟨adding operator⟩⟨term⟩ | ⟨simple arithmetic expression⟩⟨adding operator⟩⟨term⟩
⟨if clause⟩ ::= **if** ⟨Boolean expression⟩ **then**
⟨arithmetic expression⟩ ::= ⟨simple arithmetic expression⟩ | ⟨if clause⟩⟨simple arithmetic expression⟩ **else** ⟨arithmetic expression⟩

Table 4-3. for statements—syntax

⟨for list element⟩ ::= ⟨arithmetic expression⟩ | ⟨arithmetic expression⟩ **step** ⟨arithmetic expression⟩ **until** ⟨arithmetic expression⟩ | ⟨arithmetic expression⟩ **while** ⟨Boolean expression⟩
⟨for list⟩ ::= ⟨for list element⟩ | ⟨for list⟩, ⟨for list element⟩
⟨for clause⟩ ::= **for** ⟨variable⟩ := ⟨for list⟩ **do**
⟨for statement⟩ ::= ⟨for clause⟩⟨statement⟩

Table 4-4. Assignment statements—syntax

⟨left part⟩ ::= ⟨variable⟩ :=
⟨left part list⟩ ::= ⟨left part⟩ | ⟨left part list⟩⟨left part⟩
⟨assignment statement⟩ ::= ⟨left part list⟩⟨arithmetic expression⟩

★4-8. EVALUATING ARITHMETIC STATEMENTS

In this section, it will be shown how the concept of a *push-down store* yields an algorithm for evaluating simple arithmetic expressions in ALGOL. In the next section, it will be shown how the same concept gives an algorithm for compiling such expressions by translating them into assembler language.

The factor which most facilitates the evaluation or compiling process is that arithmetic operations are "ordered" by the syntax of ALGOL. This allows the use of what are called *operator precedence techniques*. The definitions of arithmetic operations in ALGOL are such that the operator $\uparrow$ has highest precedence; the operators /, $\div$, and $\times$ have second highest precedence and are equal in precedence; and the operators + and $-$ have least precedence. If we consider precedence as a binary relation $>$ and equality as a binary relation $=$, we have

$$\uparrow \; > \; \div \; = \; \times \; = \; / \; > \; + \; = \; -$$

The fact that these operators are so ordered, plus the left-to-right rule,[1] leads to a simple set of steps for evaluating arithmetic expressions. First let us limit our language to variables of single letters, the above arithmetic operators, and left and right parentheses. We further consider $-$ as a binary operator only, not allowing $-b + a$. The use of $-$ as both a unary and binary operator is easily taken care of but we do not digress to describe the procedure.

The procedure to be given makes use of what computer programmers and designers call a *push-down store* or a *stack*. This is a memory which operates on a "last-in-first-out" basis. A stack memory consists of an ordered set of locations in the memory. This can be imagined as a vertical stack of cards with a marker at the bottom. The stack is started empty. The first item placed in the memory goes onto the marker. If a second item is put in, this goes on top of the first item. When an item is retrieved from the memory, or "read," it is always the last item placed in the memory. Thus, if the memory stack $S1$ is started empty and we put the number 16 in followed by the number 36 and then read from the memory, the number 36 will be read first. If we read again, the number 16

[1] As was mentioned in Sec. 4-2, this means performing operations from left to right among operators of equal precedence.

will be read. If we start with the empty stack $S1$ and then place the numbers 6 and then 26 in the stack and then read one number, we will read 26. If we then place 42 and 34 in the stack and read three numbers, we will read 34, 42, and then 6 in that order.

We implement our algorithm for evaluating arithmetic expressions in ALGOL, using two push-down stores or stacks called $S1$ and $S2$. The algorithm assigns variable symbols to $S1$ and operation symbols, or operators, to $S2$, as follows.

ALGORITHMS FOR EVALUATING ARITHMETIC EXPRESSIONS

1. Read the input string from left to right.
2. Read the first value of a variable into $S1$ and the first operator into $S2$.
3. Always read each new variable on the top of $S1$.
4. Each time an operator is read
   (a) If the operator has precedence over the operator currently on top of $S2$, place the new operator on top of $S2$.
   (b) Otherwise (i.e., if operator read has equal or lower precedence), apply the operator on top of $S2$ to the two elements on $S1$. Place the result on top of $S1$, remove the operator from $S2$, and repeat step 4.
5. When the last variable has been read
   (a) If $S2$ is empty, the program terminates.
   (b) Otherwise (i.e., if application of the preceding rules does not leave $S2$ empty), apply the top operator of $S2$ to the top two variables on $S1$, storing the result on $S1$.
   (c) Repeat step 5 until $S2$ is empty.

The above algorithm will result in the evaluation of a given arithmetic expression. (In fact, if a precedence relation is defined on a set of binary operators, this algorithm will result in the evaluation of an expression written using these operators; this applies to Boolean operators or to any other type of operator that might be defined.)

Consider the evaluation of $a \times b \times c + d \times e \times f$. First, a is placed on $S1$ and $\times$ is placed on $S2$; then b is read and placed on $S1$, $\times$ is read and placed on $S2$, and then c is read and placed on $S1$. The situation is

$S1$	$S2$
c	$\times$
b	$\times$
a	

When the $+$ operator is read, it yields precedence to the $\times$, so the $\times$ operator is applied to c and b, and the product stored on top of the stack. The situation is then

$S1$	$S2$
$c \times b$	$\times$
a	

Since $S2$ is not empty, the operator $\times$ is applied to $c \times b$ and a, giving $c \times b \times a$. Then $+$ is placed on $S2$; the situation is (using commutativity to write $a \times b \times c$ for $c \times b \times a$):

$S1$	$S2$
$a \times b \times c$	$+$

Now d is placed on $S1$ and an $\times$ is read. Since d is multiplied by e and the product placed on $S1$, the situation is

$S1$	$S2$
$d \times e$	$+$
$a \times b \times c$	

Finally the last $\times$ is read with the result that the top of $S1$ is multiplied by f and the product is placed on top of $S1$; the situation now is

$S1$	$S2$
$d \times e \times f$	$+$
$a \times b \times c$	

Since S_2 is not empty, the operator symbol is applied to the top two items in $S1$, giving

$S1$	$S2$
$a \times b \times c + d \times e \times f$	

Thus, in conclusion, the evaluation takes place by "parsing" the original ALGOL expression $a \times b \times c + d \times e \times f$ according to the parsing tree displayed below. The syntactic rules of ALGOL and the left-to-right parsing rule make no other interpretation possible.

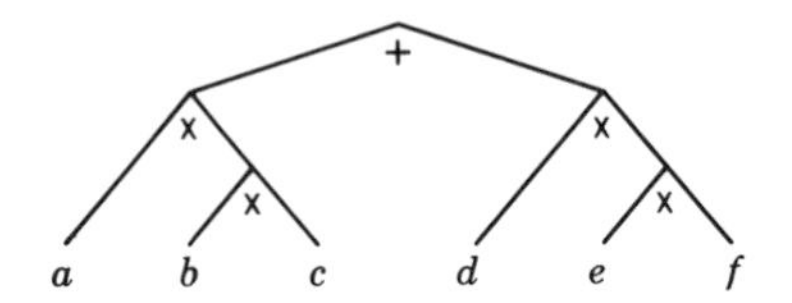

★4-9. COMPILING ARITHMETIC EXPRESSIONS

The preceding algorithm evaluates one arithmetic expression in ALGOL form. We shall now use this algorithm to compile a code in assembler language which will evaluate *any* expression of this sort. Let us use the following four typical machine instructions:

CLA X	Places the number stored at location X in the accumulator (Clear and add X).
ADD X	Adds the number at location X in the memory to the current contents of the memory.
MUL X	Multiplies the number in the accumulator by the number at location X in the memory and places the product in the accumulator.
STO X	Stores the number in the accumulator in location X in the memory.

We restrict our binary operators to + and × and define two *generators;* thus

Operator	*Generator*	*Comments*
+	CLA X ADD Y STO Z	X is the address on top of $S1$, Y is the next address down; these are removed and replaced by the new address Z
×	CLA X MUL Y STO Z	Same as above

We now compile by following the rules given in the algorithm, except that instead of values for the variables, the compiler actually reads the variables as letters and assigns an address to each variable; it is this address which is placed in the stack. In the sequel that follows, then, we associate with a given variable A the address (A). Then each time an operator is applied, the appropriate generator is added to the list of machine coding by the compiler.

Let us compile $A + B \times C$. First A is read, and next the address (A) is assigned and placed on $S1$; after this the + is put on $S2$. Then B is read, the address (B) is assigned, and then × and C are read, resulting in

$S1$	$S2$
(C)	
(B)	+
$(A)\times$	

Since $\times$ has precedence over $+$, we apply it, and the compiler puts out the machine code for the generator; this gives

```
CLA  (C)
MUL  (B)
STO  (D)
```

Here (D) is a new address replacing (C) and (B) on the stack, which now reads

$S1$	$S2$
(D)	$+$
(A)	

Now the generator for $+$ is added, giving

```
CLA  (C)
MUL  (B)
STO  (D)
CLA  (D)
ADD  (A)
STO  (E)
```

If values for A, B, and C are stored in the addresses (A), (B), and (C), the above machine language will store $A + B \times C$ in E.

Adding generators for the other arithmetic operators will enable us to evaluate all nonparenthesized arithmetic expressions. Adding the rules that (is placed on top of $S2$ when read and that) means to process all operators in $S2$ until some (is read will enable the use of parentheses.

The above is a very simple example of how some compilers work; syntax-driven compilers use the syntax of the language to determine which generators from some set should be added to the list of machine codes which will comprise the final compiled program. The subject is very complex, however, and the above example simply demonstrates some of the principles involved in the translation.

The precedence relations which exist among arithmetic operations in the absence of parentheses make it possible to compile arithmetic expressions easily. However, for other parts of the ALGOL grammar and for context-free languages in general, the going is not so easy. It is necessary to identify each of the syntactic classes which comprise a string in order for the compiler to intelligently turn out machine code. But in most cases there is no direct way to do this. The hypothesis that a given part of a string is in a given syntactic class must be checked by studying the adjacent parts of the string to verify that a consistent parsing of the entire string is possible. There are various approaches to this. The compiler can run down a tree-like structure, noting the location of each branch node and returning to it if a potential parsing does not work out, or all possible options can be tried when the choice of syntax class is not clear. For further discussions of the complicated problems which can arise, see the books by Chomsky and Ginsburg in the list of references.

EXERCISES C

1. Show how to derive the expression $A := B + C**D$ from the Backus normal form rules for ALGOL.
2. Show how the number 10.98 can be derived from Table 4-1.
3. Explain the statement $A := B \times 3$ in Sec. 4-3 using Table 4-4.
4. (a) Derive a finite-state machine which will accept only strings of the form

   01, 001, 0001, 00001, . . . and 011, 0011, 00011, . . .

5. Derive a finite-state machine which will print a 1 for only the strings 0101,0110, . . . and all strings of the form 10,110,1110, . . .
6. Convert the state tables for the machines in Exercises 4 and 5 to rules in Backus normal form.
7. Design a push-down store machine which will print a 1 for only strings of the form αbcd, where α is a string of 0's and 1's and b, c, and d are special tape symbols.

REFERENCES

BOTTENBRUCH, H.: "Structure and Use of ALGOL 60," *J. ACM*, **9**:161–221 (1962).
CHOMSKY, N.: "Syntactic Structures," Mouton and Co., 1957.
DIJKSTRA, E. W.: "A Primer of ALGOL 60 Programming," Academic Press, 1962.
FLOYD, R. W.: "The Syntax of Programming Languages—a Survey in Programming Systems and Languages," Saul Rosen (ed.), McGraw-Hill, 1967.
GINSBURG, S.: "The Mathematical Theory of Context-free Languages," McGraw-Hill, 1966.
KNUTH, D.: "The Art of Computer Programming," 9 vols., Addison-Wesley, in press.
MC CRACKEN, D. D.: "A Guide to ALGOL Programming," Wiley, 1962.
NAUR, P.: "A Course of ALGOL 60 Programming," Regnecentralen, 1961.
NICOL, K.: "Elementary Programming and ALGOL," McGraw-Hill, 1968.
RUTISHAUSER, H.: "Description of ALGOL 60," Springer, 1967.

Boolean Algebras

5-1. INTRODUCTION

The concept of a Boolean algebra was defined informally in Sec. 1-2 and related there to various examples of Boolean algebras such as power sets. For convenience, we repeat this definition formally here. Most of the next two chapters will be devoted to applying it systematically.

Definition. A *Boolean algebra* $B = [A,\wedge,\vee,',O,I]$ is a set A with two binary operations $(\wedge,\vee)$, two universal bounds (O,I), and one unary operation $'$ such that for all $x,y,z \in A$,

L1.	$x \wedge x = x \qquad x \vee x = x$	(Idempotent)
L2.	$x \wedge y = y \wedge x \qquad x \vee y = y \vee x$	(Commutative)
L3.	$x \wedge (y \wedge z) = (x \wedge y) \wedge z$ $x \vee (y \vee z) = (x \vee y) \vee z$	(Associative)
L4.	$x \wedge (x \vee y) = x \qquad x \vee (x \wedge y) = x$	(Absorption)
L5*a*.	$x \wedge [y \vee (x \wedge z)] = (x \wedge y) \vee (x \wedge z)$	(Modularity)

L5b. $x \vee [y \wedge (x \vee z)] = (x \vee y) \wedge (x \vee z)$

L6. $x \wedge (y \vee z) = (x \wedge y) \vee (x \wedge z)$

$x \vee (y \wedge z) = (x \vee y) \wedge (x \vee z)$ (Distributive)

L7. $x \wedge O = O \qquad x \vee O = x$

$x \wedge I = x \qquad x \vee I = I$ (Universal bounds)

L8. $x \wedge x' = O \qquad x \vee x' = I$ (Complements)

L9. $(x')' = x$ (Involution)

L10. $(x \wedge y)' = x' \vee y' \qquad (x \vee y)' = x' \wedge y'$ (de Morgan)

Observe carefully the axiomatic form of the preceding definition. Having first postulated five operations explicitly (two binary, one unary, two zero-ary), we then listed 21 identities, grouped into 10 axioms, or *postulates*. An algebraic system qualifies as a Boolean algebra if and only if it has two binary, one unary, and two zero-ary operations which satisfy the specified identities. (Of course, it really doesn't matter which particular symbols are used to signify the operations in question, but the notation given above is that most widely used in algebra today.)

The symbols $\wedge$ and $\vee$ are called "wedge" and "vee"; $x \wedge y$ is called the "meet" or "product" of x and y, while $x \vee y$ is called their "join" or "sum."

The simplest nontrivial Boolean algebra is provided by Example 1 in Sec. 1-2, which we present again in terms corresponding to our formal definition.

Example 1. The *two-element* Boolean algebra is $\mathbf{B} = [\{0,1\},\wedge,\vee,',0,1]$, where $\wedge$ means *the lesser of*, $\vee$ means *the greater of*, and $'$ means *the opposite of*, so that for example, $0 \wedge 1 = 0$, $0 \vee 1 = 1$, and $0' = 1$.

A trivial Boolean algebra is $[\{0\},\wedge,\vee,',0,0]$ with

$$0 \wedge 0 = 0 \vee 0 = 0' = 0$$

$0 = 0$, and $I = 0$. With this definition, all identities of L1 to L10 hold trivially, with both sides 0.

Our main concern in this chapter will be with *finite* Boolean algebras. As we shall show in Chap. 9, these are all given (up to isomorphism) by the following set of examples (see Theorem 1 of Chap. 1).

Example 2. For any positive integer n, the 2^n-*element* Boolean algebra $\mathbf{B}^n = [\mathcal{P}(\mathbf{n}),\cap,\cup,',\varnothing,\mathbf{n}]$ consists of the power set of all subsets of the set $\mathbf{n} = \{1, \ldots ,n\}$, with $\wedge$, $\vee$, and $'$ taken as set intersection, set union, and set complement, with $\varnothing$ as the empty set, and with $I = \mathbf{n}$.

In proving theorems about Boolean algebras in general, it is essential to avoid assuming anything except the identities postulated in L1–L10 and results which have been proved from them. Only in this way can we be sure that the theorems will apply to *all* Boolean algebras, no matter whether their elements represent sets, logical propositions, or mathematical descriptions of electronic hardware such as the gating networks to be described in Sec. 5-4. This *postulational approach* will be adhered to in the present chapter; it is standard in *modern algebra.*

The list of 21 identities in postulates L1 to L10 is highly redundant. For example, postulates L1, L5, L9, and L10 are consequences of the remaining six postulates. We prove first that L1 is redundant.

Lemma 1. (*Dedekind.*) *The idempotent laws of* L1 *follow from the absorption laws of* L4.

Proof. Setting $y = x \wedge x$ in the first identity of L4, we get

$$x \wedge [x \vee (x \wedge x)] = x$$

By the second identity of L4 (with $y = x$), the expression in square brackets is simply x; hence $x \wedge [x \vee (x \wedge x)] = x \wedge x$. It follows that $x = x \wedge x$. The identity $x = x \vee x$ can be proved in the same way, interchanging $\wedge$ and $\vee$.

The last remark is a special case of the following fundamental metamathematical[1] Duality Principle, closely related to that for partial ordering defined in Sec. 2-5.

Theorem 1. *Any true theorem about Boolean algebras whose statement involves only the three operations* $\wedge$, $\vee$, *and* $'$ *remains true if* $\wedge$ *and* $\vee$ *are interchanged throughout.*

Proof. The set of identities which can be used in proofs remains invariant when these two binary operations are interchanged in each of the postulates L1–L10. Hence every proof can be dualized by systematic interchange of $\wedge$ and $\vee$.

Likewise, we can show that postulate L5 is redundant; indeed, it will not be referred to again until Chap. 9 and was only included here for completeness.

Lemma 2. *Postulate* L5 *follows from* L1–L4 *and* L6.

Proof. By the duality principle of Theorem 1, it is sufficient to derive the first identity of L5. This we do as follows:

$$\begin{aligned} x \wedge [y \vee (x \wedge z)] &= x \wedge [(y \vee x) \wedge (y \vee z)] = [x \wedge (y \vee x)] \wedge (y \vee z) \\ &= x \wedge (y \vee z) = (x \wedge y) \vee (x \wedge z) \end{aligned}$$

using successively L6, L3, L4, and L6.

[1] A "theorem about theorems" is called *metamathematical,* just as a verbal analysis of languages is called *metalinguistic.*

In Chap. 1, we showed that the set $\mathbf{2}^{\mathbf{X}}$ of all binary-valued functions from any set X to $\mathbf{2} = \{0,1\}$ is a Boolean algebra. We now examine in detail the particular Boolean algebra obtained in this way by choosing X as the Boolean algebra $\mathbf{2}^{\mathbf{2}}$ of Chap. 1.

Example 3. Let F_2 consist of all the $2^4 = 16$ possible functions f: $\{00,01,10,11\} \to \{0,1\}$. In the notation of Chap. 1, this is the set of all functions f: $\mathbf{2}^{\mathbf{2}} \to \mathbf{2}$; it is itself the Boolean algebra $\mathbf{2}^{\mathbf{4}}$. At each $\mathbf{x} \in X = \mathbf{2}^{\mathbf{2}}$ of the domain, the "complement" f' of each function f takes on the opposite value from f; thus

$$f'(\mathbf{x}) = 1 - f(\mathbf{x}) \qquad \text{for all } \mathbf{x} \in \mathbf{2}^{\mathbf{2}} \tag{1}$$

where the minus sign has its usual meaning. For any two such functions e and f, the "join" $j = e \vee f$ and the "meet" $m = e \wedge f$ take on, respectively, the greater and the lesser of the values assumed by e and f at each $\mathbf{x}$, and we have

$$j(\mathbf{x}) = \max\,\{e(\mathbf{x}),f(\mathbf{x})\} \qquad m(\mathbf{x}) = \min\,\{e(\mathbf{x}),f(\mathbf{x})\} \tag{2}$$

Two examples of complements of functions, and four examples of joins and meets of functions, are given in the partial list of elements of F_2 in Table 5-1. The remaining four elements of F_2 are the constant functions 0 and 1, and the functions α and β specified below:

Table 5-1. Partial list of elements of F_2

	g	h	g'	h'	$g \wedge h$	$g \wedge h'$	$g' \wedge h$	$g' \wedge h'$	$g \vee h$	$g \vee h'$	$g' \vee h$	$g' \vee h'$
00	0	0	1	1	0	0	0	1	0	1	1	1
01	0	1	1	0	0	0	1	0	1	0	1	1
10	1	0	0	1	0	1	0	0	1	1	0	1
11	1	1	0	0	1	0	0	0	1	1	1	0

	0	1	α	β
00	0	1	0	1
01	0	1	1	0
10	0	1	1	0
11	0	1	0	1

$$\text{where} \begin{cases} 0 = g \wedge g' = h \wedge h' \qquad 1 = g \vee g' = h \vee h' \\ \alpha = (g \wedge h') \vee (g' \wedge h) = (g \vee h) \wedge (g' \vee h') \\ \beta = (g \wedge h) \vee (g' \wedge h') = (g \vee h') \wedge (g' \vee h) = \alpha' \end{cases}$$

By inspecting the preceding table, we obtain Lemma 2.

Lemma 2. *The Boolean algebra F_2 is "generated" by g and h as specified in the preceding tables; that is, every element of F_2 can be expressed as a Boolean combination of the two generators g and h.*

We shall see in Sec. 5-10 that F_2 is the "free" Boolean algebra with two generators; that is, any Boolean algebra generated by two elements is an *epimorphic image* of F_2.

The particular Boolean function defined by the expression (Boolean polynomial)

$$\alpha = (g \wedge h') \vee (g' \wedge h) = g + h \tag{3}$$

above is called the *symmetric difference* of the elements g and h. One writes $\beta(g,h) = g + h$ because β *adds modulo 2* the values of the functions g and h. More generally, the same is true if S and T are any subsets of a set U, with characteristic functions e_S and e_T; we have

$$e_{S+T} = e_S + e_T \pmod 2 \qquad S + T = (S \cap T') \cup (S' \cap T) \tag{3'}$$

The Boolean algebra F_2 of Example 1 also has a natural interpretation in terms of the Euler circles (disks) of Fig. 5-1. These circles partition

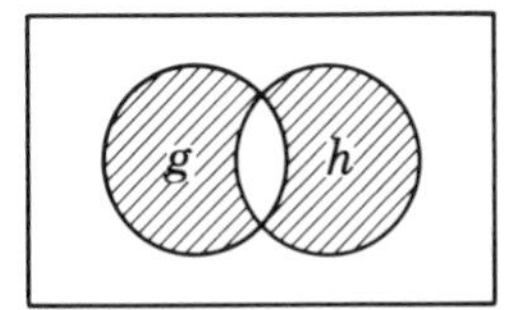

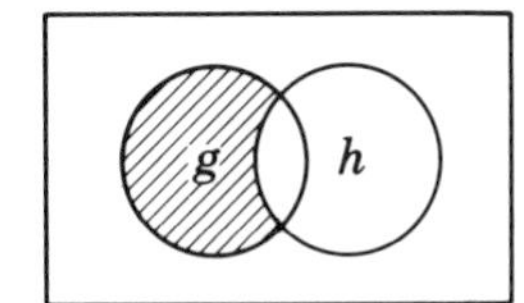

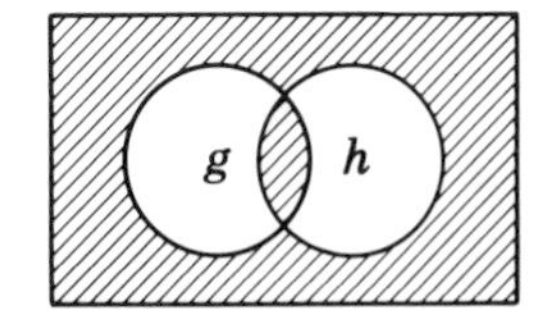

Fig. 5-1. Euler circle diagrams for F_2.

the plane into four regions; the 16 elements of F_2 are mapped in a natural way onto the field of sets (Chap. 1), consisting of the 16 unions of subsets of these four regions (see Exercise A4).

5-2. ORDER

The postulates for Boolean algebra listed in Sec. 5-1 had a special form; they were *identities* on the three *operations* $\wedge$, $\vee$, and $'$. Alternatively, we can give a very different set of postulates for Boolean algebra in terms of the properties of a single (partial) *order* relation. We shall describe such a set of postulates in Chap. 9; here we shall simply indicate the most important connections between the three Boolean operations and this order relation.

Intuitively, this order relation is associated with the Boolean operations $\wedge$, $\vee$, and $'$ in the same way that the *inclusion* relation $S \subset T$ between sets is associated with the operations $S \cap T$, $S \cup T$, and S'.

For instance, $S \subset T$ is equivalent to each of the four conditions

$$S \cap T = S \qquad S \cup T = T \qquad S \cap T' = \varnothing \qquad S' \cup T = I$$

We prove in Lemma 2 below from postulates L1 to L10 that this result holds in any Boolean algebra. We begin the proof with Lemma 1.

Lemma 1. *In any Boolean algebra*

$$a \wedge x = O \quad \text{and} \quad a \vee x = I \quad \text{imply} \quad x = a' \tag{4}$$

This is true since the hypotheses of (4) imply the successive equalities

$$\begin{aligned} x &= x \wedge I = x \wedge (a \vee a') = (x \wedge a) \vee (x \wedge a') = O \vee (x \wedge a') \\ &= (a \wedge a') \vee (x \wedge a') = (a \vee x) \wedge a' = I \wedge a' = a' \end{aligned}$$

by L7, L8, L6, hypothesis, L8, L6 and L2, hypothesis, and L7, respectively. This proves Lemma 1.

Setting $a = I$ and $x = O$, we infer as a corollary (since $I \wedge O = O$ and $I \vee O = I$ by L7 and L2) that

$$O' = I \quad \text{and} \quad I' = O \quad \text{in any Boolean algebra} \tag{5}$$

Lemma 2. *In any Boolean algebra A, the four equations which follow are mutually equivalent:*

$$a \wedge b = a \qquad a \vee b = b \qquad a' \vee b = I \qquad a \wedge b' = O \tag{6}$$

Proof. If $a \wedge b = a$, then $a \vee b = (a \wedge b) \vee b = b$ by L4. Next, if $a \vee b = b$, then

$$a' \vee b = a' \vee (a \vee b) = (a' \vee a) \vee b = I \vee b = I$$

Again, by de Morgan's laws of L10, $a' \vee b = I$ implies

$$a \wedge b' = (a' \vee b)' = I' = O$$

using (5). Finally, $a \wedge b' = O$ implies

$$a \wedge b = (a \wedge b) \vee O = (a \wedge b) \vee (a \wedge b') = a \wedge (b \vee b') = a \wedge I = a$$

which completes the cycle of implications.

Definition. In any Boolean algebra A, we define $a \leqq b$ to mean that one (or, equivalently, all) of the four equations of (6) holds.

The inclusion relation is basic in Boolean algebra because we can define all the Boolean operations in terms of it. This fact will be demonstrated in full in Chap. 9. Thus, O and I are definable by their universal bounds properties

$$O \leqq a \leqq I \qquad \text{for all } a \in A \tag{7}$$

For, these relations are equivalent by (6) to $O \vee a = a$ and $a \vee I = I$, which holds by L7 (and L2). (This proof shows that the other two identities of L7 are redundant.)

Theorem 2. *In any Boolean algebra* $\mathcal{B} = [A, \wedge, \vee, ']$, *the relation* $a \leqq b$ *is a partial ordering of* A. *Moreover, in terms of this partial ordering, we have*

$$a \wedge b = \text{glb } \{a,b\} \qquad \textit{and} \qquad a \vee b = \text{lub } \{a,b\} \tag{8}$$

Proof. By L1, $a \wedge a = a$ and so by definition of inclusion, $a \leqq a$ for all $a \in A$. Again, if $a \leqq b$ and $b \leqq a$, then $a \wedge b = a$ and $b \wedge a = b$. This implies $a = b$ by the commutative law (L2) and transitivity of equality. Finally, $a \leqq b$ and $b \leqq c$ mean that $a \wedge b = a$ and $b \wedge c = b$, implying $a = a \wedge b = a \wedge (b \wedge c) = (a \wedge b) \wedge c = a \wedge c$ by L3. Therefore, they imply $a \leqq c$, which proves that $\leqq$ is a partial ordering of A.

It remains to prove (8). By Theorem 1 on duality, each of the equations $a \wedge b = \text{glb } \{a,b\}$ and $a \vee b = \text{lub } \{a,b\}$ implies the other; hence it is sufficient to prove that $a \wedge b = \text{glb } \{a,b\}$. To prove this, we first observe that since $a \wedge (a \wedge b) = (a \wedge a) \wedge b = a \wedge b$ by L3 and L1, $a \wedge b \leqq a$ by (6). Similarly, $a \wedge b = (a \wedge b) \wedge b$, and so $a \wedge b \leqq b$. This shows that $a \wedge b$ is a lower bound of a and of b.

Finally, let c be *any* lower bound of a and b. Then, using L3 and the hypothesis on c (twice), we obtain

$$(a \wedge b) \wedge c = a \wedge (b \wedge c) = a \wedge c = c$$

and so $a \wedge b \geqq c$. This shows that $a \wedge b$ is the (unique, by Sec. 2-4) *greatest* lower bound of a and b.

We can use the result of Theorem 2 to describe the Boolean algebra $\mathbf{2}^4$ of Example 3 by its order diagram, shown in Fig. 5-2. The shaded elements are g and h; $p = g \wedge h$, $q = g \wedge h'$, $r = g' \wedge h$, and $s = g' \wedge h'$.

Lemma 3. (*Isotonicity.*) *In any Boolean algebra* A

$$b \leqq c \textit{ implies } a \wedge b \leqq a \wedge c \textit{ and } a \vee b \leqq a \vee c \textit{ for all } a \in A \tag{9}$$

Proof. Since $b \leqq c$, $b = b \wedge c$, and thus

$$a \wedge b = a \wedge (b \wedge c) = (a \wedge a) \wedge (b \wedge c) = (a \wedge b) \wedge (a \wedge c)$$

where we have used freely the commutativity and associativity of $\wedge$, as well as L4. Hence, by definition, $a \wedge b \leqq a \wedge c$. The proof that $a \vee b \leqq a \vee c$ is dual.

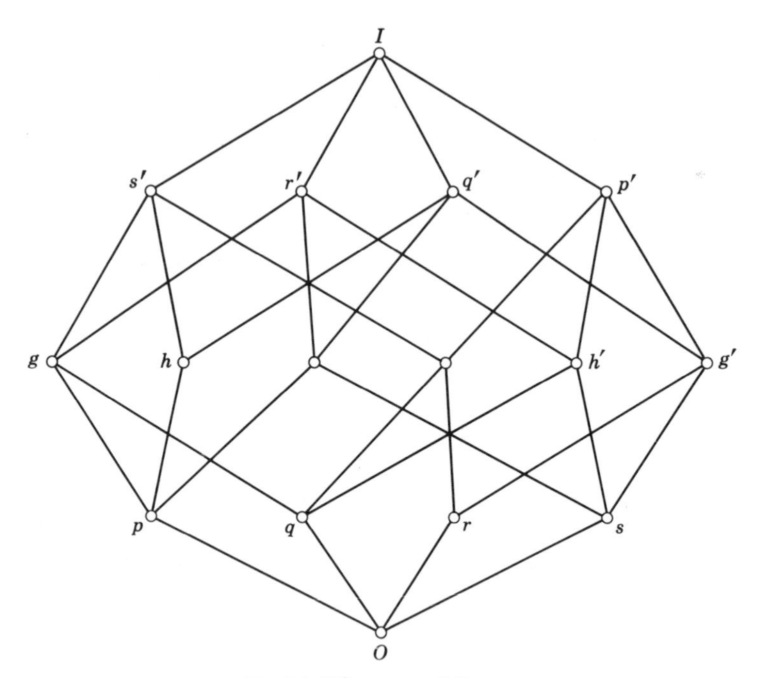

Fig. 5-2. Diagrams of F_2.

5-3. BOOLEAN POLYNOMIALS

In general, a *Boolean polynomial* is any expression which can be built up recursively by repeated application of the operations $\wedge$, $\vee$, and $'$ to some set of symbols $x_1, \ldots, x_n$. In the Backus normal form of Sec. 4-9, letting BOOLEPOLY represent our class, we can write

$$\text{BOOLEPOLY} ::= \langle\text{letter}\rangle \mid \langle\text{BOOLEPOLY} \wedge \text{BOOLEPOLY}\rangle \mid \langle\text{BOOLEPOLY} \vee \text{BOOLEPOLY}\rangle \mid \langle\text{BOOLEPOLY}'\rangle$$

This gives us a set of "production rules" for writing Boolean polynomials as algebraic expressions.

Clearly, any Boolean polynomial $F(x_1, \ldots, x_n)$ defines a function $F\colon \mathcal{B}^n \to \mathcal{B}$ on any Boolean algebra $\mathcal{B}$, which can be computed by evaluating the polynomial. Indeed, this result has the following partial converse (whose proof may be hard to follow at this stage).

Theorem 3. *Every function $f\colon \mathcal{B}^n \to \mathcal{B}$ is a Boolean combination of the coordinate functions $\delta_i\colon \mathbf{x} \mapsto x_i$, this combination being defined by some Boolean polynomial $F(\delta_1, \ldots, \delta_n)$.*

Proof. By definition, each δ_i associates the value x_i with each $\mathbf{x} = (x_1, \ldots, x_n) \in \mathcal{B}^n$. On the other hand, to any $\mathbf{v} \in \mathbf{2}^n$, we can assign the Boolean polynomial function

$$p_{\mathbf{v}} = y_1 \wedge \cdots \wedge y_n \qquad \text{where } y_i = \begin{cases} x_i & \text{if } v_i = 1 \\ x_i' & \text{if } v_i = 0 \end{cases}$$

This has the characteristic property that

$$p_{\mathbf{v}}(\mathbf{v}) = 1, \qquad \text{and} \qquad p_{\mathbf{v}}(\mathbf{w}) = 0 \qquad \text{if } \mathbf{w} \neq \mathbf{v} \text{ in } \mathbf{2}^n$$

We now use the obvious bijection $\mathcal{B}^n \leftrightarrow \mathbf{2}^n$ to assign to any given $f: \mathcal{B}^n \to \mathcal{B}$ the set $S(f)$ of all $\mathbf{v} \in \mathbf{2}^n$ such that $f(\mathbf{v}) = f(v_1, \ldots, v_n) = 1$. Consider the function

$$F(x_1, \ldots, x_n) = \bigvee_{S(f)} p_{\mathbf{v}}$$

By the characteristic property of $p_{\mathbf{v}}$,

$$F(\mathbf{v}) = \begin{cases} 1 & \text{if } f(\mathbf{v}) = 1 \\ 0 & \text{otherwise, i.e., if } f(\mathbf{v}) = 0 \end{cases}$$

This is the desired conclusion.

An *identity* is an equation $F(x_1, \ldots, x_n) = G(x_1, \ldots, x_n)$ which is valid for all $x_1, \ldots, x_n$ in any Boolean algebra. Our postulates L1 to L10 list 21 such identities (with $n = 1,2,3$) which hold by definition in all Boolean algebras. These identities bring out forcibly the *redundancy* of the expressions of the symbolic language defined by BOOLEPOLY. One of the main objectives of this chapter will be to eliminate the ambiguity which would otherwise result, by constructing a systematic algorithm which will reduce every Boolean polynomial F to a simple *canonical form* $\tilde{F}$, such that F and G always represent the same function from $\mathcal{B}^n$ to $\mathcal{B}$ if and only if $\tilde{F}$ and $\tilde{G}$ are the same symbolic expressions. This algorithm is closely related to the construction used to prove Theorem 3.

This canonical form thus gives a systematic process for determining in a finite number of steps, for any two given Boolean polynomials F and G, whether or not $F = G$ is an identity, i.e., whether or not F and G are always equal as functions, on any Boolean algebra. The problem of finding such a process is called the *word problem*, which we shall thus solve for Boolean algebras in this chapter.

Table 5-2 solves the word problem for $n = 1$; it shows that any Boolean function of p alone is equal to either p, p', O, or I. There are just four different Boolean functions of one variable p. The table can be deduced from our postulates; the table of complements follows from (5) and L9, the diagonal entries in the other two tables are from L1, and the off-diagonal entries are correct by L7 and L8.

Table 5-2. Boolean functions of p

$'$	
O	I
p	p'
p'	p
I	O

$\wedge$	O	p	p'	I
O	O	O	O	O
p	O	p	O	p
p'	O	O	p'	p'
I	O	p	p'	I

$\vee$	O	p	p'	I
O	O	p	p'	I
p	p	p	I	I
p'	p'	I	p'	I
I	I	I	I	I

Likewise, the Euler circles of Fig. 5-1 suggest that there are just $2^4 = 16$ nonequivalent Boolean functions of two variables g and h, an impression which is corroborated by the function tables of Sec. 5-1 and which is correct. When these tables are rearranged in lexicographic order of the sets of function values, as we have done in Table 5-3, we have a simple solution of the word problem for $n = 2$; namely, given $F(g,h)$ and $G(g,h)$, to test for $F = G$, simply evaluate the functions which they give in Example 1. F and G are the same if and only if these functions are identical; moreover, Table 5-3 gives a simple and short notation for the functions of each equivalence class.

Table 5-3. Canonical forms for $F(g,h)$

0000	0	0100	$g' \wedge h$	1000	$g' \wedge h'$	1100	g'
0001	$g \wedge h$	0101	h	1001	$g' + h$	1101	$g' \vee h$
0010	$g \wedge h'$	0110	$g + h$	1010	h'	1110	$g' \vee h'$
0011	g	0111	$g \vee h$	1011	$g \vee h'$	1111	1

Likewise, the Venn diagram of Fig. 1-1*b* suggests that there are exactly $256 = 2^{2^3}$ different (i.e., nonequivalent) Boolean functions of three variables, and that these correspond in a natural way to the subsets of the set of eight regions into which the disks decompose the rectangle U. This is also true (Theorem 9), but it is less easy to pick a shortest form for each equivalence class when three variables are involved. This shortest-form problem depends on the choice of operations to which individual symbols, or "characters," are ascribed, i.e., on the "alphabet" or "vocabulary" used to describe operations. In Table 5-3, this alphabet includes $+$ (addition mod 2) as well as the usual Boolean operations $\wedge$, $\vee$, $'$.

In Chap. 6, the shortest-form problem will be attacked for general n using only $\wedge$, $\vee$, $'$, and, indeed, "joins-of-meets" of a rather special form, well adapted to electronic design. In the present chapter, our main objective will be to solve the word problem for Boolean polynomials in n symbols. In the rest of the present section, we shall lay the groundwork for solving this problem, first proving the following result.

Lemma 1. *The join or meet of a finite number of elements in a Boolean algebra A depends only on the set of elements involved, and not on the order in which the elements are listed or combined or on the number of repetitions.*

A more formal statement is in terms of the following identities, valid for any finite sets S and T of elements of any Boolean algebra:

$$\bigwedge_S a_i \wedge \bigwedge_T a_j = \bigwedge_{S \cup T} a_k \qquad \bigvee_S a_i \vee \bigvee_T a_j = \bigvee_{S \cup T} a_k \tag{10}$$

Here and below, the expressions $\bigwedge_S a_i$ and $\bigvee_S b_i$, S a set, signify glb S and lub S, respectively. Thus, if $S = \{a_1, \ldots, a_n\}$ is finite,

$$\bigwedge_S a_i = a_1 \wedge \cdots \wedge a_n$$

and $\bigvee_S a_i = a_1 \vee \cdots \vee a_n$. By the Duality Principle (Theorem 1), it suffices to prove the first of these identities. Actually, both identities of (10) are obvious consequences of L1 and the generalized commutative and associative law of Sec. 1-10 (see also Sec. 2-11); if any letter a_h occurs both in S and in T, then we can permute the terms on the left side of (10) so as to make the two occurrences adjacent and use L1 to replace $a_h \wedge a_h$ by a_h.

Lemma 2. *For any finite sets of indices, S and T,*

$$\left(\bigvee_S a_i\right) \wedge \left(\bigvee_T b_j\right) = \bigvee_{S \times T} (a_i \wedge b_j) \tag{11}$$

and, dually,

$$\left(\bigwedge_S a_i\right) \vee \left(\bigwedge_T b_j\right) = \bigwedge_{S \times T} (a_i \vee b_j) \tag{11'}$$

The proof depends only on the commutative, associative, and distributive laws (L2, L3, and L6) and is just like that of the general distributive law of Sec. 1-8.

We now generalize the result of (4).

Lemma 3. *If $a \wedge x = a \wedge y$ and $a \vee x = a \vee y$, then $x = y$.*

Proof. By L4 and L6, using L2 and substituting equals for equals freely, we have

$$\begin{aligned} x &= x \wedge (x \vee a) = x \wedge (y \vee a) = (x \wedge y) \vee (x \wedge a) \qquad \text{(by L6)} \\ &= (x \wedge y) \vee (y \wedge a) = y \wedge (a \vee x) = y \wedge (a \vee y) = y \end{aligned}$$

We next come to a very important generalization of de Morgan's laws (L10), which also utilizes the involution law (L9).

Lemma 4. *To find the complement of any Boolean polynomial, interchange* $\vee$ *and* $\wedge$ *(vees and wedges) throughout, prime each unprimed letter, and unprime each primed letter.*

Explanation. Thus, the complement of $(x \wedge y') \vee z'$ is $(x' \vee y) \wedge z$. (We assume that double primes have been eliminated by L9 and that primed parentheses have been eliminated by induction.)

Proof. We use induction on the number n of letters appearing in the expression, counting repetitions. The lemma is true for $n = 1$, since $(x)' = x'$ and $(x')' = x$. Otherwise, since no parentheses are primed, we can write the polynomial as $f = p \wedge q$ or $f = p \vee q$. By de Morgan's laws, we conclude that $f' = p' \vee q'$ or $f' = p' \wedge q'$, as the case may be. By induction on n, we can apply Lemma 4 to p and q; if we do this, the effect on f is as stated.

EXERCISES A

ASSUME ONLY POSTULATES L1 TO L10 IN EXERCISES 1 TO 6.

1. Prove the following identities:
   (a) For the Boolean sum ("symmetric difference") $a + b$:
   $$g + h' = g' + h \qquad a + b = a' + b' \qquad a + (b + c) = (a + b) + c$$
   (b) $(x \wedge y') \vee (x' \wedge y) = (x \vee y) \wedge (x' \vee y')$.
2. (a) Write down the dual of the identity of Exercise 1*b*.
   (b) Prove the second identity of L5, without assuming Theorem 1.
3. Show that in F_2, with the notation of the text,
   $$\alpha \wedge \beta = 0 \qquad \text{and} \qquad \alpha \vee \beta = I, \qquad \text{whence} \qquad \beta = \alpha'$$
4. Writing $x \wedge y' = x - y$, show that all 16 Boolean polynomials in x and y are given by $0, x, y, x \wedge y, x \vee y, x - y, y - x, x + y$, and their complements.
5. Prove that $(x \wedge y) \vee (y \wedge z) \vee (z \vee y) = (x \vee y) \wedge (y \vee z) \wedge (z \vee x)$, using L1–L6 alone.
6. Prove that no nontrivial (finite) Boolean algebra can have an odd number of elements. (*Hint:* Pair x and x'.)
7. Show that the (trivial) one-element Boolean algebra $\{0\}$ satisfies axioms L1–L10.
8. Prove by induction on n that $\left(\bigwedge_{i=1}^{n} x_i\right)' = \bigvee_{i=1}^{n} x_i'$.
9. Prove identity (10) in detail. (*Hint:* Use induction on the number of elements in S and in T.)
10. Prove that identities (11)–(11′) are true in any Boolean algebra.

11. Assuming only L1–L4, show that each of the two identities of L5 is equivalent to the following self-dual identical implication:

    L5*: If $x \leqq z$, then $x \vee (y \wedge z) = (x \vee y) \wedge z$

12. Relate the Duality Principle of this chapter to the Duality Principle of Sec. 2-4.

5-4. BLOCK DIAGRAMS FOR GATING NETWORKS

In Chap. 3, we noted that present-day digital computers had *bistable* memory elements, and that it was convenient to designate the two stable states of any element by the symbols 0 and 1, bearing in mind that "on" and "off," and T and F (for true and false) are also used.

In this section, we shall consider a special kind of logical network occurring in digital computers, designed to produce reliably (from simultaneous input signals entering through n input leads from memory elements in states $X_1, X_2, \ldots, X_n$) one or more specified binary outputs $F(X_1, X_2, \ldots, X_n)$, so as to realize in each output some specified function of the inputs. Such a network is called a *gating network*, or *gate*.[1]

Clearly, there are 2^n different combinations of signals from n binary inputs, and thus exactly 2^{2^n} effectively different input-output functions $f: \mathbf{2}^n \rightarrow \mathbf{2}$ can be realized by gates. As we showed in Theorem 3, these are all Boolean functions, and Boolean algebra provides a very convenient symbolism for describing them (see again Sec. 5-1, where the case $n = 1$ was described in detail).

It is suggestive to represent gating networks as (acyclic) *labeled directed graphs* (see Sec. 2-11), whose arcs represent electrical conductors and whose nodes are gates. Such graphs are called *block diagrams*. Thus,

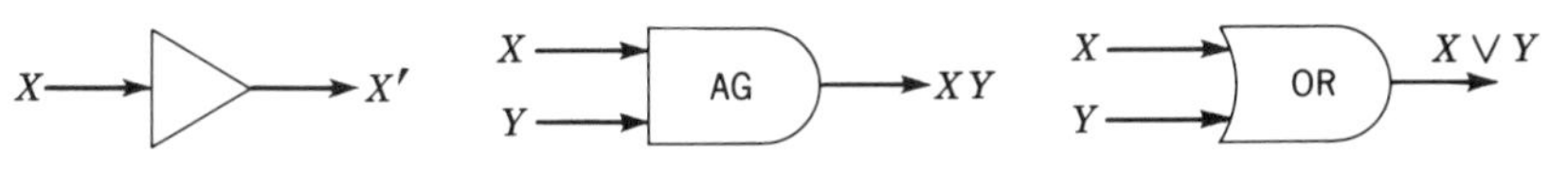

Fig. 5-3. Gates.

the three basic Boolean operations of $'$, $\wedge$, and $\vee$ are affected by three kinds of gates called an *inverter*, an *AND gate*, and an *OR gate*, respectively. In block diagrams, these kinds of gates are represented symbolically by the shapes[2] drawn in Fig. 5-3.

[1] As contrasted with a *sequential network*, in which one or more delay elements are included (see Sec. 6-9).

[2] These are the shapes specified by the United States of America Standards Institute and the Military Standards Specifications. These organizations use $+$ where we use $\vee$ and omit wedges $\wedge$; we shall often do this in Chap. 6.

By definition, an inverter has the following effect: if the input signal **X** is at level 0 (see Sec. 3-2), then the output signal has the level 1, and vice versa. Thus, if the input is **X**, the output is designated **X′**.

Also shown in Fig. 5-4 are block-diagram symbols of AND gates and OR gates which yield binary operations. The AND gate in Fig. 5-4 is shown with two inputs X and Y; the output for this gate is designated[1] as XY, and the output from the gate will be a 1, if both inputs to the gate represent 1. In Fig. 5-4 the block-diagram symbol for an OR gate is also shown. This gate again has two inputs X and Y and an output signal which is designated $X \vee Y$, indicating the output signal from this gate will be a 1 unless both of the input signals are 0's.

Single gates such as those shown in Fig. 5-4 and combinations of gates which are interconnected are examples of gating networds. The

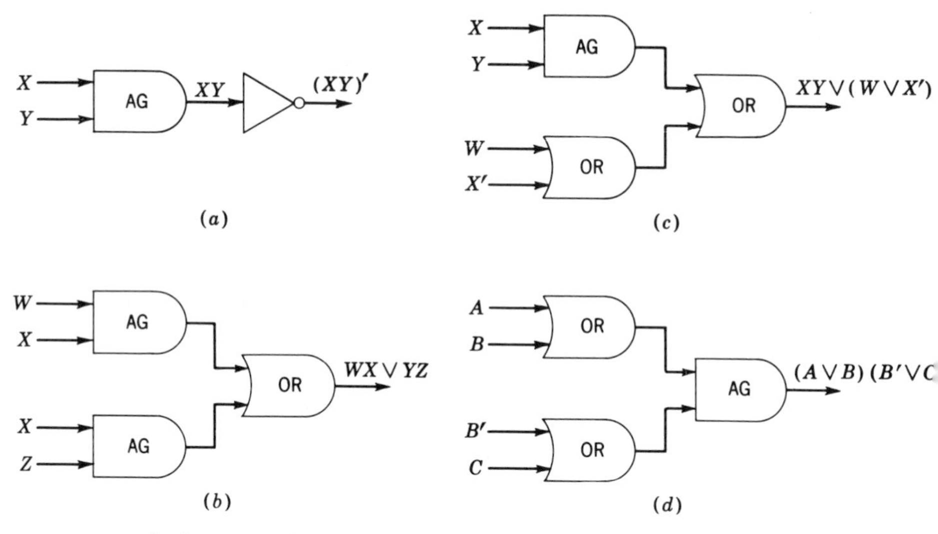

Fig. 5-4. Gating networks. (*a*) AND gates to inverter; (*b*) AND-to-OR network; (*c*) gate network; (*d*) OR-to-AND gate network.

operations of any such network can be described using Boolean algebra, provided no two gates have their outputs connected together and the input to any gate is never connected to an output from the same gate, even if such an output passes through several other gates before being connected back to the input. (Such interconnections are called *loops* and in many cases leave outputs of the logical network indeterminate.)

Figure 5-4 shows several combinations of AND gates, OR gates, and inverters along with the corresponding descriptions of the operations of these gates in the language of Boolean algebra. Figure 5-4*a* shows an

[1] XY is a commonly used shorter form of $X \wedge Y$; thus XYZ means $X \wedge Y \wedge Z$, etc.

AND gate with two inputs X and Y. The output of this gate is connected to an inverter, the output of which is designated by $(XY)'$. The output from this gate combination will be a 1 if either X or Y represent a 0. Figure 5-4*b* depicts a circuit consisting of two AND gates and an OR gate. This particular configuration is called an *AND-to-OR gate network* for obvious reasons. Again, the output function is described in terms of Boolean algebra; it is $WX \vee YZ$. Therefore, if (W AND X) OR (Y AND Z) are 1's, the output from the gating network will represent 1 but will otherwise be 0.

The rules for converting gate configurations into Boolean algebra expressions should be evident from these examples. Further examples are shown in Figs. 5-4*c* and 5-4*d*. The ability to analyze gating networks in this manner is important for the maintenance and operations analysis of digital computers. The internal operations of gating networks in computers are described by means of block diagrams composed of symbols of the type shown, plus descriptions of the outputs of these gating networks, expressed as Boolean functions of the input variables.

The particular gates shown, AND and OR gates, quite often have more than two inputs. Since the $\vee$ and $\wedge$ operations are commutative and associative, it is possible to describe the output functions yielded

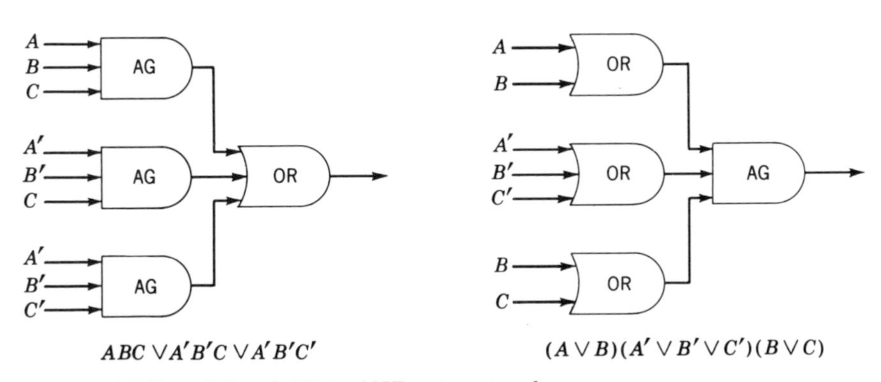

Fig. 5-5. AND-to-OR and OR-to-AND gate networks.

by AND and OR gates unambiguously without the use of brackets. Figure 5-5 shows this, giving the output of an OR-and-AND gate network using the $\vee$ and $\wedge$ operations, and also of an AND-to-OR network with each gate having three inputs.

There is a systematic technique for determining the Boolean function realized by a given combinational network. This technique is based upon a *table of combinations*. For any given combinational network, we will have some finite number, say n, of input variables and therefore 2^n different input values. It is then possible to write down each output value of the network (which corresponds to the value of the Boolean algebra expression), given the values of the input variables. A table listing the output value beside each possible set of input values will then completely describe the operation of the network. Figure 5-6 shows a *table of combinations* for a logic network consisting of an AND gate, an OR gate, and an inverter with three input variables X_1, X_2, and X_3; the values at the outputs of the AND gate, the inverter, and the OR gate are all shown on the table of combinations. Calculation of each row of the table is

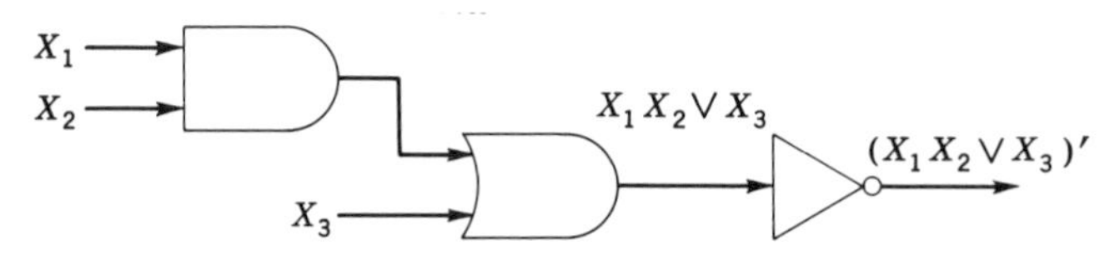

Inputs

X_1 X_2 X_3	X_1X_2	$X_1X_2 \vee X_3$	$(X_1X_2 \vee X_3)'$
0 0 0	0	0	1
0 0 1	0	1	0
0 1 0	0	0	1
0 1 1	0	1	0
1 0 0	0	0	1
1 0 1	0	1	0
1 1 0	1	1	0
1 1 1	1	1	0

Fig. 5-6.

straightforward, consisting of an evaluation of Boolean expressions. Analysis of a given gating network is therefore straightforward; one derives the Boolean algebra expression for the network, and this expression is then evaluated for each combination of values of the input variables.

5-5. CONNECTIONS WITH LOGIC

Boolean algebra applies to logic as well as to sets. Thus, let $p,q,r, \ldots$ stand for any properties of objects (red, round, soft, etc.). One can combine properties symbolically by giving the following interpretations to the symbols $\wedge$, $\vee$, and $'$:

$p \wedge q$	means	*p and q*
$p \vee q$	means	*p or q*
p'	means	*not p*

These conventions are directly related to those for sets. Thus, if p and q are any specified properties and we let $S(p)$ denote the set of all objects (for a given class U) which have property p, then we have

$$
\begin{aligned}
S(p \vee q) &= S(p) \cup S(q) \\
S(p \wedge q) &= S(p) \cap S(q) \\
S(p') &= [S(p)]'
\end{aligned}
\tag{12}
$$

In other words, the assignment $p \mapsto S(p)$ behaves like[1] a *morphism* of the Boolean algebras concerned (see p. 158).

Note that Boolean algebra distinguishes clearly between the "inclusive or" $p \vee q$ (sometimes written "and/or") and the "exclusive or" $p + q = (p \wedge q') \vee (p' \wedge q)$, meaning p or q but not both. Ordinary English is ambiguous about this.

A second area of application of Boolean algebra to logic concerns statements, or *propositions*, also denoted $p,q,r, \ldots$. As in the case of electronic signals in a digital computer, propositions are conceived of in logic as capable of assuming precisely one of two truth values, true or false (T or F). It is usual to let "$p = q$" mean that proposition p is logically equivalent to proposition q (in more familiar symbols, that $p \Leftrightarrow q$). Likewise, "p implies q," which is another way of saying "if p, then q" (or $p \Rightarrow q$), is logically equivalent to $p' \vee q$ ("either q is true, or p is false").

The above examples show that the everyday language of logic is highly redundant; there are many Boolean combinations of statements which have the same meaning. There are also statements which are logical *tautologies*, i.e., true by virtue of their logical structure alone. The simplest such tautology is $p \vee p'$ (read "either p or not p"). Tautologies such as the preceding can be recognized in (Boolean) algebraic form by the property that their value is I, no matter what values p,q, etc., have.

Another important tautology is

$$[(p \Rightarrow q) \wedge (q \Rightarrow r)] \Rightarrow (p \Rightarrow r) \tag{13}$$

In other words, the relation $\Rightarrow$ is transitive. To prove that (13) is a tautology, replace each implication symbol by its definition, and use Lemma 4 of Sec. 5-3 to obtain

$$
\begin{aligned}
[(p' \vee q) \wedge (q' \vee r)]' \vee (p' \vee r) &= (p \wedge q') \vee (q \wedge r') \vee p' \vee r \\
&= [(p \wedge q') \vee p'] \vee [(q \wedge r') \vee r] \qquad \text{(Lemma 1, Sec. 5-3)} \\
&= [(p \vee p') \wedge (q' \vee p')] \vee [(q \vee r) \wedge (r' \vee r)] \qquad \text{(L6)} \\
&= (q' \vee p') \vee (q \vee r) = q' \vee q \vee p' \vee r = I \vee p' \vee r = I
\end{aligned}
$$

The reader should satisfy himself that our postulates L1–L10 for Boolean algebra are indeed laws of intuitive logic not only for *sets*, but

[1] We say "behaves like" because the Boolean algebra of "properties" has no precise mathematical definition; properties do not form a well-defined set.

also for *properties* of objects (usually written as adjectives), and for *propositions.* It is this fundamental observation and its applications by Boole and others which initiated modern symbolic logic.

We shall consider symbolic logic further in Chap. 14.

EXERCISES B

1. Show that $(p \Rightarrow q) \Rightarrow (q' \Rightarrow p')$ follows from the definitions of Sec. 5-5.
2. Prove that the following are tautologies of Boolean algebra:
   (a) $(p \vee p) \to p$
   (b) $p \to (p \vee q)$
   (c) $(p \to q) \to r = (p \to r) \to q$
   (d) $(p \to q) \to [(r \vee p) \to (r \vee q)]$, regardless of r
3. In logic, a statement which is never true is called an *absurdity*. Show that the following are absurdities:
   (a) $q' \wedge p \wedge (p \to q)$
   (b) $(p' \to p) \wedge (p \to p')$
4. Show that $p \to p'$ is *not* an absurdity, and give an example to show that this convention makes sense.
5. (a) Show that $\phi(g,h) = g + h = (g \wedge h') \vee (g' \wedge h)$ adds mod 2 the values of the characteristic functions $e_G = g$ and $e_H = h$ of any two sets G and H.
   (b) Show that, in Fig. 5-1, the shaded region is $g + h$.

5-6. LOGICAL CAPABILITIES OF ALGOL

In Sec. 4-6, we gave a preliminary account of the conditional **if** · · · **then** · · · **else** format of ALGOL and of the related **go to** conditional capability. The use of conditional expressions is of the utmost importance to the programmer because it permits him to sequence the operations performed by a computer in a way which depends on the values of the variables involved.

We now present a systematic account of this and other logical capabilities of ALGOL, using the Boolean notation introduced earlier in this chapter. A major advantage of ALGOL over FORTRAN lies in the power of the Boolean expressions available; these greatly simplify the formulation of complicated logical conditions.

In ALGOL, Boolean variables and expressions assume just one of the values **true** or **false** at any given instant. Boolean variables are announced as such by a type declaration of the forms

boolean $A, B, Y, Z;$

which says that the variables A, B, Y, and Z are to have Boolean values. One commonly obtains Boolean variables from *relations* between arithmetic expressions; in ALGOL, these are defined from any of the following *relational operators:*

$$< \quad \leq \quad = \quad \geq \quad > \quad \neq$$

For example, if X has been declared **boolean,** the statement

$$X := 2 \uparrow 4 < 15;$$

assigns the value **false** to X. Again, if Y has been declared **boolean,** the statement

$$Y := X \times 2 < A + B;$$

assigns the value **true** to Y if $2X$ is indeed less than $A + B$ when the statement is evaluated and the value **false** if $2X \geqq A + B$.

From Boolean variables arising, for example, from relations, one can obtain other Boolean expressions in ALGOL by using one or more of the following five *logical operators:* $\neg$, which is a unary operator standing for "not," and the binary $\wedge$, $\vee$, $\supset$, $\equiv$. Table 5-4 gives the function tables for these operators; they symbolize not, and, or, implies, and is logically equivalent to, respectively, where F stands for **false** and T for **true.**

Table 5-4. Values of logical operators

	$\neg$	$\wedge$	T	F	$\vee$	T	F	$\supset$	T	F	$\equiv$	T	F
T	F	T	T	F	T	T	T	T	T	T	T	T	F
F	T	F	F	F	F	T	F	F	F	T	F	F	T

In order to avoid the use of unnecessary parentheses, ALGOL uses the following precedence scheme. Arithmetic operators are highest in rank and are evaluated first, in the order of precedence described in Chap. 4. All relational operators have the same rank and are evaluated after the arithmetic operators. Boolean operators are evaluated last, and their order of precedence is $\neg$, $\wedge$, $\vee$, $\supset$, $\equiv$. Thus, $\neg A \vee B$ means $(\neg A) \vee B$, and $\neg A \vee B \wedge \neg C$ means $(\neg A) \vee (B \wedge (\neg C))$.

The basic forms of *conditional statements* in ALGOL are

if B **then** S;
if B **then** S_1 **else** S_2;

In the above, B is a Boolean expression, and S, S_1, and S_2 are statements. In the first case, if B evaluates to **true,** then the statement S is executed; otherwise statement S is skipped, and the next statement is executed. In the second case if B evaluates to **true,** then statement S_1 is executed (and S_2 is skipped); otherwise statement S_2 is executed, and S_1 is skipped.

As an example of the first form, we have

$$\textbf{if } X \geq Y \vee Y \geq Z \textbf{ then } W := W \uparrow 2;$$

which will square W if X is greater than or equal to Y or if Y is greater than or equal to Z.

As an example of the second form, we have

$$\textbf{if } W = Y \vee W = Z \textbf{ then } X := X \uparrow 2 \textbf{ else } X := X \uparrow 3;$$

This statement will square X if W equals Y or if W equals Z; otherwise it will cube X. Another such example is

$$A := \textbf{if } J > 6 \textbf{ then } 4 \textbf{ else } 5;$$

This statement says "if the variable J has a value greater than 6, give A the value 4; otherwise give A the value 5." Again, if W, X, Y, and Z are all Boolean variables, the statement

$$W := 4 \textbf{ if } X \wedge Y \vee X \wedge Z \textbf{ then } Y \wedge Z \textbf{ else } X;$$

gives a Boolean value to W.

go to statements in ALGOL permit one to order the execution of other statements and to repeat sequences of statements. To use this capability, one must *label* the statement referred to by a string of letters, numbers, or blanks which precedes a statement and is separated from it by a colon. Thus

```
L:      X := A + B;
1L2:    X := A ↑ B;
53:     X := A/B;
056:    X := A/B;
```

Labels are in effect just *names* for the statements following the colon. Thus, L is the code name for the statement "set X equal to $A + B$," and $1L2$ a code name for "set X equal to A^B." If a label is strictly numerical, leading zeros are ignored; thus 056 is the same label as 56.

The statement **go to** M tells the compiler to execute the statement with label M next

```
        Z := 3;
        X := 5;
        Y := 6;
        go to B4;
        Z := X + Y;
B4:     Z := Z + X;
```

After the above sequence is executed, Z will have value 8 since the program omits the statement $Z := X + Y$; skipping to the last statement.

Labels also permit one to use statements of the form

if A **then go to** L;

where A is a Boolean expression and L is a label. Such a statement says, in effect, "if A has the value **true,** execute the statement with label L next; if A has the value **false,** then simply execute the following statement in the program." For example, the sequence of labels and statements for a given **real** X,

```
A1: W := 2 × X;
if W < 5₁₀9 then go to A1;
end
```

gives to W the value equal to the largest number $2^n X$ not exceeding 5 billion.

Example 4. The following ALGOL program computes a table of

$$F(x) = \begin{cases} 17.3 - (x+1)^x & x < 5 \\ 19.4/(1+x^2) & x \geq 5 \end{cases}$$

for x ranging from 0 to 10 in steps of 0.1:

```
       begin real x, F;
       x := 0;
back:  F := if x < 5 then 17.3 − (x + 1) ↑ x else 19.4/(1 + x ↑ 2);
       print (x,F);
       if x < 10 then begin x := x + .1; go to back end
       end
```

It is assumed that the effect of executing the print statement (which is not an ALGOL 60 statement) is to print out the values of the variables x and F. A somewhat simpler version of the above program is the following, using the **for** statement:

```
begin real x, F;
for x := 0 step .1 until 10 do
    begin F := if x < 5 then 17.3 −(x + 1) ↑ x else 19.4/(1 + x ↑ 2);
          print (x,F)
    end
end
```

5-7. BOOLEAN APPLICATIONS

The logical capabilities of ALGOL make it easy to computerize many computations of Boolean algebra. In this section, we shall illustrate this with some simple applications.

Example 5. In the power set $\mathcal{P}(\mathbf{n})$, it is easy to program Boolean operations by specifying each subset S of $\mathbf{n} = \{1,2, \ldots ,n\}$ as a one-dimensional **boolean array,** through its characteristic function $e_S(k)$ which is written without subscripts in ALGOL as $eS(k)$.

$$eS(k) = \begin{cases} \textbf{true} & \text{if } k \in S \\ \textbf{false} & \text{if } k \notin S \end{cases} \tag{14}$$

Thus, using the familiar Boolean formulas

$$e_{S \cap T} = e_S \wedge e_T \qquad \text{and} \qquad e_{S \cup T} = e_S \vee e_T$$

We can get a printout of the elements of $S \cap T$ by writing the ALGOL instruction

for k := **step** 1 **until** n **do**
if $eS \wedge eT(k)$ = **true then** print k;

To get printouts of the elements of $S \cup T$ or of the elements of S', listed in their natural order, we can proceed similarly.

Similarly, one can perform Boolean operations on binary relations by identifying each binary relation ρ on $\mathbf{m}$ to $\mathbf{n}$ as a two-dimensional **boolean array,** in which the 1's of the relation matrix of ρ are given the Boolean value **true,** and the 0's are given the value **false** (Secs. 2-1 and 2-2). To *compose* (multiply) binary relations is almost as easy.

Example 6. Let ρ be a binary relation on a set $X = \{x_1, \ldots ,x_n\}$, and let $R = \|r_{ij}\|$ be its relation matrix. Or let $\vec{G} = \vec{G}(\rho)$ be the simple *directed graph* with nodes $x_i, \ldots ,x_n$, and let $r_{ij} = 1$ mean that there is a (unique) arc from x_i to x_j. The two-dimensional Boolean array $B(\rho) = \|b_{ij}\|$, with

$$b_{ij} = \begin{cases} \textbf{true} & \text{if there is an arc from } x_i \text{ to } x_j \\ \textbf{false} & \text{otherwise} \end{cases}$$

then describes the graph $\vec{G}$. The matrix R (or $\|b_{ij}\|$) is its *adjacency matrix.*

As in Chap. 2, the Boolean matrix $\|s_{ij}\|$ corresponding to ρ^2 is easily obtained from B by

$$s_{ij} = \bigvee_{k=1}^{n} (b_{ik} \wedge b_{kj})$$

In terms of the graph $\vec{G}(\rho)$, s_{ij} = **true** if and only if there is a path of length exactly 2 from node x_i to node x_j.

Assume that the $N \times N$ ($N \leq 50$) Boolean array R contains the relation matrix for a relation ρ (1 = **true**, 0 = **false**). The following ALGOL program computes the relation matrix for the relation ρ^2:

```
begin boolean array R[1:50,1:50]; S[1:50,1:50];
      integer i,j,k,N;
for   i := 1 step 1 until N do
for   j := 1 step 1 until N do
      begin S[i,j] := false;
            for k := 1 step 1 until N do
            S[i,j] := S[i,j] ∨ (R[i,k] ∧ R[k,j])
      end
end
```

Example 7. Suppose that we are given the $n \times n$ relation matrix $R = \|r_{ij}\|$ of a graph with n nodes as a (symmetric, irreflexive) two-dimensional *Boolean array*, and that we want to know whether or not the graph is connected.

We could proceed by the method described in Chap. 2 and compute the relation matrix

$$e \vee R \vee R^2 \vee \cdots \vee R^n \tag{15}$$

which is J, if and only if the graph is connected. However, this would require n^3 multiplications each time $R^{k+1} = R^kR$ was computed, hence about n^4 Boolean multiplications in all. The following procedure is much more economical of computing time.

Clearly, the graph is connected if and only if there is a simple path from node 1 to every other node. Hence, it suffices to use the shortest-path algorithm of Sec. 2-10, computing the following sets S_k, T_k, and U_k.

First, set $S_0 = \{1\} = T_0 = U_0$. Then compute recursively (e.g., by an ALGOL program)

$$T_{k+1} = S_kR \qquad S_{k+1} = T_{k+1} \cap U'_k \qquad U_{k+1} = S_{k+1} \cup U_k$$

Here

$$S_kR = \{j \in \mathbf{n} \mid iR_j \text{ for some } i \in S_k\}$$

its characteristic function is given by the assignment $j \mapsto \bigvee_{i \in S_k} r_{ij}$.

Clearly, T_{k+1} is the set of all nodes adjacent to some node of S_k; S_{k+1} is the set of nodes $h \in \mathbf{n}$ such that the shortest (simple) path from 1 to h has length $k + 1$, that is, of nodes whose *distance* from 1 is $k + 1$. Likewise, U_{k+1} is the set of nodes whose distance from 1 is at most $k + 1$. The process terminates when $S_m = \emptyset$ is the empty set. Clearly, $m \leq n$, and the graph is connected if and only if U_m includes all nodes.

Since the S_k are mutually exclusive, this procedure requires only about n^2 Boolean multiplications and so is very much quicker (consider the case $n = 50$).

EXERCISES C

1. Write an ALGOL program which will enumerate the elements of $S \cup T$, for arbitrary $S \subset \mathbf{n}$, $T \subset \mathbf{n}$.
2. Write an ALGOL program which will enumerate $S' = \neg S$ for any $S \subset \mathbf{n}$.
3. If $A = \|a_{ij}\|$ and $B = \|b_{ij}\|$ are $m \times n$ and $n \times r$ matrices (arrays) of Boolean elements, then $AB = \|c_{ij}\|$, where $c_{ij} = \bigvee_{\mathbf{n}} a_{ik}b_{kj}$. Write an ALGOL program which will compute AB.
4. Justify the statement, "A two-dimensional array of type **boolean** is a relation matrix."
5. Write an ALGOL program to compute the real roots (if any) of the quadratic equation

$$4x^2 + e^{\pi/\sqrt{2}} + \ln 80 = 0$$

6. Write an ALGOL program to compute

$$F(x) = \begin{cases} 3125 - x^x & \text{if } x < 5 \\ (x - 5)/(1 + x^2) & \text{if } x \geqq 5 \end{cases}$$

for x ranging from 1 to 10 in steps of 0.1.

★7. Let R be the relation (connectivity) matrix of a directed graph $\vec{G}$ with nodes 1,2, . . . ,50, and let $N \geqq 1$. Write an ALGOL program that computes a matrix S such that $S_{ij} = 1$ (**true**) if and only if there is a path in $\vec{G}$ from node i to node j of length $\leq N$. The overall program organization is as follows:

```
begin integers N, . . . ;
      Boolean array R[1:50,1:50],S[1:50,1:50,1:50], . . . ;
Read (N, matrix R);
      .
      .
      .
Print (N, matrix R, matrix S)
end
```

As indicated above, we need not be concerned with the specific format of the input-output instructions. These will depend on the particular ALGOL processor used and the input-output devices available.

5-8. BOOLEAN SUBALGEBRAS

In any Boolean algebra $\mathcal{B} = [A, \wedge, \vee, ', O, I]$, we define a *Boolean subalgebra* to be a subset $S \subset A$ which:

1. Contains O and I
2. Contains with any x, its complement x'
3. Contains with any x and y, their meet $x \wedge y$ and the join $x \vee y$

By L7, given any $x \in A$ with $O < x < I$, the four elements O, x, x', and I constitute a Boolean subalgebra of A, the subalgebra *generated* by x (or x'). Thus, in Fig. 5-7, the shaded elements form a subalgebra of the

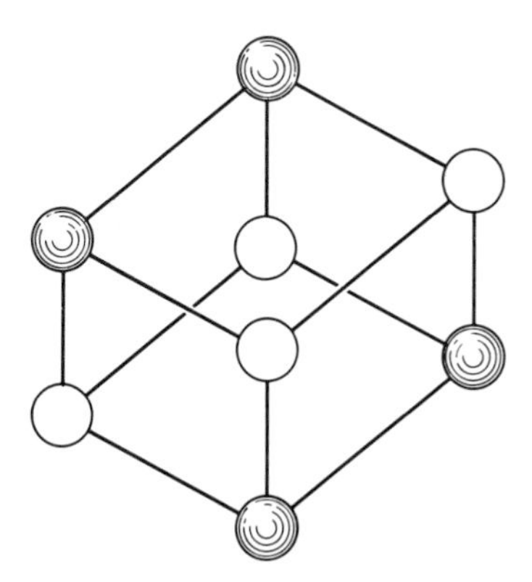

Fig. 5-7.

Boolean algebra $\mathcal{B} = [\mathcal{P}(\mathbf{3}), \cap, \cup, ', \phi, \mathbf{3}]$. Moreover, the elements O and I form another subalgebra.

In general, as the word "subalgebra" suggests, we have the following theorem.

Theorem 4. *Any Boolean subalgebra S of a Boolean algebra is itself a Boolean algebra $\mathcal{B} = [S, \wedge, \vee, ', O, I]$ under the operations of $\mathcal{B}$.*

This is true for the simple reason that each of the identities L1 to L10 holds in S a fortiori because it holds in $\mathcal{B}$ and its terms are all in $S \subset A$.

Example 8. As was observed in Chap. 1, the power set $\mathcal{P}(I)$ of any set I is a Boolean algebra under the operations $\cap$ (intersection), $\cup$ (set-union), and complement. The Boolean subalgebras of $\mathcal{P}(I)$ are therefore the (nonvoid) families $F \subset \mathcal{P}(I)$ of sets $X, Y, Z, \ldots \subset I$ which contain with any set X its complement X' and which contain with any X and Y also $X \cap Y$ and $X \cup Y$. Such families of sets are called *fields* of sets.

In general, if a subalgebra S of a Boolean algebra A contains an element x, then it must contain x', O, and I, as explained above. But as noted above, the quadruple x,x',O,I always is a subalgebra; it is thus the smallest subalgebra of A which contains x and is therefore called the Boolean subalgebra of A *generated* by x.

We saw in Sec. 5-1 that the Boolean algebra $F_2 = \mathbf{2}^4$ was generated by two elements f and g (see also Fig. 5-2). We now generalize this result to one valid for any exponent which is a power of 2.

Theorem 5. *The Boolean algebra* $\mathcal{P}(\mathbf{2}^r) = B^{2^r}$ *is generated by r elements* $X_1, \ldots, X_r$.

Proof. Let $\mathbf{2}^r$ be the set of all r-bit binary words, $\mathbf{w} = w_1 \cdots w_r$ (each $w_i = 0$ or 1), and let $X_i \subset \mathbf{2}^r$ be the set of all those r-bit binary words which have a 1 in the ith place. Thus, for $r = 3$, we set

$$\begin{aligned} X_1 &= \{100,101,110,111\} \\ X_2 &= \{010,011,110,111\} \\ X_3 &= \{001,011,101,111\} \end{aligned}$$

For each individual n-bit binary word $\mathbf{w}$, we have

$$\mathbf{w} = \bigcap_{i=1}^{r} Y_i(\mathbf{w}) \qquad \text{where } Y_i(\mathbf{w}) = \begin{cases} X_i & \text{if } w_i = 1 \\ X_i' & \text{if } w_i = 0 \end{cases} \tag{16}$$

Since every element of $\mathcal{P}(\mathbf{2}^r)$ considered as a subset $\mathbf{2}^r$ is just some set (join) of such words (possibly the void set $\varnothing = X_i \cap X_i'$), we are done.

Intervals. Given $a \leqq b$ in a poset A, the *interval* $[a,b]$ is defined to be the set of all $x \in A$ such that $a \leqq x \leqq b$.

Thus, in any Boolean algebra A, $[O,I] = A$.

Lemma 1. *If* $x \in [a,b]$ *and* $y \in [a,b]$, *then* $x \wedge y$ *and* $x \vee y$ *belong to* $[a,b]$.

This is true because by hypothesis, a is a lower bound and b an upper bound to $\{x,y\}$; hence a must be contained in the *greatest* lower bound (and any upper bound), whereas b must contain the *least* upper bound (and any lower bound) of $\{x,y\}$.

We now define a *lattice* to be any algebra $[L, \wedge, \vee]$ in which L1–L4 hold, and a *distributive* lattice to be a lattice in which the distributive laws L6 hold. Lattices will be studied systematically in Chap. 9. For the moment, we note only the following consequences of Lemma 1.

Theorem 6. *Any interval* $[a,b]$ *of a Boolean algebra* A *is a distributive lattice under the operations of* A.

Proof. By Lemma 1, $[a,b]$ contains with any x and y also $x \wedge y$ and $x \vee y$. That the operations $\wedge$ and $\vee$ in $[a,b]$ satisfy identities L1–L4 and L6 now follows as in the proof of Theorem 4, because they are restrictions to a sublattice of operations which satisfy L1–L4 and L6 in A.

Observe that the interval $[a,b]$ is *not* a Boolean subalgebra of A except in the trivial case $a = O$, $b = I$ that the interval is all of A; in other cases, $[a,b]$ fails to contain a' or b'. However, we can define the *relative* complement of x in the interval $[a,b]$ as $x^* = (a \vee x') \wedge b$ and then prove the following theorem.

Theorem 7. *Given the interval* $[a,b]$ *in a Boolean algebra* A, *the algebraic system* $[[a,b], \wedge, \vee, {}^*, a, b]$ *is a Boolean algebra.*

We leave the proof to the reader as an exercise (Exercise D7 below).

5-9. DISJUNCTIVE NORMAL FORM

In this section, we shall complete the solution of the "word problem" for Boolean expressions (Boolean polynomials). Its solution is greatly helped by having at hand the concept of disjointness, which applies not only to Boolean algebras and lattices, but even to posets in which x and y are called disjoint when O is the only lower bound to $\{x,y\}$.

Definition. In any poset with universal lower bound O, two elements x and y are said to be *disjoint* when $x \wedge y = O$.

Thus, in a Boolean algebra A, complementary elements a and a' are always disjoint. Indeed, the set of all elements disjoint from any $a \in A$ constitutes precisely the interval $[O,a']$:

$$a \wedge x = O \qquad \text{if and only if } x \leqq a' \tag{17}$$

More generally, in any distributive lattice,

$$a \wedge x = O \qquad \text{and} \qquad a \wedge y = O \text{ imply} \qquad a \wedge (x \vee y) = O \tag{18}$$

for $a \wedge (x \vee y) = (a \wedge x) \vee (a \wedge y) = O \vee O = O$, using L6.

Theorem 8. *Let* $p_1, \ldots, p_m$ *be pairwise disjoint elements of a distributive lattice* L *with* O. *Define the function* $\theta\colon \mathcal{P}(\mathbf{m}) \to L$ *as the mapping which takes each subset* $S \subset \mathbf{m}$ *into* $\theta(S) = \bigvee_{i \in S} p_i$ *and* $\varnothing$ *into* O. *Then*

$$\Big(\bigvee_{i\in S} p_i\Big) \vee \Big(\bigvee_{j\in T} p_j\Big) = \bigvee_{S\cup T} p_k \tag{19}$$

and

$$\Big(\bigvee_{i\in S} p_i\Big) \wedge \Big(\bigvee_{j\in T} p_j\Big) = \bigvee_{S\cap T} p_k \tag{19'}$$

Explanation. We set $\bigvee_{\varnothing} p_i = O$, as also follows from the definition of $\bigvee$ as least upper bound.

Proof. By the commutative, associative, and idempotent laws extended by induction as in Sec. 5-3, we have (19).

Likewise, by the general distributive law of Sec. 1-8, we have for any $S \subset \mathbf{m}$, $T \subset \mathbf{m}$

$$\left(\bigvee_{i \in S} p_i\right) \wedge \left(\bigvee_{j \in T} p_j\right) = \bigvee_{S \times T} (p_i \wedge p_j)$$

However, since $p_i \wedge p_j = O$ if $i \neq j$ by disjointness, one can simply omit all terms $p_i \wedge p_j$ with $i \neq j$ above, and so

$$\bigvee_{S \times T} (p_i \wedge p_j) = \bigvee_{S \cap T} p_k \qquad \text{if the } p_i \text{ are disjoint}$$

Combining the preceding equalities (by the transitivity of equality), we have (19′), completing the proof of Theorem 8.

It is now straightforward to prove our main result. To express it in compact form, we shall let $\mathbf{v} = v_1 \cdots v_n$ and $\mathbf{w} = w_1 \cdots w_n$ signify words of n binary digits, each 0 or 1. To each such n-bit word there corresponds a unique Boolean polynomial $p(\mathbf{w}) = \bigwedge_{\mathbf{n}} z_i$, where z_i is y_i if $w_i = 1$ and y_i' if $w_i = 0$. This gives 2^n elements $p(\mathbf{w})$; our main result will be that every Boolean polynomial $f(y_1, \ldots, y_n)$ can be expressed as a join of some set of the $p(\mathbf{w})$.

Theorem 9. *Each Boolean expression in $Y = \{y_1, \ldots, y_n\}$ can be put into the disjunctive normal form*

$$\bigvee_S p(\mathbf{w}) \qquad p(\mathbf{w}) = \bigwedge_{\mathbf{n}} z_i \qquad z_i = \begin{cases} y_i & \text{if } w_i = 1 \\ y_i' & \text{if } w_i = 0 \end{cases} \tag{20}$$

in one and only one way.

Proof. By Theorem 8, the set of all joins of the $p(\mathbf{w})$ is closed under $\wedge$ and $\vee$ since the $p(\mathbf{w})$ are disjoint [this since if $\mathbf{v} \neq \mathbf{w}$, then some $v_i \neq w_i$, and so $p(\mathbf{v}) \subset y_i$ and $p(\mathbf{w}) \subset y_i'$, or vice versa, whence $p(\mathbf{v}) \wedge p(\mathbf{w}) = O$]. Furthermore, if S and S' are complementary in the set of all n-bit binary words, then by (19)

$$\bigvee_S p(\mathbf{w}) \vee \bigvee_{S'} p(\mathbf{w}) = \bigvee_W p(\mathbf{w}) = I$$

since

$$I = I \wedge \cdots \wedge I = (y_1 \vee y_i') \wedge \cdots \wedge (y_n \vee y_n') = \bigvee_W p(\mathbf{w})$$

again by the generalized distributive law. Hence, by the uniqueness of complements in a Boolean algebra, $\bigvee_{S'} p(\mathbf{w})$ is the complement of $\bigvee_S p(\mathbf{w})$, and the set of Boolean expressions in disjoint canonical form (20) is a Boolean subalgebra of the given Boolean algebra.

It remains to show that all y_i are in the set (20); if so, they obviously generate it. To show this, note that

$$y_1 = y_1 \wedge I \wedge \cdots \wedge I = y_1 \wedge (y_2 \vee y_2') \wedge \cdots \wedge (y_n \vee y_n')$$

Expanding the last expression into 2^{n-1} terms by the generalized distributive law again, we have

$$y_1 = \bigvee_{S(1)} p(\mathbf{w}) \qquad S(1) = \{\mathbf{w} \mid w_1 = 1\}$$

Similarly, we can prove that every $y_i = \bigvee_{S(i)} p(\mathbf{w})$, which completes the proof of Theorem 9.

It is less obvious that each of the 2^{2^n} expressions (20) represents a different function. However, this fact is a corollary of Theorem 5. The example $\mathcal{P}(r)$ of this theorem shows that no use of identities L1–L10, however ingenious, can ever prove that two formally distinct $\bigvee_S p(\mathbf{w})$ represent the same function; such a proof would have to apply to this $\mathcal{P}(r)$. This proves the following important corollary of Theorems 5 and 9.

Corollary. *Each Boolean expression $p(y_1, \ldots, y_n)$ can be reduced to one and only one disjunctive normal form* (20) *by repeated use of identities* L1–L10.

It follows that $\mathbf{B}^{2^n}$ is the *free Boolean algebra* with n generators. We shall give this fact a deeper interpretation in the next section.

EXERCISES D

1. Show that one cannot assign to each Boolean polynomial in six symbols a different 48-bit word as "name," and that to give each Boolean polynomial in 20 symbols a different name would require more than 10^6 bits for each name.
2. Show that to assign to each Boolean polynomial in six symbols a fixed address would require more than 2 million 64-bit words of storage.

3. Show that $g + h' = g' + h$ and $g + h = g' + h'$.
4. Describe a Boolean isomorphism from $F_2 \cong \mathbf{2}^4$ to sets of the regions of Fig. 5-1, which maps g and h into the two disks shown there.
5. Prove that for any Boolean polynomial $p(x,y)$

$$p(x,y) = [x \wedge p(I,y)] \vee [x' \wedge p(O,y)]$$

and

$$p(x,y) = [p(I,I) \wedge x \wedge y] \vee [p(I,O) \wedge x \wedge y'] \vee [p(O,I) \wedge x' \wedge y] \vee [p(O,O) \wedge x' \wedge y']$$

6. Establish a bijection between the Boolean subalgebras of $\mathbf{2}^n$ and the partitions of the set of its "atoms" (i.e., elements which cover O).

★7. (a) Show that, if $[a,b]$ is any interval at a Boolean algebra A, then $[[a,b],\wedge,\vee,a,b]$ is a distributive lattice.
(b) Show that the operation $a \mapsto a^*$ satisfies conditions L7–L10.
(c) Conclude that Theorem 7 is true.

★5-10. DIRECT PRODUCTS; MORPHISMS

Given any two Boolean algebras A and B, we can construct their *direct product* $A \times B$ as follows. The elements of $A \times B$ are the ordered pairs (a,b) with arbitrary $a \in A$ and $b \in B$. (Thus, as a set, $A \times B$ is just the Cartesian product of A and B.) The operations (binary, unary, zero-ary) of $A \times B$ are defined by

$$\begin{gathered}(a_1,b_1) \vee (a_2,b_2) = (a_1 \vee a_2,\ b_1 \vee b_2)\\ (a_1,b_1) \wedge (a_2,b_2) = (a_1 \wedge a_2,\ b_1 \wedge b_2)\\ (a,b)' = (a',b') \qquad O = (O_A,O_B) \qquad I = (I_A,I_B)\end{gathered} \tag{21}$$

It is immediately obvious that $A \times B$ is a Boolean algebra under the operations defined by (21), because each of the identities L1 to L10 can be verified separately in the first and second component of the expressions in question.

We next define a *Boolean morphism* to be a function $\theta: A \to B$ which has Boolean algebras for domain and codomain, such that

$$\theta(x \wedge y) = \theta(x) \wedge \theta(y) \qquad \theta(x \vee y) = \theta(x) \vee \theta(y) \qquad \theta(x') = [\theta(x)]'$$

for all $x,y \in A$. These identities imply

$$\theta(O_A) = \theta(x \wedge x') = \theta(x) \wedge \theta(x') = \theta(x) \wedge [\theta(x)]' = O_B$$

and, dually, $\theta(I_A) = I_B$.

As in Sec. 2-6, a Boolean morphism which is one-one is called a (Boolean) *mono*morphism; one which is onto is called a Boolean *epi*morphism, and one which is a bijection is called a Boolean *iso*morphism.

Clearly, the assignment $p_A\colon (a,b) \mapsto a$ and $p_B\colon (a,b) \mapsto b$ are Boolean epimorphisms $p_A\colon A \times B \to A$ and $p_B\colon A \times B \to B$.

Theorem 10. *A bijection between Boolean algebras is a Boolean isomorphism if (and only if) it preserves the inclusion relation.*

Proof. If the inclusion relation is preserved, Theorem 2 shows that the definitions of glb, lub, universal bounds, and complement are preserved. The "only if" is trivial.

It is a corollary that any finite Boolean algebra is determined up to Boolean isomorphism by its diagram (the directed graph of its covering relation), but such diagrams are not very helpful when the number of elements is 32 or more.

Note again that if $\mathcal{P}(U)$ and $\mathcal{P}(V)$ are Boolean algebras defined as power sets (under the Boolean operations of intersection, union, and complement), then the mapping $S \sqcup T \to (S,T)$ for variable $S \subset U$, $T \subset V$ defines an isomorphism $\mathcal{P}(U \sqcup V) \cong \mathcal{P}(U) \times \mathcal{P}(V)$.

Free Boolean Algebra. Finally, let $G = \{g_1, \ldots, g_r\}$ be the generators of the Boolean algebra $\mathcal{P}(\mathbf{2}^r) = \mathbf{B}^{2^r}$, which were denoted by $X_1, \ldots, X_r$ in Theorem 5. Let A be any Boolean algebra, and let $f\colon G \to A$ be any mapping from G to a subset of A; we may designate $f(g_i)$ by y_i.

Then the mapping $S \mapsto \bigvee_S p_i$ of Theorem 8 is a *morphism* of joins and meets for $S \subset \mathcal{P}(G)$, from the Boolean algebra $\mathcal{P}(\mathbf{2}^r)$ to A. As was shown in the proof of Theorem 9, it is also a morphism for complements. Therefore, it is a Boolean morphism.

We have sketched a proof of the following result.

Theorem 11. *Any mapping $g_i \mapsto y_i$ from the generators of the Boolean algebra $\mathcal{P}(\mathbf{2}^r)$ of Theorem 5 to elements of a Boolean algebra A can be extended to a Boolean morphism $\mu\colon \mathcal{P}(\mathbf{2}^r) \to A$.*

In general, an algebraic system with r generators having the property of Theorem 10 is called a *free algebra* with r generators. We have thus shown that $\mathcal{P}(\mathbf{2}^r) \cong \mathbf{B}^{2^r}$ is a free Boolean algebra with r generators.

EXERCISES E

1. Find the disjunctive normal form for the following Boolean polynomials in x,y,z:

   (a) y (b) $xy \vee z'$ (c) $x + yz'$

2. Simplify the Boolean expression $(xy \vee x)(z' \vee zx)$.

3. Prove that $xy \vee xz \vee zy' = xy \vee z'y'$.
4. Show that each of the two distributive laws of postulate L6 implies the other.
5. Let $\mathcal{B}$ be a Boolean algebra, and let the elements $a_1, \ldots, a_n$ of $\mathcal{B}$ be disjoint with union $\bigvee_{i=1}^{n} a_i = I$. Show that the mapping $\mu: S \mapsto \bigvee_{S} a_i$ is a Boolean monomorphism from the power set $\mathcal{P}(\mathbf{n})$ of all $S \subset \mathbf{n}$ to $\mathcal{B}$.
6. Prove in detail that F_2 is "free" on two generators.
7. Classify the elements of F_3 into equivalence classes
   (a) Under the group of permutations of x,y,z.
   (b) Under the group generated by permutations of x,y,z and the correspondences $x = x'$, $y = y'$, $z = z'$.
8. Let $\mathcal{A} = [A,\wedge,\vee,']$ and $\mathcal{B} = [B,\wedge',\vee,']$ be Boolean algebras, and let $\theta: A \to B$ be a Boolean morphism. Show that the inverse image $\theta^{-1}(O)$ in A of $O \in B$ contains
   (a) $O \in A$
   (b) with any $a \in A$, all $x \leqq a$ in A
   (c) with a_1 and a_2, also $a_1 \vee a_2$. (Such a subset of A is called an *ideal*, the *kernel* of θ.)
9. Show that if J is any ideal of A, then there exists a Boolean algebra $\mathcal{C} = [C,\wedge,\vee,']$ and a Boolean epimorphism $A \to C$ with kernel J.

REFERENCES

HOHN, F. E.: "Applied Boolean Algebra," Macmillan, 1960.
WHITESITT, J. E.: "Boolean Algebra and its Applications," Addison-Wesley, 1961.

Optimization and Computer Design

6-1. INTRODUCTION

An important application of Boolean algebra has been to the analysis and design of digital computers, especially of the combinational or gating circuits described in Chap. 5. The problem of optimizing the design of these circuits is essentially a problem of Boolean algebra. This will be explained in Secs. 6-3 and 6-4, where we shall describe a general technique for constructing gating circuits from a few simple components to realize any given Boolean function.

These sections will be preceded by a section on the general concept of optimization (Sec. 6-2), a technique for optimizing a class of simple "multistage decision processes" which can be formulated as problems of finding shortest paths in directed graphs whose arcs have well-defined *lengths* (or "costs").

After this introductory material will come three sections (6-5 to 6-7), which present some useful techniques for *economizing* the construction of gating circuits. There are many ways to construct a gating circuit which will realize a given Boolean function $f(x_1, \ldots, x_n)$ of binary inputs. The problem is to simplify the design by reducing the number of components used, thus making the design more economical and less complicated. This is called *minimization* in the literature, and the procedures are called *minimization procedures.*

In Secs. 6-8 and 6-9, we turn our attention to the more complicated problem of designing *sequential* machines. These are machines containing both memory elements and gates. A procedure is given for realizing any finite-state machine.

6-2. OPTIMIZATION

In most engineering design, economy is a basic consideration. Commonly, one wishes to *minimize cost* subject to various performance requirements. The study of the mathematical techniques which facilitate cost minimization subject to given performance requirements or, dually, which helps one to obtain *maximum performance* with given resources, comprises an important area of applied mathematics known as *optimization theory.*

The general idea of optimization and a mathematical technique for achieving optimality are illustrated by multistage decision problems of the following kind.

Suppose we have some *process* which proceeds in discrete *stages,* representing *transitions* from one *state* of the process to the next. Such processes can be represented schematically by *directed graphs* (or *flow charts*), whose nodes represent the different states of completion of the processes and whose arcs represent the possible transitions from one state of completion to the next, each *labeled* with its *cost.* The problem is to *minimize the total cost*—i.e., the sum of the costs from a given *initial state* to the *final state* (when the process is *completed*).

If we interpret cost as *length,* we see that the *optimal decision* amounts mathematically to finding a *shortest path,* i.e., one in which the sum of the costs (lengths) of the arcs traversed (the total cost) is minimized, from a given initial node to a specified final node. Call such a path an *optimal path.* In the special case in which all arcs have the same cost this problem has already been described in Sec. 2-10.

Example 1. To illustrate the relevant technique, we now find the path of least cost from the starting state $s_{0,0}$ to the final state $s_{5,0}$ in the labeled directed graph of Fig. 6-1. This graph has only 14 states, and 14 different paths from state $s_{0,0}$ to state $s_{0,5}$. Hence one could find the optimal path by computing the cost of each path and selecting one with minimum cost.

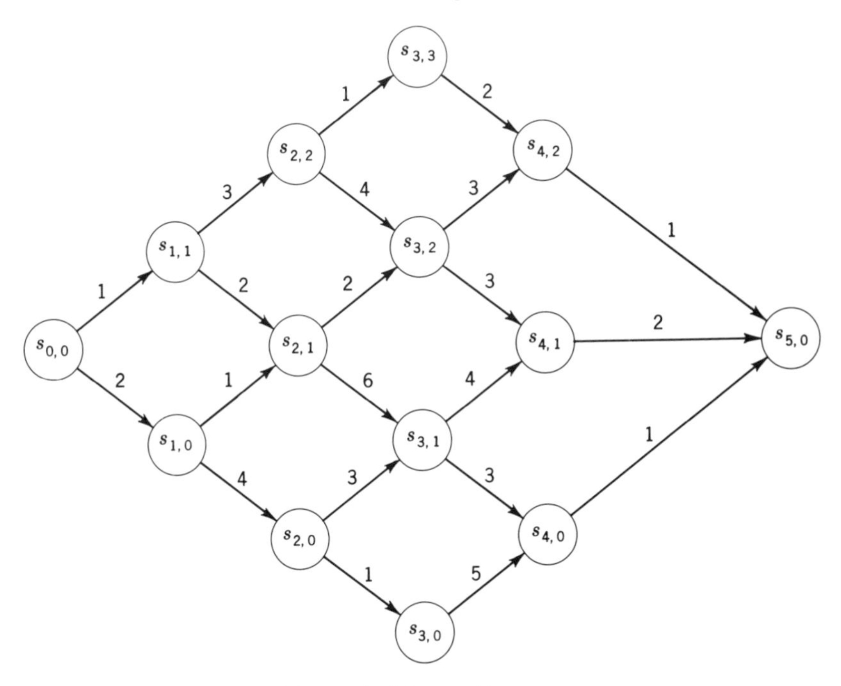

Fig. 6-1. A graph for a multistage decision problem.

However, this procedure becomes impractical with large graphs because there are so many alternative paths to compare. A more efficient procedure can be found, based on the following so-called *Optimality Principle.*

Theorem 1. (*Optimality Principle.*) *Any subpath of an optimal path is optimal.*

The proof is obvious. If $\vec{P}$ is any optimal path with successive arcs $\alpha_i = \overrightarrow{s_{i-1}s_i}$ $(i = 1, \ldots, n)$ from s_0 to s_n and $\vec{Q} = [s_j, s_{j+1}, \ldots, s_k]$ is any subpath (segment) of $\vec{P}$ the cost (length) of which can be reduced by replacing it with

$$\vec{R} = [\tilde{s}_h = s_j, \tilde{s}_{h+1}, \ldots, \tilde{s}_l = s_k]$$

then the new path

$$\vec{s} = [s_0, \ldots, s_{j-1}, \tilde{s}_h, \ldots, \tilde{s}_l, s_{k+1}, \ldots, s_n]$$

constructed by "splicing in" $\vec{R}$ in place of $\vec{Q}$, will cost less (be shorter) than the original path $\vec{P}$ so that $\vec{P}$ cannot have been optimal.

Corollary. *In a directed graph whose arcs all have positive costs (lengths), every optimal path is simple.*

We now apply the preceding Optimality Principle to the problem of Fig. 6-1, considering first the optimal subpaths from each state in stage 4 to $s_{5,0}$, then from stage 3 to $s_{5,0}$, from stage 2 to $s_{5,0}$, from stage 1 to $s_{5,0}$, and finally from $s_{0,0}$ to $s_{5,0}$.

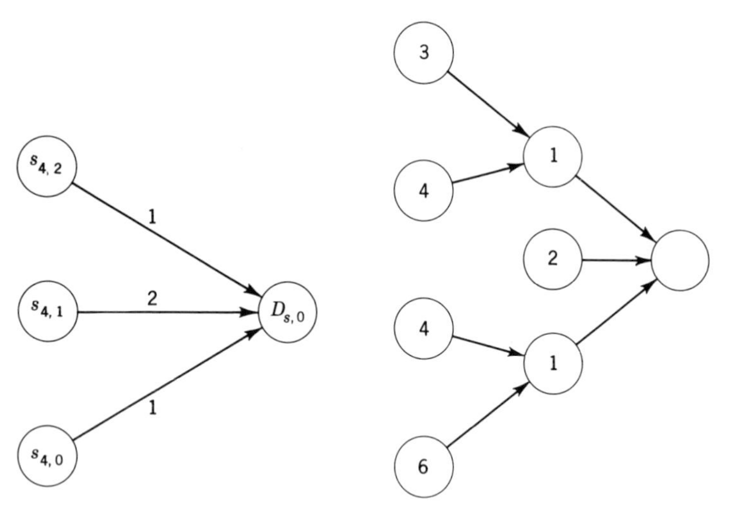

Fig. 6-2. Graph for determining the values of M for stage 4 of the problem.

Fig. 6-3. Graph for determining the values of M for stage 3 of the problem.

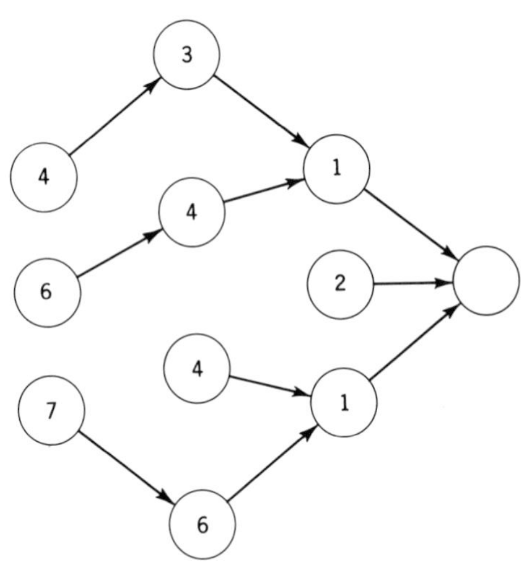

Fig. 6-4. Graph for determining M for stage 2 of the problem.

Define the function $M(s_{i,j})$ as the cost of the path of least weight leading from state $s_{i,j}$ to state $s_{5,0}$. Some costs are

$$M(s_{5,0}) = 0 \qquad M(s_{4,2}) = 1 \qquad M(s_{4,1}) = 2 \qquad M(s_{4,0}) = 1$$

Given these costs for the states in the fourth stage of the process, we now wish to calculate the cost of M for each state in stage 3. This can be done by drawing the subgraph formed by the decisions (choices of paths) made thus far, although for our first step this set of decisions is forced.

The new graph is shown in Fig. 6-2, and the value of each $M(s_{i,k})$ is now used as a label for $s_{i,k}$. The values for each $M(s_{i-1,\,k})$ can then be determined by simply examining each arc from $s_{i-1,\,k}$ to some $s_{i,j}$, adding its cost to $M(s_{i,j})$, and then choosing the minimum for all j. This will give a minimal value for any path from $s_{5,0}$ to $s_{i-1,\,k}$. For instance, there are two arcs from $s_{3,2}$ to a state in stage 4 for the graph in Fig. 6-1. The cost of the upper arc is 3, and this goes to $s_{4,2}$ where $M(s_{4,2}) = +1$; this path to $s_{5,0}$ would cost 4. The lower arc is to $s_{4,1}$, where $M(s_{4,1}) = 2$, and the arc from $s_{2,2}$ to $s_{3,1}$ costs 3 for a total of 5. The upper path is therefore less costly and should be chosen so that the new $M(s_{3,2}) = 4$. The resulting subgraph for the third stage is shown in Fig. 6-3.

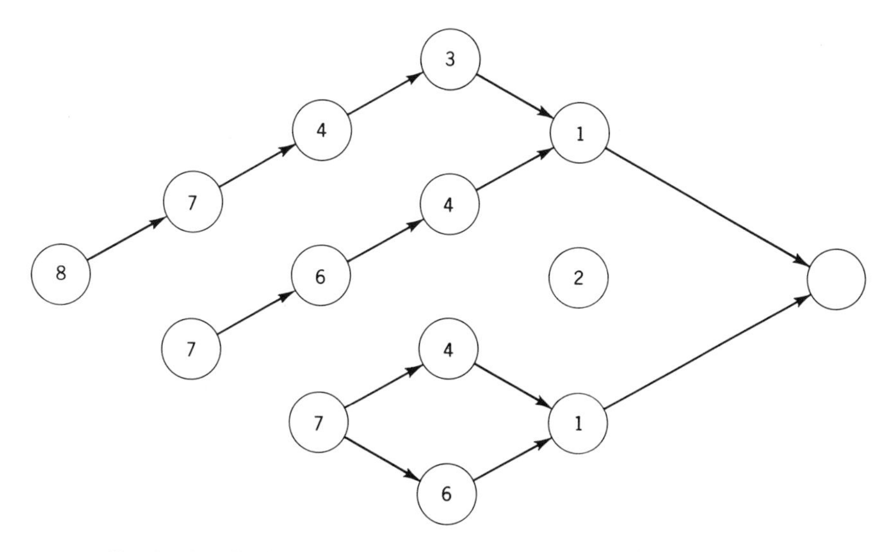

Fig. 6-5. Graph of optimal solution.

The new graph and values for the $M(s_{2,k})$ is shown in Fig. 6-4, and the final graph is shown in Fig. 6-5. This graph shows that the optimal policy involves following the top path in the graph.

6-3. COMPUTERIZING OPTIMIZATION

In this section we consider another example and then describe a general computer program for finding optimal paths in finite directed graphs with specified arc lengths (costs).

Example 2. Another multistage decision problem is shown in graph form in Fig. 6-6. Here the problem is simply to find a path of least cost

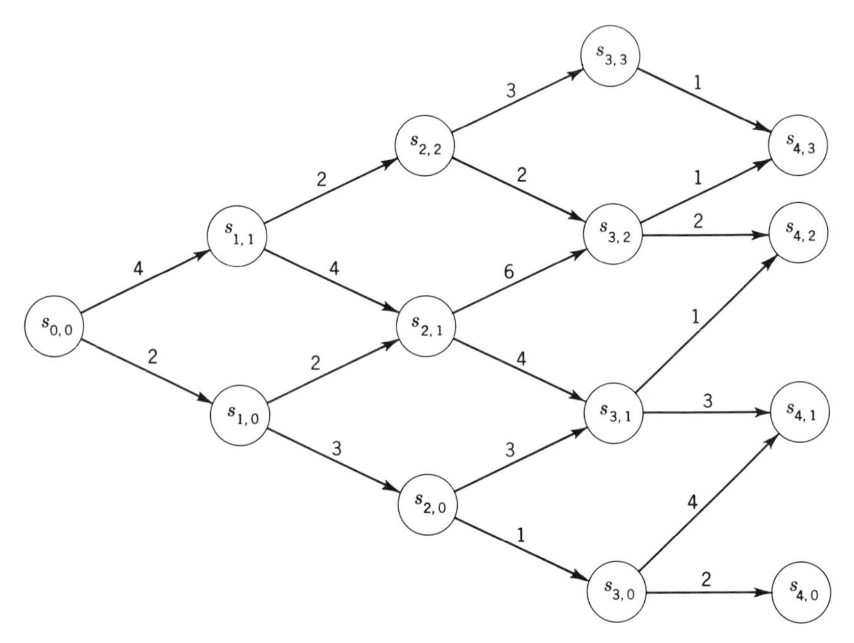

Fig. 6-6. A finite-state problem with a terminal stage.

leading from $s_{0,0}$ to any state in stage 4 (to any $s_{4,k}$). This problem is solved by calculating the optimal paths from each state in stage 3 to any state in stage 4 and then calculating the optimal paths from each state in stage 2 to any state in stage 4. This is then repeated for stage 1 and stage 0.

Define $M(s_{i,k})$ as the length of the optimal path from every $s_{i,k}$ to $s_{4,j}$ for $i = 3,2,1,0$. We have, trivially, $M(4,k) = 0$ for all $k = 0,1,2,3$. The values of $M(s_{3,k})$ are also obvious; that is, $M(s_{3,0}) = 2$, $M(s_{3,1}) = 1$, $M(s_{3,2}) = 1$, $M(s_{3,3}) = 1$, and the new subgraph is shown in Fig. 6-7*a*.

The subgraph for the values of $M(s_{2,k})$ is shown in Fig. 6-7*b*, and finally, the graph showing the optimal policy is in Fig. 6-8.

The technique sketched above can greatly reduce the number of steps required to find optimal policies for large but finite multistage decision processes. Many such processes arise in transportation theory, e.g., in optimizing routing and flow and in locating distribution centers for

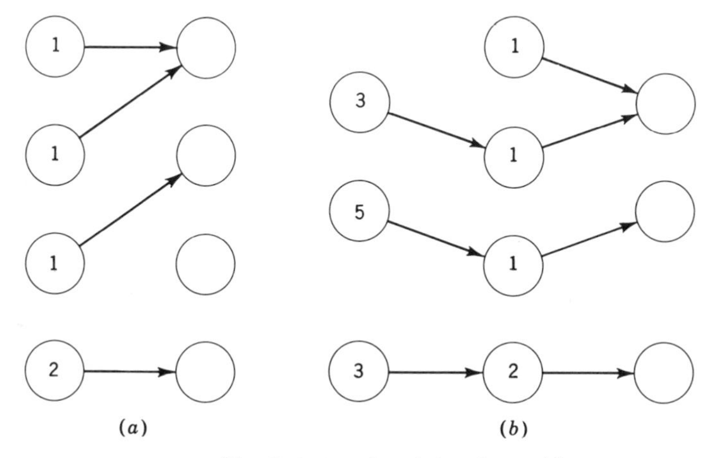

Fig. 6-7. The first steps in solving the problem.

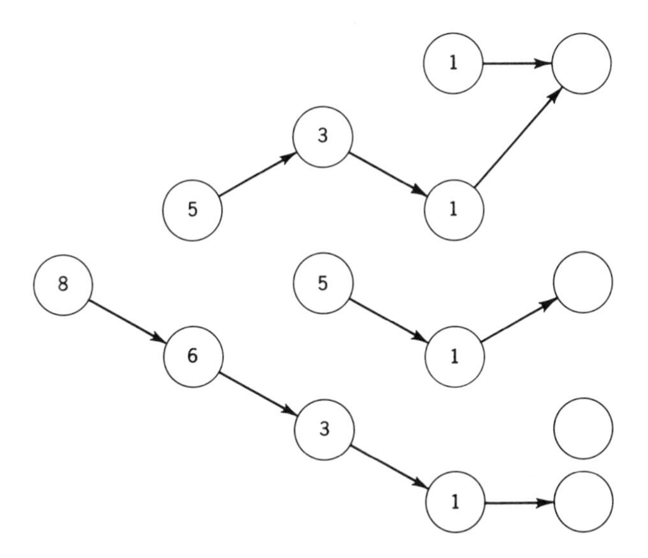

Fig. 6-8. The solution graph for Fig. 6-6.

commodities so as to minimize the cost or the maximum time for distribution. Others concern resource allocation and arrangement in multistage production processes, where the objective may be to maximize the output.[1]

Similar considerations can be used to find shortest or quickest paths in the calculus of variations, where the Optimality Principle is invoked in deriving the Euler-Lagrange equations. Optimal control problems provide

[1] For discussions of many such problems, see Richard E. Bellman, "Dynamic Programming," Princeton University Press, 1957.

a still more extensive area of application for the ideas involved.[1] Difference approximations to such continuum problems provide an intermediate class of problems solvable by computer.

General Case. We conclude this section by describing an easily programmed generalization of the technique applied in Examples 1 and 2, which can be used to find the shortest path from node a to node b in any directed graph $\vec{G}$ whose arcs α_i have specified positive lengths $\lambda(\alpha_i)$. Using the Optimality Principle, one can proceed as follows (since only simple paths can be shortest, one can also ignore loops).

Stage 0. List all nodes a_i. Set $\lambda(a) = 0$, and set all other $\lambda(a_i) = \infty$ (intended to represent the cost (length) of the optimal path from a to a_i *known at this stage*).

Stage 1. List all arcs $\overrightarrow{ax_i} = \xi_i$ issuing from a. For each x_i in turn, compare $\lambda(\xi_i)$ with $\lambda(x_i)$. If $\lambda(\xi_i) \geqq \lambda(x_i)$, move on to x_{i+1}. If $\lambda(\xi_i) < \lambda(x_i)$, replace $\lambda(x_i)$ by $\lambda(\xi_i)$, and record the path $\overrightarrow{ax_i}$ as the shortest path from a to x_i *found so far*.

Stage 2. For each node $x_i \neq a$ considered in stage 1, list all arcs $\overrightarrow{x_iy_j} = \eta_{ij}$ issuing from x_i. For each η_{ij} in turn, compute $\mu_{ij} = \lambda_i + \lambda(\eta_{ij})$. If $\mu_{ij} \geqq \lambda(y_j)$, move on to the next arc η_{ij}. If $\mu_{ij} < \lambda(y_j)$, replace $\lambda(y_j)$ by μ_{ij}, and record the path $\overrightarrow{ax_iy_i}$ as the shortest (optimal) path from a to y_j *found so far*.

We now proceed recursively, changing notation.

Stage n. For each node a_i for which $\lambda(a_i)$ was diminished at the $(n-1)$st stage, list all arcs $\overrightarrow{a_iz_k} = \zeta_{ik}$ issuing from a_i. For each ζ_{ik} in turn, compute $\nu_{ik} = \lambda(a_i) + \lambda(\zeta_{ik})$, and compare ν_{ik} with $\lambda(z_k)$. If $\nu_{ik} \geqq \lambda(z_k)$, move on to the next arc ζ_{ik} (either $\zeta_{i,\,k+1}$ or $\zeta_{i+1,1}$). If $\nu_{ik} < \lambda(z_k)$, replace $\lambda(z_k)$ by ν_{ik}, and record the new *optimal path* $\overrightarrow{a \cdots a_iz_k}$ *from* a *to* z_k *found so far* (alternatively, this will be the *first* path found from a to z_k having total length $\leqq \nu_{ik}$).

For some first mth stage, at most equal to the number of nodes in $\vec{G}$, no new cheaper (shorter) path of length n will be found. At this stage, every $\lambda(a_i)$ will be optimal.

We now present an ALGOL program for the above optimal-path algorithm. For a directed graph $\vec{G}$ with N nodes $1,2, \ldots ,N$, the input

[1] See E. B. Lee and L. A. Markus, "Foundations of Optimal Control Theory," Wiley, 1967.

data to the program is N and an $N \times N$ matrix Z where Z_{ij} is the length of the arc from node i to node j. If there is no arc from node i to node j, then $Z_{ij} = \infty$ (actually some very large positive number within range of the computer, say 10^{20}). It is assumed that the initial node is 1, and the program computes the length of the optimal path from node 1 to node k for $k = 2,3, \ldots ,N$. The following notation is used:

L_k = length of a current optimal path from node 1 to node k.
m_k = number of nodes, excluding the initial node 1 and final node k, in a current optimal path from node 1 to node k.
P_k = array of successive nodes $P_{k_1},P_{k_2}, \ldots ,P_{k_{m(k)}}$ in a current optimal path from node 1 to node k, again excluding the initial node 1 and final node k.
d = array of nodes $d_1,d_2, \ldots ,d_r$ for which the optimal length L_{d_i} was diminished at the previous iteration.

```
begin integer N, i, j, k, q, r, r, temp;
      integer array m[1:50], d[1:50], d temp[1:50], P[1:50, 1:50];
      real nu; real array Z[1:50, 1:50], L[1:50];
      for i := 2 step 1 until N do
          begin L[i] :=₁₀ + 20; m[i] := 0; d[i] := i end
      r := N − 1;
A:    r temp := 0;
      for q := 1 step 1 until r do
      begin i := d[q];
            for k := 2 step 1 until N do
            begin nu := L[i] + Z[i,k]
                  if nu < L[k] then
                     begin L[k] := nu; m[k] := m[i] + 1;
                           for j := 1 step 1 until m[i] do
                               P[k.j] := P[i.j];
                           P[k,m[k]] := i;
                           r temp := r temp + 1;
                           d temp [r temp] := k
                     end
            end
      end
      if r temp > 0 then
            begin r := r temp;
                  for i := 1 step 1 until r do d[i] := d temp[i];
                  go to A
            end
end
```

After execution of the above, L_k is the length of a shortest (optimal) path from the starting node 1 to node k. If $L_k \geqq 10^{20}$, then no path has been found from node 1 to node k. The intermediate successive nodes of any optimal path from node 1 to node k are given by the kth row of the array P.

EXERCISES A

1. Find the path of least length (cost) through the following labeled directed graph (numbers represent arc lengths):

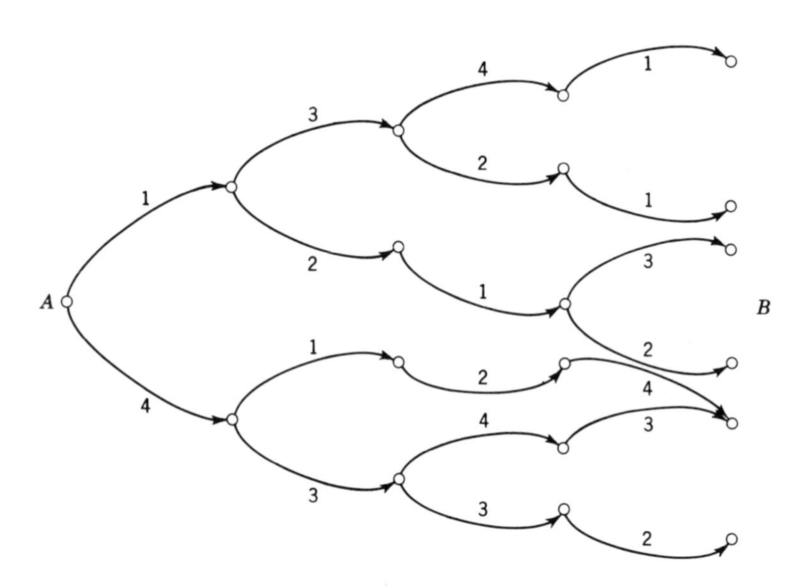

2. Find the shortest path from A to B on the following labeled graph:

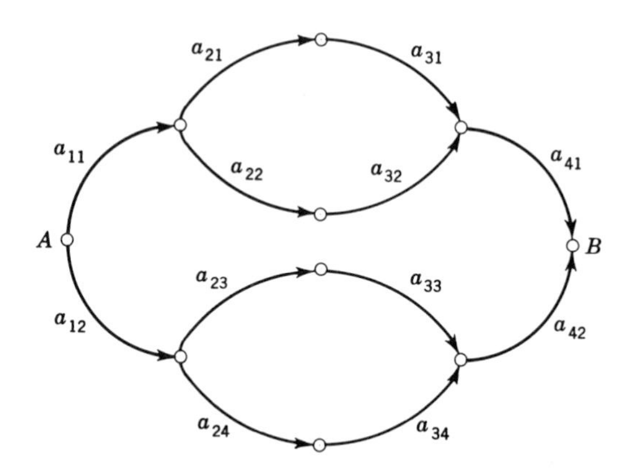

3. Write an ALGOL program that finds the shortest path from A to B in the following labeled graph:

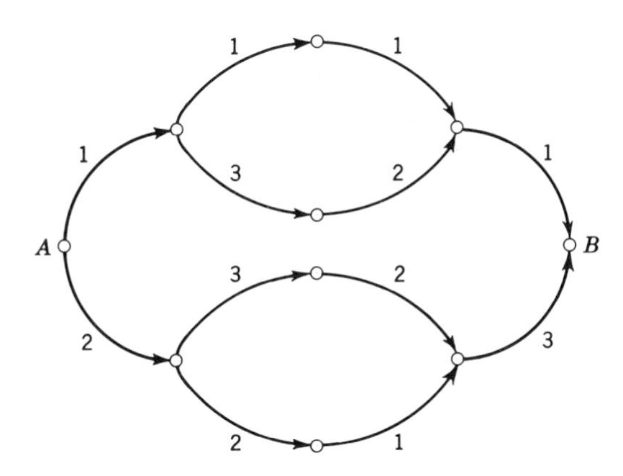

6-4. LOGIC DESIGN

Because of the connections between Boolean algebra and logic, the design of gating circuits (or combinational networks) in a computer is referred to by computer designers as the problem of *logic design*. There exists a straightforward procedure for designing a gating circuit to realize a given Boolean function $f(x) = f(x_1, \ldots, x_r)$ of r Boolean input variables, that is, a given Boolean function $f\colon \mathbf{2}^r \to \mathbf{2}$. We shall now describe this procedure in some detail; it requires the use of only AND gates, OR gates, and inverters.

Some comment should be made concerning notation and terminology. Computer logic designers[1] have developed a notation and terminology which is different from that commonly used by mathematicians. The binary operation symbol + is used by such designers to denote what has been written $\vee$ in Chap. 5. Thus, to the logic designer $0 + 0 = 0$, and $0 + 1 = 1 + 0 = 1 + 1 = 1$. Algebraists, however, commonly use the symbol + in Boolean algebra as follows:

$$0 + 0 = 0, \qquad 1 + 1 = 0$$

and

$$0 + 1 = 1 + 0 = 1$$

Logic designers use the symbol $\oplus$ to denote this binary operation of addition mod 2, so that for them $1 \oplus 1 = 0$.

The principal reason for this use by logic designers of the symbol + for what we have designated $\vee$ is that the logic diagrams for computers are maintained on magnetic tape and printed on commercial line printers

[1] Computer designers refer to gating circuits as constituting the *computer logic;* they refer to the design of such circuits as *switching circuit theory*.

which do not have the symbol $\vee$. Logic designers therefore use $\cdot$ or juxtaposition for $\wedge$, writing for instance abc or $a \cdot b \cdot c$ for $a \wedge b \wedge c$; logic designers also use $+$ for the binary operation which we have designated $\vee$; thus $a + b$ is written instead of $a \vee b$.

In this chapter, we shall use juxtaposition as a shortened form where the binary operation symbol $\wedge$ might normally appear, and use $\vee$ as in Chap. 5. The convenient terminology *product term* instead of *meet term* and *sum-term* instead of *join term* will also be used; this is part of a very convenient overall terminology, which is used by logic designers and which is embodied in the following definitions.

Definition. A *literal* is a variable or the complement of a variable. Thus, x, x', y, a', and b are five different literals.

Definition. A *product term* is the meet of a set of literals. A *sum-term* is the join of a set of literals. Thus ab, $ab'c'$, xyz, and $z'x'$ are four different product terms, and $(a \vee b \vee c)$, $(a' \vee b')$, $(a \vee b' \vee c)$, and $(x \vee y)$ are four different sum-terms.

Definition. A *sum-of-products expression* is a sum (join) of a set of product terms. A *product-of-sums expression* is a product of sum-terms. Thus $ab \vee c$, $ab \vee a'b'$, and $ac \vee abc \vee a'c'$ are three different sum-of-products expressions, and $(a \vee b)(c \vee d)$, $(a \vee b)(a' \vee b')$, $(a \vee b \vee c)(a' \vee b' \vee c)$, and $(a \vee b' \vee c')$, are four different product-of-sums expressions.

We now give a technique for designing a gating network which will realize any given function of n binary "input" variables. This technique follows exactly the same procedure as was used to prove Theorem 3 in Chap. 5, which asserts the existence of a Boolean sum-of-products expression equivalent to any binary function $f\colon \mathbf{2^n} \to \mathbf{2}$ of n binary input variables.

First, a table of combinations listing the value of the function for each input value is formed. Figure 6-9 will be used as our example.

Next, for each set of values for the variables, i.e., for each row of the table of combinations, a product term is formed which has value 1 for these particular values of the variables. Thus, for the row in Fig. 6-9 where $A = 0$, $B = 1$, and $C = 0$, we form $A'BC'$. For the row where $A = 1$, $B = 1$, and $C = 0$, we form ABC'. These terms are listed in a column entitled "product terms."

Another column is then formed headed "sum-terms." This column contains the complement of each product term in the same row. Thus, in the row with product term $AB'C'$ we write the sum-term $(A' \vee B \vee C)$, and in the row with product term $A'BC'$ we write the sum term $(A \vee B' \vee C)$. (These terms will be used to derive product-of-sums expressions.)

We then form a sum-of-products expression for the desired function. This expression is formed by simply taking the sum of the product terms from those rows where the function has value 1.

A	B	C	$f(A,B,C)$	*Product terms*	*Sum-terms*
0	0	0	0	$A'B'C'$	$A \vee B \vee C$
0	0	1	1	$A'B'C$	$A \vee B \vee C'$
0	1	0	0	$A'BC'$	$A \vee B' \vee C$
0	1	1	1	$A'BC$	$A \vee B' \vee C'$
1	0	0	0	$AB'C'$	$A' \vee B \vee C'$
1	0	1	1	$AB'C$	$A' \vee B \vee C'$
1	1	0	1	ABC'	$A' \vee B' \vee C$
1	1	1	1	ABC	$A' \vee B' \vee C$

$$f(A,B,C) = A'B'C \vee A'BC' \vee AB'C \vee ABC' \vee ABC$$
$$= (A \vee B \vee C)(A \vee B' \vee C)(A' \vee B \vee C)$$

Fig. 6-9.

In Fig. 6-9, this means selecting product terms from the second, fourth, sixth, seventh, and eighth rows, yielding the product terms

$$A'B'C \qquad A'BC \qquad AB'C \qquad ABC' \qquad ABC$$

The sum-of-products expression formed from these terms will give the desired function. Thus we have

$$f(A,B,C) = A'B'C \vee A'BC \vee AB'C \vee ABC' \vee ABC$$

for the function in Fig. 6-9. Notice that this expression will have the desired value 1 if $f(A,B,C) = 1$, because the values for A, B, and C which lead to 1 as function value make the product term (and therefore cause the sum-of-products expression) in the same row to have value 1. This is a general procedure which works for any given function f: $\mathbf{2^n} \to \mathbf{2}$ and leads to the following result.

Theorem 2. *Any Boolean function of n variables can be expressed by a sum-of-products expression.*

The gating network which will realize this Boolean polynomial is easily formed. An AND gate is required for each product term and an OR gate with as many inputs as there are product terms. Each AND gate has its output connected as an input to the OR gate. Each AND has as inputs the literals from some product term. Figure 6-10 shows the network for the sum-of-products expression derived above.

We assume here that input lines for A, A', B, B', C, and C' are all available. If the complemented values (that is, A', B', and C') were not available, these could be formed using inverters. In most cases, however, signals for each of the variables will be available. Also, the signal or input line for A, B, and C can be used as an input to as many AND gates as is necessary.[1]

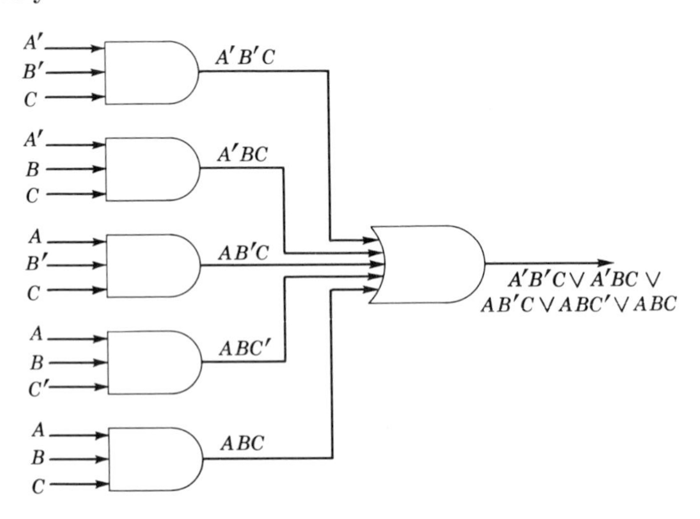

Fig. 6-10. Gating network for Fig. 6-9.

A product-of-sums expression and an OR-to-AND gate network are similarly derived. In this case we solve for $f'(A,B,C)$ (f' can be easily calculated by simply complementing each of the values for f). The problem is then solved as before, but for f'. This means selecting the terms from the rows with function values of $f'(A,B,C) = 1$; these are the rows where values of $f = 0$ appeared previously. The sum (join) of these product terms gives a sum-of-products expression for f'. For Fig. 6-9 this would give $f'(A,B,C) = (A'B'C' \vee A'BC' \vee AB'C)$. The complement of this gives f, since $f'' = f$. Thus

$$\begin{aligned} f &= (A'B'C' \vee A'BC' \vee AB'C')' \\ &= (A \vee B \vee C)(A \vee B' \vee C)(A' \vee B \vee C) \end{aligned}$$

using De Morgan's law. This proves the following theorem.

[1] Each time a signal **a** is used as in input to a gate, the "load" on the signal is increased by some fixed amount. A given signal can be "overloaded," but there are ways to alleviate this. In any case, the problem can be attacked only in the context of some specific technology and will not be treated in this book.

Theorem 3. *Any Boolean function of n variables can be expressed in product-of-sums form.*

The above procedure can be mechanized by simply forming the product (meet) of those sum-terms in Fig. 6-9 where values of 0 appear for $f(A,B,C)$. This gives the same expression as above. Implementing this expression requires as many OR gates as sum-terms and an AND gate with as many inputs as there are OR gates.

The OR-to-AND gate network for the problem in Fig. 6-9 appears in Fig. 6-11.

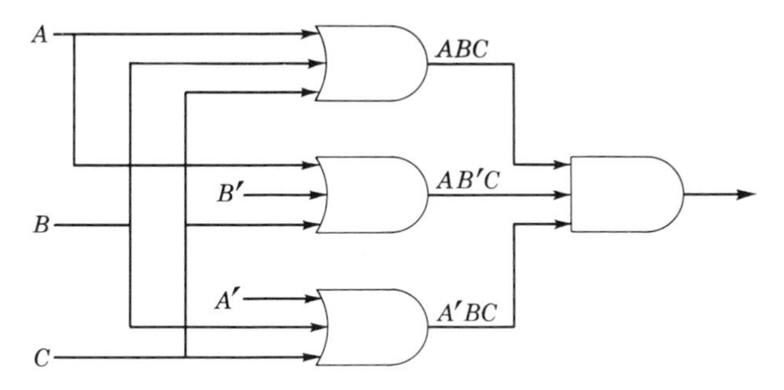

Fig. 6-11. OR-to-AND gate network.

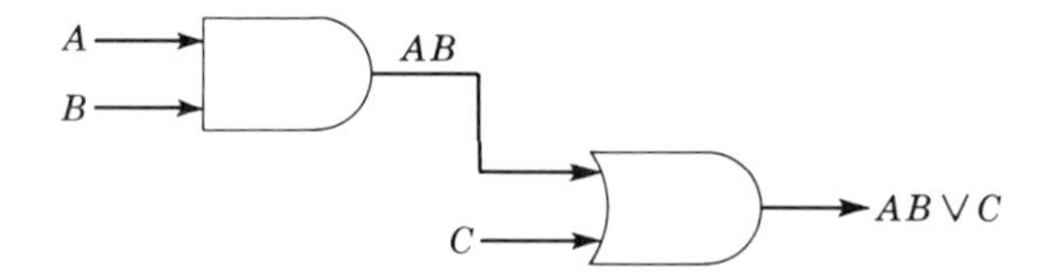

Fig. 6-12. Derivation of gating network from table of combinations.

The objection to the preceding design method is that it is uneconomical. The number of gates used and the complexity of the network can be greatly reduced using the theorems of Boolean algebra. For instance, Fig. 6-12 shows a combinational network which will yield the same function as those of the other two shown.

6-5. NAND GATES AND NOR GATES

The preceding section dealt with two binary operations, join (sum) and meet (product), and a unary operation of complementation. It was shown that any binary-valued function of n variables can be described by a Boolean polynomial using only these three operations. The question naturally arises as to whether all three of these operations are necessary. This is quite simply answered by De Morgan's laws. Product and complementation are sufficient, for $(X'Y')' = X \vee Y$, so sums can be manufactured from products and complements. Similarly, $(X' \vee Y')' = XY$, so products can be made from sums and complementation. Given any expression in products, sums, and complements, therefore, we can eliminate all the product operations by sums and complements in a systematic way, or we can replace sums similarly.

The next natural question concerns whether some single operation might not be universal; that is, can perhaps all Boolean functions of n variables be described by using repeatedly just a single binary operation? Two such universal binary operations exist: the Pierce operation, usually denoted by $\downarrow$ and defined by $x \downarrow y = x'y'$, and the Sheffer "stroke" operation, usually denoted by $|$ and defined by $z \mid y = z' \vee y'$. To show that the $\downarrow$ operation is universal, we simply manufacture complements as follows: $x' = x \downarrow x$ and (since complements have become available) we can now write $xy = x' \downarrow y'$. Dual arguments show that $|$ is universal.

There are two very popular electronic gates called *NOR gates* and *NAND gates* which realize these binary Pierce and Sheffer operations. Figure 6-13 shows the block-diagram symbols for these gates. The gates

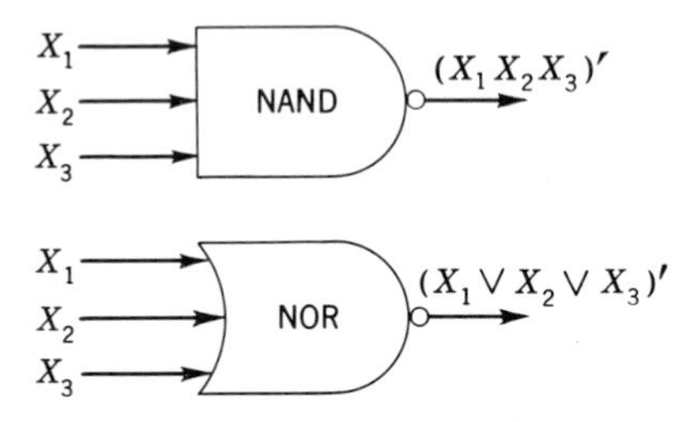

Fig. 6-13. Universal gates.

do not realize simple operations or variables when connected in parallel, however, nor could they, for neither operation is associative;

$$(x \downarrow y) \downarrow z \neq x \downarrow (y \downarrow z)$$

as is seen by considering $x = 0$, $y = 1$, and $z = 1$, for instance. Therefore, we cannot write $x \downarrow y \downarrow z$ unambiguously, and the operation of a gate which would physically realize such an expression would be ill-defined.

The NOR gate with n inputs yields by definition a function of n variables which we shall denote as $P(x_1,x_2, \ldots ,x_n) = x_1' \cdot x_2' \cdots x_n'$, and similarly for NAND gates $N(x_1,x_2, \ldots ,x_n) = x_1' \vee x_2' \vee \cdots \vee x_n'$.

Now consider Fig. 6-14, which shows a two-level NAND-to-NAND configuration and a two-level NOR-to-NOR configuration. Notice that the NAND-to-NAND configuration yields a function corresponding exactly to a (disjunctive) sum-of-products normal form in the same literals; that is, if we replace each of the left-most NAND gates with

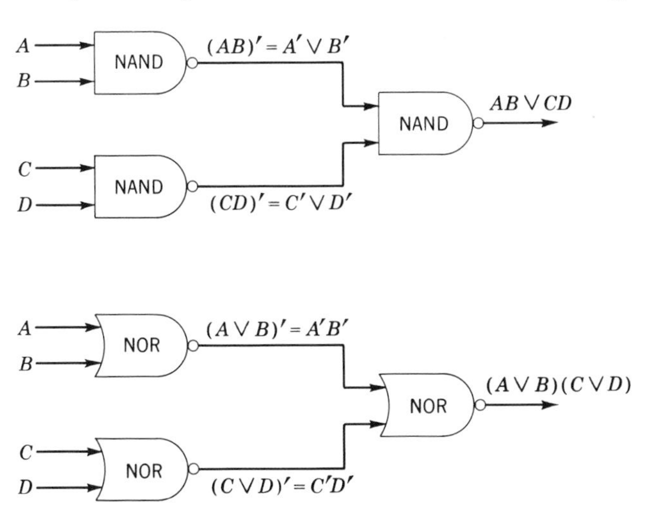

Fig. 6-14. Universal gate networks.

AND gates and the right-most NAND gate with an OR gate, a network yielding the same function will result. A similar line of reasoning shows that NOR-to-NOR configurations can always be replaced by OR-to-AND configurations.

The above procedure illustrates how one can form gate networks for an arbitrary function $f\colon \mathbf{2}^n \to \mathbf{2}$. This is the same procedure used to prove Theorem 3 in Chap. 5, which showed that the most general function $f\colon \mathbf{2^n} \to \mathbf{2}$ of the binary variables $x_1, x_2, \ldots, x_n$ could be expressed in a "join-of-meets" form

$$f = \bigvee_l Z_l \qquad \text{where} \qquad Z_l = \bigwedge_{i=1}^{n} Y_{i(l)} \qquad \text{and} \qquad Y_{i(l)} = x_i \text{ or } x_i' \quad (1)$$

We shall now give to the result just stated a new interpretation. First, we observe that in our new terminology we should rewrite the canonical form of (1) as

$$f = \sum_l \left(\prod_{i=1}^{n} Y_{i(l)} \right) \qquad \text{each } Y_{i(l)} \text{ a literal } (X_i \text{ or } X_i') \qquad (2)$$

This gives the same result again. Since each product term requires at most one AND gate and there are at most 2^n terms, this implies that any Boolean function $f: \mathbf{2}^n \to \mathbf{2}$ can be realized using at most 2^n AND gates, each having n input leads (actually, at least one less gate is required and generally far fewer are required).

Dually, every Boolean function $f: \mathbf{2}^n \to \mathbf{2}$ can be rewritten in the product-of-sums form

$$f = \prod_{l} \left(\sum_{i=1}^{n} Z_{i(l)} \right) \qquad \text{each } Z_i \text{ a literal } (X_i \text{ or } X_i') \tag{3}$$

Recall also that it was shown in Sec. 5-9 that every Boolean polynomial in elements $X_1, \ldots ,X_n$ of a Boolean algebra can be reduced to the canonical form (2) or (3) by systematic use of the postulates L1–L10 of Boolean algebra.

Observe that most of the Boolean functions described in this section are symmetric, in the sense of the following definition.

Definition. A function $f: \mathbf{2}^n \to \mathbf{2}$, that is, a Boolean function f of n variables $X_1,X_2, \ldots ,X_n$, is said to be *symmetric* when

$$f(X_1, \ldots ,X_n) = f(X_{\sigma 1}, \ldots ,X_{\sigma n})$$

for any permutation[1] σ of the X_i. It is called *$\pm$-symmetric* when it is symmetric in the variables $Y_i, Y_2, \ldots ,Y_n$, where each Y_i is X_i or X_i'. It is called *partially symmetric* when it is symmetric in some subset of $n \geqq 2$ of its variables.

An important symmetric Boolean function is

$$\alpha(a,b,c) = a + b + c = abc \vee ab'c' \vee a'bc' \vee a'b'c$$

depicted in the Venn diagram below. Regarded as a subset of $\mathcal{P}(\mathbf{2}^3)$, it

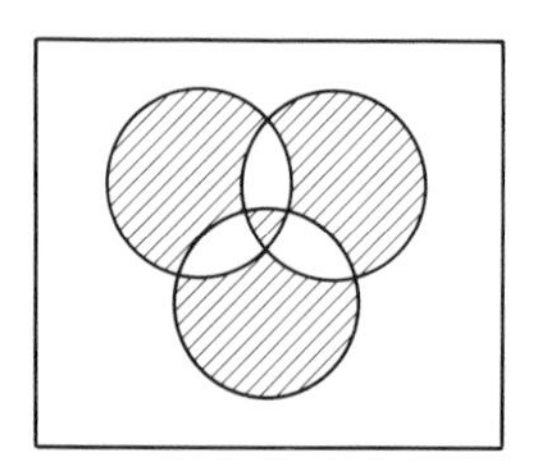

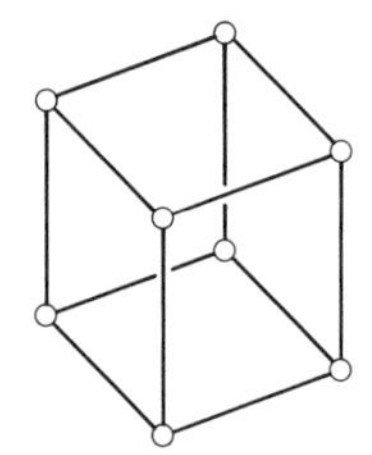

corresponds to the set of all "red" vertices in the bipartite graph of $\mathbf{2}^3$ shown above. One can define

$$a + b + c + d = (a + b + c) + d = (a + b) + (c + d)$$

[1] For a general discussion of permutations $\sigma: \mathbf{n} \to \mathbf{n}$, see Sec. 7-9.

similarly, as well as the Boolean sum (mod 2) of n variables. The circuits required to realize these functions require a surprisingly large number of components.

6-6. THE MINIMIZATION PROBLEM

The basic minimization problem for optimizing gate network design can now be couched in rational terms: given a Boolean polynomial, how can we generate an equivalent expression which will be in some sense "simplest"?

Definition. A *canonical product-of-sums expression* in n variables consists of the product of a set of sum-terms, each sum-term containing exactly one occurrence of each of the n variables and with no sum-term repeated in the expression. This class of expressions was given the name "disjunctive canonical form" in Chap. 5.

The techniques used to derive the expression in the previous section lead to canonical expressions.

We give two criteria for minimality of expressions in sum-of-products form; dual criteria will apply for product-of-sums form. For the reasons stated earlier,[1] these criteria neglect the cost of inverting (complementation).

Criterion I. *An expression Φ in sum-of-products form is minimal according to the criterion of fewest literals when no other expression Ψ which is equivalent to Φ contains fewer occurrences of literals.*

Criterion II. *An expression Φ in sum-of-products form is minimal according to the criterion of fewest terms when no expression Ψ equivalent to Φ has fewer product terms and if no expression Φ containing the same number of product terms contains fewer occurrences of literals.*

These two criteria make good sense from an implementation viewpoint: the first gives the least inputs and AND gates, whereas the second gives the least number of gates. Two other minimum cost criteria are also often used: "least number of literals plus terms," and "least diodes to construct." But both of these follow easily from our two criteria and the same basic minimization techniques apply to them. Also, we shall study only minimization of sums-of-products expressions. Dual techniques can be used for product-of-sums expressions.

[1] Flip-flops usually have two complementary outputs, so that a variable *and* its complement are usually available.

Definition. An expression Ψ *implies* an expression Φ when no assignment of values to the variables in Ψ and Φ makes Ψ have value 1 and Φ have value 0. A product term α which implies f is called an *implicant* of f.

Theorem 4. *Given a product term α in a sum-of-products expression Φ for a function f, then α implies f.*

Explanation. This theorem asserts that the product term $\alpha = ab'c'$ implies $ab'c' \vee a'b'c$, for instance, and that ab implies $ab \vee ab'c' \vee a'b'$.

Proof. If f does not have the value 0 for all assignments of values to the variables in f, then Φ must contain at least one product term. If we consider a product term β in Φ, then any assignment of values to the variables in β which make $\beta = 1$ will cause Φ to have value 1.

Now, if we seek a minimal expression Φ for a function f and Φ must be in sum-of-products form, then Φ must consist of sum-of-products terms each of which implies f, for Φ must certainly imply f.

Definition. A *prime implicant* for a function f is a product term α which implies f, but which does not imply f if any literal in α is deleted.

In effect, a prime implicant is a product term which is the "shortest" of all product terms which imply Φ and contain the same literals.

There is a simple test that can be used to determine whether or not a product term α implies an expression Φ. Simply assign values 1 to unprimed and 0 to primed literals in α; this makes α have value 1. Then assign the same values to the variables in Φ, and see whether or not Φ has value 1, that is, whether $\Phi = 1$ regardless of the values of the other literals in Φ.

For instance, $\alpha = xy$ implies $\Phi = xyz \vee xyz'$ because, setting $x = 1$ and $y = 1$, we have $\Phi = 1 \cdot 1 \cdot z \vee 1 \cdot 1 \cdot z' = z \vee z' = 1$. Also, $\beta = x'y$ implies $\Psi = wx'yz \vee wx'yz' \vee w'x'yz \vee w'x'yz'$ because, making $x = 0$ and $y = 1$, we have

$$\begin{aligned}\Psi &= w \cdot 1 \cdot 1 \cdot z \vee w \cdot 1 \cdot 1 \cdot z' \vee w' \cdot 1 \cdot 1 \cdot z \vee w' \cdot 1 \cdot 1 \cdot z' \\ &= wz \vee wz' \vee w'z \vee w'z' = w(z \vee z') \vee w'(z \vee z') = w \vee w'\end{aligned}$$

In both of the examples of the preceding paragraph, the product term given as α is a prime implicant because shortening either α or β by deleting a literal will cause the term to no longer imply the expression.

Theorem 5. *A function f is equivalent to a sum-of-products expression which is the sum of all the prime implicants of f.*

Proof. Certainly, f is implied by each prime implicant, and therefore by their sum. Then, any assignment of values of the variables of f which makes f true will make some product term β which implies f true. Hence, by simply shortening β until it is a prime implicant, we can generate sufficient prime implicants so that their sum will take the value 1 whenever f has value 1.

Definition. An expression Φ in sum-of-products form equivalent to a function f is *irredundant* when (*i*) each product term of Φ is a prime implicant and (*ii*) no term in Φ can be deleted without destroying the relation $\Phi = f$.

Theorem 6. *Any expression Φ which is minimal according to the criterion for fewest literals or according to the criterion of fewest terms is irredundant.*

Proof. If we could strike a literal from one of the product terms from Φ, then either the resulting expression would not be equivalent to Φ, or Φ would not be irredundant. If we could strike a variable or term from Φ and maintain equivalence, certainly the original expression would not be minimal since the shortened expression would cause it to fail to satisfy (*i*) or (*ii*).

To find an irredundant expression for a function f, generate a complete set of prime implicants, and then select from these prime implicants a subset whose sum is irredundant.

Identifying irredundant forms is not as easy as it might appear at first; consider $\psi = ab' \vee ac' \vee b'c$. This is not irredundant because, although each term is a prime implicant, the term ab' can be deleted without changing the function represented by ψ. (We could test to see if a given term is superfluous by seeing if the term implies the expression with the term removed, for if $\alpha \vee \Phi = \Phi$, then α implies Φ and α can be deleted.)

Theorem 6 reduces our problem to the following: (*i*) to generate systematically the set of all prime implicants, (*ii*) to generate from these all irredundant expressions, (*iii*) to choose from these particular expression(s) all which satisfy the chosen criterion for minimality.

First we present a technique for generating the prime implicants.

Definition. A product term α *subsumes* a product term β when every literal in β appears in α. For instance, abc subsumes ab, $ac'd'$ subsumes ad', ab subsumes ab, etc.

Definition. A product term α is a *completion of β with respect to Φ* when α subsumes β and each variable in Φ appears in α.

Thus, $abc'd'$, $abcd'$, $abc'd$, and $abcd$ are all completions of ab with respect to $\Phi = abc' + d + cd + ab$.

Theorem 7. *If Φ is a canonical sum-of-products expression for a function f, then a product term α implies f if all completions of α are in Φ.*

If some term β is a completion of α and does not appear in Φ, then we assign values to the variables in Φ such that $\beta = 1$; then $\alpha = 1$, and $\Phi = 0$ so α does not imply Φ.

Thus, for ac' to imply a function f in variables a, b, c, and d, the terms $ab'c'd'$, $ab'c'd$, $abc'd'$, and $abc'd$ must all appear in the canonical disjunctive form Φ for f.

6-7. PROCEDURE FOR DERIVING PRIME IMPLICANTS

A systematic technique for deriving all the prime implicants of a given Boolean function f will now be given. Starting with a canonical sum-of-products form Φ for the function f, apply the rule $\gamma x \vee \gamma x' = \gamma$, where γ is a product term and x is a variable in all possible ways to the terms of Φ. After making all possible applications of this rule, discard all subsuming terms since these cannot be prime implicants.

Example 6. Let us apply the above technique to the canonical sum-of-products expression $\Phi = abcd \vee ab'cd' \vee ab'cd \vee abcd' \vee a'b'cd'$.

Applying the rule to the first and third terms gives acd; similary, the first and fourth terms yield abc, the second and third terms yield $ab'c$, etc., until we form a list of implicants

$abcd$	acd	ac
$ab'cd'$	abc	
$ab'cd$	$ab'c$	
$abcd'$	$b'cd'$	
$a'b'cd'$	acd'	

The only two terms in this list which are not subsumed by some other term are ac and $b'cd'$. These are the two prime implicants, and it happens that $ac + b'cd'$ is also a minimal expression for Φ, although there is generally a second step in selecting prime implicants.

The above procedure is due to Quine; McCluskey has organized this procedure and made it more efficient as follows:

Step 1. If the variables are ordered, represent the literals in product terms by the binary digits 0 and 1, and represent the absence of literals by dashes. For instance, represent the expression $\Phi = abc \vee ac' \vee ad'$ by $111- \vee 1-0- \vee 1--0$, and the expression $\psi = ad' \vee a'c \vee abcd$ by $1--0 \vee 0-1- \vee 1111$.

Step 2. Next, partition the binary n-tuples representing the meet terms into equivalence classes, where each equivalence class contains only n-tuples having the same number of 1's, and arrange these equivalence classes into a list so that the number of 1's increases as the list progresses downward. This gives a table as follows:

$$\Phi = abc'd \vee abc'd \vee ab'c'd' \vee ab'c'd' \vee a'b'cd' \vee a'b'c'd \vee a'b'c'd'$$

0000
0001
0010
1000
1100
1001
1101

Step 3. It is now only necessary to attempt matches (application of $\gamma x \vee \gamma x' = \gamma$) in adjacent equivalence classes, and if we partition the terms which result into classes where all the members of a given class were derived from matches from the same two classes, then again only adjacent classes need be considered. We further check, using a $\checkmark$, each term when it is matched with another term (continuing to use checked terms for matches when possible, however). This technique indicates terms which subsume others and are therefore not prime implicants; it is known as the *Quine-McCluskey technique.* Thus, we have

✓0000	✓000−	−00−	The prime implicants are
✓0001	00−0	1−0−	$a'b'd'$, $b'c'$, and ac'
✓0010	✓−000		
✓1000	✓−001		
✓1100	✓1−00		
✓1001	✓100−		
✓1101	✓110−		
	✓1−01		

We have now found all prime implicants. Notice that the rule for matching only applies when the dashes in two terms are the same in number and in the same position and the binary digits representing literals differ in only one position. This greatly facilitates the matching process.

Step 4. Finally, choose a subset from the set of all prime implicants for an expression Φ, whose join will be implied by Φ and which will satisfy a chosen criterion for minimality.

This problem can be examined by means of *prime implicant tables* such as those shown in Fig. 6-15.

	$a'b'c'd'$	$a'b'c'd$	$a'b'cd'$	$ab'c'd'$	$abc'd'$	$ab'c'd$	$abc'd$
$a'b'd'$	×		×				
$b'c'$	×	×		×		×	
ac'				×	×	×	×

	0000	0001	0010	1000	1100	1001	1101
00–0	×		×				
–00–	×	×		×		×	
1–0–				×	×	×	×

Fig. 6-15. Prime implicant tables for Example 6.

The rule for forming a prime implicant table is straightforward. One lists the terms of the canonical expansion Φ as column headings and the prime implicants as row headings, and places crosses at each intersection where the term of Φ subsumes the prime implicant at the intersection.

Definition. A term α *covers* a term β when each literal in α is in β.

Notice that "covers" is "in the opposite direction" to "subsumes." If α subsumes β, then β covers α.

In order for our new expression ψ (which will consist of a sum of prime implicants) to be implied by Φ, every term in Φ must be covered by some term in ψ. Otherwise, we could assign values to some term γ not covered which would make γ (and therefore Φ) have value 1 while ψ had value 0.

From the viewpoint of the prime implicant table, this observation means that a subset of the column headings must be selected so that at least one x will be in each column.

This fact makes the following rule for solution categorical: if a single x lies in some column, the prime implicant lying in the same row *must* be selected because it is the only term in that column covering the term of the canonical expansion. The set of all such terms which must be selected in this way is called the *core* for the expansion.

In Fig. 6-15 this forces the selection of all three prime implicants because $a'b'c'd'$ is covered only by $b'c'$, $a'b'cd$ is covered only by $a'b'd'$, and $abc'd'$ and $abc'd$ are covered only by ac'. All these prime implicants therefore comprise the core, and the minimal expression by any criterion is $a'b'd' \vee b'c' \vee ac'$.

In most cases removal of the core will still leave terms in Φ to be covered. In some cases removal of the core and the canonical expression terms covered by the core causes some prime implicants to cover none of the remaining terms. Such terms imply the core, are *absolutely eliminable,* and will appear in no minimal expression. An example of this is shown in Fig. 6-16, where the removal of $a'b$ and $b'c'd'$, the core, and then the canonical expansion terms $a'bc'd'$, $a'bc'd$, $a'bcd'$, and $a'b'c'd'$ leads to the table of Fig. 6-16*b*. Notice there that $a'c'd'$ covers no remaining term, and so there is no point in considering this term further (it is absolutely eliminable).

		0000	0100	1000	0011	0101	0110	1010	1011	0111
$a'b$	01−−		×			×	×			×
$a'c'd'$	0−00	×	×							
$b'c'd'$	−000	×		×						
$ab'c$	101−							×	×	
$b'cd$	−011				×				×	
$a'cd$	0−11				×					×

(*a*)

	0011	1010	1011
0−00			
101−		×	×
−011	×		×
0−11	×		

(*b*)

Fig. 6.16.

Figure 6-16*b* indicates the remaining choices which can be made; they require that $ab'c$ be chosen and that either $b'cd$ or $a'cd$ be chosen. There are therefore two irredundant expressions

$$a'b \vee b'c'd' \vee ab'c \vee b'cd \qquad \text{and} \qquad a'b \vee b'c'd' \vee ab'c \vee a'cd$$

Both of these are minimal by either criterion I or criterion II.

There is a nice technique for systematically listing all redundant covers of a table of this sort. Consider Fig. 17 where we have simply

	a	*b*	*c*	*d*	*e*
A	×		×	×	
B		×	×		
C	×		×		×
D		×		×	

Fig. 6-17. Sample prime-implicant table.

listed the prime implicants as A, B, C, and D and the canonical expression terms as a, b, c, d, and e. Using Boolean algebra we can describe this table by saying that a can be covered by $A \vee C$ and b can be covered by $B \vee D$, and c can be covered by $A \vee B \vee C$, etc. Written out this is

$$(A \vee C)(B \vee D)(A \vee B \vee C)(A \vee B)(B \vee C)(B \vee D)$$

If this expression is multiplied out and the simple rule for subsuming used which is $\alpha\beta \vee \alpha = \alpha$, where α and β are product terms, then each product term in the resulting expression will cover the table. In Fig. 6-16, we have, after multiplication and the elimination of subsuming terms

$$AB \vee CD$$

This indicates that "A and B" or "B and D" or "A and C" can be selected and a cover will result.

EXERCISES B

1. Find a minimal sum-of-products expression for each of the following:
   (a) $f(a,b,c) = \sum (0,3,5,6)$, where $0 = a'b'c'$, $3 = a'bc$, $5 = ab'c$,
   (b) $f(a,b,c) = \sum (0,3,4,6)$
   (c) $f(a,b,c,d) = \sum (0,1,2,3,4,6,7,8,9,11,15)$
   (d) $f(a,b,c,d,e) = \sum (0,1,3,7,8,9,10,11,14,15,16,18,22,23,24,25,26,30)$
2. Design a two-output gating network with outputs Z_1 and Z_2 using AND gates and OR gates only, for the two Boolean algebra expressions which follow. Assume that the signals A, B, and C are from flip-flop outputs so that A, A', B, B', C, and C' are available. Share gates if and when possible. Use as few gates as possible regardless of the number of inputs to each gate. Draw a block diagram for the network.

$$Z_1 = ABC + A'BC + AB'C + A'B'C + ABC'$$
$$Z_2 = AB'C + AB'C' + A'B'C + A'B'C' + ABC'$$

3. (a) Select a set of α_j which will cover the β_i in the following prime implicant table:

	β_1	β_2	β_3	β_4	β_5	β_6	β_7	β_8
α_1	×	×	×	×	×			
α_2	×					×		
α_3		×	×		×		×	×
α_4			×	×				×
α_5				×		×	×	

   (b) Explain and prove the following rule for solving prime implicant tables: if a column C_i contains crosses in each row in which another column C_j contains crosses, then column C_i can be eliminated.
4. Design two-level AND-to-OR gate networks for the following two Boolean polynomial expressions. Minimize the total number of gates used, sharing gates

between expressions when possible and considering the network as a single two-output network rather than as two separate single-output networks.

$$f_1(w,x,y,z) = wx'y'z' + wx'y'z + wx'yz' + wx'yz + wxy'z' + w'zy'z$$
$$f_2(w,x,y,z) = wx'yz' + wx'yz + wxyz' + wxyz + wxy'z' + w'xy'z$$

5. (a) Determine a minimal product-of-sums expression for the Boolean expression

$$f = b'd \vee abd \vee a'bc'd \vee a'bcd'$$

   (b) Draw a minimal two-level gating network for f using only NOR gates.

6. (a) Using a prime implicant technique, derive a minimal sum-of-products and product-of-sums expression for

$$f(a,b,c,d) = \sum (1,12,13,14,15)$$

   (b) Draw the following two-level gating networks for the preceding f, using only

   (1) AND-to-OR (3) NAND-to-NAND
   (2) OR-to-AND (4) NOR-to-NOR

7. Consider the design of the following *selected network:* a_1, a_2, and a_3 are a set of control or *selection* variables, with the property that $a_ia_j = 0$ $(i \neq j)$ and $a_1 + a_2 + a_3 = 1$; x_1, x_2, and x_3 are a set of variables, and it is desired to realize the Boolean function

$$f(a_1,a_2,a_3,x_1,x_2,x_3) = x_i \qquad \text{if } a_i = 1$$

   that is, the function f which has the value x_i when the corresponding selection variable a_i is "on." Design two-level (1) AND-to-OR and (2) OR-to-AND logical networks for the above.

★8. Let $\mathfrak{A}$ be an algorithm which produces a minimal sum-of-products expression f for arbitrary Boolean function F. Schematically, let $\mathfrak{A}(F) = f$. Let $\mathfrak{A}'$ be the following composite algorithm: (1) write F', (2) produce $\mathfrak{A}(F')$, (3) and using de Morgan's laws, write $[\mathfrak{A}(F')]' = \mathfrak{A}'(F) = g$. Prove that g is a minimal product-of-sums expression equivalent to F.

★9. Using only AND gates, OR gates, and inverters, determine an upper bound to the number of components required to realize a general Boolean function (a) of two variables, (b) of three variables, and (c) of r variables.

★6-8. CONSENSUS TAKING

The previous section has described the problem of minimizing a Boolean algebra expression, and outlined a technique whereby any expression can be systematically reduced. There are refinements to this technique for forming and selecting prime implicants, and several further improvements in the basic procedure are discussed in the exercises.

Definition. The *consensus* of two product terms α and β is defined if exactly one variable in α appears complemented in β. The consensus of α and β, written $\sigma(\alpha,\beta)$, is formed by deleting the variable which is opposed in α and β from both α and β, and then forming the product of the remaining literals while striking out duplications of literals.

Thus, the consensus of abc and $ab'c'$, xyz and $x'y'$, or xyz and xz is not defined. The consensus of abc and $bc'd$ is abd, the consensus of xyz and $xy'z$ is xz, likewise, $\sigma(x'y'z, x'yw) = wx'z$, etc.

Theorem 8. *If α and β are terms of an expression Φ, then* $\Phi \vee \sigma(\alpha,\beta) = \Phi$.

Proof. Notice that $\sigma(\alpha,\beta)$ implies $a \vee \beta$ because letting the variable that appears opposed in α and β be written as x, α can be written as πx and β as $\gamma x'$; hence we have $\sigma(\alpha,\beta) = \pi\gamma$, and making $\pi\gamma = 1$ gives

$$\alpha \vee \beta = (1 \cdot x) \vee (1 \cdot x') = 1$$

which proves the implication.

Theorem 9. *Given an expression Φ in sum-of-products form, a complete set of prime implicants for Φ can be found by repeatedly performing the two following operations on Φ as long as possible:*

(*i*) *If a clause α subsumes a clause β, drop α.*
(*ii*) *If for some two terms λ and π a consensus is defined and this consensus subsumes no terms already present, add $\sigma(\lambda,\pi)$ to the expression.*

Example 7. Consider $yz \vee y'z' \vee x'u \vee wyz \vee xwz' \vee uwxy$. The consensus of wyz and xwz' is wxy, which is subsumed by $uwxy$. This gives $yz \vee y'z \vee x'v \vee wyz \vee xwz' \vee wxy$, and thus

$$\sigma(x'u,xwz') = uwz' \qquad \text{and} \qquad \sigma(x'u,wxy) = uwy$$

giving

$$yz \vee y'z \vee x'u \vee wyz \vee uwz' \vee uwy \vee xwz' \vee wxy$$

which is a complete list of prime implicants for the expression.

Proof. There are three things to be proved:

(1) If some prime implicant is not among the terms of expression Φ, a consensus term can be found and added to Φ.
(2) If some term α in Φ is not a prime implicant, then either
   (a) α subsumes a term prime implicant term already in Φ or
   (b) a consensus term not present can be formed and added [i.e., either (a) or (b) above can be formed].
(3) Steps (*i*) and (*ii*) in Theorem 9 cannot be performed indefinitely.

The hard part is to prove (1). Let us consider an expression Φ in sum-of-products form for a function f. Suppose a prime implicant γ exists which differs from all terms in Φ. We must show it is possible to form a consensus term which can be added to the set of terms comprising Φ. Certainly γ, which is a prime implicant, subsumes no terms in Φ, and therefore at least one term (possibly γ itself) exists which subsumes γ and subsumes no term of Φ. Now, by adjoining literals of Φ to γ, we can form a term δ of maximum length which subsumes γ *and* subsumes no term of Φ. Notice δ cannot be subsumed by γ since all completions of a prime implicant of a function f imply f and therefore subsume some term of any normal expression for f.

Let us select a variable not in δ, say x; then we form δx and $\delta x'$, both of which will subsume γ and both of which will subsume some term of Φ, as δ was picked to be the maximum length. Let the two terms of Φ which are subsumed by δx and $\delta x'$ be named α and β. Then α and β will contain x and x' as literals and do not otherwise differ, in that no variable Φ appears complemented in one term and uncomplemented in the other (also, α and β are not x and x', because then $\alpha \vee \beta$ would equal 1, as would f). We therefore form the consensus of α and β, and this consensus will be subsumed by δ, which is subsumed by no term in Φ.

The proof of (2) is straightforward. A term α of Φ, not a prime implicant, must subsume some prime implicant ξ; if ξ is a term of Φ, then ξ can be immediately marked out of Φ; if ξ is not a prime implicant, then (1) applies, and a consensus term can be formed.

As far as part (3) is concerned, there are only a finite number of different meet terms which can be formed in $2n$ literals with no literal repeated. If a term α appears in Φ, then α can be dropped only if some term β is subsumed by α, β can be dropped only if some term γ is subsumed by it, etc., and all terms in any sequence of this type will be subsumed by α so α cannot reappear. Also, another α cannot be added as α would subsume it. Further, consensus taking can generate no terms in other than the $2n$ literals, and so the process must terminate after a finite number of steps.

6-9. FLIP–FLOPS

The present section lays the basis for realizing physically any finite-state automaton (as defined in Chap. 3) by a suitable combination of gating networks and simple bistable memory elements. The next section will show how to do this in detail.

Figure 6-18*a* shows a block diagram for a particular type of electronic circuit called a *flip-flop*. The simplest flip-flop has a single input line K and two output lines, labeled L and L' in Fig. 6-18*a*. The input value from K at time i, which we call $K(i)$, is stored in the flip-flop B until time $i + 1$, at which time (after a short delay), the flip-flop assumes the value $K(i + 1)$ until time $i + 2$, and so on. Thus the "state" of the flip-flop B at time $i + 1$ is that of K at time i, as indicated in Fig. 6-18*b*.

Thus, such a flip-flop simply "remembers" its input from the previous pulse, changing its state only when forced to by a new and different clock pulse at some later time.

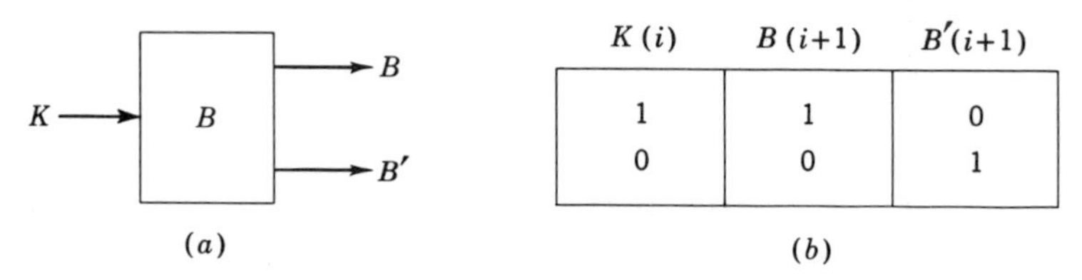

Fig. 6-18. Function of a flip-flop.

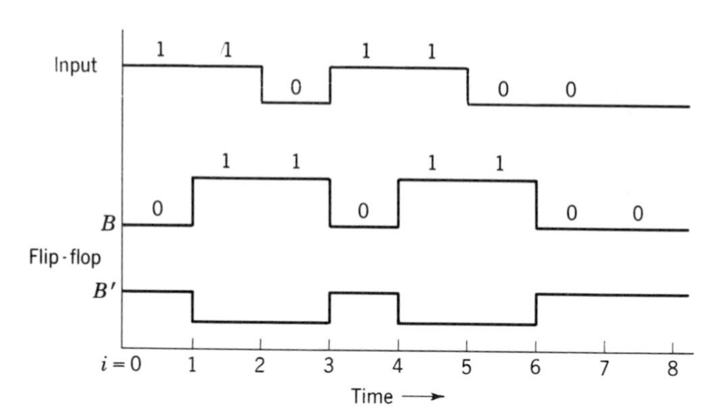

Fig. 6-19. Flip-flop operation.

There are two opposite outputs from the flip-flop B. That which we have labeled B gives at time (clock pulse) $i + 1$ the signal $K(i)$, and that labeled B' gives at time $i + 1$ the opposite (complementary) signal.

Figure 6-19 shows a possible input-output sequence for a flip-flop.

Flip-flops are the most used active memory elements in present-day digital computers. There are many input arrangements for flip-flops, and the input scheme given here is about the simplest of these. Interested readers may find the other input configurations in the literature.[1]

The use of flip-flops in sequential network design is illustrated in Fig. 6-20. This consists of a gating network to which are connected p

[1] See, for example, T. C. Bartee, "Digital Computer Fundamentals," 2d ed., McGraw-Hill, 1966.

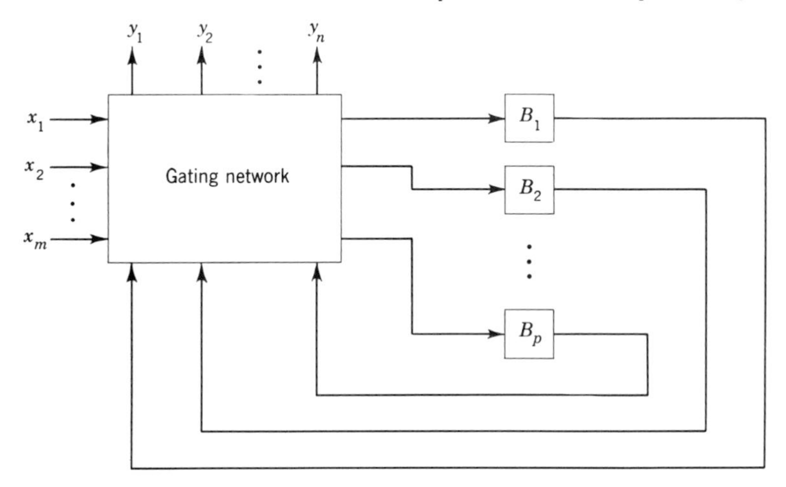

Fig. 6-20. Logical arrangement for sequential machines.

flip-flops $B_1, \ldots, B_p$ by input *and* output leads, as well as m other (binary) inputs $x_1, \ldots, x_m$ and n outputs $y_1, \ldots, y_n$.

The outputs from the flip-flops then go into the gating network, so the gating network has $m + p$ inputs, m from the input signals and p from the flip-flops. The outputs $y_1, \ldots, y_n$ are also taken from the gating network, so the network has $n + p$ outputs in all.

6-10. SEQUENTIAL MACHINE DESIGN

Electronic networks of the type depicted in Fig. 6-20 can be designed to realize physically the state function of any finite-state machine. More precisely, we have the following theorem.

Theorem 10. *The state function* ν: $A \times S \rightarrow S$ *of any finite-state machine can be realized by a sequential network consisting of inverters, AND gates, OR gates, and flip-flops.*

Proof. Let us assume a state table with q states (or a state graph with q states). This defines a function δ from $A \times S$ into S.

We would now like to represent the internal states as binary p-tuples. If the sequential machine is to have q internal states, we must have $2^p \geqq q$, where p is the number of places in a p-tuple or the number of flip-flops. Values can be given to each of the 2^p different p-tuples by using the binary number system and letting the value of a given set of B_i be $\sum_{i=0}^{q-1} B_i 2^i$. We can then encode q, letting the subscript on the s_i equal this value. For instance, a machine with five internal states $s_0, s_1, \ldots, s_4$ can be encoded with 3-tuples as follows:

	B_0	B_1	B_2
s_0	0	0	0
s_1	1	0	0
s_2	0	1	0
s_3	1	1	0
s_4	0	0	1

If need be, encode X and Y in a similar manner. This is only necessary when input and output symbols are defined abstractly. In most practical cases X and Y are given in terms of binary values; the following concrete example is typical.

Present state	*Next state* 0	1	*Output* 0	1
s_0	s_0	s_1	1	0
s_1	s_2	s_1	0	1
s_2	s_0	s_1	1	1

(*a*)

	B_0	B_1
s_0	0	0
s_1	1	0
s_2	0	1

(*b*)

Fig. 6-21. State table to be realized.

Example 8. Let the state table be that given in Fig. 6-21*a*. Let the representation be as in Fig. 6-21*b*. Let the representation be as in Fig. 6-21*b*, so that s_0 is represented by $B_0 = 0$, $B_1 = 0$; s_1 by $B_0 = 1$, $B_1 = 0$; etc. The state table defines a partial function δ, as follows:

X	B_0	B_1		Y	B_0^*	B_1^*
0	0	0	$\mapsto$	1	0	0
0	1	0	$\mapsto$	0	0	1
0	1	1	$\mapsto$	1	0	0
1	0	0	$\mapsto$	0	1	0
1	1	1	$\mapsto$	1	1	0
1	0	1	$\mapsto$	1	1	0

The above function can be implemented physically as three single-output combinational networks, using the techniques described in previous sections. The Boolean equations are

$$\begin{aligned} Y &= X'B_0'B_1' \vee X'B_0B_1 \vee XB_0'B_1 \\ B_0^* &= XB_0'B_1' \vee XB_0B_1 \vee XB_0'B_1 \\ B_1^* &= X'B_0B_1' \end{aligned}$$

Here B_0^* and B_1^* are the next-state values to be taken by the flip-flops which store the values of the states. If we substitute the gating network described above in the box marked "Gating network" in Fig. 6-20, a machine will result which will physically realize the state table in Fig. 6-21.

Example 9. Consider the problem of designing a sequential machine which will add two binary nonnegative integers C and D presented with the least significant digits first. Thus, if $C(0) = 0$, $C(1) = 1$, $C(2) = 1$, and $C(3) = 0$ [$i > 3$, $C(i) = 0$], then C represents the integer 6. Our output will be called Y, and we would like $Y = C + D$. Two internal states are needed (to remember the "carry" from digit to digit). A state table is shown in Fig. 6-22.

	Next state CD				*Output* CD			
$C_i d_i =$	00	01	10	11	00	01	10	11
s_0	s_0	s_0	s_0	s_1	0	1	1	0
s_1	s_0	s_1	s_1	s_1	1	0	0	1

Fig. 6-22. State table for binary adders.

Our functions $\nu: C \times D \times S \to S$ and $\zeta: C \times D \times S \to Z$ can be easily read off from Fig. 6-22; the result is as follows:

C	D	S		Z	S
0	0	s_0	$\mapsto$	0	s_0
0	0	s_1	$\mapsto$	1	s_0
0	1	s_0	$\mapsto$	1	s_0
0	1	s_1	$\mapsto$	0	s_1
1	0	s_0	$\mapsto$	1	s_0
1	0	s_1	$\mapsto$	0	s_1
1	1	s_0	$\mapsto$	0	s_1
1	1	s_1	$\mapsto$	1	s_1

To convert these functions to Boolean expressions in sum-of-products form, we let $s_0 = 0$ and $s_1 = 1$ and then use a flip-flop B to store the internal state. This gives

$$Y = C'D'B \vee C'DB' \vee CD'B' \vee CDB$$
$$B = CDB' \vee CD'B \vee C'DB \vee CDB$$

A block diagram of a physical network for a finite-state sequential machine which will add two binary numbers is shown in Fig. 6-23. This

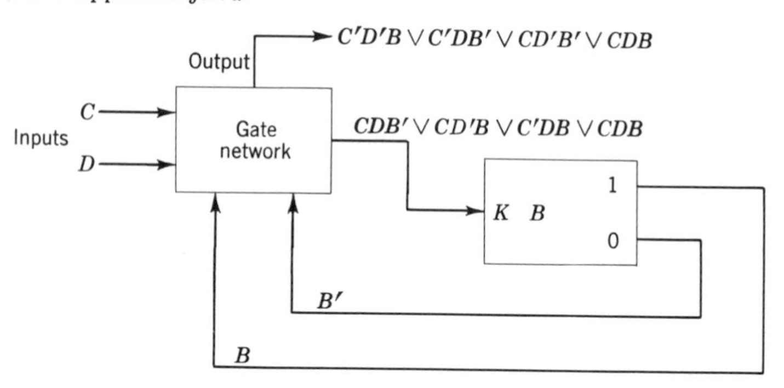

Fig. 6-23. Serial-binary adder.

is called a *serial-binary adder.* The "carry" flip-flop B should be started in its 0 state.

The above has assumed an arbitrary coding of the values from the s_i into the B flip-flop. The optimization problem of coding so that the combination network which results is minimal is an unsolved problem in switching theory.

What has been presented is an algorithm whereby an arbitrary state table or state graph can be converted into a design for a real computer. The deterministic finite-state automata of Chap. 3 are thus constructible, and adding tape reading and writing devices to our sequential machine will give us the finite-state machines previously discussed.

By simply adding output lines which control tape motion, an arbitrary Turing machine can also be constructed, provided that one can add segments to the potentially infinite piece of tape at either end, as required.

EXERCISES C

1. (a) Draw the state diagram for a finite-state machine which will retain an output of 0 until the single-input line $X(t)$ has four consecutive input values of 1; that is, $X(t) = 0$ or 1, t is an integral-valued variable $t = 0,1,2, \ldots, n, \ldots$, and $Z(t)$, the output of the machine, is to be a 0 unless $X(k) = X(k+1) = X(k+2) = X(k+3) = 1$ for some k, in which case $Z(k+3)$ should be a 1 as should all $Z(k+3+j)$ where $j \geq 0$. Further, if the machine ever receives two consecutive 0's, that is, if $X(t) = X(t+1) = 0$ for some $t \geq 0$ (even if this pair of 0's precedes a sequence of three 1's), the machine is to never have an output of 1.

   (b) Design the machine using AND gates, OR gates, and $R - S$ flip-flops.

2. (a) Design a sequential machine with two input lines X_1 and X_2 and one output line Z such that the output line will have a 1 only after the sequence (0,1), (1,0), (1,1) appears on the input lines X_1 and X_2. The machine is to be started with $Z = 0$ and $(X_1,X_2) = (0,0)$. Use flip-flops, AND gates, and OR gates. Once the Z output line takes the value 1, it is to remain a 1 until the machine

is restarted. Draw a state table for the machine, minimize the number of states, and then assign values to the internal states, finally deriving the expressions for the combinational network.

3. Design a sequential circuit, using two-level AND-to-OR logic and flip-flops, that gives an output of 1 if the circuit receives a 0 followed by an odd number of consecutive 1's finally followed by a 0; that is, the circuit *recognizes* or *accepts* the set

$$S = \{01^{2n+1}0 \mid n \geq 0\}$$

4. Find minimal expressions for the following:
   (a) $f(a,b,c) = \sum (0,3,5)$
   (b) $f(a,b,c,d) = \sum (0,1,3,5,6,9,10)$
   (c) $f(a,b,c,d,e) = \sum (0,2,4,6,7,8,10,12,14,30)$
5. Is the binary relation of compatibility between product terms in a finite mixed number of variables (a) reflexive? (b) antireflexive? (c) symmetric? (d) antisymmetric? (e) transitive?
6. Design a sequential circuit using only AND gates, OR gates, and flip-flops, which realizes the following state diagram:

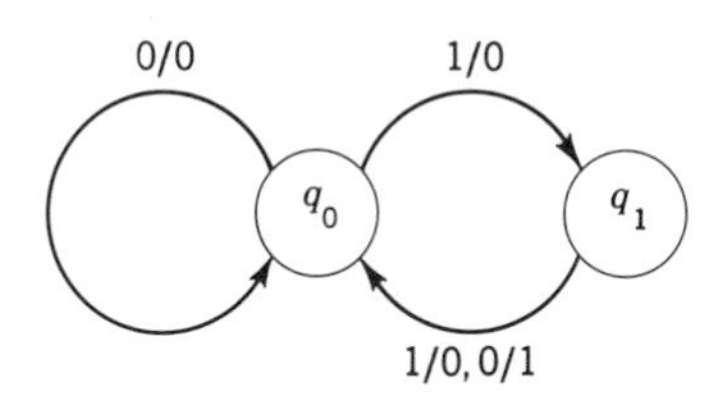

7. (a) Design a sequential circuit using only NAND gates and flip-flops, which gives an output of 1 precisely when the circuit receives an even number of consecutive 1's followed by a single 0; that is, the machine *accepts* or *recognizes* the set

$$S = \{1^{2n}0 \mid n \in \mathbf{N}\} \qquad \{\overbrace{11 \cdots 11}^{n \text{ times}}\ 0\}$$

   (b) Explain why there is no finite-state machine that accepts the set $S = \{1^n0^n \mid n \in N\}$.

Monoids and Groups

7-1. BINARY ALGEBRAS

This chapter will be mainly devoted to groups. In much the same way as Boolean algebra describes the behavior of *sets* under intersection, union, and complementation, groups describe the combination of *bijections* under composition. Such bijections include especially the *symmetries* of geometrical and algebraic configurations.

Before studying groups per se, we shall make a preliminary analysis of binary operations in general.

Definition. A *binary operation* on a set S is a function $\beta: S^2 \to S$. In other words, a binary operation is a rule which assigns to any two elements x and y a "value" $\beta(x,y)$, the effect of performing the specified operation β.

It is often helpful to think of the operation as multiplication and to write $x \circ y$ or $x \beta y$ or just xy instead of $\beta(x,y)$. However, one must remember that binary operations in general need be neither commutative nor associative. Both $xy \neq yx$ and $x(yz) \neq (xy)z$ are possible until proved impossible; that is, we may have $\beta(x,y) \neq \beta(y,x)$ and

$$\beta(x,\beta(y,z)) \neq \beta(\beta(x,y),z)$$

Definition. A *binary algebra* $[S,\beta]$ is a set S with a binary operation $\beta: S^2 \to S$.

When β is associative, the binary algebra $[S,\beta]$ is called a *semigroup*. An element $1_l \in S$ such that $\beta(1_l,x) = x$ for all $x \in S$ is called a *left identity;* an element 1_r such that $\beta(x,1_r) = x1_r = x$ for all $x \in S$ is called a *right identity.* A semigroup which contains a special identity element 1 such that $x1 = 1x = x$ for all $x \in S$ is called a *monoid.* This gives the following definition.

Definition. A *monoid* $[S,\beta]$ is a set S with a binary operation $\beta(x,y) = xy$ which satisfies

M1. $x(yz) = (xy)z$ for all $x,y,z \in S$. (Associative)
M2. For some $1 \in S$, $1x = x1 = x$ for all $x \in S$.

Note that M2 only asserts the existence of an identity 1; it is unique since if 1 and 1′ are both identity elements, then $1 = 1 \cdot 1' = 1'$, where the first equation holds because 1′ is an identity, and the second holds because 1 is an identity. If this were not true, we would postulate the presence of a zero-ary operation "take 1," change M2 to $1x = x1$ for all $x \in S$, and write $[S,\beta,1]$ instead of $[S,\beta]$.

Example 1. From any Boolean algebra $[A,\wedge,\vee,',0,I]$, one can construct two (dual[1]) monoids $[A,\wedge]$ and $[A,\vee]$, with identities 0 and I, respectively.

Example 2. For any set X, the set X^X of all functions $f: X \to X$ (unary operations of X) is a monoid $[X^X,\circ]$ under left composition. It is also a monoid under right composition, since for all $f,g,h \in X^X$

$$f \diamondsuit (g \diamondsuit h) = (h \circ g) \circ f = h \circ (g \circ f) = (f \diamondsuit g) \diamondsuit h \qquad (1)$$

$$1_X \diamondsuit f = f \circ 1_X = f = 1_X \circ f = f \diamondsuit 1_X \qquad \text{all } f \in X^X \qquad (2)$$

For instance, if $X = \mathbf{2} = \{0,1\}$, let the four possible functions $f \in \mathbf{2}^{\mathbf{2}}$ be labeled as follows:

$1: 0 \mapsto 0$	$r: 0 \mapsto 1$	$z_0: 0 \mapsto 0$	$z_1: 0 \mapsto 1$
$1 \mapsto 1$	$1 \mapsto 0$	$1 \mapsto 0$	$1 \mapsto 1$

[1] See Secs. 2-5 and 5-2.

Then the monoids $[\mathbf{2}^2,\circ]$ and $[\mathbf{2}^2,\diamondsuit]$ have the following multiplication tables:

$\circ$	1	r	z_0	z_1
1	1	r	z_0	z_1
r	r	1	z_1	z_0
z_0	z_0	z_0	z_0	z_0
z_1	z_1	z_1	z_1	z_1

$\diamondsuit$	1	r	z_0	z_1
1	1	r	z_0	z_1
r	r	1	z_0	z_1
z_0	z_0	z_1	z_0	z_1
z_1	z_1	z_0	z_0	z_1

Note that the monoids of Example 1 are *commutative* (since $x \wedge y = y \wedge x$ and $x \vee y = y \vee x$ for all $x,y \in A$), whereas the monoids of Example 2 are noncommutative except in the trivial case that X is a singleton. Referring to Theorem 2 of Chap. 2, we have also the following example.

Example 3. For any set X, the set Rel (X) of all binary relations on X is a (noncommutative) monoid under right composition, with the equality relation as the identity.

It is natural to ask whether a monoid (or any other binary algebra) can have more than one identity. The answer is no; more precisely, we have the following lemma.

Lemma 1. *If a binary algebra has a left identity and a right identity, then each of these is unique, and the two are equal to the same two-sided identity.*

Proof. Let 1_l and 1_r be assumed to be a left and a right identity, respectively. Then $1_l 1_r = 1_r$ since 1_l is a left identity, while $1_l 1_r = 1$ since 1_r is a right identity. Therefore, by transitivity (and symmetry) of equality, $1_r = 1_l$, and both are the same and unique.

Definition. In a binary algebra $[S,\beta]$, a *left zero* is an element 0_l such that $0_l x = 0_l$ for all $x \in S$; a *right zero* is an element 0_r such that $x0_r = 0_r$ for all $x \in S$. A *zero* is an element which is both a left zero and a right zero.

Thus, in Example 1, O is a zero for $[A, \wedge]$, and I is a zero for $[A, \vee]$. In Example 2, the functions z_0 and z_1 act as left zeros for left composition and right zeros for right composition. In Example 3, the null relation whose graph $S(0)$ is the empty set $\varnothing$ is a two-sided zero.

Lemma 2. *If a binary algebra has a left zero 0_l and a right zero 0_r, then each of these is unique, and the two are the same two-sided zero.*

Proof. Since 0_l is a left zero, $0_l0_r = 0_l$, and since 0_r is a right zero, $0_l0_r = 0_r$; hence $0_l = 0_r$. Hence the two are the same two-sided zero. Repeating the argument, every left or right zero must equal this element.

7-2. CYCLIC MONOIDS; SUBMONOIDS

In a monoid M, the powers of any element $a \in M$ are defined for $n \in \mathbf{N}$ by recursion as follows:

$$a^0 = 1,\ a^1 = a,\ a^2 = aa,\ \ldots\ ,a^{n+1} = a^n a \tag{3}$$

A monoid is called *cyclic* when it consists of the powers c^n of some one of its elements c (is "generated" by c).

Theorem 1. *In a monoid, $a^m a^n = a^{m+n}$ for all $m,n \in \mathbf{N}$.*

The proof goes for each fixed $m \in \mathbf{N}$ by induction on n. Let $P_m(n)$ be the proposition that $a^m a^n = a^{m+n}$. Then $P_m(0)$ holds trivially, since $a^m a^0 = a^m 1 = a^m$. Likewise, granted that $P_m(n)$ is true, we have successively

$$a^m a^{n+1} = a^m(a^n a) = (a^m a^n)a = a^{m+n}a = a^{m+n+1}$$

where the four equations follow from (3), associativity, the truth of $P_m(n)$, and (3), respectively.

Corollary. *Any cyclic monoid is commutative.*

Example 4. Let f: $\mathbf{4} \to \mathbf{4}$ be the function whose assignments are indicated in Fig. 7-1*a*. Then the powers $f^0 = 1, f^1 = f, f^2 = f \circ f$, and $f^3 = f^2 \circ f$ of f are all different, but $f^4 = f$. Hence the cyclic monoid of powers of f has the multiplication table of Fig. 7-1*b* (see also Sec. 2-6).

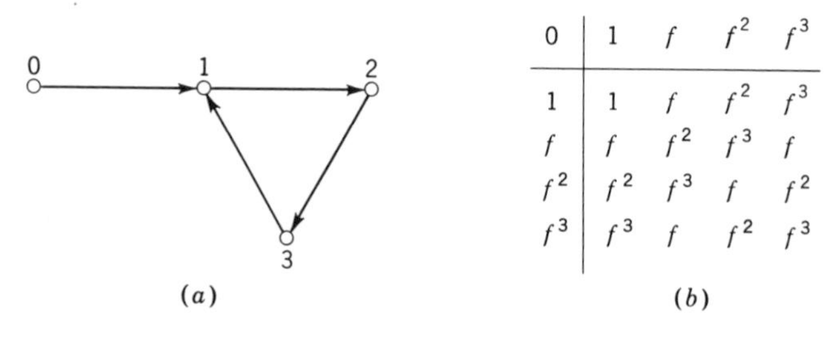

$\circ$	1	f	f^2	f^3
1	1	f	f^2	f^3
f	f	f^2	f^3	f
f^2	f^2	f^3	f	f^2
f^3	f^3	f	f^2	f^3

Fig. 7-1. Example of cyclic monoid.

Example 5. Under addition, the nonnegative integers form a cyclic monoid $\mathbf{N}$ with identity 0 generated by 1.

Although the sudden switch from a multiplicative to an additive notation may come as a shock to the reader, it should be obvious upon reflection that $+(x,y) = x + y$ is just as good a binary operation (function) from $\mathbf{N} \times \mathbf{N}$ to $\mathbf{N}$ as $\cdot\,(x,y) = xy$. The latter, incidentally, defines a noncyclic monoid $[\mathbf{N}, \cdot\,]$ with identity 1 and zero 0. Clearly $[\mathbf{N},+]$ has the identity 0 and no zero (although $+\infty$, if adjoined, would act like one).

Now let C be any cyclic monoid with generator c. By Theorem 1, multiplication by c (on either side) carries c^r into c^{r+1}. We signify this by defining $f_c\colon C \to C$ as the function $c \mapsto c^{r+1}$. We now consider, as in Sec. 2-7, the action f_c and its powers $f_c^h\colon c^r \to c^{r+h}$ on

$$C = \{f_c^h(1)\} = \{1^c, f_c(1), f_c^2(1), \ldots, f_c^h(1), \ldots\}$$

If the c^r are all different, then $C \cong \mathbf{N}$ by Theorem 1. Otherwise there will be some least $s \in \mathbf{P}$ such that $c^s = c^m$ for some $m < s$ (that is, $C = \{1_C, c, \ldots, c^{s-1}\}$); hence s is the *order* (cardinality) of C. Moreover, using induction on j, one can prove the formula $c^ic^j = c^{\phi(i,j)}$, with

$$\phi(i,j) = i + j - kn \tag{4}$$

where $n = s - m$ and $k \in \mathbf{N}$ is the smallest integer such that $k > (i + j - s)/n$. (Thus, if $i + j < s$, then $k = 0$.) In summary, we have proved the following result.

Lemma. *Every infinite cyclic monoid C is isomorphic with* $[\mathbf{N},+]$. *Every finite cyclic monoid of order s is isomorphic for some nonnegative integers $m < s$ and $n = s - m$, with the binary algebra (monoid)* $[\mathbf{s}^*,+]$, *where $\mathbf{s}^*$ is the set $\{0,1, \ldots, s - 1\}$ with multiplication defined by the rule* (4). *We shall refer to this monoid as $C_{m,n}$, noting that it is characterized up to isomorphism by the integers m and n.*

The proof depends on the results of Sec. 2-7, where it was shown that each $C_{m,n}$ can be represented by a directed graph $G_{m,n}$ consisting of an initial segment of "tail" of length m followed by a cycle or "loop" of length n, as shown in Fig. 7-2. For any $x \in C_{m,n}$, multiplication by the

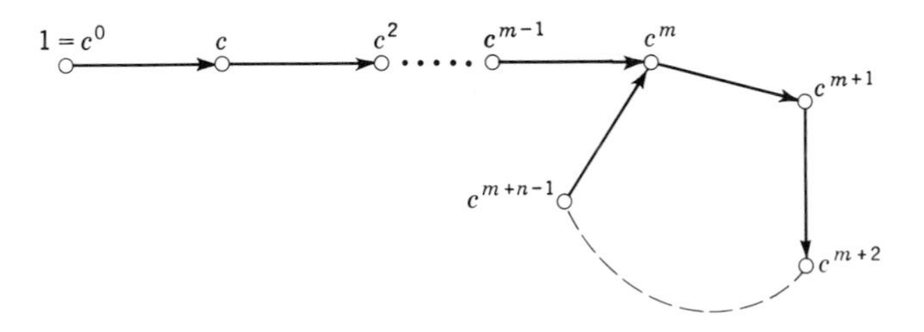

Fig. 7-2.

generator c is represented graphically in $G_{m,n}$ by a one-step transition from the node x, and in general, multiplication by c^l is represented graphically by an l-step transition from the node x. Now it is evident that an l-step transition from any node in the loop of $\vec{G}_{m,n}$ ends up where it started if and only if the loop is traversed an integral number of times. With respect to $C_{m,n}$ this means that

$$c^i c^l = c^i \qquad \text{if and only if } n \mid l (m \leq i < m + n)$$

Hence, choosing for l that one of the n consecutive integers $\{m, m + 1, \ldots, m + n - 1\}$ which is divisible by n and setting $i = l$ above, we have $c^l c^l = c^l$. We conclude the following results.

Theorem 2. *Every finite cyclic monoid $C_{m,n}$ $(m \neq 0)$ contains precisely one idempotent besides the identity. If $m = 0$, then the identity is the only idempotent.*

Corollary 1. *Every finite cyclic semigroup contains at least one idempotent (a semigroup C is cyclic when $C = \{c^r\}$, $r \in \mathbf{P}$).*

A much more interesting result is the following corollary.

Corollary 2. *Let M be a finite monoid. Then for each $x \in M$, x^n is an idempotent for some $n \in \mathbf{P}$.*

Proof. It is easily verified that any $x \in M$ generates a cyclic semigroup; that is, the set $\{x^n\}$ with $n = 1,2,3, \ldots$ constitutes a cyclic semigroup. The desired result now follows immediately from Corollary 1.

Universality. Theorem 1 exhibits the special monoid $[\mathbf{N},+]$ as *universal* in the following sense. For any map of its generator 1 into any other monoid $[M, \cdot\,]$, the assignment $\theta\colon n \mapsto a^n$ carries the binary operation $+$ in $\mathbf{N}$ into the binary operation in M because we have

$$(m + n)\theta = a^{m+n} = a^m \cdot a^n$$

by Theorem 1. As in Sec. 2-6, such mappings are called *morphisms*. We can describe the content of Theorem 1 also by Fig. 7-3. Let $\theta \times \theta$ be the assignment $(m,n) \to (m\theta,n\theta) = (a^m,a^n)$. Then $(\theta \times \theta)\diamondsuit = + \diamondsuit\, \theta$; the diagram is "commutative" in the usual sense that all paths between the same endpoints have the same composites.

The reader can verify that $[\mathbf{P},+]$ has just the same universality property for semigroups. To describe the situation, one says that $[\mathbf{N},+]$

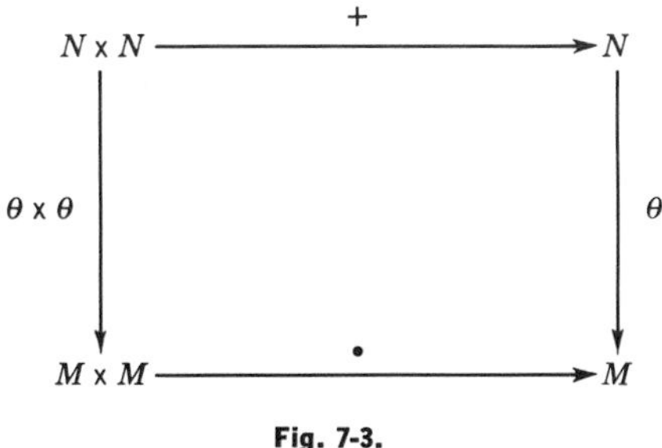

Fig. 7-3.

is the "free" monoid with one generator, and that $[\mathbf{P}, +]$ is the "free" semigroup with one generator.

It is easy to prove that such universal monoids or semigroups are unique up to isomorphism, for if A and B are two such universals with generators a and b, respectively, then the maps $a \mapsto b$ and $b \mapsto a$ of these generators can be extended to morphisms $\alpha: A \to B$ and $\beta: B \to A$ whose composites are $\alpha \diamondsuit \beta = 1_A$ and $\beta \diamondsuit \alpha = 1_B$, respectively. It follows that α and β are isomorphisms, as claimed.

The preceding argument applies to algebraic systems with any number of generators.

7-3. GROUPS

An element a in a monoid M with identity 1 is called *left*-invertible when $xa = 1$ for some $x \in M$, and *right*-invertible when $ay = 1$ for some $y \in M$. An element which is both left- and right-invertible is called *invertible.* We shall now see that the properties of invertibility for functions (i.e., for elements in monoids X^X), which we studied in Sec. 1-5, hold for monoids in general.

Lemma 1. *An invertible element a of a monoid M has a two-sided inverse a^{-1}, such that $aa^{-1} = a^{-1}a = 1$. It has no other left or right inverse.*

Proof. Let b_l and b_r be left and right inverses of a in M, respectively. Then

$$b_l = b_l(ab_r) = (b_la)b_r = b_r$$

Hence we can set $a^{-1} = b_l = b_r$. Similarly, for any left inverse x of a, 1 implies $x = x(aa^{-1}) = (xa)a^{-1} = a^{-1}$, and likewise for right inverses.

Monoids in which every element is invertible are called *groups;* finite groups constitute the most deeply studied class of binary algebras. We shall therefore now define the group concept in full detail.

Definition. A *group* is a monoid in which every element is invertible. Thus a group is an algebra $[G, \cdot\,] = [G, \cdot\,,1,^{-1}]$ in which

G1. $x(yz) = (xy)z$ for all $x,y,z \in G$. (Associative)
G2. $1x = x1 = x$ for all $x \in G$. (Identity)
G3. $xx^{-1} = x^{-1}x = 1$ for all $x \in G$. (Inverses)

Among finite monoids, groups have a very simple alternative characterization.

Theorem 3. *A finite monoid M is a group if and only if* 1 *is its only idempotent.*

Proof. If 1 is the only idempotent of M, then by Corollary 2 of Theorem 2 every $x \in M$ has an inverse $x^{n-1} \in M$ with $x^{n-1}x = xx^{n-1} = 1$; hence M is a group. Conversely, if M is a group with identity 1 and e is idempotent, then for any right inverse e^{-1} of e

$$e = e1 = e(ee^{-1}) = (ee)e^{-1} = ee^{-1} = 1$$

Hence 1 is the only idempotent of M.

We now define a *submonoid* as a subset S of a monoid $M = [M,\beta]$ such that (*i*) $1 \in S$, and (*ii*) $x \in S$ and $y \in X$ imply $xy = \beta(x,y) \in S$. Evidently, every element a of any monoid M generates a cyclic submonoid of M consisting of the powers of a.

Lemma 2. *In any monoid M, the left-invertible and right-invertible elements form submonoids L and R of M.*

Proof. Clearly, $1 \in L$ and $1 \in R$, since $11 = 1$. Again, if $xa = 1$ and $yb = 1$, then

$$(yx)(ab) = y(xa)b = y1b = yb = 1$$

Hence L is a submonoid. A similar discussion applies to R.

The invertibles of M constitute a submonoid $L \cap R$. Moreover the sequence of equations displayed above proves that

$$(ab)^{-1} = b^{-1}a^{-1} \qquad \text{for any invertibles } a,b \in M \tag{5}$$

Similar formulas hold for left-invertibles and for right-invertibles, but with the qualification that one-sided inverses are not in general unique.

Example 6. Let $[\mathbf{Z}_n, \cdot\,]$ be the multiplicative monoid of integers mod n. The submonoid (group) of invertibles consists of the positive integers $k < n$ with k relatively prime to n [in symbols, with $(k,n) = 1$].

In any semigroup S, an important role is played by the set of elements e such that $ee = e$, the *idempotents* of S. If e is an idempotent in a monoid M, then (by induction) $e^n = e$ for all $n \in \mathbf{P}$; hence the cyclic submonoid of M generated by e consists of 1 and e alone. The following result is also easy.

Lemma 3. *In any commutative monoid M, the idempotents form a submonoid.*

Proof. Clearly $1^2 = 1$ in any monoid. Moreover, in a commutative monoid, $e^2 = e$ and $f^2 = f$ together imply

$$(ef)^2 = (ef)(ef) = e(fe)f = e(ef)f = (ee)(ff) = ef$$

More generally, the preceding calculation shows that if any two idempotents e and f of a monoid or semigroup are permutable (that is, if $ef = fe$), then their product is an idempotent.

Definition. In a binary algebra $[S, \cdot\,]$, $a \in S$ is

(1) Right-cancellative when $xa = ya$ implies $x = y$.
(1′) Left-cancellative when $ax = ay$ implies $x = y$.

Lemma 4. *In a monoid M, any right-invertible element is right-cancellative, and any left-invertible element is left-cancellative.*

Lemma 5. *In any finite monoid M*

(2) *Any right-invertible is left-cancellative.*
(2′) *Any left-invertible is right-cancellative.*

Corollary. *In a finite monoid M, any right inverse or left inverse is unique and necessarily a two-sided inverse.*

The proofs are left to the reader.

We now present an ALGOL program which uses the result of Theorem 3. The 50×50 array T contains the multiplication table for a monoid $[M,\mu]$, where $M = \{1,2, \ldots ,N\} (N \leq 50)$, $\mu: M \times M \to M$: $(k,j) \to T[i,j]$, and $1 \in M$ satisfies $T[1,j] = T[j,1] = j$ for all $j \in S$. The ALGOL program given below the Boolean variable GROUP has the value *true* if the monoid $[M,\mu]$ specified by T is a group; it has the value *false* otherwise.

```
begin integer array T[1:50, 1:50];
      integer i, N; boolean GROUP;
for i = 2 step 1 until N do
      if T[i,i] = i then begin GROUP := false; go to F end;
GROUP := true:
F: end
```

EXERCISES A

1. Show that any semigroup S can be extended to a monoid by adjoining an identity element.
2. Show that if z is a left zero of a semigroup S, then so are all its left multiples $xz(x \in S)$.
3. Show that if the set T contains more than one element, then the monoid of all transformations of T has more than one left zero, but no right zero.
4. Show that in the monoid of Example 3, multiplication is noncommutative and $fg \neq gf$ for two suitably chosen elements.
5. Show that a left zero cannot have a left inverse in a monoid containing more than one element.
6. Show that the mapping $g_a: x \to xa$ (a fixed, x variable) is one-one if and only if a has a right inverse.
7. Let S be any finite semigroup.
   (a) Show that an element $a \in S$ has a left inverse if and only if the transformation $f_a: x \to ax$ is one-one.
   (b) Show that an element $a \in S$ has a left inverse if and only if it has a right inverse.
8. Exhibit a semigroup with left identity and right zero which is not a monoid.
★9. Define the free unary algebra with two generators (see Sec. 2-7), and show that it is $[\mathbf{N},+] \times [\mathbf{N},+]$.
★10. (a) Define the free "binary algebra" with one generator, F_1^2. (*Hint:* See the diagram of Fig. 7-4.)
   (b) Prove that F_1^2 is isomorphic to the set of all strings α, $(\alpha\alpha)$, $((\alpha\alpha)\alpha)$, $(\alpha(\alpha\alpha))$, . . . consisting of all "well-formed strings" of α's and parentheses), (, under the operation of juxtaposition within parentheses.

7-4. MORPHISMS; DIRECT PRODUCTS

In this section, we shall develop further the concept of morphism for semigroups and monoids and shall discuss its relation to direct products.

Definition. A *morphism of semigroups* is a function $\theta: S \to T$ whose domain and codomain are both semigroups, and which carries the semigroup operation of S into that of T. Thus, if the operation is multiplication, the condition is

$$\theta(ss') = \theta(s)\theta(s') \qquad \text{for all } s,s' \in S \tag{6}$$

If S and T are monoids with identities 1_S and 1_T, respectively, then $\theta: S \to T$ is a *morphism of monoids* when (6) holds, and also

$$\theta(1_S) = 1_T \tag{7}$$

As in Sec. 2-6, a morphism which is onto is called an *epimorphism*, a morphism which is one-one is called a *monomorphism*, and a morphism which is bijective is called an *isomorphism*. We will now show that for S and T (any monoids) the notions of *isomorphism of semigroups* and *isomorphism of monoids* are equivalent to each other.

Note that the operations in S and T may be symbolized in different ways. Thus, the mapping $x \mapsto e^x$ defines a morphism θ: $[\mathbf{R},+] \to [\mathbf{R}^+, \cdot\,]$ of semigroups (and of groups) from the additive group of real numbers to the multiplicative group of all positive real numbers. Note also the characterization of the morphism condition (6) in terms of the following mapping diagram for the domain $[S,\circ]$ and codomain $[T,*]$.

$$\begin{array}{ccc} S \times S & \xrightarrow{\circ} & S \\ {\scriptstyle \theta\times\theta}\downarrow & & \downarrow{\scriptstyle \theta} \\ T \times T & \xrightarrow{*} & T \end{array}$$

Lemma 1. *If S and T are any two monoids, then a bijection $\beta: S \leftrightarrow T$ is a morphism of monoids if and only if it is a morphism of semigroups.*

Proof. Let $\beta^{-1}(1_T) = s \in S$ so that by definition

$$\beta(ss') = \beta(s)\beta(s') = 1_T\beta(s') = \beta(s') \in T \qquad \text{for all } s' \in S$$

Since β is a bijection, this implies that

$$ss' = \beta^{-1}(\beta(ss')) = \beta^{-1}(\beta(s')) = s' \qquad \text{for all } s' \in S$$

Similarly, $s's = s'$ for all $s' \in S$, whence $s = 1_S$ by Lemma 1 of Sec. 7-1. In summary, $\beta^{-1}(1_T) = 1_S$; therefore, $\beta(1_S) = 1_T$, and β is a morphism of monoids.

Lemma 2. *The set of all morphisms of any monoid $M = [S, \cdot\,]$ is a submonoid of S^S.*

Proof. Substituting into (6), we have

$$\theta'(\theta(ss')) = \theta'(\theta(s)\theta(s')) = \theta'(\theta(s))\theta'(\theta(s'))$$

Lemma 3. *The inverse θ^{-1} of any isomorphism $\theta: S \to T$ of semigroups (monoids) is itself an isomorphism of semigroups (monoids).*

Proof. For any $t,t' \in T$, there exists $s,s' \in S$ with $t = \theta(s)$ and $t' = \theta(s')$. By (1), then,

$$\theta^{-1}(tt') = \theta^{-1}(\theta(s)\theta(s')) = \theta^{-1}(\theta(ss')) = ss' = \theta^{-1}(t)\theta^{-1}(t')$$

for any $t,t' \in T$.

We now define two further variants of the fundamental morphism concept. A morphism from a semigroup (or other algebraic system) to itself is called an *endomorphism*. And an endomorphism which is bijective, i.e., an isomorphism of an algebraic system with itself, is called an *automorphism.*

Lemma 4. *The automorphisms of any semigroup S form a group called the automorphism group of S.*

Proof. Trivially, 1_S is an automorphism of S. By Lemma 1, the composite of any two automorphisms of S is an automorphism of S. And by Lemma 2, the inverse of any automorphism of S is an automorphism of S.

Lemma 5. *Let $\theta: M \to N$ be any morphism of monoids. Then the image under θ of any submonoid $S \subset M$ is a submonoid $\theta(S)$ of N, and the inverse image $\theta^*(T)$ of any submonoid $T \subset N$ is a submonoid of M.*

The proof follows almost immediately from the relevant definitions and will be left to the reader. Similar results hold for morphisms of semigroups. Indeed, at the postulational level, the main difference between semigroups and monoids is that the intersection of two nonvoid subsemigroups of a given semigroup may be void, whereas all submonoids of a given monoid S contain 1_S.

A sharper result is the following.

Theorem 4. *Let $M = [S, \cdot, 1]$ be any monoid, and let $\theta: S \to T$ be any surjection of M onto a binary system $[T,\beta]$ with the morphism property*

$$\theta(s,s') = \beta(\theta(s),\theta(s')) \qquad \text{in } T \tag{8}$$

Then $[T,\beta]$ is a semigroup, and $[T,\beta,\theta(1)]$ is a monoid. Moreover, if M is a group, then so is $[T,\beta]$.

Proof. For any elements $t_i \in T$ $(i = 1,2,3)$, there exist elements $s_i \in S$ with $\theta(s_i) = t_i$, since θ is a surjection.

$\beta(t_1,\beta(t_2,t_3)) = \theta(s_1(s_2s_3)) = \theta((s_1s_2)s_3) = \beta(\beta(t_1,t_2)t_3)$, applying (8) twice. Hence β is associative; that is, $[T,\beta]$ is a semigroup. Likewise, writing e for $\theta(1) \in T$, we have

$$\beta(e,t_i) = \theta(1s_i) = \theta(s_i) = t_i \qquad \text{for all } t_i$$

Hence e is a left identity. Likewise, e is a right identity, completing the proof.

We now prove a "representation theorem" which shows that the postulates for monoids imply all the general properties of functions under composition.

Theorem 5. *Any monoid M is monomorphic to the monoid M^M of all functions $f\colon M \to M$ under right or left composition (Cayley's theorem).*

Proof. Consider the mapping $\mu\colon s \mapsto f_s$, where f_s is a function $x \to xs$ from M to M. By associativity, $x(st) = (xs)t$ for all $x \in M$; hence $f_{st} = f_s \diamond f_t$, and μ is a morphism (carrying products into right composites). Since $f_s = f_t$ implies $s = 1s = 1t = t$ for 1 the identity of M, it is a monomorphism. The mapping $\nu\colon s \mapsto g_s$, where $g_s(x) = sx$ for all $x \in M$, is likewise a monomorphism of monoids, carrying products into left composites $g_{st} = g_s \circ g_t$.

Example 7. Let M be the monoid which consists of an identity 1 and r left zeros $z_1, \ldots, z_r$, so that $z_iz_j = z_i$ for all $i,j \in r$. Then the monomorphism μ of Theorem 5 maps 1 onto 1_M and each z_k onto $\phi_k \in M_M$, where ϕ_k takes 1 to z_k and leaves each $z_j \in M$ fixed.

Another interesting example is furnished by the case $X = \mathbf{2}$ of Example 2. By definition, the four functions 1, r, z_0, z_1 have the following action on the set $\{0,1\}$:

	1	r	z_0	z_1
0	0	1	0	1
1	1	0	0	1

Let multiplication in M signify right composition. Then Theorem 5 represents the elements of M monomorphically as functions acting on a set of four points 1, r, z_0, z_1, and not on the set $\mathbf{2} = \{0,1\}$.

Definition. The *direct product* $S \times T$ of two semigroups S and T is the set of all couples (s,t), multiplied component by component (termwise) according to the rule

$$(s,t)(s',t') = (ss',tt') \qquad \text{for all } s,s \in S \text{ and } t,t' \in T \tag{9}$$

If S and T are monoids, then so is $S \times T$, with identity $(1_S,1_T)$. In this case, $S \times T$ contains *submonoids* $S_1 \cong S$ and $T_1 \cong T$, consisting of all elements of the form $(s,1_T)$ resp. $(1_S,t)$. In any case, the mappings $(s,t) \mapsto s$, and $(s,t) \mapsto t$ are epimorphisms $S \times T \to S$ and $S \times T \to T$.

For monoids, we can summarize the preceding remarks in the diagram of Fig. 7-4a, in which μ and ν are epimorphisms and μ' and ν' are monomorphisms of monoids. More generally for semigroups, we also have an evident isomorphism $S \times (T \times U) \cong (S \times T) \times U$ defined by $(s,(t,u)) \mapsto ((s,t),u)$.

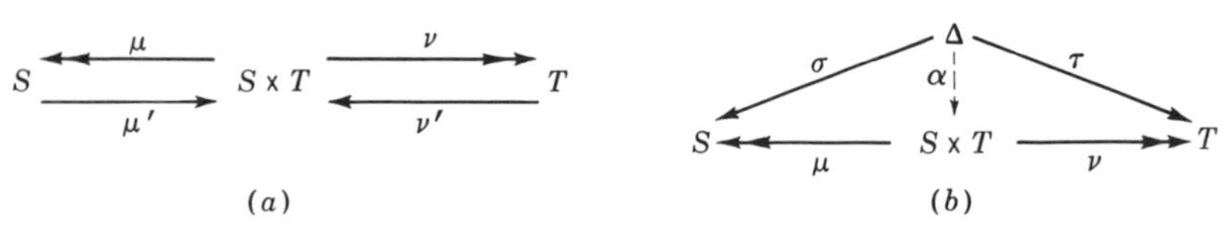

Fig. 7-4.

Another basic property of direct products is displayed in the diagram of Fig. 7-4b. If $\sigma: A \to S$ and $\tau: A \to T$ are any morphisms of semigroups, then the assignment $a \to (\sigma(a),\tau(a))$ defines a morphism $\alpha: A \to S \times T$ which makes the diagram of Fig. 7-4b commutative, and there is no other morphism $\beta: A \to S \times T$ with these properties.

7-5. EXAMPLES OF GROUPS; POSTULATES

After recalling some familiar groups (which are not always thought of as groups), we shall begin a systematic study of abstract groups by recalling their definition by *postulates* and proving from these postulates some elementary properties valid in all groups.

Many groups are familiar from elementary algebra; most of them are *commutative* (or Abelian) in the sense that in them $xy = yx$ for all x,y. Thus we have the following example.

Example 8. The nonzero complex numbers $z = x + y\sqrt{-1}$ with $x,y \in \mathbf{R}$ and not both zero, form a commutative group $[\mathbf{C}^*, \cdot]$ under multiplication.

Likewise, the nonzero real numbers and the nonzero rational numbers form groups under multiplication; but the nonzero integers do not. Many other familiar commutative groups from algebra have an operation commonly written as addition (and the group identity correspondingly written as zero). Thus we have the following example.

Example 9. The integers form a commutative group $[\mathbf{Z},+]$ under addition.

Example 10. The complex numbers form a commutative group $[\mathbf{C},+]$ under addition.

Many other groups arise as sets of *symmetries* of algebraic or geometrical configurations. Among these, one of the most basic is the group of all symmetries of an unstructured set of n elements.

Example 11. For any set X, the set of all bijections $\beta: X \to X$ of X is a group. If X has n elements, this group is called the *symmetric group* of degree n and often denoted S_n; it has $n!$ elements.

Example 12. The *dihedral group* Δ_n consists of the symmetries of a regular polygon Π_n of n sides.

To visualize the symmetries of Π_n, stand it on a horizontal side and place the origin at its center 0. Figure 7-5 shows the (typical) cases $n = 3$

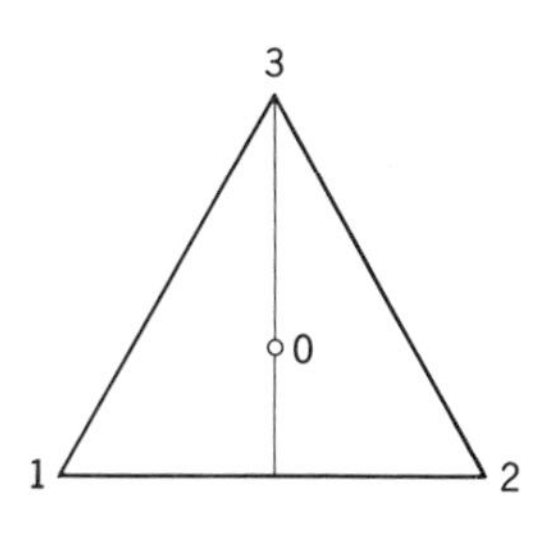

Fig. 7-5.

and $n = 4$. Let α denote the (counterclockwise) angle made by an arbitrary vector through the center with the resulting vertical axis of symmetry. Clearly, rotation R through $(360/n)°$ is a symmetry of Π_n, and so are its different powers $R^k (k \in \mathbf{n})$. So is reflection V in the vertical axis of symmetry and in $n - 1$ other symmetry axes through 0, making angles of $(180/n)°$ with the vertical and giving $2n$ symmetries in all.

The actions of the R^k and V on any angle α with the vertical are clearly given by

$$R^k(\alpha) = \alpha + (k/n)360° \qquad V(\alpha) = -\alpha$$

From these formulas one can compute

$$R^n = V^2 = 1$$

$$R^kV = VR^{-k} \qquad \text{takes } \alpha \text{ to } -\alpha + \frac{k}{n}360°$$

However, there is no need to compute a complete multiplication table for the dihedral group Π_n; one can easily deduce this table from the following three equations:

$$R^n = 1 \qquad V^2 = 1 \qquad RV = VR^{-1} \tag{10}$$

where I is the identity transformation. Every element can be written R^k or VR^k for some $K = 0,1, \ldots, n-1$. We have the multiplication rules

$$R^hR^k = R^{h+k} \text{ (or } R^{h+k-n} \text{ if } h + k \geqq n) \qquad (VR^n)R^k = VR^{h+k} \tag{10'}$$

$$R^h(VR^k) = (R^hV)R^k = (VR^{-h})R^k = \begin{cases} VR^{k-h} & \text{if } k \geqq h \\ VR^{n+k-h} & \text{if } k < h \end{cases} \tag{10''}$$

and $(VR^h)(VR^k) = R^{k-h}$ or R^{n+k-h} similarly.

In order to have a list of postulates for groups for convenient reference, we next recall their definition from Sec. 7-3, in somewhat expanded form.

Definition. A (multiplicative) *group* $G = [G, \cdot\,]$ is a set G with a binary *associative* multiplication, which contains an *identity* element 1 such that

$$x1 = 1x = x \qquad \text{for all } x \in G \tag{11}$$

and in which each element x has an *inverse* x^{-1} such that

$$xx^{-1} = x^{-1}x = 1 \tag{12}$$

Now for some general properties of groups. We recall from Sec. 7-1 that in any monoid, and hence in any group, the element 1 of (11) is unique, so that uniqueness need not be formally postulated. Moreover, as in Sec. 7-3, for a given $x \in G$ the element x^{-1} in (12) is also unique, since if y is any other such element (for given x), then

$$y = y1 = y(xx^{-1}) = (yx)x^{-1} = 1x^{-1} = x^{-1}$$

There are also some simple rules which are valid in any group, but *not* in monoids generally. Among these rules, the following cancellation laws are basic.

Rule 1. *The elements of any group G satisfy the left and right cancellation laws*

$$ax = ay \textit{ implies } x = y \qquad \textit{and} \qquad xa = ya \textit{ implies } x = y \tag{13}$$

We now prove the left cancellation law. If $ax = ay$, then for the element a^{-1} given by (12), we have

$$x = 1x = (a^{-1}a)x = a^{-1}(ax) = a^{-1}(ay) = (a^{-1}a)y = 1y = y$$

(The student should justify specifically each step in the preceding chain of equalities and then observe that by induction and the transitivity of equality, the final conclusion $x = y$ follows.)

Note that the cancellation law implies the unicity of right and left inverses as the special cases $ax = ay = 1$ and $xa = ya = 1$.

Rule 2. *For given elements a and b in a group G, the equations* $xa = b$ *and* $ay = b$ *have unique solutions* $x = ba^{-1}$ *and* $y = a^{-1}b$, *respectively.*

Proof. Clearly $x = ba^{-1}$ is a solution of $xa = b$, since

$$(ba^{-1})a = b(a^{-1}a) = b1 = b$$

Conversely, if $xa = b$, then

$$x = xe = x(aa^{-1}) = (xa)a^{-1} = ba^{-1}$$

Hence there is no other solution. The proof for $ay = b$ is similar.

As a corollary of Rules 1 and 2, it follows that each *left translation* $\phi_a = c \mapsto ax$ of a group G is a bijection of G, and so is each *right translation* $\psi_a = x \mapsto xa$. The name "translation" is used here because in the special case that G is the additive group of all ordered pairs (x,y) of real numbers under addition, the mapping

$$\tau_{a,b}: (x,y) \rightarrow (x + a,\ y + b) = (x,y) + (a,b)$$

defines a translation of the (x,y)-plane (the complex plane of Example 4) in the usual geometrical sense.

Rule 3. *For* 1 *the identity element and for any a and b*

$$1^{-1} = 1 \qquad \textit{and} \qquad (ab)^{-1} = b^{-1}a^{-1} \tag{14}$$

Proof. Since 1 is the left identity, $11 = 1$. This states that 1 is the (unique) left inverse of 1 and also its right inverse. Likewise,

$$(ab)(b^{-1}a^{-1}) = a(bb^{-1})a^{-1} = a1a^{-1} = aa^{-1} = 1$$

This proves that $b^{-1}a^{-1}$ is a right inverse of ab; the proof that it is a left inverse is similar.

Alternative Postulate Systems. By refining the arguments presented above, one can derive various alternative postulate systems for groups, which have less redundancy. Two such alternative postulate systems are described by the following statements:

1. Any semigroup $[S, \cdot\,]$ which has a *left* identity e such that $ex = x$ for all $x \in S$ and a *left* inverse x^{-1} such that $x^{-1}x = e$ for each x is necessarily a group.
2. Let S be a semigroup in which for any $a,b \in S$, there exist elements $x,y \in S'$ such that $xa = b$ and $ay = b$; then S is a group.

7-6. SUBGROUPS

A subgroup of a group G is defined to be a subset $S \subset G$ which is itself a group under the restriction to S of multiplication in G. This definition requires that $x,y \in S$ imply $xy \in S$. Since 1 is the only idempotent of G, it also requires that $1 \in S$. Finally, since x^{-1} is the only solution in G of the equation $xy = 1$ and 1 is the only possible identity of S, S cannot be a subgroup of G unless it contains with any x, also x^{-1}.

Conversely, let $S \subset G$ satisfy the three preceding conditions. Then multiplication in S is automatically associative, 1 acts as an identity, and every $x \in S$ is invertible (in S). This proves the following theorem.

Theorem 6. *A subset S of a group G is a subgroup of G if and only if (i) $1 \in S$, (ii) $x \in S$ implies $x^{-1} \in S$, and (iii) $x \in S$ and $y \in X$ imply $xy \in S$.*

In groups, the definition of power in Sec. 7-2 can be extended to arbitrary integral exponents as follows:

$$a^m = aaa \cdots a \text{ (to } m \text{ factors)} \qquad a^0 = 1 \qquad a^{-m} = (a^{-1})^m \tag{15}$$

With this definition, Theorem 1 has the following extension.

Theorem 7. *If G is a group, then*

$$a^r a^s = a^{r+s} \qquad (a^r)^s = a^{rs} \qquad \textit{for all } a \in G \tag{16}$$

Proof. For $r,s \in \mathbf{N}$, the first identity of (16) restates Theorem 1. For $r = -m$ and $s = -n$ both negative, we have by (14)

$$a^r a^s = (a^{-1})^m (a^{-1})^n = (a^{-1})^{m+n} = a^{-(m+n)} = a^{r+s}$$

There remains the case in which the exponents have opposite signs, say $r = m$ and $s = -n$, with $m > 0$ and $n > 0$. Thus

$$a^m a^{-n} = a^m (a^{-1})^n = (\overbrace{a \cdots a}^{m \text{ terms}})(\overbrace{a^{-1} \cdots a^{-1}}^{n \text{ terms}})$$

By the associative law we can cancel successive a's against the inverses a^{-1}. If $m \geqq n$, there remains a^{m-n}, whereas if $m < n$, there remain $n - m$ factors a^{-1}, giving $(a^{-1})^{n-m}$ or $a^{-(n-m)}$. In both cases we have the desired law $a^m a^{-n} = a^{m+(-n)}$.

The second half of (16) is even easier to establish. If $s > 0$, then by the first half of (16):

$$a^r a^r \cdots a^r \text{ (to } s \text{ factors)} = a^{r+r+\cdots+r} = a^{rs}$$

If $s < 0$, we can make a similar expansion, noting that $(a^r)^{-1} = a^{-r}$ whether r is positive, zero, or negative. If $s = 0$, the result is obvious.

Corollary. *In any group G, the powers of any fixed element a constitute a subgroup of G.*

Definition. A group C which consists of the powers a^r of a single element a is called *cyclic,* and is said to be *generated* by a. The cardinality (or order) of the cyclic subgroup $\{a^r \mid r \in \mathbf{Z}\}$ of any element a in any group G is called the *order* of a.

Note that $[\mathbf{Z}, +]$ is an (additive) cyclic group, but not a cyclic monoid. This cyclic group is infinite; the other cyclic groups are the additive groups $\mathbf{Z}_m$ of the integers mod m and are also cyclic when considered as monoids (see Chap. 2).

Definition. A morphism of groups is a function $\mu: G \to H$ with groups as domain and codomain, which satisfies (*i*) $\mu(1_G) = 1_H$, (*ii*) for all $g \in G$, $\mu(g^{-1}) = [\mu(g)]^{-1}$ in H, and (*iii*) for all $g, g' \in G$,

$$\mu(gg') = \mu(g)\mu(g') \tag{17}$$

We shall now show that conditions (*i*) and (*ii*) in the preceding definition are redundant.

Lemma. *If the function f in the preceding definition satisfies condition (iii) then it also satisfies conditions (i) and (ii).*

Proof. Since $\mu(1_G)\mu(1_G) = \mu(1_G1_G) = \mu(1_G)$, clearly $\mu(1_G)$ is the only idempotent element of H, which is 1_H; likewise, for any $g \in G$, since

$$1_H = \mu(1_G) = \mu(gg^{-1}) = \mu(g)\mu(g^{-1})$$

$\mu(g^{-1})$ is the inverse of $\mu(g)$ in H.

The preceding lemma asserts that if G and H are groups, then any morphism $\mu: G \to H$ of semigroups is also one of groups. Note that its analog is not true for monoids. If M is any monoid with two idempotents 1_M and $e \neq 1_M$, then the assignment $x \mapsto e$ for all $x \in M$ is a morphism of semigroups, but not one of monoids.

Theorem 8. *If $\mu: G \to H$ is any morphism of groups, then (i) the image $\mu(S)$ of any subgroup of G is a subgroup of H, (ii) the inverse image $\mu^{-1}(T)$ of any subgroup $T \subset H$ is a subgroup of G.*

Proof. Since $\mu(1_G)\mu(1_G) = \mu(1_G 1_G) = \mu(1_G)$, then $\mu(1_G)$ is the only idempotent element of H, namely, 1_H. Likewise, since

$$1 = \mu(1) = \mu(xx^{-1}) = \mu(x)\mu(x^{-1})$$

$\mu(x^{-1})$ is the unique inverse of $\mu(x)$ in H. Conclusion (*i*) is a corollary of Theorem 6. Conclusion (*ii*) also follows from Theorem 6; thus $1_G \in \mu^{-1}(T)$ since $\mu(1_G) = 1_H$ and $1_H \in T$. $\mu(x) \in T$ and $\mu(y) \in T$ imply

$$\mu(xy) = \mu(x)\mu(y) \in T$$

and so $xy \in \mu^{-1}(T)$ by (14); and $\mu(x) \in T$ implies $\mu(x^{-1}) = [\mu(x)]^{-1} \in T$ by Theorem 6 and the lemma. Hence $\mu^{-1}(T)$ satisfies the conditions of Theorem 6.

Corollary. *If $\theta: G \to H$ is an epimorphism of semigroups and G is a group, then so is H.*

Proof. Since $\theta(1_G g) = \theta(1_G)\theta(g)$ and $\theta(g 1_G) = \theta(g)\theta(1_G)$ for all $\theta(g) \in H$, $\theta(1_G)$ is an identity for H. Similarly, for any $\theta(g) \in H$, $\theta(g^{-1})$ is a two-sided inverse of $\theta(g)$; we omit the details. Even the associativity of multiplication in H follows from that of multiplication in G, and need not be assumed.

Finally, we prove a result which states that $[\mathbf{Z},+]$ is the *free group* with one generator.

Theorem 9. *If G is a group and $a \in G$, then the assignment $\mu(1) = a$ can be extended to a unique morphism $\mu: [\mathbf{Z},+] \to G$.*

Proof. By induction on positive and negative powers, any extension to such a morphism must make $\mu(m) = a^m$ for all $m \in \mathbf{Z}$, by (14); hence there is at most one extension to a morphism. Conversely, by the first part of (15) this assignment is such a morphism, which therefore exists.

EXERCISES B

1. List the eight symmetries of the square, describing the geometrical effect of each one.
2. Show that the group of symmetries of an equilateral triangle (under composition) is isomorphic with the group of all invertible functions $f \in \mathbf{3} \to \mathbf{3}$.
3. (a) Show that the cyclic groups of orders 5,6 and 14 have 4,2 and 6 generators, respectively.
   (b) Show that the groups of automorphisms of the preceding groups are all cyclic.
4. (a) Show that the group Aut $[\mathbf{Z},+]$ of all automorphisms of the additive group $[\mathbf{Z},+]$ of the integers has order 2.
   (b) Show that the automorphisms of $[\mathbf{Z}_8,+]$ form a group of order 4 which is not cyclic.

5. Determine all isomorphisms between $[\mathbf{Z}_4,+]$ and $[\mathbf{Z}_5^*,\cdot]$. (*Explanation:* $\mathbf{Z}_5^* = \{1,2,3,4\}$, mod 5.)
6. Show that the group of symmetries of a regular tetrahedron is isomorphic with the group of all permutations of $\{1,2,3,4\}$.
7. Construct an epimorphism from the "affine" group of all linear transformations $x \mapsto ax + b$ ($a \neq 0$, b real) onto $[\mathbf{Z}_2,+]$.
8. Show that the endomorphisms of any group form a monoid under composition.

IN EXERCISE 10, LET $\tilde{M} = [\{0,1\},\cdot]$ BE THE TWO-ELEMENT MONOID WHOSE MULTIPLICATION TABLE IS GIVEN BELOW.

$\cdot$	0	1
0	0	0
1	0	1

9. (a) Let $M = [\mathbf{2}^{\mathbf{2}}, \circ]$ be the monoid of all functions $f: \mathbf{2} \to \mathbf{2}$ under left composition. Show that the function $\theta: M \to \tilde{M}$ which carries bijections of $\mathbf{2}$ into $1 \in M$, and all other f into 0, is an epimorphism of M onto $\tilde{M}$.
   (b) Construct an analogous epimorphism from $M_n = [\mathbf{n}^{\mathbf{n}}, \diamond]$ onto $\tilde{M}$.
10. (a) For M any *finite* monoid not a group, let M^* be the set of its invertibles Show that the assignment $\theta: M \to \tilde{M}$ defined by $\theta(M^*) = 1$, $\theta(M - M^*) = 0$, is an epimorphism.
    ★(b) Show that the preceding result is not necessarily true for infinite monoids. (*Hint:* Consider the monoid of all functions from $\mathbf{N}$ to $\mathbf{N}$.)

7-7. ABELIAN GROUPS

Many important groups are commutative; such groups are called *Abelian groups*. In particular, algebraic systems which have an *addition* operation (e.g., the set of all real $n \times n$ matrices) are almost always Abelian groups with respect to this addition. In such additive (Abelian) groups, it is most suggestive to write the binary group operation as $+$, the group identity as 0, and the inverse of an element x and $-x$.

We shall derive some special properties of Abelian groups in this section, using the above additive notation. This notation makes the defining conditions for Abelian groups assume the form of the following identities, which can be taken as axioms for Abelian groups:

$$x + y = y + x \tag{18a}$$
$$x + (y + z) = (x + y) + z \tag{18b}$$
$$x + 0 = 0 + x = x \tag{18c}$$
$$x + (-x) = (-x) + x = 0 \tag{18d}$$

Moreover, since integral powers a^r ($r \in Z$) of any element a of an Abelian group go into multiples ra in the additive notation, the laws of exponents (15) become

$$(r + s)a = ra + sa \text{ and } s(ra) = (sr)a \qquad \text{for any } r,s \in \mathbf{Z} \tag{19}$$

Thus, formally, they become distributive and associative laws. Moreover, the third law of exponents, $(ab)^r = a^r b^r$, which is not valid for general a,b in *non*Abelian groups, is valid in any Abelian group A. In the additive notation, the third law of exponents becomes

$$r(a + b) = ra + rb \qquad \text{for all } r \in \mathbf{Z} \text{ and } a,b \in A \tag{20}$$

Equation (20), which can be proved by induction, can be restated in terms of morphisms as follows. Define an endomorphism of an Abelian group A as a morphism $A \to A$.

Theorem 11. *For any integer r, the mapping $a \mapsto ra$ is an endomorphism in any Abelian group.*

Corollary. *In any Abelian group A, for any positive integer r, the set rA of all ra ($r \in \mathbf{Z}$, $a \in A$) and the set $A:r$ of all $a \in A$ such that $ra = 0$ are subgroups of A.*

It is a further corollary of (20) that in any Abelian group A, the subset of all elements of finite order is a subgroup. This subgroup is called the *torsion subgroup* of A.

Now for more examples of Abelian groups. Evidently [Eq. (15)], any cyclic group generated by a single element is Abelian. From such cyclic groups, one can construct many others as *direct products* (Sec. 7-4). We have, almost trivially, the following theorem.

Theorem 12. *The direct product $G \times H$ of any two groups G and H is a group. Any direct product of Abelian groups is Abelian.*

Proof. Just as for monoids, $(1_G, 1_H)$ is an identity for $G \times H$. Moreover, any $(g,h) \in G \times H$ has the inverse (g^{-1},h^{-1}) in $G \times H$ (we use the multiplicative notation). Finally, if $gg' = g'g$ for all $g,g' \in G$ and $hh' = h'h$ for all $h,h' \in H$, then

$$(g,h)(g',h') = (gg',hh') = (g'g,h'h) = (g',h')(g,h)$$

for all (g,h) and $(g'h')$ in $G \times H$.

Example 13. For any prime p, the direct product $Z_p \times \cdots \times Z_p$ of n copies of the cyclic group $Z_p = [\mathbf{Z}_p,+]$ is called the *elementary Abelian group* of order p^n. In particular, the elementary Abelian group of order 2^n is related to the Boolean algebra of order 2^n as a special case of the following construction.

Example 14. If A is any Boolean algebra, then $G = [A,+]$ is an Abelian group under the following self-dual addition operation

$a + b = (a \wedge b') \vee (a' \wedge b)$ of Sec. 5-1. In G, $0 \in A$ is the identity, and each element is its own inverse, $x + x = 0$ for all $x \in A$.

It is easily verified that $C_2 \times C_3 \cong C_6$; this follows since $(1,1) \in C_2 \times C_3$ has additive order 6 and generates all of $C_2 \times C_3$. More generally, let m and n be any two relatively prime positive integers; that is, let $(m,n) = 1$ so that m and n have no common factors other than ± 1. Then $C_m \times C_n \cong C_{mn}$ is cyclic with generator $(1,1)$ of order mn, since $(r,r) = (0,0)$ if and only if m and n are common divisors of r.

Therefore, any direct product of cyclic groups is a direct product of cyclic groups of prime-power orders. The Fundamental Theorem on Abelian groups states that conversely any *finite* Abelian group is a direct product of cyclic groups of prime-power orders, and that the set of these prime-power orders is unique. This fact makes it easy to list *all* the Abelian groups of any finite order.

EXERCISES C

1. (a) Show that if $a^2 = 1$ for all a in a group G, then G is commutative.
   (b) Show that the same is true in any monoid.
2. Prove carefully and rigorously, in full detail, that every subgroup of a cyclic group is cyclic.
3. Show that a group G is Abelian if and only if the assignment $a \mapsto a^2$ is an endomorphism of G.
4. Show that a group G is Abelian if and only if $a \mapsto a^{-1}$ is an endomorphism of G.
5. How many Abelian groups are there of order 36? of order 300? of order p^6 (is p a prime)?
6. (a) Show that, in any additive Abelian group G, for any $n \in \mathbf{P}$, the set $0{:}\, n = \{x \in G \mid nx = 0\}$ is a subgroup.
   (b) Show that, in the group of symmetries of the square, the set of all elements of order 2 is not a subgroup.
7. In a Boolean algebra A, let $a + b = (a \wedge b') \vee (a' \wedge b)$. Show that $[A, +]$ is an Abelian group in which every element has order 2.
8. Show that, for $F = [\mathbf{Z}_2, +][\mathbf{Z}_2, +]$ the 4-group, Aut F is the symmetric group of degree 3.
9. Define an *anti*automorphism of a semigroup S as a bijection $0{:}\, x \mapsto x'$ such that $(xy)' = y'x'$.
   (a) Show that the monoid $\mathbf{n}^{\mathbf{n}}$ of all functions $f{:}\, \mathbf{n} \to \mathbf{n}$ has no nontrivial antiautomorphism. (*Hint:* Consider its left and right zeros.)
   ★(b) Show that every finite group with $\gamma' > 2$ elements has a nontrivial antiautomorphism. (You may assume the fundamental theorem on Abelian groups.)

7-8. GROUPS ACTING ON SETS

We now introduce a fundamental idea, that of a group acting on a set. If $f{:}\, X \to X$ is any bijection of a set, then its positive and negative powers

f^r and f^{-r} constitute, with the identity $1_X = f^0$, a cyclic group acting on X. More precisely, as in Sec. 7-2, we have

$$f^s(f^r(x)) = f^{r+s}(x) \qquad \text{for all } x \in S \text{ and } r,s \in \mathbf{Z} \tag{21}$$

so that

$$f^r \diamond f^s = f^s \diamond f^r = f^{r+s} \qquad \text{for all } r,s \in \mathbf{Z}$$

This defines two morphisms, μ and $\nu = \mu^{\text{opp}}$ from the *additive* group $[\mathbf{Z},+]$ to the *multiplicative* symmetric group of all bijections of X, under right and left composition, respectively.

Changing emphasis, one may also say that the mapping $r \mapsto f^r$ defines an *action* (or representation by permutations; see Sec. 7-9) of the group $[\mathbf{Z},+]$ on the set X. This can be generalized from $\mathbf{Z}$ to any multiplicative monoid $[M, \cdot\,]$ by making the following definition.

Definition. Let M be any (multiplicative) monoid, and let X be any set. A function $\mu: M \to X^X$ is called an *action* of M on X when $\mu(1) = 1_X$ and either

$$\mu(rs) = \mu(r) \diamond \mu(s) \qquad \text{for all } r,s \in M \tag{22}$$

or

$$\mu(rs) = \mu(r) \circ \mu(s) \qquad \text{for all } r,s \in M \tag{22$'$}$$

More precisely, (22) is said to define a *right* action of M on X, and (22′) is said to define a *left* action of M on X.

In other words, a right or left action of M on X is a morphism μ of monoids from M to X^X under right or left composition, respectively. If $M = G$ is a group, then since

$$\mu(g) \diamond \mu(g^{-1}) = \mu(gg^{-1}) = \mu(1) = 1_X \qquad \text{and} \qquad \mu(g^{-1}) \diamond \mu(g) = 1_X$$

all mappings $\mu(g)$ of any right or left action must be bijections. For any function $f: X \to X$, the assignment $r \mapsto f^r$ defines an action of the additive monoid $[\mathbf{N},+]$ on X^X, with $f^\theta = 1_X$.

Any action $\mu: G \to X^X$ of a group G on a set X defines a very important *equivalence relation* γ on X as follows:

$$x \,\gamma\, y \qquad \text{means that} \qquad \mu_g(x) = y \qquad \text{for some } g \in G \tag{23}$$

where μ_g is written for $\mu(g)$.

In words, some transformation $\mu_g: X \to X$ belonging to G carries x into y. The relation defined by (23) is clearly reflexive, since $1_X(x) = x$ and $\mu_1 = 1_X$. It is symmetric because $\mu_g(x) = y$ implies $\mu_g - 1(y) = x$ [since $\mu_{g-1} = (\mu_g)^{-1}$], and it is transitive because $\mu_g(x) = y$ and $\mu_h(y) = z$

imply $\mu_{gh}(x) = (\mu_g \diamond \mu_h)(x) = z$ for any right action and

$$\mu_{hg}(x) = (\mu_h \cdot \mu_g)(x) = \mu_h(\mu_g(x)) = \mu_h(y) = z \qquad hg \in G$$

for any left action of G on X. This proves the following result.

Theorem 13. *For any action of any group G on a set X,* (23) *defines an equivalence relation on X.*

The equivalence classes of X under the equivalence relation associated by (23) with any action of a group G on X are called *sets of transitivity* for this action. The action is called *transitive* when there is only one such equivalence class, so that $x \gamma y$ for all $x,y \in X$ (i.e., when for any $x,y \in X$, $\mu_g(x) = y$ for some $g \in G$).

For example, Cayley's representation of a group G by its right translations is a transitive right action of G on its own elements; for any $g \in G$, $\mu_g(x) = xg$. The left translations of G define an analogous left action of G on G, with $\nu_g(x) = gx$ for all $x,y \in G$. These actions are *simply* transitive, in the sense that for each $x,y \in G$ there is exactly one $g \in G$ with $xg = y$ and exactly one $h \in G$ with $hx = y$.

Likewise, the *Euclidean group* of all rigid motions of the plane into itself is transitive on points; the associated equivalence relation on triangles defined through (23) by the action of this group on them is called in geometry the relation of *congruence*. The subgroup of all parallel translations of the plane is simply transitive on its points; this representation of $[\mathbf{R},+]^2$ is really a special case of Cayley's theorem. By (23), two straight lines are equivalent under this group if and only if they are parallel. Finally, a polyhedron is regular if and only if its isometries are transitive on its vertices and faces.

7-9. PERMUTATIONS

A bijection $\beta: X \leftrightarrow X$ of a finite set to itself is usually called a *permutation* of X. By definition, the powers of such a β form a cyclic group of permutations. Consider now the action of this cyclic group β^r as a morphism of monoids from $[\mathbf{Z},+]$ to X^X. The sets of transitivity for this action are called the *cycles* of β. It follows from Theorem 13 that X is the disjoint sum of its cycles.

The standard way to specify a permutation β of (say) $\mathbf{n} = \{1, \ldots, n\}$ consists of listing in parentheses first the cycle

$$\gamma_1 = (1, 1\beta, 1\beta^2, \ldots, 1\beta^{m1-1})$$

where 1β is written for $\beta(1)$ and $1\beta^2$ for $\beta(\beta(1))$, etc.

Then the (disjoint) cycle γ_2 beginning with the first $j_2 \in \mathbf{n}$ omitted by γ_1, then the cycle γ_3 beginning with the first element $j_3 \in \mathbf{n}$ omitted by γ_1 and γ_2 (that is, $\gamma_1 \sqcup \gamma_2$), and so on, until all $\mathbf{n}$ is exhausted. The one-element cycles can be omitted; these are just the $k \in \mathbf{n}$ which are left untouched by β. Commas between symbols can also be omitted if no confusion will result.

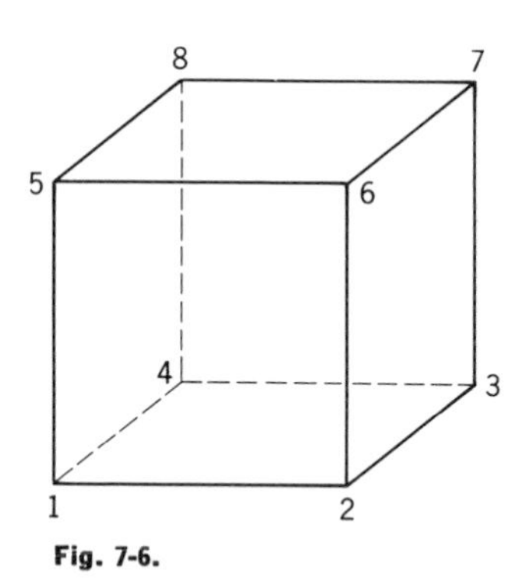

Fig. 7-6.

Example 16. The symmetries of the cube displayed in Fig. 7-6 induce 48 different permutations of its vertices. Thus, counterclockwise rotation through 90° around its vertical axis induces the product

$$(1234)(5678) = (5678)(1234)$$

Rotation through 120° around the diagonal axis $\overrightarrow{17}$ induces the permutation (245)(386), and so on.

Note that *disjoint cycles are permutable;* in symbols, $\gamma_k\gamma_l = \gamma_l\gamma_k$ for all k,l. This makes the order in which the γ_k are written down immaterial. A cycle of length 2 is called a *transposition*, and we have the following result.

Lemma. *Every permutation β of $\mathbf{n}$ can be written as a product of m transpositions for some $m \leqq n(n-1)/2$.*

Proof. In the list $1\beta^{-1}, 2\beta^{-1}, 3\beta^{-1}, \ldots, n\beta^{-1}$, transpose $n = (n\beta)\beta^{-1}$ to the right, one place at a time, as long as possible. It will reach the right end after $n - n\beta \leqq n - 1$ transpositions. Now repeat the process for $n - 1$, then for $n - 2$, and so on. After at most

$$m = (n-1) + (n-2) + \cdots + 1 = \frac{n(n-1)}{2}$$

such transpositions, we will have moved each $k\beta^{-1}$ to the kth position, thus effecting $\beta: k\beta^{-1} \mapsto k$.

Even and Odd Permutations. A permutation ϕ of $\mathbf{n}$ is called *even* or *odd* according to whether the number of pairs $(i\phi, j\phi)$ with $i,j \in \mathbf{n}$ and $i < j$ whose order is reversed by ϕ [so that $\phi(j) < \phi(i)$] is even or odd. Writing

$$K_n = \prod_{\substack{i<j \\ i,j \in \mathbf{n}}} (j - i) = 1^{n-1} 2^{n-2} 3^{n-3} \cdots n - 1 = \prod_{k=1}^{n-1} k! \qquad (24)$$

since the set of all *unordered* pairs (i,j) with $i \neq j$ is unaffected by ϕ, we see that clearly $\prod_{i<j} (j\phi - i\phi) = \phi(K_n)$ is K_n or $-K_n$ according to whether ϕ is even or odd. Here each pair $k \neq l$ occurs exactly once as either some $i\phi < j\phi$ or some $j\phi < i\phi$, $i < j$. In the first case, we write sgn $\phi = +1$; in the second, we write sgn $\phi = -1$.

Moreover, each *transposition* $\tau = (i,j)$ is odd; τ reverses the sign of $j - i$, and also of all $(j - k)$ and $(k - i)$ with $i < k < j$, giving an odd number $2(j - i) - 1$ of sign changes in all. It follows that $\tau(K_n) = -K_n$, and more generally that $\tau(\phi(K_n)) = -\phi(K_n)$, since the effect of ϕ is just to list the $i \in \mathbf{n}$ in some new order. Using the lemma and this last result repeatedly, we see that ϕ is even or odd according to whether ϕ is the product of an even or an odd number of transpositions. As a corollary we have the following.

Theorem 14. *The function defined by the assignment $\phi \mapsto$ sgn ϕ is an epimorphism for multiplication. The even permutations form a subgroup of S_n.*

The preceding subgroup is called the *alternating group* of degree n and is denoted A_n.

EXERCISES D

1. Let $\alpha = (12)(354)$. List the powers of α, in the standard "product of disjoint cycles" notation.
2. Show that any 3-cycle $\gamma = (p\ q\ r)$ is the commutator $\alpha = \sigma^{-1}\tau^{-1}\sigma\tau$ of two suitable 2-cycles ("transpositions").
3. (a) Express (1234567) as a product of (not necessarily disjoint) 3-cycles.
   (b) Express (1234)(56) and (1234)(5678) as products of 3-cycles.
4. ★ Show that every even permutation can be expressed as a product of (not necessarily disjoint) 3-cycles.
5. Show that the relations $s^2 = t^2 = (st)^3 = 1$ determine a group isomorphic to the symmetric group of degree 3.

6. Show that the symmetric group S_n is generated by the cyclic permutations (12) and $(23\ .\ .\ .\ n)$.

★7. Interpret the set of functions $f(x) = (ax + b)/(cx + d)$, with a,b,c,d real and $ad - bc = 1$, as a group acting on the disjoint sum $R \sqcup \{\infty\}$.

7-10. LAGRANGE'S THEOREM

Now let G be any group, and let S be any subgroup of G. Consider the right action of S on G via right translations $x \mapsto xs$ $(s \in S)$. The identity $x(ss') = (xs)s'$ shows that this is indeed a right action of S on G (a morphism $S \to G^G$ for right composition), the restriction to S of the Cayley right representation. Note that $1 \in S$ induces the identity map 1_G, since $x1 = x$ for all $x \in G$.

For this *right* action, the set of transitivity containing any given $g \in G$ consists by definition of all gs with $s \in S$. This set is denoted gS and is called the *left* coset of S containing g. When applied to this case, Theorem 13 gives the following result.

Lemma 1. *The left cosets gS of any subgroup S is a given group G are the subsets of a partition of G.*

In other words, distinct left cosets are nonoverlapping, and the left cosets of S exhaust G. We can also verify this result independently, as follows. If gS and hS have an element $u = gs = hs'$ in common $(s,s' \in S)$, then $gs'' = hs's^{-1}s''$ for any $s'' \in S$; hence $gS \subset hS$ since $s's^{-1}s'' \in S$. Likewise, $hS \subset gS$, which proves that overlapping left cosets are identical. Since $g = g1 \in gS$, the left cosets of S exhaust G.

Now define the *index* $[G:S]$ of S in G as the number of distinct left cosets of S in G. Trivially, $[S:1] = o(S)$ is the order of S. Since left translation of G by any $g \in G$ is a bijection of G, each left coset gS of S has precisely $o(S) = [S:1]$ elements. Combining this observation with Lemma 1, we have Lagrange's theorem in the following form.

Theorem 15. *For any subgroup S of a group G,*

$$[G:1] = [G:S][S:1] \tag{25}$$

In words, the order of G is the order of S multiplied by the index of S in G.

Corollary 1. *Every element of a finite group G has an order which divides the order of G.*

This is true because the order of the element is by definition the order of the cyclic subgroup which it generates, and this divides $o(G)$ by Theorem 15.

Corollary 2. *Every group of prime order p is cyclic.*

In this case, the order of any element c not the identity must exceed 1 and be a divisor of p; hence the order of c must be p. Thus c must generate the group.

Corollary 3. *Let G be a nontrivial group. Then G has no proper subgroups if and only if $o(G)$ is a prime.*

We omit the proof.

Example 17. In the dihedral group Δ_n, let S be the subgroup consisting of 1 and reflection in an axis through the vertex 1. Then for any $g \in \Delta_n$, Sg consists of all symmetries which carry vertex 1 into its image $g(1)$ under g. Likewise, the left coset gS consists of all symmetries which carry the same vertex into vertex 1 as g does. Finally, the set $g^{-1}Sg$ of all $g^{-1}sg$ with $s \in S$ consists of 1 and its reflection in the axis through the vertex $g(1)$.

We now digress to prove another corollary which will be essential for the theory of Galois fields in Chap. 12.

Theorem 16. *Let G be a noncyclic Abelian group of finite order g. Then there is a proper divisor h of g such that $x^h = 0$ for all $x \in G$.*

Proof. Express $g = p_1^{e_1}p_2^{e_2} \cdots p_r^{e_r}$ as the product of powers of its different prime factors $p_1 < p_2 < \cdots < p_r$. Let e_i' be the largest integer such that G contains an element g_i of order $p_i^{e_i'}$. Then $e_i' = e_i$ by Theorem 15; hence there are only two possible cases.

Case 1. Every $e_i' = e_i$. In this case, G contains elements $y_1, \ldots, y_r$ of orders $p_1^{e_1}, \ldots, p_r^{e_r}$. Since the y_i commute, their product $y = y_1 \cdots y_r$ must have order $p_1^{e_1} \cdots p_r^{e_r}$; hence G is cyclic in this case.

Case 2. Some $e_i' < e_i$, whence $h = \Pi p_i^{e_i}$ is a proper divisor of g. Then let the order of any given $x \in G$ be $p_1^{c_1} \cdots p_r^{c_r}$, where, of course, the c_i depend on x. Setting $n_i = \prod_{j \neq i} p_j^{c_j}$, we get an element $y_i = x^{n_i}$ of order $p_i^{c_i}$; hence $c_i = e_i'$. Substituting back, we see that the order of x divides h in this case, where h is a proper divisor of g independent of x.

EXERCISES E

1. (a) In S_5, find a 3-cycle γ such that $\gamma^{-1}(123)\gamma = (124)$.
   (b) Infer that, if $n \geqq 5$, all 3-cycles are conjugate in A_n (the alternating group of degree n).

THE COMMUTATOR SUBGROUP $G' = [G,G]$ OF A GROUP G IS THE SUBGROUP OF G GENERATED BY ITS COMMUTATORS $a^{-1}b^{-1}ab$.

2. (a) Show that $G' = 1$ if G is Abelian.
   (b) Show that $G' \lhd G$, and in fact that $\alpha(G') = G'$ for any automorphism α of G.
   (c) Show that G/G' is Abelian.
   (d) Show that, for $N \lhd G$, G/N is Abelian if and only if $N \supset G'$.
3. Show that, for all n, the commutator group of S_n is A_n. (*Hint:* Use Exercise D4 and Exercise E1 above.)

★4. Let $H \lhd A_n$ with $n \geq 5$.
   (a) Show that, if H contains a 3-cycle, then $H = A_n$. (*Hint:* Use Exercise D4 and Exercise E1 above.)
   (b) Show that, unless $H = 1$, H contains a 3-cycle.
   (c) Show that A_n is simple for all $n \geq 5$.
5. How many nonisomorphic labeled squares are there whose vertices are colored red, yellow, or blue? Display one representative of each type.
6. Same question for labeled hexagons?

★7. For the subgroup S of a group G, define the relation $x\ E_S\ y$ to mean that $xy^{-1} \in S$.
   (*i*) Show that E_S is an equivalence relation on G.
   (*ii*) Describe the equivalence classes for the partition defined by this relation.
   (*iii*) When does the relation E_S have the Substitution Property (*a*) for the unary operation $x \mapsto x^{-1}$? (*b*) for the operation xy?

7-11. NORMAL SUBGROUPS

Given an element a of a group G, the mapping $C_a: x \mapsto a^{-1}xa$ is called *right conjugation* by a; since $a^{-1}(xy)a = (a^{-1}xa)(a^{-1}ya)$, C_a is an endomorphism of G (a morphism $C_a: G \to G$). More generally, this is true if a is any invertible in any monoid.

Moreover, the mapping $\gamma: a \mapsto C_a$ defines a morphism $\gamma: G \to G^G$ since $C_b(C_a(x)) = b^{-1}(a^{-1}xa)b = C_{ab}(x)$ for all $a,b,x \in G$. Therefore, the morphism C_a is a bijection of G with inverse $C_a{}^{-1}$ and thus is an automorphism of G. Collecting the above results, we have the following theorem.

Theorem 17. *In any group G, right conjugation by any $a \in G$ defines an automorphism $C_a: x \mapsto a^{-1}xa$ of G; moreover, the mapping $a \mapsto C_a$ is a right action of G on itself.*

Definition. An automorphism of the form $x \mapsto a^{-1}xa$ is called an *inner automorphism* of the group (or monoid) G.

Clearly, an Abelian group has the trivial automorphism 1_G for its only inner automorphism, but every non-Abelian group has nontrivial inner automorphisms.

The equivalence classes of a group G under the action of its inner automorphisms are called *conjugacy classes* of G; note that since

$bxb^{-1} = (b^{-1})^{-1}x(b^{-1})$, the meaning of inner automorphism and conjugacy class is the same, whether right or left conjugation is used.

Definition. A subgroup S of a group G is called *normal* (or self-conjugate) in G, a relation written $S \lhd G$, when $g^{-1}Sg = S$ for all $g \in G$.

Lemma 1. *Each of the following conditions on a subgroup S of a group G is necessary and sufficient for S to be normal: (i) $Sg = gS$ for all $g \in G$, and (ii) $g^{-1}Sg \subset S$ for all $g \in G$.*

Proof. If $S \lhd G$, then $S = g^{-1}Sg$ for all $g \in G$, whence

$$gS = gg^{-1}Sg = Sg$$

for all $g \in G$; hence $S \lhd G$ implies condition 1. Next, condition (*i*) trivially implies condition (*ii*). Finally, condition (*ii*) implies $S \lhd G$ because if $g^{-1}Sg \subset S$ for all $g \in G$, then $(g^{-1})^{-1}Sg^{-1} \subset S$ also, whence

$$S = g^{-1}(g^{-1})^{-1}Sg^{-1}g \subset g^{-1}Sg$$

Since $g^{-1}Sg \subset S$ and $S \subset g^{-1}Sg$ imply $g^{-1}Sg = S$, the proof that $(S \lhd G) \Rightarrow (i) \Rightarrow (ii) \Rightarrow (S \lhd G)$ is complete.

Note that, as a consequence of Lemma 1 (*i*), one need not talk about "right" or "left" cosets of a normal subgroup; they are the same and can be called simply "cosets."

Theorem 18. *Let $\theta: G \to H$ be any morphism of groups. Then the inverse image $\theta^{-1}(1_H) = K$ of 1_H under θ is a normal subgroup of G.*

Proof. Evidently, $x \in K$ and $y \in K$ imply

$$\theta(xy) = \theta(x)\theta(y) = 1_H 1_H = 1_H$$

and $\theta(x^{-1}) = [\theta(x)]^{-1} = 1_H{}^{-1} = 1_H$; hence K is a subgroup. Likewise, $x \in K$ and $g \in G$ imply

$$\theta(g^{-1}xg) = [\theta(g)]^{-1}\theta(x)\theta(g) = [\theta(g)]^{-1}1_H\theta(g) = [\theta(g)]^{-1}\theta(g) = 1_H$$

Hence K is a normal subgroup of G.

Definition. The inverse image of 1_H under any morphism $\theta: G \to H$ of groups is called the *kernel* of θ.

Quotient Groups. The converse of Theorem 18 is also true and gives a powerful method for constructing groups. But the proof of this converse uses a symbolic calculus which we have not studied before.

Calculus of Complexes. Given any two nonvoid subsets or "complexes" S and T in a monoid M, we define their product as the set ST of all products st with $S \in S$ and $t \in T$; this is also nonvoid. The following observation is often useful.

Lemma. *The complexes of any monoid M themselves form a monoid $\mathcal{P}(M) - \{\varnothing\}$ under multiplication.*

Thus $(ST)U = S(TU)$ is the set of all products stu with $s \in S$, $t \in T$, $u \in U$, and the singleton set $\{1\}$ as an identity in $\mathcal{P}(M) - \{\varnothing\}$. Letting S^{-1} stand for the set of all inverses s^{-1} $(s \in S)$, it is, however, not true that $SS^{-1} = 1$ in a group G; indeed, $SS^{-1} \subset S$ if and only if S is a subgroup of G.

Theorem 19. *Let N be any normal subgroup of the group G. Then there exists an epimorphism $\theta: G \to H$ with kernel N onto some group H. Moreover, H is determined to within isomorphism by N.*

Proof. Consider the mapping $\theta: x \mapsto Nx$ from G to cosets of N. Multiply two cosets by the calculus of complexes: let $(Nx)(Nx')$ be the set of products yy' with $y \in Nx$, $y \in Nx'$. Then $(Nx)(Nx') = Nxx'$, as is easily verified using the calculus of complexes:

$$(Nx)(Nx') = N(xN)x' = N(Nx)x' = NNxx' = Nxx'$$

In other words, multiplication is single-valued, and θ is an epimorphism of semigroups. Since G is a group, it follows by the corollary of Theorem 6 that so is H, the set of Nx with the multiplication defined above.

Conversely, if $\theta: G \to H$ is an epimorphism with kernel N, then $\theta(x) = \theta(y)$ in H if and only if

$$\theta(xy^{-1}) = \theta(x)\theta(y^{-1}) = \theta(x)[\theta(y)]^{-1} = \theta(y)[\theta(y)]^{-1} = 1_H$$

Hence the condition is $xy^{-1} \in N$, $x \in Ny$, or $Nx = Ny$; the equivalence relation associated with θ as in Secs. 2-3 and 2-4 partitions G into the different (left or right, by Lemma 1) cosets of N in G. Consider the bijection b between the cosets Nx and the $h \in H$ so defined. Since

$$\theta(xx') = \theta(x)\theta(x')$$

the product $N(xx') = (Nx)(Nx')$ defined in the preceding paragraph corresponds under b to the product hh', where $h = b(Nx)$ and $h' = b(Nx')$; that is, b is an isomorphism, as claimed.

Example 18. In the additive group $[\mathbf{Z},+]$, let (n) be the set of multiples of $n \in \mathbf{P}$. Then $\mathbf{Z}/(n)$ is the additive group of the integers mod n.

Example 19. Each permutation $\pi \in S_n$ of the n digits $1, \ldots, n$ carries the polynomial $P = \prod_{i<j} (x_i - x_j)$ into itself or into $-P$. Hence S_n acts on the pair $\{P,-P\}$. This defines an epimorphism $S_n \to (\pm 1)$, from S_n to a multiplicative copy of the cyclic group of order 2. The kernel of this epimorphism is, by definition, the alternating group $A_n \lhd S_n$, consisting of all even permutations of the digits $1, \ldots, n$.

EXERCISES F

1. In the symmetric group S_5, list the conjugates of $(12)(34)$.
2. Show that, in the symmetric group S_n, two permutations represented in "product of disjoint cycles" form are conjugate if and only if they have the same number of conjugates of each length.
3. Enumerate the conjugacy classes of S_4, and specify the cardinality of each.
4. Same question for S_5.
5. Show that if N is a maximal normal subgroup of a group G, then G/N is simple.
6. Show that a subgroup S of a group G is normal if it contains all the conjugates of some set of generators.
7. (a) Describe the six symmetries of a cube with one vertex held fixed.
   (b) Show that the cube has exactly 48 symmetries.
8. Show that a regular octahedron has exactly 48 symmetries, and relate them to the symmetries of a cube.
9. Show that the translations $(x,y) \mapsto (x + a, y + b)$ of the (x,y)-plane form a normal subgroup of the "affine" group of all transformations of the form

$$(x,y) \mapsto (\alpha x + \beta y + a, \gamma x + \delta y + b), \qquad \alpha\delta \neq \beta\gamma$$

10. (a) Prove that if n is not divisible by a prime p, then $n^{p-1} \equiv 1 \pmod{p}$. (*Hint:* Consider the multiplicative group $[\mathbf{Z}_p - \{0\}, \cdot]$.)
    (b) Prove that, if p is a prime, then $n^p \equiv n \pmod{p}$ for all $n \in \mathbf{Z}$. (*Hint:* Use (a) and consider two cases.)
11. Let $2^r + 1$ be a prime p.
    (a) Prove that, in the group $[\mathbf{Z}_p - \{0\}, \cdot]$, the order of 2 is $2r$.
    (b) Using Exercise 10, infer that $2r$ is a divisor of $p - 1 = 2^r$.
    (c) Conclude that $r = 2^s$ is a power of 2 ($p = 2^{2^s} + 1$). (Numbers of the form $2^{2^s} + 1$ are called Fermat numbers.)

REFERENCES

CHEVALLEY, C.: "Fundamental Concepts of Algebra," Academic, 1956.
HALL, M.: "The Theory of Groups," Macmillan, 1960.
HIGMAN, B.: "Applied Group-theoretic and Matrix Methods," Dover, 1955.
LJAPIN, S.: "Theory of Semigroups," *Am. Math. Soc. Translations*, 1963.
WEYL, H.: "Symmetry," Princeton University Press, 1952.
WIELANDT, H.: "Finite Permutation Groups," Academic, 1964.

Binary Group Codes

8-1. INTRODUCTION

This chapter will describe and explain one technique for solving an important problem of communications—that of encoding and decoding digital information for reliable transmission through communications *channels* with "noise." Typically, one wishes to transmit a *message* which consists of a *string* (finite sequence) of characters from some finite "alphabet." As was explained in Chaps. 3 and 4, this alphabet can consist simply of 0 and 1, or it can consist of small and/or capital Latin letters, or all Arabic numerals, or both (*aphanumeric* characters). Thus, a message can consist of printed English text (in which case punctuation marks and blanks are also characters). The characters may also be numbers from a file to be transmitted from one computer to another, or numbers giving the results of measurements by instruments in a space vehicle. The strings of characters might then consist of telemetry frames, which are organized sequences of such measurements.

In any event, *data communication* consists of the transmission of characters from some finite alphabet through some communications channel. In almost every case of interest, imperfections in this communications channel (whether it be wires or cables as in the telephone system or space as in satellite communications systems) will cause a finite probability of error q that a given transmitted character will be incorrectly received by the receiver. For long messages, even what appears to be a low probability of error, say 10^{-6} per character, is not tolerable. For instance, in computer-to-computer systems this figure is unacceptable.

As an example, consider one computer communicating with another computer via an orbiting satellite communications link. In this case, the alphabet is normally binary-valued, and the channel consists of the space separating the transmitter attached to the first computer to the satellite and then the space separating the retransmitted signal from the satellite to the second computer. This channel distorts and attenuates electromagnetic signals representing 0's and 1's, and as a result the receiver may occasionally err in interpreting the transmitted signals. Such a system will be especially sensitive to noise introduced by sun spots, bad weather, etc.

Binary Symmetric Channels. Generally, designs for communication systems are analyzed mathematically using simplified models for the channel. For a binary-valued alphabet, whose signals may be designated by 0 and 1, the simplest, reasonably realistic model is called the *binary symmetric channel*. It is the most frequently used model and the only model which will be used in this chapter.

Let binary signals, represented symbolically by 0's and 1's, be transmitted sequentially through a channel to a receiver, as indicated in Fig. 8-1. Let the probability that a given signal will be read correctly by the

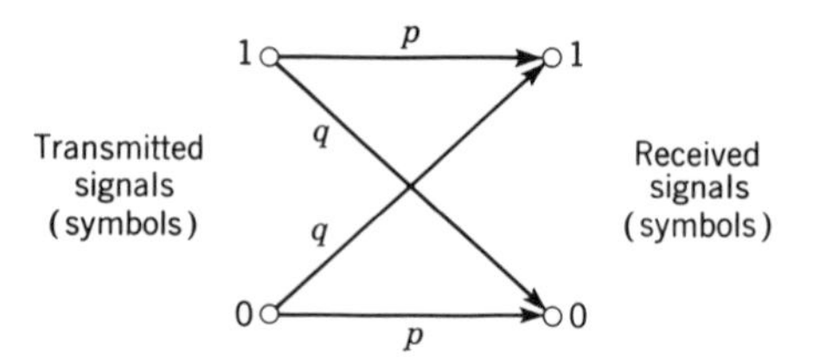

Fig. 8-1. Graph of transition probabilities for a binary symmetric channel.

receiver be p, and let the complementary probability that it will be incorrectly read be $q = 1 - p$. Finally, let it be assumed that errors in transmitting successive digits occur independently, in the sense of the following definition.

Definition. Let $\mathcal{E}_1, \mathcal{E}_2, \mathcal{E}_3, \ldots$ be a sequence of trials (events), and let T be any property which each $\mathcal{E}_i$ may or may not have. Let p_i be the probability that $\mathcal{E}_i$ will have property T, and let $q_i = 1 - p_i$ be the complementary probability that it will not. Then the trials are *independent* with respect to property T when, for any two disjoint subsets I and J of the trials, the probability that all $\mathcal{E}_i \in I$ will have property T *and* no $\mathcal{E}_i \in J$ will have property T is the product $\prod_I p_i \prod_J q_i$.

The preceding condition is algebraically equivalent to the following statement: Considering only those trials which satisfy the condition that all $\mathcal{E}_i \in I$ have property T and no $\mathcal{E}_j \in J$ has property T, the *conditional probability* that $\mathcal{E}_k$ ($k \notin I \sqcup J$) will have property T is p_k.

To fix ideas, suppose that the chance of incorrect transmission of a single binary symbol 0 or 1 (*bit* of information) is $q = 1\% = 0.01$, and that a valuable message 10,000 symbols long must be correctly transmitted. Then a straightforward symbol-by-symbol transmission without repetition would have a negligible chance, only

$$P_0 = (1 - 0.01)^{10,000} \simeq 10^{-4.4} < 0.004\%$$

of being transmitted without error.

The preceding numerical result is a special case of a classic algebraic formula of Bernoulli, valid under our assumption that the probabilities of error for different signals are independent (uncorrelated).[1] Letting $\binom{n}{k}$ denote as usual the binomial coefficient

$$\binom{n}{k} = \frac{n!}{(k!)(n-k)!} = \frac{n(n-1) \cdots (n-k+1)}{1 \cdot 2 \cdot 3 \cdots k}$$

we have the following basic result.

Theorem 1. *The probability that precisely k errors will occur in transmitting an n-bit message through a binary symmetric channel is precisely*

$$P_k = \binom{n}{k} p^{n-k} q^k = \binom{n}{k} p^{n-k}(1-p)^k = \binom{n}{k}(1-q)^{n-k} q^k \qquad (1)$$

The magnitude of the binomial coefficient $\binom{n}{k}$ can be accurately estimated for large k and n by Stirling's formula. A study of this magnitude leads to the Law of Large Numbers, which asserts that for large n, the fraction of errors will actually be very near to p. Thus, for $p = 0.99$

[1] For a detailed derivation of Bernoulli's formula (1), one can consult Abramson, Fano, or almost any introductory text on probability theory.

and a message 10,000 binary digits long, as above, the Law of Large Numbers asserts that about 100 digits will be read wrong and indeed, that between 50 and 150 errors will occur with virtual certainty (this is again because of our assumption that the probabilities of error for individual letters are independent).

The preceding example illustrates a very general situation in digital data transmission, which is depicted in the block diagram in Fig. 8-2.

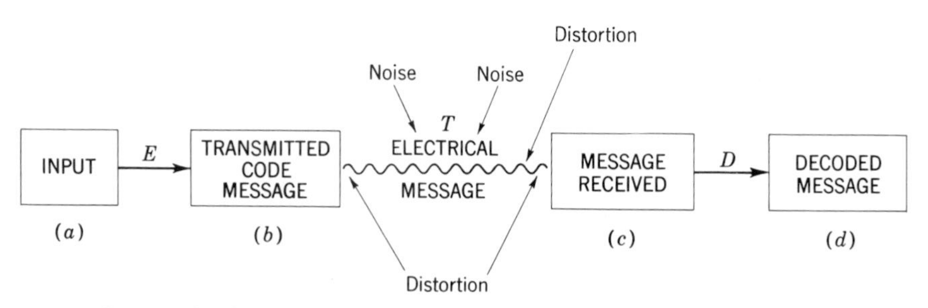

Fig. 8-2. Communications channel block diagram.

In each case, there is the information for transmission through the channel, the addition of noise to the signal transmitted through the channel, reception of the signals plus noise at the receiver, and translation of these signals plus noise back into strings of characters, some of which may contain errors due to the effects of the noise on the receiver.

The general problem facing the designer of such a communications system is as follows. Because of the physical limitations of the system (such as total transmitter power available, signal attenuation in the communications channel, and disturbing sources), the designer cannot hope to receive faithfully at the receiver an exact reproduction (even of diminished amplitude) of the signal transmitted. If the signals are grossly distorted, the receiver will make errors in "guessing" which characters were transmitted. The designer must either accept these errors or invent schemes for reducing them.

8-2. ENCODING AND DECODING

In many information processing systems, such as those arising in communication with satellites and in the operation of large-scale computers, errors can be very expensive. Hence it is important to make them extremely improbable.

This chapter will be mainly devoted to efficient ways for improving the reliability of information transmission by *systematic codes* of various kinds. Most of these efficient codes are *group codes,* the understanding of which depends on Lagrange's theorem (Sec. 7-10).

The basic idea used in all systematic codes may be described as follows. One first maps the sequence of characters to be transmitted into a longer sequence of the same characters (usually 0's and 1's) by an *encoding* scheme. By inspecting the additional information in the extra characters, the receiver is then able to detect and/or correct errors which have been introduced by noise during transmission. The longer message received is then mapped onto a sequence of characters of the original (correct) length by a *decoding* scheme.

Definition. An (m,n)-*code* for a binary message consists of an encoding scheme $E: \mathbf{2}^m \to \mathbf{2}^n$ and a decoding scheme $D: \mathbf{2}^n \to \mathbf{2}^m$, where $\mathbf{2}^n$ is the set of all binary n-tuples. Its objective is to make $H = E \diamond T \diamond D$ the identity function with probability very near 1, where T is the "error function" due to the channel (see Fig. 8-2). Since the encoding and decoding take place under controlled conditions, they may be assumed to be *error-free*. Therefore, our mathematical model of a communications system will have the nature indicated in Fig. 8-3, with deterministic E,D.

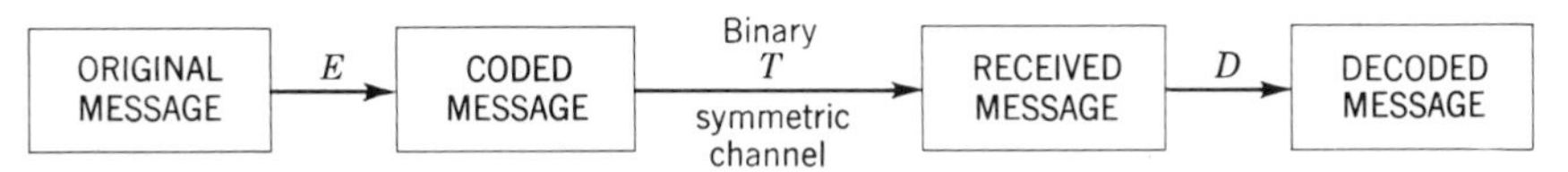

Fig. 8-3. Mathematical model of communications system.

In general, codes are classified into *error-correcting codes* in which, as suggested above, an attempt is made to correct errors which occur because of noise in the transmission channel, and *error-detecting codes*, whose purpose is simply to detect the presence of errors whose nonrecognition could be harmful. We shall now give two examples to illustrate this distinction.

Example 1. A simple error-detecting code is the following *parity-check scheme*, applicable to message blocks $\mathbf{a} = (a_1, \ldots, a_m) = a_1 \cdots a_m$. of any fixed length m. The encoding function

$$E: (a_1, \ldots, a_m) = a_1 \cdots a_m = \mathbf{a} \mapsto \mathbf{b} = b_1 \cdots b_{m+1}$$

is defined by

$$b_i = a_i \qquad \text{for} \qquad i = 1, \ldots, m \tag{2}$$

$$b_{m+1} = \begin{cases} 0 & \text{if } \sum_{i=1}^{m} a_i \text{ is even} \\ 1 & \text{if } \sum_{i=1}^{m} a_i \text{ is odd} \end{cases} \tag{2'}$$

Thus, for example, if $m = 2$, then E makes the assignments $00 \mapsto 000$, $01 \mapsto 011$, $10 \mapsto 101$, $11 \mapsto 110$. From the definitions (2) and (2′) it follows that the sum of the digits of any code word $\mathbf{b} = E\mathbf{a}$ is even.

The corresponding decoding scheme $D: \mathbf{b} \mapsto \mathbf{c}$ is defined by

$$b_i = c_i \qquad \text{for} \qquad i = 1, \ldots, m \tag{2''}$$

If $\sum_{i=1}^{m+1} c_i$ is odd, the receiver knows that *a transmission error must have occurred*. Even if the parity check sum $\sum c_i$ is even, of course, one cannot conclude with certainty that the message was transmitted correctly. However, assuming the model of a binary symmetric channel with a small error probability of q per digit, the fraction of messages which are incorrectly received is $q^3 + 3q^2p + 3qp^2$ (three, two, or one error). Of these, for our three-digit code, only double errors go undetected. Hence the fraction of errors which pass undetected is only

$$\frac{3q^2p}{q^3 + 3q^2p + 3qp^2} < \frac{q}{q+p} = q$$

The probability that an error will *not* be detected is less than q.

We next analyze in more detail an example of an error-*correcting* code. Although error correction which simply "repeats the message" is very inefficient (far from optimal), it initially appears attractive because of its simplicity.

Example 2. Consider the binary symmetric channel of Sec. 8-1 for transmitting strings of binary digits. The following $(m,3m)$ *triple-repetition code* is sometimes useful. The message is *encoded* by breaking it up into m-character "blocks," each of which is transmitted three times. The message received (after distortion by noise) is then *decoded* by breaking it up into $3m$-character blocks which should be repetitive, and then assigning to the ith digit whichever value (0 or 1) occurs at least twice (a majority of times).[1]

With this code, the probability that any one digit will be received correctly all three times is p^3; the probability of incorrect reception the first of the three times and correct reception the other two times is p^2q. The probability of incorrect reception the second or third time and correct reception the other two is similarly $p^2q + p^2q = 2p^2q$. Thus, the probability of no error is p^3, and the probability of exactly one error is $3p^2q$, giving a probability of $p^3 + 3p^2q$ that a single digit will be read correctly and a probability of $3pq^2 + q^3$ that it will be read incorrectly.

[1] If the received digits $c_ic_{i+m}c_{i+2m}$ consist of two or more 1's, then c_i is assumed to be a 1; otherwise, c_i is assumed a 0.

Now, suppose that the probability of error per digit would have been $q = 0.1$ without the code. Then each digit in any block will be read correctly all three times with the probability of 0.729 and read twice correctly with a probability of 0.243. It will be read incorrectly twice with a probability of 0.027 and read incorrectly three times with a probability of 0.001. This shows that the $(m,3m)$ code of Example 2 reduces the probability of error per digit from 10 to 2.8%.

Likewise, a code in which each block of m letters is encoded by quintuple repetition (repeating it five times) and then decoded by "majority rule" will have a probability of error of only $q^5 + 5q^4p + 10q^3p^2 = 0.00856$, or less than 1%. Consequently, the probability of transmitting correctly a *message block* 10 units long will be increased from $(0.9)^{10} \simeq 0.35$ to $(0.972)^{10} \simeq 74\%$ by the above triple-repetition code and to $(0.99144)^{10} \simeq 91.5\%$ by the majority rule five-times repetition code.

In conclusion, we note that the triple-repetition code above *corrects single errors* in message blocks; for instance, one must transmit nine digits for each three message digits (bits), which makes the transmission *three times as slow* (although the decoded message is much more reliable). In Example 4, we shall show how to achieve single-error correction with a (3,6) code, without reducing reliability.† This will increase the efficiency by one-third.

EXERCISES A

1. Show that if the binary symmetric channel of Fig. 8-1 is used to transmit two-bit words, then the probability of messages received is described by the following matrix:

Transmitted message	00	01	10	11
00	p^2	pq	pq	q^2
01	pq	p^2	q^2	pq
10	pq	q^2	p^2	pq
11	q^2	pq	pq	p^2

2. The following array is the top of what is called *Pascal's triangle.*

```
            1
          1   1
        1   2   1
      1   3   3   1
    1   4   6   4   1
  1   5  10  10   5   1
  . . . . . . . . . . .
```

   (a) Add three rows to the array.
   (b) Explain its relation to the Bernoulli formula.

† The triple-redundancy code will correct some cases of multiple errors.

3. If a code with 14-digit code words is transmitted through the binary symmetric channel of Fig. 8-1:
   (a) What is the probability that exactly five errors will occur during transmission of a word?
   (b) What is the probability that five or less errors will occur?
   (c) How many words are within distance 4 of a given code word?
4. Assume that a message of three binary digits is transmitted through a binary symmetric channel whose probability of error varies with time, and that the ith digit is transmitted correctly with probability p_i $(i = 1,2,3)$.
   (a) Show that this matrix $T = \|t_{hl}\|$ is symmetric.
   (b) Show that T is a stochastic matrix—i.e., a matrix with nonnegative real entries, such that $\sum p_{hl} = 1$ for all h.

8-3. BLOCK CODES

The two codes described above are examples of block codes. By definition, a *block code* takes a fixed number m of information characters, and replaces it by an encoded larger number of characters n (a "coded word" of some fixed length n) to be transmitted. We shall consider only such block codes below.[1]

For reasons of equipment simplicity and reliability, and physical channel characteristics, most data communication systems transmit sequences of signals representing binary symbols. As a result, we can imagine our transmitted character sequences to consist of 0's and 1's. As a result, block codes are defined by an *encoding function E* and a *decoding function D*, where, as in Sec. 8-2,

$$E\colon \mathbf{2}^m \to \mathbf{2}^n \qquad D\colon \mathbf{2}^n \to \mathbf{2}^m \qquad \mathbf{m} \leqq \mathbf{n}\dagger \tag{3}$$

that is, E takes each of the m-digit message words into an n-digit code word and D takes each n-digit code word into an m-digit message word. Any such code is called a (binary) (m,n) *block code*. As before, if 1 is the identity function, we must have $E \diamond D = D \circ E = 1$ so that the message will read correctly if there are no errors in transmission; D must be a one-sided inverse of E.

For instance, in a (15,20) block code, the message is broken up into 15-digit segments, each of which is encoded into a 20-digit code word, which is transmitted. The receiver will now receive 20-digit words, some of whose digits may be in error. The function D will now map the received

[1] There are also "convolution" or "sequential" codes which intersperse check characters with message characters, in a way which depends on the entire previous message. The encoding techniques are like those for block codes, but we shall not study them here.

† The choice $m = n$ is useful only for a cipher code, used for messages whose decoding (deciphering) function is kept secret.

word, possibly including errors, into some m-digit word, which may still contain errors.

Against this background, one optimization problem for error-correcting codes in a binary symmetric channel can be stated as follows: given m and n, find D and E which minimize the probability of a message containing errors after E has operated.

Distance between Words. The key concept of coding theory is that of the distance between code words. Each n-digit binary word

$$\mathbf{a} = (a_1, \ldots, a_n)$$

can be considered as a vertex (node) of the *hypercube* of all binary words of length n, which was already considered as a directed graph in Sec. 2-8. In this hypercube, the *distance* between two words $\mathbf{a}$ and $\mathbf{b}$ is just the number of digits i with $a_i \neq b_i$, and two words are adjacent when they differ in just one digit.

We now reconsider this hypercube as an (elementary, Abelian) *binary group* under the "exclusive" or *sum* $\mathbf{a} + \mathbf{b}$. Since every element is of order 2, this is also the difference $\mathbf{a} - \mathbf{b}$.

Definition. The *weight* $w(\mathbf{a})$ of a word $\mathbf{a} = (a_1, \ldots, a_n)$ is the number of 1's in it. The *distance* $d(\mathbf{a},\mathbf{b})$ between two binary words of length m is the weight of their exclusive-or sum, that is, $d(\mathbf{a},\mathbf{b}) = w(\mathbf{a} + \mathbf{b})$. Hence the distance $d(\mathbf{a},\mathbf{b})$ is the number of values of i such that $a_i \neq b_i$. Thus, $w(0101) = 2$, $w(1101) = 3$, $d(1011,1111) = 1$, $d(0000,0011) = 2$, and $d(1101,1101) = 0$. Note that the distance between pairs of points $\mathbf{a}$ and $\mathbf{b}$ is unchanged by any "translation" $\mathbf{x} \mapsto \mathbf{x} + \mathbf{c}$. This fact is evident since

$$d(\mathbf{a} + \mathbf{c}, \mathbf{b} + \mathbf{c}) = w(\mathbf{a} + \mathbf{b} + \mathbf{c} + \mathbf{c}) = w(\mathbf{a} + \mathbf{b}) = d(\mathbf{a},\mathbf{b})$$

Let a given message word to be encoded be the m-tuple $\mathbf{v} = (v_1,v_2, \ldots ,v_m)$, and let the corresponding encoded code word be $\mathbf{b} = (b_1,b_2, \ldots ,b_n)$ for our (m,n) code. If we assume a binary symmetric channel as in Fig. 8-2, the probability that every digit in a given code word will be correctly received is p^n. The probability that exactly one error will occur is $np^{n-1}q$. The probability of exactly two errors is $\binom{n}{2} p^{n-2}q^2$ and, in general, the probability of exactly k errors is $\binom{n}{k} p^{n-k}q^k$. The probability that l or fewer errors will occur is

$$p^n + \binom{n}{1} p^{n-1}q + \binom{n}{2} p^{n-2}q^2 + \cdots + \binom{n}{l} p^{n-l}q^l \tag{4}$$

Let us now relate the distance function $d(\mathbf{b},\mathbf{b}^*)$ to error probabilities. The probability that the code word $\mathbf{b}$ will be changed into the code word

$\mathbf{b}^*$ is simply $p^{n-d(b,b^*)} \cdot q^{d(b,b^*)}$. Thus the probability that 0011 will be turned into 1011 is p^3q.

Notice that to *detect* all single-error patterns, the code words must have a minimum distance of 2 between any two words $\mathbf{a}$ and $\mathbf{b}$. If, for instance, 011 and 001 are both code words, then an error in the second position will change one word into the other and make it impossible to detect whether or not an error has occurred.

Theorem 2. *For a code to detect all sets of k or fewer errors, it is necessary and sufficient that the minimum distance between any two code words be $k + 1$ or more.*

If we have an error-detecting code with a minimum distance of $k + 1$ between any two code words, then the probability that a word will be received erroneously without being detected is at most

$$q^n + \binom{n}{1} pq^{n-1} + \cdots + \binom{n}{k+1} p^{n-k-1}q^{k+1} \tag{5}$$

For small q and adequate k, this is about $\binom{n}{k+1} p^{n-k-1}q^{k+1}$; moreover, the actual probability (in a binary symmetric channel) of errors escaping undetected will be much less than the above value, as we shall see.

Theorem 3. *For a code $[D,E]$ to correct all sets of k or fewer errors, it is necessary that the minimum distance between code words be at least $2k + 1$.*

Conversely, when this condition is fulfilled, one can always construct D so as to correct all such error vectors. For example, for the simple (1,3) code with E: $0 \mapsto 000$, $1 \mapsto 111$, we can use the decoding function

$$\begin{array}{ll} 000 \mapsto 0 & 111 \mapsto 1 \\ 001 \mapsto 0 & 011 \mapsto 1 \\ 010 \mapsto 0 & 101 \mapsto 1 \\ 100 \mapsto 0 & 110 \mapsto 1 \end{array}$$

at the receiver, and any single error will be corrected.

More generally, we have, by Theorem 1, the following result.

Theorem 4. *For a code which automatically corrects k or fewer errors, with a binary symmetric channel the probability of incorrect interpretation of an n-bit message is at most*

$$\binom{n}{k+1} p^{n-k-1}q^{k+1} + \cdots + \binom{n}{1} pq^{n-1} + q^n \tag{6}$$

The probability of *correct* decoding is, correspondingly, at least

$$p^n + \binom{n}{1} p^{n-1}q + \cdots + \binom{n}{k} p^{n-k}q^k \tag{7}$$

Actually, we may do better, for k is our minimum distance. However, few codes based on algebraic decoding procedures (rather than tables) do appreciably better.

It will be convenient to consider error n-tuples $\mathbf{e} = e_1e_2 \cdots e_n$ and to represent our system as follows. We encode by mapping a given m-bit message word $\mathbf{a} = a_1a_2 \cdots a_m$ into a code word $\mathbf{b} = b_1b_2 \cdots b_n$. The channel then "adds" the *error n-tuple* $e_1e_2 \cdots e_n$, giving $b_1 + e_1$, $b_2 + e_2, \ldots, b_n + e_n$, and we designate this as $\mathbf{r} = r_1r_2 \cdots r_n$, where $r_i = b_i + e_i$. An error-*correcting* system would then map $r_1r_2 \cdots r_n$ into the nearest code word $b_1b_2 \cdots b_n$. An error-*detecting* system would simply see if $\mathbf{r} = r_1r_2 \cdots r_n$ was a code word and would say an error had occurred if this was not the case.

As an example, suppose that our message is $\mathbf{a} = 01$, that our encoder maps 01 into $\mathbf{b} = 0110$, and that the channel generates an error pattern $\mathbf{e} = 0010$. The received word is then $\mathbf{r} = 0100$; an error-*correcting* system would map $\mathbf{r} = 0100$ back into 0110 and finally into the correct message 01.

The consideration of error n-tuples $\mathbf{e}$ and their effect on code words is also illuminating when error detection and error correction are considered.

When a code is single-error detecting, no error n-tuple with a single 1, that is, no error n-tuple $\mathbf{e}$ of weight 1, can be added to a code word $\mathbf{b}$ so as to make (ambiguously) $\mathbf{b}^* = \mathbf{b} + \mathbf{e}$ some other code word $\mathbf{b}^* \neq \mathbf{b}$. For example, consider the (2,3) parity-check code, with encoding scheme

$$E: 00 \mapsto 000, \quad 10 \mapsto 101, \quad 01 \mapsto 011, \quad 11 \mapsto 110$$

With B the set of code words 000, 011, 101, 110, no error word 000, 001, 010, 100 can change one of these code words into another, so any occurrence of just one error will be detected (more precisely, one or three errors will be detected).

A two-error detecting code is a code where the addition of no error n-tuple with one or two 1's will change one code word into another.

Example 3. The following (2,5) encoding scheme is two-error detecting.

$$E: \quad \begin{array}{ll} 00 \mapsto 00000 = \mathbf{b}^1 & 01 \mapsto 01011 = \mathbf{b}^2 \\ 10 \mapsto 10101 = \mathbf{b}^3 & 11 \mapsto 11110 = \mathbf{b}^4 \end{array}$$

This encoding scheme can be given single-error correcting capability because all code words are separated by at least three units. Since $d(\mathbf{b}^i,\mathbf{b}^j) \geqq 3$ for any $\mathbf{b}^i \neq \mathbf{b}^j$, a single error will leave any transmitted message a unit distance from only one code word, the right one. For no two distinct code words $\mathbf{b}$ and $\mathbf{b}^*$ do there exist error patterns $\mathbf{e}$ and $\mathbf{e}^*$, each containing a single 1 (that is, with $w(\mathbf{e}) = w(\mathbf{e}^*) = 1$), such that $\mathbf{b} + \mathbf{e} = \mathbf{b}^* + \mathbf{e}^*$.

Thus, a decoding scheme which simply moves the received word into the nearest code word [i.e., that minimizes $d(\mathbf{b}^i,\mathbf{c})$] will implement a single-error correcting code. Thus, assuming a binary symmetric channel, transmission errors will be correctly decoded with probability of at least $p^5 + 5p^4q$.

Notice that an error pattern $\mathbf{e} = 00 \cdots 0$ indicates that no errors occur, $\mathbf{e} = 100 \cdots 0$ indicates an error in the first position, etc. The most probable error pattern which can occur is $\mathbf{e} = 00 \cdots 0$. The probability of this is p^n; clearly, $p^{n-k}q^k$, the probability of a particular error pattern of $n - k$ 0's and k 1's, is always less than p^n for $k > 0$. Similarly, using induction, one can show that $p^{n-1}q$, the probability of a given error pattern with a single 1, is greater than the error probability for any error pattern with more than a single 1.

This argument can be repeated until we deduce that it is best, from a probabilistic viewpoint, to assume that an error pattern with the least possible number of 1's has occurred; this corresponds to picking the code word which minimizes $d(\mathbf{b},\mathbf{c}.)$

EXERCISES B

1. Show that a binary code with minimum distance 7 is capable of detecting any pattern of six or fewer errors, or of correcting any pattern of three or less errors.
2. If a single parity-check code is used for eight message digit blocks, an (8,9) code results. For the binary symmetric channel, what is the probability that an error will go undetected in a nine-digit block?
3. Suppose that a (4,5) single parity-check code is used for error detection and a (4,12) triplicating single-digit code is used for the binary symmetric channel. Compare the probability of erroneous digits escaping for $p = 0.9$ and $q = 0.1$.
4. Often error correction and error detection are mixed in the same code. A code with minimum distance 4 can be used for single-error correction and double-error detection. If the code is a (4,8) code, analyze the probability of an error passing through the system for the binary symmetric channel.

8-4. MATRIX ENCODING TECHNIQUES

In the preceding sections, we have specified various encoding schemes by means of "dictionaries" which list beside each message word $\mathbf{a}$ the corresponding code word $\mathbf{b} = E[\mathbf{a}]$. Such lists become very long for codes with many code words and make impractical demands on storage capacity.

We shall now describe a systematic algebraic technique for encoding binary words by *matrix multiplication*,[1] which makes much smaller demands on storage. Let E be an $m \times n$ *encoding matrix* (or *generator matrix*) of 0's and 1's. We write $E = \|e_{ij}\|$, where e_{ij} is the entry in the ith row and jth column. Letting $+$ stand for addition mod 2 (Boolean sum), we can describe encoding by the following equation:

$$b_j = a_1e_{1j} + a_2e_{2j} + \cdots + a_me_{mj} = \sum_{i=1}^{m} a_ie_{ij} \qquad j = 1, \ldots, n \qquad (8)$$

Evidently, formula (8) defines encoding for an (m,n) code.

Example 4. Consider the following 3×6 encoding matrix:

$$E = \begin{bmatrix} 1 & 0 & 0 & 1 & 1 & 0 \\ 0 & 1 & 0 & 0 & 1 & 1 \\ 0 & 0 & 1 & 1 & 1 & 1 \end{bmatrix}$$

If $\mathbf{a} = 100$, then for $\mathbf{b} = \mathbf{a}E$, (8) reduces to $b_j = e_{1j}$ (dropping terms equal to zero); the code word for $\mathbf{a}^1 = 100$ is thus the first row of the matrix E. Likewise, the code word for $\mathbf{a}^2 = 010$ is the second row of E, and the code word for $\mathbf{a}^4 = 001$ is the third row of E.

We can describe the complete (3,6) encoding defined by the matrix E by the following list ("dictionary"):

$$\begin{aligned} \mathbf{a}^0 &= 000 \mapsto 000000 \\ \mathbf{a}^1 &= 100 \mapsto 100110 \\ \mathbf{a}^2 &= 010 \mapsto 010011 \\ \mathbf{a}^3 &= 110 \mapsto 110101 \\ \mathbf{a}^4 &= 001 \mapsto 001111 \\ \mathbf{a}^5 &= 101 \mapsto 101001 \\ \mathbf{a}^6 &= 011 \mapsto 011100 \\ \mathbf{a}^7 &= 111 \mapsto 111010 \end{aligned}$$

This list illustrates an advantage of encoding by matrix multiplication—one need only store m code words and not 2^m code words. The preceding discussion applies generally.

For a code to be useful, clearly all code words must be different. This requirement can be met by the simple device of letting the $m \times m$ square submatrix consisting of the first m columns of E be I_m, the $m \times m$ identity matrix having 1's along the main diagonal (upper left to lower right) and 0's elsewhere, as in the example chosen. We omit the proof.

[1] In order to write words horizontally (as "row vectors"), we will be multiplying by matrices on the *right*.

Group Addition. We now recall from Example 13 of Sec. 7-7 that the set of all message words $\mathbf{a} = a_1 \cdots a_m$ of m binary digits forms an *Abelian group* $\mathbf{Z}_2{}^m$ under addition of digits mod 2. The code words $\mathbf{b} = b_1 \cdots b_n$ form a similar Abelian group.

Now suppose that $\mathbf{a} = \mathbf{a}' + \mathbf{a}''$ in the Abelian group of all message words, which means that $a_i \equiv a_i' + a_i''$ (mod 2) for $i = 1, \ldots, m$. For $\mathbf{b} = \mathbf{a}E$, $\mathbf{b}' = \mathbf{a}'E$, $\mathbf{b}'' = \mathbf{a}''E$, we clearly have (all in the field $\mathbf{Z}_2$)

$$\begin{aligned} b_j &\equiv \sum_{i=1}^{m} a_i e_{ij} \equiv \sum_{i=1}^{m} (a_i' + a_i'')e_{ij} \\ &\equiv \sum_{i=1}^{m} a_i' e_{ij} + \sum_{i=1}^{m} a_i'' e_{ij} \equiv b_j' + b_j'' \end{aligned}$$

all mod 2. In summary, we have proved that

$$E[\mathbf{a}] = \mathbf{b} = \mathbf{b}' + \mathbf{b}'' = E[\mathbf{a}'] + E[\mathbf{a}''] \tag{9}$$

In terminology of Chap. 7, this conclusion can be restated as follows.

Theorem 5. *For any $m \times n$ encoding matrix E with an $m \times m$ submatrix I_m, the (m,n) code $\mathbf{a} \mapsto \mathbf{a}E = E[\mathbf{a}]$ is a monomorphism from the additive group of message words to the additive group of code words.*

8-5. GROUP CODES

Theorem 5 shows that (binary) matrix codes are group codes, in the sense of the following definition.

Definition. When the code words in a block code form an additive group, the code is called a *group code*. The following result is of considerable help in determining the error-detecting or error-correcting capabilities of group codes.

Theorem 6. *For a group code with code words $\mathbf{b}^0, \mathbf{b}^1, \ldots, \mathbf{b}^{2^m-1}$, the minimum distance $d(\mathbf{b}^i, \mathbf{b}^j)$ (where $\mathbf{b}^i \neq \mathbf{b}^j$) between any two code words is equal to the minimum weight $w(\mathbf{b}^k)$ for some nonzero code word.*

Proof. The code words $\mathbf{b}^j$ form an additive group. Since the minimum distance is $d(\mathbf{b}^i, \mathbf{b}^j)$ for some $\mathbf{b}^i$ and $\mathbf{b}^j$ ($\mathbf{b}^i \neq \mathbf{b}^j$) and since $\mathbf{b}^i + \mathbf{b}^j$ is also a code word, say $\mathbf{b}^k$, one has $w(\mathbf{b}^k) = d(\mathbf{b}^i, \mathbf{b}^j)$ by definition of weight and distance, proving the theorem.

For the group code in Example 4, there are four code words of weight 3. Therefore, the minimum distance is 3, and the code will correct all occurrences of single errors or detect any double error.

It is natural to ask what error patterns will pass undetected. In group codes, these are simply those corresponding to nonzero code words. For instance, for Example 4, the error pattern $e = 100110$ will change any code word into another code word because given any $\mathbf{b}^i$, $\mathbf{b}^i + \mathbf{e} = \mathbf{b}^j$ for some $\mathbf{b}^j$, and every code word will be changed into another by such an error pattern.

This rule also enables us to determine the probability of making an error (if error detection is the objective) since if we sum the probabilities of the error patterns consisting of the nonzero code words, this sum gives us the probability the transmitted code word will be changed into another code word, thus foiling the error-detecting system. In Example 4, the probability of an error escaping undetected is $4p^3q^3 + 3p^2q^4$.

Optimal Decoding. We now consider the problem of optimizing the *decoding* of a given group encoding scheme $E: \mathbf{a} \mapsto \mathbf{a}E$ with an $m \times n$ encoding matrix E (with binary entries $e_{ij} \in \mathbf{Z}_2$); that is, we shall consider the problem of minimizing the probability that an error will be made (i.e., the probability that $D[\mathbf{a}E] \neq \mathbf{a}$). We will solve this problem under two assumptions: (1) that all message words are equally probable, and (2) that communication is through a binary symmetric channel.

The decoding scheme is dependent on a decoding table which lists all possible words which can be received. The decoding table is constructed using Lagrange's theorem (Sec. 7-10). Since the code words form a subgroup $\mathbf{B}$ of the set of all receivable words $\mathbf{C}$, we can construct a table of $\mathbf{C}$ modulo $\mathbf{B}$ by listing $\mathbf{B}$ as a first row, placing $\mathbf{0}$ in the left-most position, thus

$$\mathbf{0} \quad \mathbf{b}^1 \quad \mathbf{b}^2, \quad \ldots, \quad \mathbf{b}^{2^m-1}$$

Then, some word in $\mathbf{c}^i \in \mathbf{C}$ which is not a $\mathbf{b}^j \in \mathbf{B}$ is selected and a new row or *coset* is formed by adding $\mathbf{c}^i$ to each $\mathbf{b}^j$ as follows

$$\begin{array}{lllll} \mathbf{0} & \mathbf{b}^1 & \mathbf{b}^2 & , \ldots, & \mathbf{b}^{2^m-1} \\ \mathbf{0} + \mathbf{c}^i & \mathbf{b}^1 + \mathbf{c}^i & \mathbf{b}^2 + \mathbf{c}^i & & \mathbf{b}^{2^m-1} + \mathbf{c}^i \end{array}$$

This row is then rewritten, if necessary, with the element $\mathbf{b}^j + \mathbf{c}^i$ of least weight (i.e., so that no element $\mathbf{b}^k + \mathbf{c}^i$ has less weight) in the left-most position. This is called the *coset leader*. Call this coset leader $\mathbf{c}^1$ and rewrite the row as follows

$$\begin{array}{lllll} \mathbf{0} & \mathbf{b}^1 & \mathbf{b}^2 & & \mathbf{b}^{2^m-1} \\ \mathbf{c}^1 & \mathbf{b}^1 + \mathbf{c}^1 & \mathbf{b}^2 + \mathbf{c}^1, & \ldots, & \mathbf{b}^{2^m-1} + \mathbf{c}^1 \end{array}$$

Now select some other element not in the above two rows and form a third row, then select the word of least weight in that row and rewrite the row with this element as coset leader calling it $\mathbf{c}^2$. Continue this until each word in $\mathbf{c}$ is somewhere in the table.

Lagrange's theorem guarantees that if the above procedure is followed, no word will be duplicated in the table. That is, consider a row or coset $\mathbf{c}^i + B$ which is $\{\mathbf{c}^i + \mathbf{b}\} : \mathbf{b} \in B$, then each element $\mathbf{c} \in \mathbf{C}$ is uniquely representable as $\mathbf{c}^j + \mathbf{b}^k$ for some coset leader $\mathbf{c}^j$ and code word $\mathbf{b}^k$.

The final decoding table has the form

$$\begin{array}{llllll}
\mathbf{0} & \mathbf{b}^1 & \mathbf{b}^2 & , & \ldots, & \mathbf{b}^{2^m-1} \\
\mathbf{c}^1 & \mathbf{b}^1 + \mathbf{c}^1 & \mathbf{b}^2 + \mathbf{c}^1 & , & \ldots, & \mathbf{b}^{2^m-1} + \mathbf{c}^1 \\
\mathbf{c}^2 & \mathbf{b}^1 + \mathbf{c}^2 & \mathbf{b}^2 + \mathbf{c}^2 & , & \ldots, & \mathbf{b}^{2^m-1} + \mathbf{c}^2 \\
\ldots & \ldots & \ldots & & \ldots & \ldots \\
\mathbf{c}^{2^{n-m}-1} & \mathbf{b}^1 + \mathbf{c}^{2^{n-m}-1} & \mathbf{b}^2 + \mathbf{c}^{2^{n-m}-1}, & & \ldots, & \mathbf{b}^{2^m-1} + \mathbf{c}^{2^{n-m}-1}
\end{array}$$

To decode, a received word $\mathbf{c} = \mathbf{b}^i + \mathbf{c}^j$ is located in the table and the code word $\mathbf{b}^i$ at the head of the column is taken to be the code word that was transmitted.

For B itself, this coset leader will be $\mathbf{e}^0 = 0$. By Lagrange's theorem, C will be the disjoint sum of the cosets $\mathbf{e}^i + B$. Indeed, as was shown in the proof of Lagrange's theorem, a sharper result holds.

Lemma 1. *In any group code, binary or not, each* $\mathbf{c} \in C$ *can be uniquely expressed as the sum* $\mathbf{c} = \mathbf{e}^i + \mathbf{b}$ *of a code word* $\mathbf{b} \in B = \mathbf{a}E$ *and a coset leader* $\mathbf{e}^i$.

8-6. DECODING TABLES

The above can be made more tangible by constructing a specific *decoding table D*. A decoding table for the (3,6) code of Example 4 is the following:

Row of code words →	000000	100110	010011	011100	001111	101001	110101	111010
	100000	000110	110011	111100	101111	001001	010101	011010
	010000	110110	000011	001100	011111	111001	100101	101010
	001000	101110	011011	010100	000111	100001	111101	110010
	000100	100010	010111	011000	001011	101101	110001	111110
	000010	100100	010001	011110	001101	111011	110110	111000
	000001	100111	010000	011101	001110	101000	110100	111011
	000101	100011	010110	011001	001010	101100	110000	111111
	↑ Column of coset leaders							

To *decode* a received word $\mathbf{b}^j + \mathbf{e}$, the n-tuple is located in the table, and the code word at the head of the column is then selected and taken to be the transmitted code word. For instance, if 110011 is received, the code word transmitted is taken to be 010011; if 100101 is received, the code word is taken to be 110101; if 110101 is received, we assume 110101 was transmitted, etc.

Just what error patterns[1] will such a group code correct? It will correct all error patterns which occur as coset leaders and no others. The error pattern 000001 will always be corrected, whereas the error pattern 010001 will always lead to an error after decoding. This code will correct all occurrences of single errors and also the double error 000101, which is the only coset leader of weight $w > 1$. We therefore have the following theorem.

Theorem 7. *Group encoding with decoding by coset leaders corrects precisely those error patterns which are coset leaders.*

For a given binary symmetric channel, if a sum is formed of the probabilities of the occurrences of each error pattern (including the all-zero error pattern) which occurs as a coset leader, the value of this sum will give the probability that a transmitted code word will be received and correctly decoded. In Example 4, the probability of correct transmission of a 6-tuple is $p^6 + 6p^5q + p^4q^2$.

The question as to whether or not this is a good scheme can be answered as follows.

Theorem 8. *Let a table be formed in which each coset of the subgroup of all code words is listed under the subgroup of code words and in which each coset leader is the word of least weight in its coset. This leads to a decoding scheme where the distance between the code word at the head of each column is less than or equal to the distance to any other code word.*

In other words, decoding by the coset leader nearest **0** selects a "closest" code word in the sense that no other code word will have less distance. Recall that the selection each time of the code word having least distance is *optimal* in that it minimizes the probability of error for the binary symmetric channel.

Proof. Suppose that we transmit $\mathbf{b}^i$ and receive $\mathbf{b}^i + \mathbf{e}$. [The distance of $\mathbf{b}^i + \mathbf{e}$ from $\mathbf{b}^i$ is $w(e) = D(\mathbf{b}^i + \mathbf{e}, \mathbf{b}^i)$.] Now, if we add $\mathbf{e}$ to each $\mathbf{b}^i$ in the code, we form a coset or row in the table; by then picking the

[1] By definition, an error pattern is a sequence giving the characteristic function (Chap. 1) of the set of erroneous digits.

Standard Array

	Coset leaders ↓								
Row of code words	l_0 = 000000	100110	010011	001101	110101	101011	011110	111000	
	l_1 = 000001	100111	010010	001100	110100	101010	011111	111001	Each
	l_2 = 000010	100100	010001	001111	110111	101001	011100	111010	row
	l_3 = 000100	100010	010111	001001	110001	101111	011010	111100	is a
	l_4 = 001000	101110	011011	000101	111101	100011	010110	110000	coset
	l_5 = 010000	110110	000011	011101	100101	111011	001110	101000	
	l_6 = 100000	000110	110011	101101	010101	001011	111110	011000	
	l_7 = 100001	000111	110010	101100	010100	001010	'11111	011001	

$$\text{Encoding matrix is } E = \begin{bmatrix} 100110 \\ 010011 \\ 001101 \end{bmatrix} \qquad \mathbf{a}^j E = \mathbf{b}^j$$

$$\text{Parity-check matrix is } F = \begin{bmatrix} 101100 \\ 110010 \\ 011001 \end{bmatrix}, \qquad \mathbf{b}^j F^T = 0 \qquad EF^T = 0$$

Fig. 8-4. A (3,6) code, decoding table for Example 5.

n-tuple of least weight in this coset as coset leader, we form a row of the table and effectively minimize $d(\mathbf{b}^i + \mathbf{e}, \mathbf{b}^j)$ for each $\mathbf{b}^j$ and hence for $\mathbf{b}^i$, our transmitted code word. Notice this holds if $\mathbf{e}$ is another code word $\mathbf{b}^k$ or an error pattern $\mathbf{e}$ which contains more errors than the code will correct, as well as for correctable patterns. The procedure moves a received n-tuple into a nearest code word in all cases.

Geometrical Interpretation. The notion of *optimality* mentioned above has a simple geometric interpretation. In any (m,n) binary code, the *encoding* scheme E defines an injection (one-one mapping) from the set of 2^m message words to the *hypercube* $\mathbf{C} = \mathbf{2}^n$ of all n-bit binary words. In the *graph* whose nodes and links are the vertices and edges of this cube, the distance between two words has its usual meaning for graphs.

Each *decoding* scheme D maps $\mathbf{2}^n$ back onto $\mathbf{2}^m$, so that the antecedents of each m-bit binary word form an equivalence class which contains exactly one code word $E[\mathbf{a}]$. To correct every error of weight w or less, each equivalence class must contain the "ball" of radius w and center $E[\mathbf{a}]$, consisting of all words (vertices) whose distance from the code word in the class is w or less; there are $1 + n + \binom{n}{2} + \cdots + \binom{n}{w}$ such vertices. A code in which each equivalence class contains these vertices and no others is called *perfect*. It is very hard to construct perfect codes.

Indeed, for an (m,n) binary block code to be perfect, it is necessary that $1 + n + \cdots + \binom{n}{w} = 2^{n-m}$, and there are few positive integers w with $1 < w < (n - 1)/2$ for which $1 + n + \cdots + \binom{n}{w}$ is a power of 2, as this requires. Indeed, no general method is even known for constructing an (m,n) binary block code which is *optimal* in the sense that it maximizes the probability of correcting random errors. This requires "packing" as many nonoverlapping "spheres" of radius w in the n-cube as possible.

Example 5. Figure 8-4 shows a binary group code, its encoding matrix, standard array (table), and probabilities. If α_i is the number of coset leaders of weight i, then the probability of correct decoding is $P = \sum \alpha_i p^{n-i} q^i$, where α_i is the number of coset leaders of weight i.

For the above code, $\alpha_0 = 1$, $\alpha_1 = 6$, $\alpha_2 = 1$, and all other α_i are 0. The code is quasi-perfect in that it corrects all single errors, one double error, and no better (3,6) code can be found.

EXERCISES C

1. The following is the encoding matrix G for an error-detecting code:

$$G = \begin{bmatrix} 1 & 0 & 0 & 0 & 0 & 1 & 1 & 1 \\ 0 & 1 & 0 & 0 & 1 & 0 & 1 & 1 \\ 0 & 0 & 1 & 0 & 1 & 1 & 0 & 1 \\ 0 & 0 & 0 & 1 & 1 & 1 & 1 & 0 \end{bmatrix}$$

   Each code word has four message digits followed by four check digits. If used as an error-detecting code on a binary symmetric channel, what is the probability that a transmitted code word will be changed by the channel into another code word, thereby causing a mistake at the receiver?
2. If we omit the bottom row and the fourth column of the matrix G in Exercise 1, we have a three-message digit, four-check digit code.
   (a) Write this code in standard array form, and assume that all the capabilities of the code are used for error correction (more than single errors can be corrected).
   (b) What is the probability that a given code word will be in error after the decoding process?
   (c) If the code is used for error detection, what is the probability of committing an error?
3. (a) Prove that any group translation $\mathbf{x} \mapsto \mathbf{x} + \mathbf{c}$ of the group $[\mathbf{Z}_2{}^r, +]$ preserves distance between code words.
   (b) Prove that in a binary group code, every code word is at the same minimum distance from the set of the other code words.

IN AN n-CUBE, LET Δ_l BE A SET OF VERTICES NO TWO OF WHICH ARE CONNECTED BY A PATH OF LENGTH LESS THAN l.

4. Prove that if $m = 3$, then Δ_3 can contain at most two members.
5. Prove that Δ_5 can contain at most $2^{n+1}/(n^2 + n + 2)$ members.

$$\left[\textit{Hint: } \binom{n}{2} + \binom{n}{1} + \binom{n}{0} = (n^2 + n + 2)/2. \right]$$

A CODE C IS SAID TO BE *equivalent* TO A CODE C^* WHEN THERE IS A DISTANCE—PRESERVING BIJECTION BETWEEN THE CODE WORDS OF C AND THOSE OF C^*.

6. Given an encoding (generator) matrix G for a code, show that any (a) permutation of the rows of G, (b) permutation of the columns of G, or (c) addition of a scalar multiple of a row of G to another row will result in the new matrix G^* for a code which is equivalent to the code generated by G.
7. A matrix of the form $[I_m G]$, where I_m is a square identity matrix and G an $m \times l$ matrix and the two form an $m \times (m + l)$ matrix, is said to be in standard coding form. For instance, the matrix G in Exercise 1 is in standard coding form. Form a matrix G^* in standard coding form which will generate a code equivalent to that generated by G as follows:

$$G = \begin{bmatrix} 0 & 0 & 0 & 1 & 1 & 1 \\ 0 & 1 & 1 & 0 & 1 & 1 \\ 1 & 1 & 0 & 0 & 0 & 1 \end{bmatrix}$$

8. (a) Given an encoding matrix G, show that there exists a *parity-check matrix* H such that

   (*i*) If G is $m \times n$, H is $n \times (n - m)$.

   (*ii*) $GH = 0$.

   (*iii*) Each column $h_i, 1 \leq i \leq n - m$ of H is such that for no scalars θ_i, not all zero, does $\sum_{i=1}^{n-m} a_i h_i = 0$, that is, the columns of H are linearly independent.

   H is called a *parity-check matrix* because, for any code word **c** in the code generated by G, $cH = 0$.

   (b) Show that given a parity-check matrix for a code, the minimum weight of a code word (not the 0 code word) is equal to the minimum number of rows of H which can be added together to give 0.

9. Consider the three-message-digit, six-digit code word error-detecting binary group code with generator matrix

$$G = \begin{bmatrix} 1 & 0 & 0 & 0 & 1 & 1 \\ 0 & 1 & 0 & 1 & 0 & 1 \\ 0 & 0 & 1 & 1 & 1 & 1 \end{bmatrix}$$

   What is the probability that a message of six digits will be received and accepted as correct, when in fact at least one error has occurred and not been detected (assume a binary symmetric channel)?

10. Assume we have a parity-check matrix $A_{m,m+K} = I_m A_{m,k}$ for a code which will correct r errors. Show that the parity-check matrix

$$\left[\begin{array}{cc|c} I_{m,m} & A_{m,k} & 0_{m,1} \\ \hline \multicolumn{2}{c|}{1_{1,m+k}} & 1 \end{array}\right]$$

   results in a code which will correct r errors and detect $r + 1$ errors. Here I_m is a matrix

$$I_m = \overbrace{\begin{bmatrix} 1 & 0 & 0 & \cdot & \cdot & \cdot & 0 \\ 0 & 1 & 0 & 0 & \cdot & \cdot & \cdot \\ 0 & 0 & 1 & \cdot & \cdot & \cdot & \cdot \\ \cdot & \cdot & \cdot & \cdot & \cdot & \cdot & \cdot \\ \cdot & \cdot & \cdot & \cdot & \cdot & \cdot & \cdot \\ 0 & 0 & 0 & 0 & 0 & 0 & 1 \end{bmatrix}}^{m} \Bigg\} m$$

11. Consider the binary group code whose generator matrix is

$$\begin{bmatrix} 1 & 0 & 1 & 0 & 1 & 1 \\ 0 & 1 & 1 & 1 & 1 & 0 \\ 0 & 0 & 0 & 1 & 1 & 1 \end{bmatrix}$$

   (a) Find a generator matrix G in standard coding form for an equivalent code (Exercise 7 defines standard coding form).

   (b) Find the parity-check matrix H for the code in (a).

   (c) Find the code word that has 110 as information symbols. Show that it is in the row space of G and in the null space of H.

★12. Consider a group code with code words $C_0, C_1, \ldots, C_q$ ($q = 2^k - 1$) in a communications system where k-bit words are encoded into n-bit words. Form the decoding table

$$\begin{array}{cccc} C_0 & C_1 & & C_q \\ C_0 + S_1 & C_1 + S_1 & \cdots & C_q + S_1 \\ \cdots & \cdots & \cdots & \cdots \\ C_0 + S_r & C_1 + S_r & \cdots & C_q + S_r \end{array}$$

where $C_0 = 0000 \cdots 0$

where $C_0 + S_i$ has the least weight in the ith row

where $r = 2^{n-k} - 1$. Compute the above array for the code of Exercise 11.

13. Given a parity-check matrix H for a code, the product of a received word $\mathbf{r}$ and H is called a *syndrome* of $\mathbf{r}$; that is, $\mathbf{r}H$ is the syndrome of $\mathbf{r}$. Show that there is a bijection from the syndromes of a code to the coset leaders in the table in Exercise 10.

8-7. HAMMING CODES

Having studied the preceding algebraic technique for decoding a given group code, let us now examine an ingenious family of perfect codes originated by R. W. Hamming.

The Hamming codes are single-error-correcting codes (with minimum distance 3) which are *perfect* in that for any r there exists a ($m = 2^r - 1 - r$, $n = 2^r - 1$) code which corrects each single error which might occur, no other errors, and no ($m = 2^r - 1 - r$, $n = 2^r - 1$) code can possibly be constructed which will correct more than all single errors. This says, in effect, that if a decoding table is constructed as in the previous section, the coset leaders will consist of $\mathbf{0}$ and exactly the $2^r - 1$ error patterns with a single 1. In general, a code which has all error patterns with weight up to some integer k as coset leaders and no error patterns of weight more than k is said to be "perfect" because no better code for correcting random errors can be found. ("Quasi-perfect" means that some coset leaders of weight $k + 1$ occur, but not all, and no coset leaders of weight greater than $k + 1$ occur.)

Hamming codes also provide a simple decoding scheme for locating the error when a single error occurs. Hamming codes can be of length other than $2^r - 1$. They can be of arbitrary length, and the same basic encoding and decoding scheme applies. But we shall describe only the perfect codes, as the others follow in a straightforward manner and may be found in the exercises.

The procedure for forming a Hamming code is as follows:

1. Choose a positive integer r. The code words will then have $2^r - 1$ digits, and the message words will have $2^r - 1 - r$ digits.
2. In each code word $\mathbf{b} = b_1, b_2, \ldots, b_{2^r-1}$ use $b_{2^0}, b_{2^1}, b_{2^2}, \ldots, b_{2^{r-1}}$ as check digits, and place the $2^r - 1 - r$ message digits in the remaining b_j. For instance, if $r = 4$, then b_1, b_2, b_4, b_8 are used as check digits, and b_3, b_5, b_6, b_7, b_9, b_{10}, b_{11}, b_{12}, b_{13}, b_{14}, b_{15} are used for the 11 message digits.

3. Form a matrix M of $2^r - 1$ rows and r columns, where the row i is the binary number with value i. The matrices for $r = 2$, 3, and 4 are

$$M_{3,2} = \begin{bmatrix} 01 \\ 10 \\ 11 \end{bmatrix} \qquad M_{7,3} = \begin{bmatrix} 001 \\ 010 \\ 011 \\ 100 \\ 101 \\ 110 \\ 111 \end{bmatrix} \qquad M_{15,4} = \begin{bmatrix} 0001 \\ 0010 \\ 0011 \\ 0100 \\ 0101 \\ 0110 \\ 0111 \\ 1000 \\ 1001 \\ 1010 \\ 1011 \\ 1100 \\ 1101 \\ 1110 \\ 1111 \end{bmatrix}$$

4. Form $\mathbf{b}M = 0$ using the M in step 3. This gives r linear equations such as, for $r = 3$,

$$\begin{aligned} b_4 + b_5 + b_6 + b_7 &= 0 \\ b_2 + b_3 + b_6 + b_7 &= 0 \\ b_1 + b_3 + b_5 + b_7 &= 0 \end{aligned}$$

5. To encode a message, place the message digits in the correct b_j, and then make the check digits b_{2^i} ($i = 0, \ldots, r - 1$) satisfy the linear equations above. There will be exactly one b_{2^i} in each equation, making this a simple task. For example, in the equations above b_4 occurs in the first relation, b_2 in the second, and b_1 in the third.

The above procedure yields code words $\mathbf{b}$ which will have minimum weight 3 (except for the all-zero code word $\mathbf{0}$).

There is a straightforward way to decode these codes. A received vector consists of a code word $\mathbf{b}$ plus an error vector $\mathbf{e}$; now $\mathbf{b}M = 0$ so $(\mathbf{b} + \mathbf{e}) = \mathbf{b}M + \mathbf{e}M = 0 + \mathbf{e}M$. The result of the multiplication is therefore the same as the result of multiplying the error vector times M. If the error vector is 0, the result is 0, indicating that no errors have occurred. If the error vector has weight 1, that is, is of the form $\mathbf{e} = 00 \cdots 1 \cdots 00$, with the 1 in the ith position, then when this is multiplied by M, the 1 "picks out" the ith row of the matrix M, which is the binary number i, giving the number i to indicate which digit is in error in $\mathbf{b} + \mathbf{e}$. Simply changing this digit will correct the code word.

Example 6. The Hamming (4,7) code has as a code word $\mathbf{b} = 0001111$. The matrix M is

$$\begin{bmatrix} 001 \\ 010 \\ 011 \\ 100 \\ 101 \\ 110 \\ 111 \end{bmatrix}$$

Now $\mathbf{b}M = 000$. If the error vector $\mathbf{e} = 0010000$ is added to $\mathbf{b}$, we have $\mathbf{b} + \mathbf{e} = 0011111$ and $(\mathbf{b} + \mathbf{e})M = 011$, which is the binary number for 3, indicating an error in the third position. For $\mathbf{e} = 0000001$ we have $(\mathbf{b} + \mathbf{e})M = 111$, which is 7 in binary notation, indicating an error in the seventh position.

If more than one error occurs, the code will fail. For, if an error pattern which is also a code word occurs, then $(\mathbf{b} + \mathbf{e})M = \mathbf{0}$, and no errors will apparently have occurred. If an error pattern $\mathbf{e}$ with two 1's (a double error) occurs, then the code will "incorrectly" call out a single error and fail.

Hamming codes may be extended to perfect codes with minimum weight 4 by adding a parity check to each code word. This gives a single-error-correcting, double-error-detecting code.

To visualize the Hamming (4,7) code of Example 6 geometrically, consider the seven-dimensional *hypercube* $C = \mathbf{2}^7$ of all 7-bit words; it has 128 nodes and $7 \times 64 = 448$ edges. Let L be the set of coset leaders, defined to consist of $\mathbf{0}$ and its 7 immediate neighbors; the $\mathbf{e} \in C$ with one entry 1. The group translates $L + \mathbf{b}$ of L by code words $\mathbf{b} \in B$ constitute 16 nonoverlapping "corner pieces," each consisting of a single code word $\mathbf{b}$ and its immediate (adjacent) neighbors.

EXERCISES D

1. A parity-check matrix for a Hamming (4,7) code is

$$M^T = \begin{bmatrix} 0 & 0 & 0 & 1 & 1 & 1 & 1 \\ 0 & 1 & 1 & 0 & 0 & 1 & 1 \\ 1 & 0 & 1 & 0 & 1 & 0 & 1 \end{bmatrix}$$

The equations are

$$\begin{aligned} b_1 &= a_1 + a_2 + a_4 & b_4 &= a_1 & b_7 &= a_4 \\ b_2 &= a_1 + a_3 + a_4 & b_5 &= a_2 & & \\ b_3 &= a_2 + a_3 + a_4 & b_6 &= a_3 & & \end{aligned}$$

A received word $\mathbf{c} = \mathbf{b} + \mathbf{e}$ with one error may be decoded as follows: Form $\mathbf{c}M = \mathbf{s}$. As above, $\mathbf{s}$ is the *syndrome* for $\mathbf{r}$. Then $k = \sum_{i=1}^{3} s_i 2^{3-i}$ gives the location of the symbol to be complemented. Extend this process into a single-error-correcting double-error-detecting code. Then form the coset table, and give the functions for syndromes to coset leaders.

2. Let H be the parity-check matrix for a linear code. Show that the coset whose syndrome is $\mathbf{v}$ contains a vector of weight w if and only if some linear combinations of w columns of H equal $\mathbf{v}$. (Refer to Exercise 1 for a definition of syndrome.)
3. Show that in a binary group code either all the code words have even weight, or half have even weight and half odd weight. (*Hint:* Show that the code words of even weight form a subgroup.)
4. Design a single-error-correcting Hamming code which encodes 10 binary digits into 14 binary digits, listing the code words. If a code word from this code is transmitted in a system where we use the full error-correcting capabilities of the code, what is the probability a given code word will be correctly decoded after transmission through a binary symmetric channel?
5. Consider the three-message-digit, six-digit code word error-detecting binary code with generator matrix

$$G = \begin{bmatrix} 1 & 0 & 0 & 0 & 1 & 1 \\ 0 & 1 & 0 & 1 & 0 & 1 \\ 0 & 0 & 1 & 1 & 1 & 1 \end{bmatrix}$$

What is the probability that a message of six digits will be received and accepted as correctly transmitted through a binary symmetric channel, when in fact at least one error has occurred?

6. One Hamming code is described by the following matrix:

$$\begin{bmatrix} 1 & 1 & 1 & 0 & 0 & 0 & 0 \\ 1 & 0 & 0 & 1 & 1 & 0 & 0 \\ 0 & 1 & 0 & 1 & 0 & 1 & 0 \\ 1 & 1 & 0 & 1 & 0 & 0 & 1 \end{bmatrix} \qquad \begin{bmatrix} 0 & 0 & 1 \\ 0 & 1 & 0 \\ 0 & 1 & 1 \\ 1 & 0 & 0 \\ 1 & 0 & 1 \\ 1 & 1 & 0 \\ 1 & 1 & 1 \end{bmatrix}$$

(*a*) Generator matrix (*b*) Parity-check matrix

Assuming that no more than one error occurs during transmission, what was the transmitted code word vector when

(a) 0 1 1 1 1 1 0 is received?
(b) 0 0 0 1 1 1 1 is received?

7. Show that for any positive integer n greater than 1, the $(n,3n)$ "majority rule" code of Example 2 is never a perfect error-correcting code.

IN AN n-CUBE, LET Δ_l BE A SET OF VERTICES, NO TWO OF WHICH ARE CONNECTED BY A PATH OF LENGTH LESS THAN l.

8. Prove that if $l = 3$, then Δ_l can contain at most two members.

★9. Prove that Δ_5 can contain at most $2^{n+1}/(n^2 + n + 2)$ members. $\left[\textit{Hint: } \binom{n}{2} + \binom{n}{1} + \binom{n}{0} = (n^2 + n + 2)/2.\right]$

10. Show that a binary group code with minimum distance 7 is capable of detecting any pattern of six or fewer errors, and of correcting any pattern of three or fewer errors. What combinations of error correction and detection are possible?

EXERCISES 11–13 CONCERN THE BINARY GROUP CODE WITH ENCODING (GENERATOR) MATRIX.

$$G = \begin{bmatrix} 1 & 0 & 1 & 0 & 1 \\ 0 & 1 & 1 & 1 & 0 \end{bmatrix}$$

11. Write out a list of all the messages and their corresponding code words.
12. Draw a gating network for generating the parity-check digits from the transformation digits.
13. Describe the error-detection and error-correction probabilities of this code.

REFERENCES

ABRAMSON, N.: "Information Theory and Coding," McGraw-Hill, 1965.
BERLEKAMP, E. R.: "Algebraic Coding Theory," McGraw-Hill, 1968.
FANO, R.: "The Transmission of Information," M.I.T. Press, 1963.
PETERSON, W. W.: "Error-correcting Codes," M.I.T. Press, 1961.

Lattices

9-1. LATTICES AND POSETS

The notions of a partial ordering relation and of a poset were defined in Sec. 2-4, where some of their basic properties were derived. In Sec. 5-2, we observed that any Boolean algebra was a poset under its natural partial ordering $a \leq b$, this relation being defined to mean that $a \wedge b = a$ or, equivalently, that $a \vee b = b$. Relative to this partial ordering, moreover, $a \wedge b = \text{glb } \{a,b\}$, and $a \vee b = \text{lub } \{a,b\}$.

Furthermore, we observed that the preceding results were consequences of the postulates L1 to L4 for Boolean algebra, and that their proofs did not require any of the other postulates L5 to L10 listed in Sec. 5-1. Therefore, if we define a *lattice* to be an algebraic system $L = [L, \wedge, \vee]$ whose two binary operations satisfy the identities L1 to L4, as we did in Sec. 5-8, then we know that *any Boolean algebra is a lattice* with respect to the binary operations $\wedge$ and $\vee$.

The reader is advised to reread the sections mentioned above before continuing with this chapter.

In this chapter, we shall study the properties of lattices in general, and the properties of various special kinds of lattices (including Boolean lattices) in particular. Lattices arise in almost every branch of mathematics; the following two examples are typical.

Example 1. Let $V = \mathbf{R}^X$ be the set of all real-valued functions $f,g,h, \ldots$ with domain X. Define $f \leqq g$ to mean that $f(x) \leqq g(x)$ for all $x \in X$. Equivalently, using the glb and lub concepts, define $f \wedge g = h$ and $f \vee g = j$ to mean that for each $x \in X$, $h(x) = \min \{f(x),g(x)\}$ and $j(x) = \max \{f(x),g(x)\}$.

Example 2. Let G be any group; let $L(G)$ denote the set of all subgroups $S,T, \ldots$ of G; and let $S \leqq T$ mean that $S \subset T$. Clearly, $L(G)$ is a poset. Moreover, the intersection $S \cap T$ of any two subgroups S and T of G is itself a subgroup of G which contains every subgroup contained in both S and T. Therefore, $S \cap T = \text{glb}\ \{S,T\} = S \wedge T$ in the poset $L(G)$. Dually, the set $\{s_1t_1s_2t_2 \cdots\}$ of all products of elements $s_i \in S$ and $t_i \in T$ is a subgroup of G which is contained in every subgroup of G containing both S and T. Hence this set is the lub of S and T in the poset $L(G)$ and may be designated as $S \vee T = \text{lub}\ \{S,T\}$. This proves that the poset $L(G)$ is a lattice.

Examples 3 to 5 of Sec. 2-4 constitute other familiar examples of lattices. More will be described below, but first we recall from Chap. 2 the following principle.

Duality Principle. The converse of any partial ordering is itself a partial ordering.

This Duality Principle enables one to reduce the labor of proving many theorems by a factor of 2; it generalizes the Duality Principle for Boolean algebras already stated in Sec. 5-1. We shall shortly see how it applies specifically to lattices.

But we shall first define a class of relational systems even more general than the class of all posets. (The class of posets is already more general than that of lattices, by far.) The kind of relation which we have in mind is typified by the *divisibility* relation in any commutative monoid, such as the monoid $M = [\mathbf{Z}, \cdot]$ of all integers under multiplication. In such a commutative monoid M, define $a \mid b$ (in words, "a divides b") to mean that $ax = b$ for some $x \in M$.

Lemma 1. *In any commutative monoid M, the relation $a \mid b$ is reflexive and transitive.*

Proof. Since $a1 = a$, $a \mid a$ for all $a \in M$. Likewise, since $ax = b$ and $by = c$ imply $a(xy) = (ax)y = by = c$, clearly $a \mid b$ and $b \mid c$ imply $a \mid c$.

Hence the relation $|$ is a quasi-ordering relation of M, and the relational system $[M, |]$ is a quoset, in the sense of the following definition.

Definition. A *quasi-ordering* of a set S is a binary relation $\prec$ which is reflexive and transitive. A *quoset* $Q = [S, \prec]$ is a set with a quasi-ordering relation.

Clearly, any poset is a quoset, but the converse is not true because quasi-orderings need not be antisymmetric. (Thus $[\mathbf{Z}, |]$ is a quoset which is not a poset.)

Lemma 2. *In a quasi-ordered set* $Q = [S, \prec]$, *define* $x \sim y$ *when* $x \prec y$ *and* $y \prec x$. *Then*

(a) $\sim$ *is an equivalence relation on* S.
(b) *If* E *and* F *are two equivalence classes of* $\sim$, *then either* $x \prec y$ *for no* $X \in E$, $y \in F$, *or* $x \prec y$ *for all* $x \in E$, $y \in F$.
(c) *If* $E \leqq F$ *is defined to mean that* $x \prec y$ *for some (hence all)* $x \in E$, $y \in F$, *then the quoset* $S/\sim$ *is a poset.*

Proof of (a). Since $x \prec x$ for all $x \in S$, $\sim$ is reflexive.

Again, $x \sim y$ and $y \sim z$ imply $x \prec y$ and $y \prec z$ (by definition), hence $x \prec z$ by P3. Likewise, $z \prec x$, and so $x \sim z$, whence $\sim$ is transitive. It is symmetric by definition.

Proof of (b). If $x \prec y$ for some $x \in E$, $y \in F$, then

$$x_1 \prec x \prec y \prec y_1 \qquad \text{for all } x_1 \in E,\ y_1 \in F$$

whence $x_1 \prec y_1$ by transitivity.

Proof of (c). Clearly $E \sim E$ (since $x \sim x$) for all E. Again, $E \leqq F$ and $F \leqq G$ imply $x \prec y \prec z$ for all $x \in E$, $y \in F$, $z \in G$, whence $x \prec z$ by P3 on $\prec$, and thus $\in$ is transitive. Finally, $E \leqq F$ and $F \leqq E$ imply for all $x \in E$, $y \in F$ that $x \prec y$ and $y \prec x$, whence $x \sim y$ and $E = F$.

Lemma 2 has many applications, such as the following.

Example 3. In a commutative monoid, let $a \mid b$ mean that $ax = b$ for some $x \in S$. The relation $|$ quasi-orders S; elements of S are "equivalent" in the sense of Lemma 2 if and only if they are "associates" in the sense of number theory.

9-2. LATTICES AS POSETS

Assuming only axioms L1 to L4, we have already seen that the operations $\wedge$ and $\vee$ can be defined in terms of the binary relation $\leqq$, using the glb and lub concepts. In this section, we shall also prove that L1 to L4 *need not be assumed as axioms;* it suffices to make the weaker assumption that L is a poset in which any two elements have a glb and a lub. The ideas leading to the proof of this basic fact are well illustrated by the following example.

Example 4. Consider the poset $[\mathbf{P}, |]$ of all positive integers under the divisibility relation. Recall the familiar fact that the greatest common divisor $d = \gcd(m,n)$ of any two positive integers m and n has the following properties:

(*i*) $d \mid m$, $d \mid n$, and (*i'*) If $c \mid m$ and $c \mid n$, then $c \mid d$.

Likewise, the least common multiple $r = \operatorname{lcm}(m,n)$ satisfies

(*ii*) $m \mid r$, $n \mid r$, and (*ii'*) If $m \mid s$ and $n \mid s$, then $r \mid s$.

We shall now generalize these concepts.

Definition. Let $[P, \leqq]$ be any poset, and let $a,b \in P$ be given. Then an element $d \in P$ is called the *greatest lower bound,* or *meet,* of a and b (in symbols, $d = a \wedge b$) when

$$d \leqq a, \qquad d \leqq b \tag{1}$$

and

$$x \leqq a \text{ and } x \leqq b \text{ imply } x \leqq d \tag{1'}$$

Dually, the element $s \in P$ is called the *least upper bound,* or *join,* of a and b (in symbols, $d = a \vee b$) when

$$a \leqq s, \qquad b \leqq s \tag{2}$$

and

$$a \leqq x \text{ and } b \leqq x \text{ imply } s \leqq x \tag{2'}$$

The notions of *meet* and *join* defined above should not be confused with those of *least, greatest, minimal,* or *maximal.* Many of these terms are synonymous in ordered sets (alias "chains"), defined as in Sec. 2-4 to be posets in which the following condition holds.

P4. Given x and y, either $x \leqq y$ or $y \leqq x$

Specifically, in a chain, to be the meet of two or more elements is the same as being least or minimal, and dually, the join of a set is just the greatest or, equivalently, maximal, element in the set. However, this is not the case in lattices generally, as is shown in the Exercises.

We now come to our first main theorem.

Theorem 1. *In any poset P, if the meets and joins designated have values (i.e., if they exist), then*

L1.	$x \wedge x = x$	*and*	$x \vee x = x$
L2.	$x \wedge y = y \wedge x$	*and*	$x \vee y = y \vee x$
L3.	$(x \wedge y) \wedge z = x \wedge (y \wedge z)$	*and*	$(x \vee y) \vee z = x \vee (y \vee z)$
L4.	$x \wedge (x \vee y) = x$	*and*	$x \vee (x \wedge y) = x$

Proof. By the Duality Principle, which interchanges $\wedge$ and $\vee$, it suffices to prove one of the two identities in each of L1–L4; we shall prove the first. By P1, $x \leqq x$, while trivially $d \leqq x$ and $d \leqq x$ imply $d \leqq x$; hence, by definition, $x = x \wedge x$. Next, $x \wedge y = y \wedge x$ because interchanging x and y does not alter the meaning of the definition of glb $\{x,y\}$. Likewise, both $x \wedge (y \wedge z)$ and $(x \wedge y) \wedge z$ are glb $\{x,y,z\}$, where the glb of an arbitrary subset S of a poset is defined as follows:

Definition. $a =$ glb S means that

(i)	$a \leqq x$	for all $x \in S$	
(i')	$v \leqq x$	for all $x \in S$	implies $v \leqq a$.

Since things equal to the same thing are equal to each other, it follows that the two sides of the first equation of L3 are the same element. Finally, $x \leqq x \wedge (x \vee y)$ since $x \leqq x$ and $x \leqq x \vee y$, whereas if $b \leqq x$ and $b \leqq x \vee y$, then trivially $b \leqq x$; by the definition of glb, $x =$ glb $\{x, x \vee y\}$ follows, which proves the first equation of L4.

We now come to the basic definition of this chapter, which is motivated by the preceding theorem.

Definition. A *lattice* is a poset in which any two elements a and b have a glb called the *meet* $a \wedge b$ and a lub called the *join* $a \vee b$.

Of the three posets whose diagrams are shown in Fig. 9-1, the first two are important lattices; the third is not a lattice, but the fourth is a lattice (it is the Boolean algebra $\mathbf{2}^3$).

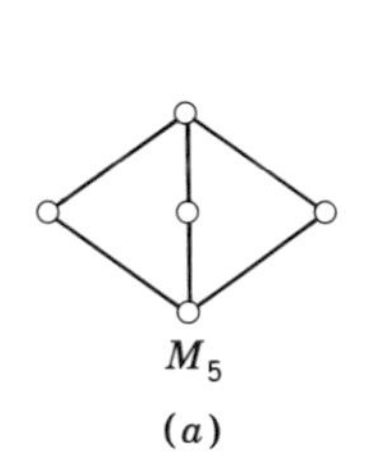
M_5

(*a*)

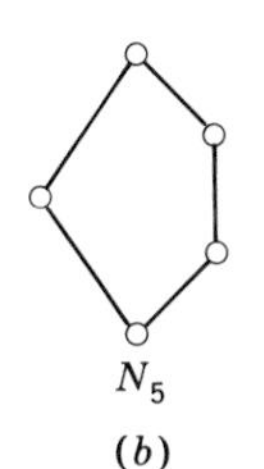
N_5

(*b*)

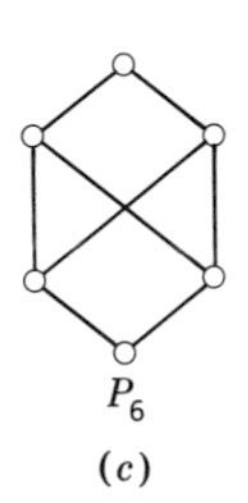
P_6

(*c*)

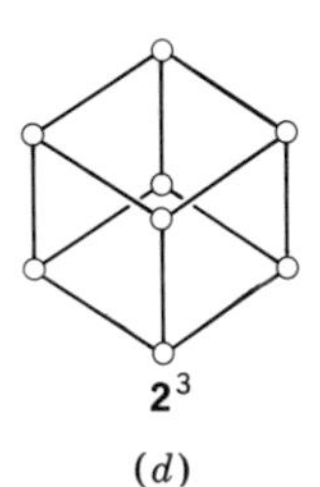
$\mathbf{2}^3$

(*d*)

Fig. 9-1. Examples of posets.

Theorem 1 has the following immediate consequences.

Corollary. *Formulas L1–L4 of Theorem* 1 *are valid identities (true for all x,y,z) in any lattice.*

Our main aim will be to see which of the results of Chap. 5 (i.e., which properties of Boolean algebras) hold also in more general lattices. Our first major step will consist in showing that the definition of lattices just given is equivalent to postulating identities L1–L4.

One can see at a glance that all the diagrams of Fig. 9-1 are posets if $a \leqq b$ is defined as in Sec. 2-4 to mean that there is a path (or "ladder") from a to b whose steps all tilt upward. The existence of glb for all pairs of elements in the first two diagrams is also obvious (using symmetry); that of lub follows by duality (each diagram goes into an "isomorphic" diagram when turned upside down).

Notice how much less work is required above than would be needed to compute and write our tables for the operations $\wedge$, $\vee$ (these would have 64 entries in the example of Fig. 9-1d) and to verify the required identities (8^3 pairs of elements would have to be compared for each side of the associative law to verify L3 for the example of Fig. 9-1d). The poset concept is indeed a great simplification.

EXERCISES A

1. Prove that semilattices can be characterized by the following two identities: (a) $xx = x$ for all x, and (b) $(xy)z = z(xy)$ for all x,y,z.
2. Prove that if the binary operations $\wedge$, $\vee$ in an algebra $[A,\wedge,\vee]$ satisfy L2–L4, then they satisfy L1.
3. Show that in Fig. 9-1, M_5 and N_5 are lattices, but that P_6 is not.
4. Prove in detail that the poset with the diagram of Fig. 9-1c is not a lattice.

IN EXERCISES 5–7, FOR THE GROUP G SPECIFIED, DRAW THE ORDER DIAGRAM OF $L(G)$, THE LATTICE OF ALL ITS SUBGROUPS.

5. (a) $G = D_6$, the dihedral group of order 6.
   (b) $G = D_8$, the dihedral group of order 8 (group of the square).
6. (a) $G = Z_2 \times Z_3$ (b) $G = Z_2 \times Z_4$
7. (a) $G = Z_{30}$ (b) $G = Z_{36}$
8. (a) Which of the posets graphed below is a lattice?
   (b) In the one which is a lattice, which elements have complements?

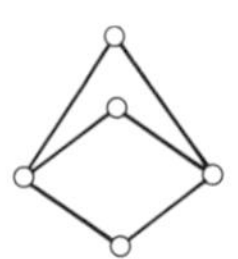

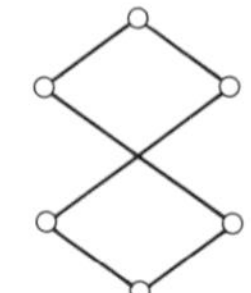

9. Show that any poset with a plane diagram and universal bounds is a lattice.
10. Let Q be any reflexive and transitive relation on a set S.
    (a) Show that the relation $E = Q \cap \ddot{Q}$, meaning that xQy and yQx, is an equivalence relation.
    (b) Show that this equivalence relation has the Substitution Property, that is, if xEy, then xQa implies yQa.
    (c) Show that $[S/E,Q]$ is a poset.

9-3. LATTICES AND SEMILATTICES

A *semilattice* is a set S with an idempotent, commutative, and associative binary operation. If the operation is indicated by juxtaposition (possibly with parentheses), the defining conditions for a semilattice are, symbolically,

$$a^2 = a \qquad ab = ba \qquad a(bc) = (ab)c \qquad \text{all } a,b,c \in S \tag{3}$$

Thus, a semilattice is a commutative semigroup whose elements are all idempotent.

Lemma. *In any semilattice, let f and g be two expressions formed from all the letters $x_1, \ldots, x_n$ (possibly with some repetitions). Then $f = g$.*

The proof is obvious.

The most interesting properties of semilattices center around the concept of divisibility. As in commutative monoids generally (see Sec. 9-1), in a semilattice S, $a \mid b$ (read, "a divides b") means that $ax = b$ for some $x \in S$.

We now compare the definition of a partial ordering relation with the following rules, valid for the divisor relation in any semilattice:

$$a \mid a \tag{4}$$

$$a \mid b \text{ and } b \mid a \text{ imply } a = b \tag{5}$$

$$a \mid b \text{ and } b \mid c \text{ imply } a \mid c \tag{6}$$

To prove (4), note that $aa = a$; to prove (5) note that $ax = b$ and $by = a$ (together with commutativity and associativity) imply

$$a = by = bby = ba = axa = aax = ax = b$$

Again, to prove (6), observe that $ax = b$ and $by = c$ imply

$$a(xy) = (ax)y = by = c$$

Finally, observe that

$$a \mid ab \qquad b \mid ab \qquad \text{while } a \mid c \text{ and } b \mid c \text{ imply } ab \mid c \tag{7}$$

that is, $ab = \text{lcm}(a,b)$ because $ab = ab$, and $ba = ab$, while $ax = c$ and $by = c$ imply $(ab)(xy) = ax(by) = cc = c$. This proves the first statement of the following theorem.

Theorem 2. *Any semilattice is a poset under its divisibility relation, and $ab = \text{lcm}(a,b)$ under this relation. Conversely, let P be a poset in which any two elements a,b have a glb $a \wedge b$; then $[P, \wedge]$ is a semilattice in which $a \mid b$ if and only if $a \geqq b$ in the poset P.*

To prove the second statement, observe that $a = a \wedge a$ trivially since $x \leqq a$ is equivalent to $x \leqq a$ and $x \leqq a$; likewise $a \wedge b = b \wedge a$ since the definition of glb (a,b) is symmetric in a and b. Further, $a \wedge (b \wedge c) = (a \wedge b) \wedge c$ stands for glb $\{a,b,c\}$. Hence $[P, \vee]$ is a semilattice. Finally if $a \geqq b$, then $a \wedge b = b$ (by transitivity), and so $a \mid b$; conversely, if $a \wedge x = b$, then $a \geqq b$ since b must be a lower bound of $\{a,x\}$.

We next recall from Sec. 9-2 the definition of a lattice as a poset in which any two elements a,b have a glb $a \wedge b$ and a lub $a \vee b$. From Theorem 2, we infer the following corollary.

Corollary. *Any lattice is a semilattice under $\wedge$ and also under $\vee$. These operations are related to $\leqq$ by the following law of consistency:*

$$a \geqq b,\ a \wedge b = b, \text{ and } a \vee b = a \text{ are equivalent} \tag{8}$$

With little more trouble, one can prove a stronger result.

Theorem 3. *A lattice is a set L of elements which satisfies axioms L1–L4; that is, it is a semilattice under each of two operations $\wedge$, $\vee$, in which the following absorption law also holds:*

$$a \wedge (a \vee b) = a \vee (a \wedge b) = a \qquad \text{for all } a,b \in L \tag{9}$$

Proof. It is obvious that any lattice has the properties stated. Conversely, from Theorem 2 and its corollary, we have all but the absorption law (9). On the other hand, $a \wedge (a \wedge b) = (a \wedge a) \wedge b = a \wedge b$ in any semilattice. By (8), this implies $a \vee (a \wedge b) = a$. The other equality of (9) follows dually.

Example 5. Let $\mathcal{P}(I)$ be the power set of all subsets of any set I, and define $S \leqq T$ to mean $S \subset T$. Then $\mathcal{P}(I)$ is a lattice, with $S \cap T = \text{glb}\,(S,T)$ and $S \cup T = \text{lub}\,(S,T)$.

9-4. SUBLATTICES; DIRECT PRODUCTS

In general, a *subalgebra* of an algebraic system A is a subsystem which is "closed" under the defining operations of A. Just as a *submonoid* of a monoid M is a subset S which contains 1 and which contains ab if it contains a and b, a *sublattice* of a lattice L is a subset $S \subset L$ such that

$$a,b \in S \qquad \text{implies } a \wedge b \in S \text{ and } a \vee b \in S \tag{10}$$

An important class of sublattices is furnished by the sublattices of the power set $\mathcal{P}(U)$ of all subsets of a given set U. A family of subsets of U which contains with any S and T also $S \cap T$ and $S \cup T$ is called a *ring of sets* (one commonly requires also that $\varnothing$ and U be in the ring). Thus a ring of sets is a sublattice of a power set.

The concept of the *direct product* $A \times B$ of two algebraic systems A, B having the same kinds of operations is only a little less obvious than that of a subalgebra. The elements of $A \times B$ are the ordered pairs (a,b) with $a \in A$, $b \in B$; thus $A \times B$ when considered as a set is simply the Cartesian product of A and B. Operations in $A \times B$ are performed independently on the two components of elements (termwise). Thus, if A, B are lattices, we define

$$\left.\begin{aligned}(a_1,b_1) \wedge (a_2,b_2) &= (a_1 \wedge a_2,\, b_1 \wedge b_2)\\ (a_1,b_1) \vee (a_2,b_2) &= (a_1 \vee a_2,\, b_1 \vee b_2)\end{aligned}\right\} \quad \text{in } A \times B \qquad \begin{aligned}(11)\\(11')\end{aligned}$$

This is again a lattice because every identity on the operations $\wedge$, $\vee$, which holds in A and B, will necessarily hold (since it holds termwise) in their direct product.

For example, the lattice of Fig. 9-1d is simply the direct product $\mathbf{2} \times \mathbf{2} \times \mathbf{2} = \mathbf{2}^3$ of three copies of the two-element lattice with diagram. This is why its diagram resembles that of a cubical box, which is the Cartesian product of three segments.

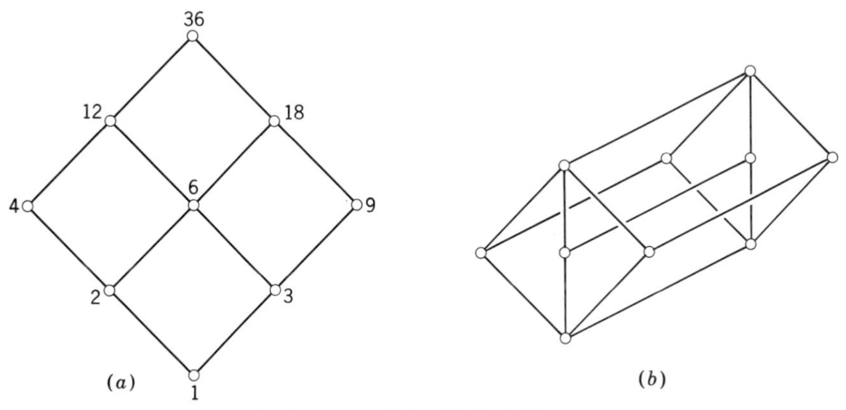

Fig. 9-2. Direct products. (a) Divisors of 36; (b) $\mathbf{2} \times M_5$.

More generally, if $\mathcal{P}(U)$ and $\mathcal{P}(V)$ are the power sets of all subsets of two given sets U and V, considered as lattices under the operations of intersection $\cap$ and set union $\cup$, then $\mathcal{P}(U \sqcup V)$ is isomorphic to the direct product of these two lattices: $\mathcal{P}(U \sqcup V) \cong \mathcal{P}(U) \times \mathcal{P}(V)$.

Again, the lattice of divisors of 36 in the poset $[\mathbf{P}, |\,]$ is the direct product $\mathbf{3}^2 = \mathbf{3} \times \mathbf{3}$ of two chains of three elements, as shown in Fig. 9-2*a*. Finally, Fig. 9-2*b* depicts the diagram of the direct product of $\mathbf{2}$ with the five-element lattice M_5 of Fig. 9-2*b*.

AND gates and OR gates. In Sec. 6-4, we showed that any Boolean polynomial function $f\colon \mathbf{2}^n \to \mathbf{2}$ could be realized by a suitable gating circuit built up from individual AND gates, OR gates, and inverters. We now suppose that neither complements of the input variables nor inverters are available and ask the following question: Which Boolean functions can be realized using AND gates and OR gates alone? In algebraic terms, we wish to know which functions $f\colon \mathbf{2}^n \to \mathbf{2}$ can be built up from the $f_i\colon \mathbf{x} = (x_1, \ldots, x_n) \mapsto x_i$, using $\wedge$ and $\vee$ alone (and not complements). Since the f_i generate under $\wedge$, $\vee$, $'$ the algebra of all such functions f (that is, by Chap. 5, all elements of the free Boolean algebra $\mathbf{2}^{2^n}$ with n generators), we are looking for the (distributive) *sublattice* of $\mathbf{2}^{2^n}$ generated by the symbols $f_1, \ldots, f_n$, without use of the complementary variables f_1'.

Since $X \wedge Y$ has the effect of putting switches X and Y in series, while $X \vee Y$ has the effect of putting them in series, the answer to this question also describes those (Boolean) functions $f\colon \mathbf{2}^n \to \mathbf{2}$ which can be realized by *series-parallel* networks.

When $n = 2$, this sublattice consists of the four elements $x, y, x \wedge y$ and $x \vee y$. When $n = 3$, it consists of 18 elements: the four self-dual elements

$$x, y, z$$

and

$$(x \wedge y) \vee (y \wedge z) \vee (z \wedge x) = (x \vee y) \wedge (y \vee z) \wedge (z \vee x)$$

and the seven elements

$$\begin{array}{cccc} x \wedge y, & y \wedge z, & z \wedge x, & x \wedge y \wedge z \\ x \wedge (y \vee z), & y \wedge (z \vee x), & z \wedge (x \vee y) & \end{array}$$

and their duals.

In Chap. 5, we showed from axioms L1–L4 that the functions $x \wedge y$ and $x \vee y$ were order-preserving (isotone) in both variables. By induction, the same is true of any lattice polynomial in any number of uncomplemented literals (whether or not the distributive law L6 holds). It is a remarkable fact that the converse of this result is also valid.

Theorem 4. *In the free Boolean algebra of all functions f: $\mathbf{2}^n \to \mathbf{2}$, the functions f_i: $(x_1, \ldots, x_n) \mapsto x_i$ generate by $\wedge$ and $\vee$ a sublattice which contains precisely those functions which are order-preserving.*

We omit the proof. For $n = 2$, the assertion is easily verified by looking at the 16 truth tables of F_2 in Example 3 of Sec. 5-1.

For $n = 3$, a direct verification for all symmetry types of Boolean polynomials is still feasible, but tedious.

EXERCISES B

1. Prove in detail that any chain is a distributive lattice.
2. Prove that a lattice is a chain if and only if all its subsets are sublattices.
3. An *interval* of a poset is a set $[a,b]$ consisting of all x such that $a \leqq x \leqq b$.
   (a) Prove that the intersection of two intervals of a lattice is void or an interval.
   (b) Prove that in a lattice, any interval is a sublattice.
4. Prove in detail that the equivalence relations on any finite set S form a lattice (under the subpartition relation) which is not a sublattice of the lattice of all binary relations on S. (*Hint:* See Sec. 2-10.)
5. Let X,Y be sets, and let f: $X \to Y$ be a function. Show that the set of images $f(S)$ of subsets $S \subset X$ is a sublattice of $\mathcal{P}(Y)$. When is it a Boolean subalgebra of $[\mathcal{P}(Y),\cap,\cup,']$?
6. Prove that in a distributive lattice L,

$$(x \wedge y) \vee (y \wedge z) \vee (z \wedge x) = (x \vee y) \wedge (y \vee z) \wedge (z \vee x)$$

   but that this is not true in the lattices M_5 and N_5 of Fig. 9-1.
7. Prove that the lattice of all normal subgroups of a group G is a sublattice of $L(G)$, the lattice of all subgroups of G.
8. Prove that in a *finite* lattice, the intervals (see Exercise 3) constitute with the void set a lattice under the inclusion relation.
9. Let $M = [S, \cdot\,]$ be a multiplicative monoid. Show that the equivalence relations on S which have the Substitution Property for multiplication constitute a sublattice of the lattice of Exercise 4.

9-5. DISTRIBUTIVE LATTICES

In the five-element lattices M_5 and N_5 of Fig. 9-1, it is easy to find triples of elements which do *not* satisfy the distributive laws of L6:

$$x \wedge (y \vee z) = (x \wedge y) \vee (x \wedge z) \tag{12}$$

$$x \vee (y \wedge z) = (x \vee y) \wedge (x \vee z) \tag{12'}$$

Therefore, these laws do *not* hold as identities in M_5 or N_5. On the other hand, we know that the distributive laws (12) and (12′) are satisfied by any sets under intersection and union; hence they are identities in the lattice $\mathcal{P}(\mathbf{3}) = \mathbf{2} \times \mathbf{2} \times \mathbf{2}$ of Fig. 9-1*d*.

In general, a lattice is called *distributive* when it satisfies the two distributive laws (12) and (12′); otherwise, it is called *nondistributive*. We have just seen that there exist both distributive and nondistributive lattices.

The distributive laws are clearly hereditary from lattice to sublattice; moreover, they hold in the power set $\mathcal{P}(U)$ of all subsets of any set U. It follows that *any ring of sets is a distributive lattice.*

Theorem 5. *In any lattice, the semidistributive laws*

$$x \wedge (y \vee z) \geqq (x \wedge y) \vee (x \wedge z) \tag{13}$$
$$x \vee (y \wedge z) \leqq (x \vee y) \wedge (x \vee z) \tag{13'}$$

hold. Moreover, (12) *implies* (12′), *and conversely.*

Proof. Since the two inequalities are dual, it suffices to prove the first. But, trivially, x is an upper bound of $x \wedge y$ and $x \wedge z$, hence of their least upper bound $(x \wedge y) \vee (x \vee z)$. Likewise, since $y \geqq x \wedge y$ and $z \geqq x \wedge z$, any upper bound of y and x must contain $(x \wedge y) \vee (x \wedge z)$; hence $y \vee z$ is an upper bound of $(x \wedge y) \vee (x \wedge z)$. This shows that $(x \wedge y) \vee (x \wedge z)$ is a *lower* bound of both x and $y \vee z$; the first displayed inequality now follows by the definition of glb.

Now we show that (12) implies (12′). From (12) we have

$$\begin{aligned}(a \vee b) \wedge (a \vee c) &= [(a \vee b) \wedge a] \vee [(a \vee b) \wedge c] \\ &= a \vee [(a \wedge c) \vee (b \wedge c)] \qquad \text{[by (9) and (12)]} \\ &= [a \vee (a \wedge c)] \vee (b \wedge c) = a \wedge (b \wedge c) \qquad \text{[by (9)]}\end{aligned}$$

where we used commutativity also in the second step. Since the implication (12′) ⇒ (12) is dual to (12) ⇒ (12′), the proof is complete.

Lemma 1. *In any distributive lattice*

$$a \wedge x = a \wedge y \quad \textit{and} \quad a \vee x = a \vee y \quad \textit{together imply } x = y \tag{14}$$

Proof. By L4, commutativity and the first equation of (14), (12′), commutativity and the second equation of (14), (12′), commutativity and the first equation of (14), and L4, respectively, we get successively

$$\begin{aligned}x = x \vee (x \wedge a) &= x \vee (y \wedge a) \\ &= (x \vee y) \wedge (x \vee a) = (y \vee x) \wedge (y \vee a) \\ &= y \vee (x \wedge a) = y \vee (y \wedge a) = y\end{aligned}$$

The case in which $a \wedge x = a \wedge y = O$, $a \vee x = a \vee y = I$ is of particular interest. In general, by a *complement* of an element a in a lattice L with universal bounds O, I is meant an element $x \in L$ such that $a \wedge x = O$ and $a \vee x = I$. The elements O and I are trivially complementary in any lattice.

Lemma 1 has the following corollary.

Corollary 1. *In a distributive lattice L, a given element a can have at most one complement.*

In such a distributive lattice, the unique complement of a given element a, when it exists, is usually written a'. Thus a' is defined in a distributive lattice by

$$a \wedge a' = O \qquad a \vee a' = I \tag{15}$$

Corollary 2. *Any isomorphism of order between Boolean algebras is an isomorphism for $\wedge$, $\vee$, and $'$.*

In most lattices, including all chains of length $n > 2$ like $\{1,2,3\}$ with $1 < 2 < 3$, there exist uncomplemented elements. In the lattice of Fig. 9-3, a is the only uncomplemented element.

Lemma 2. *In any distributive lattice, the set of all complemented elements is a sublattice.*

Proof. Let a,a' and b,b' be complementary pairs. Then

$$(a \wedge b) \wedge (a' \vee b') = (a \wedge b \wedge a') \vee (a \wedge b \wedge b') = O \vee O = O$$
$$(a \wedge b) \vee (a' \vee b') = (a \vee a' \vee b') \wedge (b \vee a' \vee b') = I \vee I = I$$

Hence $a \wedge b$ and $a' \vee b'$ are complementary; a similar argument shows that $a \vee b$ and $a' \wedge b'$ are complementary. By definition, this shows that the complemented elements form a sublattice.

In the nondistributive lattice graphed in the accompanying diagram, the complemented elements do *not* form a sublattice.

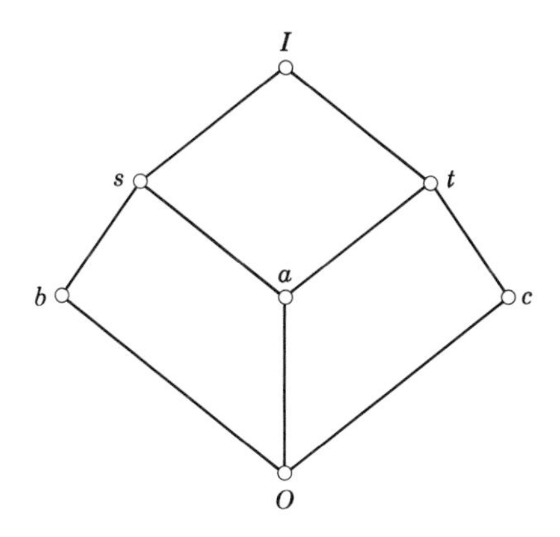

9-6. MODULAR AND GEOMETRIC LATTICES

In this section, we shall give examples of two important families of lattices which are not distributive, but which have compensating special properties.

The first of these is the family of *modular* lattices. A lattice is called *modular* when it satisfies the two identities of axiom L5 of Sec. 5-1. We now present a simplified form of this axiom.

Lemma. *In any lattice, each of the two identities of axiom L5 of Sec.* 5-1 *is equivalent to*

L5*. *If* $a \leqq c$, *then* $a \vee (b \wedge c) = (a \vee b) \wedge c$.

Proof. Since $x \leqq x \vee z$ in the second identity of L5, L5* implies that $x \vee [y \wedge (x \vee z)] = (x \vee y) \wedge (x \vee z)$, which is this identity.

Conversely, if $a \leqq c$, then $a \vee c = c$. Hence, the second identity of L5 reduces to $a \vee (b \wedge c) = (a \vee b) \wedge c$ if $a \leqq c$, which is precisely L5*. Combining this with the result of the preceding paragraph, we see that L5* is equivalent to the second identity of L5.

Finally, since L5* is self-dual, it follows from the Duality Principle that the first identity of L5 is also equivalent to L5*.

The most important class of modular lattices is described by the following theorem.

Theorem 6. *The normal subgroups of any group G form a modular lattice* $M(G)$.

Proof. It is evident that $M(G)$ is a lattice, and in fact a sublattice of the lattice $L(G)$ of all subgroups of G (see the discussion of Example 2 in Sec. 9-1). The problem is to prove that if $H \lhd G$, $K \lhd G$, and $L \lhd G$, then

$$\text{If} \quad H \leqq L \quad \text{then} \quad H \vee (K \wedge L) = (H \vee K) \wedge L \tag{16}$$

But clearly $H \leqq H \vee K$, $H \leqq L$, $K \wedge L \leqq K = H \vee K$, and $K \wedge L \leqq L$; hence H and $K \wedge L$ are lower bounds to $(H \vee K) \wedge L$, and therefore their join $H \vee (K \wedge L)$ is also such a lower bound. This proves that $H \vee (K \wedge L) \leqq (H \vee K) \wedge L$ (the argument is valid in any lattice).

To prove equality in (16), it suffices by P2 to prove that

$$(H \vee k) \wedge L \subset H \vee (K \wedge L)$$

But if $x \in (H \vee K) \wedge L$, then $x = hk = l$ ($h \in H, k \in K, l \in L$) since $H \triangleleft G$, $K \triangleleft G$ imply $H \vee K = HK$. Since $H \subset L$, however, $h \in L$ and thus $k = h^{-1}l \in L$. Therefore, $k \in K \wedge L$, and thus

$$x = hk \in H \vee (K \wedge L)$$

completing the proof.

The five-element lattice of Fig. 9-1a represents the lattice of all (normal) subgroups of the 4-group, the Abelian group of all symmetries of a rectangle.

Example 6. Consider the subgroups of the additive group of all triples of binary numbers; there are seven such subgroups of order 2 and seven of order 4. Those of order 2 include the identity element 000 and one element of the set {001,011,010,111,101,110,100}. The subgroups of order 4 consist of all three-bit words $x = (x_1,x_2,x_3)$ which satisfy

$$a_1x_1 + a_2x_2 + a_3x_3 \equiv 0 \qquad (\text{mod } 2)$$

for some $\mathbf{a} = a_1a_2a_3 \neq 000$. Listing the triples $\mathbf{a}$ in order, they are {100,101,110,110,111,011,001}. The diagram of Fig. 9-3a depicts these

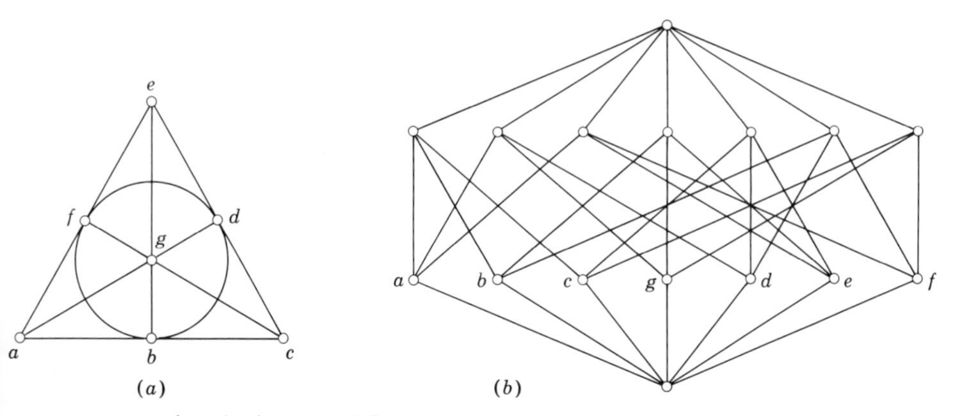

Fig. 9-3. Lattice of subgroups of $\mathbf{Z}_2{}^3$.

as represented by six segments and one circle, each of which contains three points. The diagram of Fig. 9-3b presents an order diagram for the same Fano plane.

We next turn our attention to lattices of partitions. The concept of a *partition*, which has already been studied in Chap. 2, is one of the fundamental concepts of set theory. We first sketch the proof of Theorem 7.

Theorem 7. *The set of all partitions of any given set, when partially ordered by the relation* $\pi \leqq \pi^*$ *(π is a subpartition of π^*), is a lattice defined by*

$$\pi \leqq \pi^* \qquad \text{means that if } x\,\pi\,y, \text{ then } x\,\pi^*\,y \tag{17}$$

[Note that $\leqq$ is just the relation of inclusion for binary relations, already defined in Chap. 2. For simplicity, we let π and π^* here signify the equivalence relations $E(\pi)$ and $E(\pi^*)$ associated with π and π^*, respectively.]

Proof. We define $\pi \wedge \pi_1$ by the condition

$$x(\pi \wedge \pi_1)y \qquad \text{means } x\,\pi\,y \text{ and } x\,\pi_1\,y \tag{18}$$

It is easy to verify that this $\pi \wedge \pi_1$ is an equivalence relation (reflexive, symmetric, and transitive). Again let

$$\begin{array}{c} x(\pi \vee \pi_1)y \qquad \text{means that for some } z_1, \ldots, z_{2n} \\ x\,\pi\,z_1, \quad z_1\,\pi_1\,z_2, \quad z_2\,\pi\,z_3, \quad \ldots, \quad z_{2n-1}\,\pi_1\,z_{2n}, \quad z_{2n}\,\pi\,y \end{array} \tag{18$'$}$$

The lattice of Fig. 9-4*a* depicts the symmetric partition lattice Π_4 of all partitions of a set of four elements. The corresponding lattice Π_3 is depicted in Fig. 9-1*a*.

We now prove a theorem which sharpens the result stated in Example 2.

Theorem 8. *The lattice of all subgroups of any finite group G of order n is a sublattice of the lattice Π_n of all partitions of G, considered as a set.*

Thus, Fig. 9-4*b* depicts the lattice of all subgroups in the group of symmetries of a square (the dihedral group of order 8).

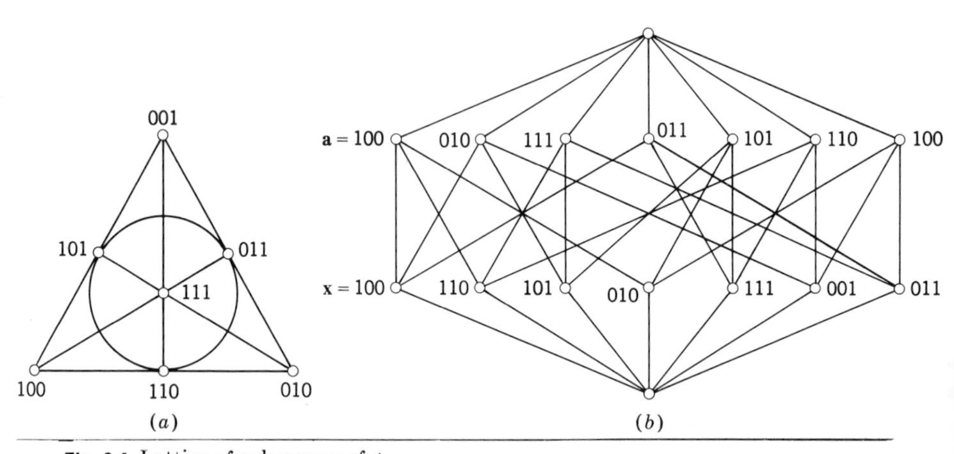

Fig. 9-4. Lattice of subgroups of Δ_4.

EXERCISES C

1. (a) Prove that if A and B are chains, then any morphism of order (Sec. 2-6) is a morphism of lattices.
   (b) How many order morphisms 1_n: $\mathbf{4} \to \mathbf{3}$ are there?
   (c) How many of these are onto (epimorphisms)?
2. Prove that 1_n is the only automorphism of $\mathbf{n}$.
3. Let P be a poset with universal bounds in which no chain has length exceeding 2.
   (a) Show that P is a lattice.
   (b) Show that unless P is the chain **3**, it has a proper automorphism (one not 1_P).
4. Let L be a distributive lattice, and let $a < b$ in L. Show that $x \mapsto (x \vee a) \wedge b$ is a lattice epimorphism which projects L onto the interval $[a,b]$.
5. Prove in detail that if f is a lattice isomorphism, then so is the bijection f^{-1}.
6. Let $h: L \to M$ be a morphism of lattices, where L has universal bounds.
   (a) Show that if $h(x') = [h(x)]'$ for one complementary pair of elements x,x' in L, then M has universal bounds, and $h(O_L) = O_M$ and $h(I_L) = I_M$.
   (b) Conversely, show that the two preceding equations imply that $h(x') = [h(x)]'$ for every complementary pair of elements $x,x' \in L$.
7. Let $h: L \twoheadrightarrow M$ be a lattice epimorphism, where L has universal bounds. Show that M has universal bounds, and that $h(x') = [h(x)]'$ in M if x has a complement in L.
8. Prove in detail that for any lattice L, the morphisms from L to itself form a monoid of transformations.
9. (a) Given distributive lattices L and M and morphisms of lattices $g: L \to M$ and $h: L \to M$, define $g \leqq h$ to mean that $g(x) \leqq h(x)$ for all $x \in L$. Show that the set of all lattice morphisms from L to M is a lattice under this relation.
   (b) Is hypothesis of distributivity needed in (a)?

★10. Show that the lattice of all Boolean subalgebras of the Boolean algebra $\mathcal{P}(\mathbf{n})$ (under inclusion) is dually isomorphic with the lattice of all equivalence relations on the set $\mathbf{n}$. (*Suggestion:* Try $n = 4$.)

11. (a) Construct an order epimorphism from 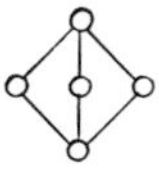 onto  .
    (b) Show that every lattice-epimorphic image of 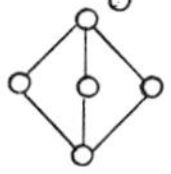 is isomorphic to 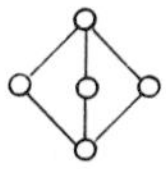 or to the one-element lattice o.
12. Show that the free Boolean algebra with two generators (see Sec. 5-1) has 24 automorphisms and 12 pairs of generators.
13. Exhibit some nontrivial epimorphisms of the lattice 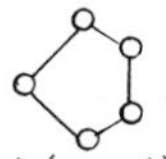 .
14. Let $P = \{x_1,x_2,x_3, \ldots ,x_N\}$ be a partially ordered set (poset). Define $\phi(1) = 1$, recursively, and define

$$\phi(i+1) = \begin{cases} i+1 & \text{if } x_{i+1} < x_{\phi(i)} \\ \phi(i) & \text{otherwise} \end{cases}$$

    Prove (by induction) that $x_{\phi(N)}$ is a minimal element of P.
15. An integer array named X contains 100 distinct integers $X[1],X[2], \ldots,$ $X[100]$. Write an ALGOL program that determines the index m of the least integer $X[m]$ of the array X.

★9-7. BOOLEAN LATTICES

It is interesting to consider Boolean algebras as relational systems $[A, \leqq]$ in which only the binary relation of inclusion is given, that is, as posets. Since in any Boolean algebra considered as a poset $\wedge$ and $\vee$ are glb and lub (alias "infimum" and "supremum"), one can say that any Boolean algebra is a poset in which every two elements have a glb and lub; that is, it is a *lattice*. We now ask how can one tell whether a given lattice $[A, \wedge, \vee]$ is associated with a Boolean algebra $[A, \wedge, \vee, ']$, in the obvious sense[1] of having the same functions $\wedge(x,y)$ and $\vee(x,y)$.

Clearly, it is necessary that the given lattice be *distributive*. It is also necessary that it be *complemented* in the following sense.

Definition. A *complemented lattice* is a lattice with universal bounds O and I in which every element a has at least one complement x, with

$$a \wedge x = O \qquad \text{and} \qquad a \vee x = I \tag{19}$$

A *Boolean lattice* is a lattice which is both complemented and distributive.

Theorem 9. *In any Boolean lattice*

$$(a')' = a \tag{20}$$

$$(a \wedge b)' = a' \vee b' \qquad (a \vee b)' = a' \wedge b' \text{ (de Morgan's laws)} \tag{21}$$

Proof. Since the relation of being complementary is symmetric, (20) follows from the unicity of complements (Lemma 1, Corollary 1, of Sec. 9-5). Hence the assignment $a \mapsto a'$ defines a function $c: A \to A$ which is its own inverse and therefore a bijection. On the other hand, c inverts order. $a \leqq b$ implies $a \wedge b' \leqq b \wedge b' = O$; hence $a \wedge b' = O$ and

$$\begin{aligned} b' &= b' \wedge I = b' \wedge (a \vee a') = (b' \wedge a) \vee (b' \wedge a') \\ &= O \vee (b' \wedge a') = b' \wedge a' \end{aligned}$$

whence $a \leqq b$ implies $b' \geqq a'$ (consistency law). It is thus a dual automorphism of A since by a dual argument $b' \geqq a'$ implies

$$b = (b')' \leqq (a')' = a$$

Therefore c interchanges glb and lub, proving (21).

Corollary. *The class of lattices which are obtained from Boolean algebras by considering only the operations $\wedge$ and $\vee$ is the same as the class of all Boolean lattices.*

[1] Equivalently, we ask the following question: when can $[A,\wedge,\vee]$ be obtained from some Boolean algebra $[A,\wedge,\vee,']$ by simply "forgetting" about the operation of complementation?

★9-8. MORPHISMS AND IDEALS

A function $\theta: L \to M$ from a lattice L to a lattice M is called a *morphism of lattices* when for all $x,y \in L$,

$$\theta(x \wedge y) = \theta(x) \wedge \theta(y) \text{ and } \theta(x \vee y) = \theta(x) \vee \theta(y) \text{ in } M \qquad (22)$$

The definition is also indicated by the commutative mapping diagram of Fig. 9-5*a*. Such a morphism of lattices is necessarily order-preserving (or "isotone") since $x \leqq y$ in L implies $\theta(x) = \theta(x \wedge y) = \theta(x) \wedge \theta(y)$, whence $\theta(x) \leqq \theta(y)$, all by the Consistency Law (8).

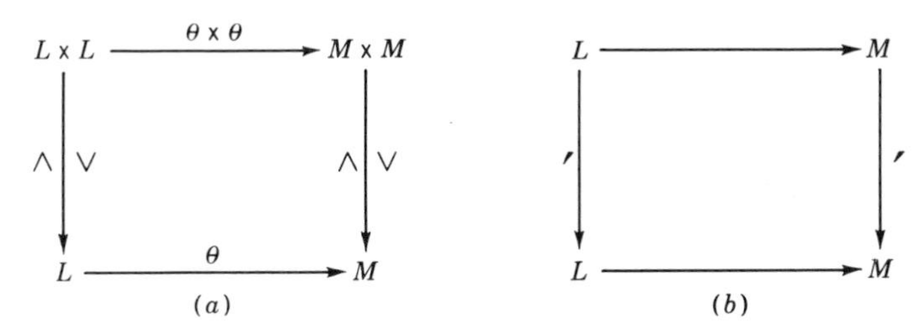

Fig. 9-5. Diagrams of lattice morphisms.

However, an *order*-preserving map is by no means necessarily a morphism of lattices. Thus for any poset $P = [S, \leqq]$, consider the mapping $\theta_P: a \mapsto \theta_P(a)$ which assigns to each $a \in S$ the set $\theta_P(a) = A \subset S$ of all $x \leqq a$ in $[S, \leqq]$. By transitivity of $\leqq$, $a \leqq b$ in S implies $\theta_P(a) \subset \theta_P(b)$ in $\mathcal{P}(S)$; hence the mapping θ_P is order-preserving (isotone) from S to its power set. Moreover, if S is a lattice, then $\theta_P(a \wedge b) = \theta_P(a) \cap \theta_P(b)$ (try to prove it). However, $\theta_P(a \vee b)$ properly contains the set union $\theta_P(a) \cup \theta_P(b)$ in general (for example, in the lattices of Figs. 9-1*a* and *b*). Hence θ_P is *not* a morphism of lattices.

Using the concept of morphism of lattices, it can be shown[1] that every finite distributive lattice is isomorphic with a ring of sets. Using the Axiom of Choice (Chap. 14), one can prove more generally that *any* distributive lattice is isomorphic with a ring of sets. Therefore, the identities which define distributive lattices completely characterize the algebraic properties of the operations $\cap$ and $\cup$ on sets. We will omit the proofs. However, it will be shown in Sec. 9-9 that every finite Boolean lattice (algebra) L is isomorphic with the power set of some set (namely, the set of its atoms).

We next define an *ideal* of a lattice L as a nonvoid subset $J \subset L$ such that (*i*) $a \in J$ and $b \in J$ imply $a \vee b \in J$, and (*ii*) $a \in J$ and $x = a$ imply $x \in J$.

[1] See, for example, Birkhoff-Mac Lane, "Survey of Modern Algebra," chap. 11, sec. 8.

Theorem 10. *Let $\theta: L \to M$ be any lattice morphism whose range includes $O \in M$. Then $\theta^{-1}(O)$ is an ideal in L.*

Proof. Write $K = \theta^{-1}(O)$ for short (it is called the *kernel* of θ). By hypothesis (the range of θ includes O), K is nonvoid. Also (*i*) holds since $\theta(a) = \theta(b) = O$ imply $\theta(a \vee b) = \theta(a) \vee \theta(b) = O \vee O = O \in M$. Finally, (*ii*) holds since $\theta(a) = 0$ and $x \leqq a$ imply $\theta(x) \leqq \theta(a) = 0$.

Boolean Morphisms. Finally, define a *Boolean morphism* $\theta: L \to M$ as a lattice morphism between Boolean algebras (lattices) which also satisfies

$$\theta(a') = [\theta(a)]' \tag{23}$$

(see Fig. 9-5*b*); that is, it is a mapping which conserves all three Boolean operations $\wedge$, $\vee$, and $'$.

The kernel of a Boolean morphism θ is the set of antecedents of $O \in M$ under θ; that is, it is $\theta^{-1}(O)$. This always contains $O \in L$ since, by (19),

$$\theta(O) = \theta(a \wedge a') = \theta(a) \wedge \theta(a') = \theta(a) \wedge [\theta(a)]' = 0 \quad \text{and} \quad a \in L$$

Corollary. *The kernel K of any Boolean morphism is an ideal because it is nonvoid since $O \in K$ as above.*

Conversely, each ideal $J \subset A$ of a Boolean algebra A is the kernel of a Boolean epimorphism $\theta: A \to B$, and $B \cong A/J$ is determined up to isomorphism by its kernel. This is most easily shown by considering A as a *commutative ring* (see Chap. 10) under the (commutative) multiplication $xy = x \wedge y$ and the addition $x + y = (x \wedge y') \vee (x' \wedge y)$.

★9-9. FINITE BOOLEAN ALGEBRAS

In this section, we shall use the concepts introduced above to show that every finite Boolean algebra is isomorphic to the power set of some finite set, and hence has 2^n elements for some integer n.

The proof depends heavily on the notions of morphism and direct product.

Lemma 1. *In any distributive lattice L, for any $a \in L$ the assignment $x \mapsto (x \wedge a, x \vee a)$ is a lattice monomorphism $\theta: L \to [O,a] \times [a,I]$.*

Proof. By the Distributive Laws L6,

$$\begin{aligned}(x \wedge y) \to &((x \wedge y) \wedge a, (x \wedge y) \vee a) \\ &= ((x \wedge a) \wedge (y \wedge a), (x \vee a) \wedge (y \vee a))\end{aligned}$$

The final expression represents, by the definition of the direct product of lattices, $\theta(x) \wedge \theta(y)$. Hence we have proved that $\theta(x \wedge y) = \theta(x) \wedge \theta(y)$, which is one of the two identities which defines a morphism. The proof of the identity $\theta(x \vee y) = \theta(x) \vee \theta(y)$ is dual.

This proves that the assignment is a morphism of lattices. It is one-one, hence a monomorphism, because $x \wedge a = y \wedge a$ and $x \vee a = y \vee a$ together imply

$$\begin{aligned} x &= x \wedge (x \vee a) = x \wedge (y \vee a) = (x \wedge y) \vee (x \wedge a) \\ &= (y \wedge x) \vee (y \wedge a) = y \wedge (x \vee a) = y \wedge (y \vee a) = y \end{aligned}$$

Lemma 2. *In any distributive lattice L, let a be complemented with complement a′. Then for any $x \in L$, there is a unique lattice morphism $\theta\colon \mathbf{3}^2 \to L$ with*

$$(2,0) \mapsto a \qquad (1,1) \mapsto x \qquad (0,2) \mapsto a'$$

as in Fig. 9-6.

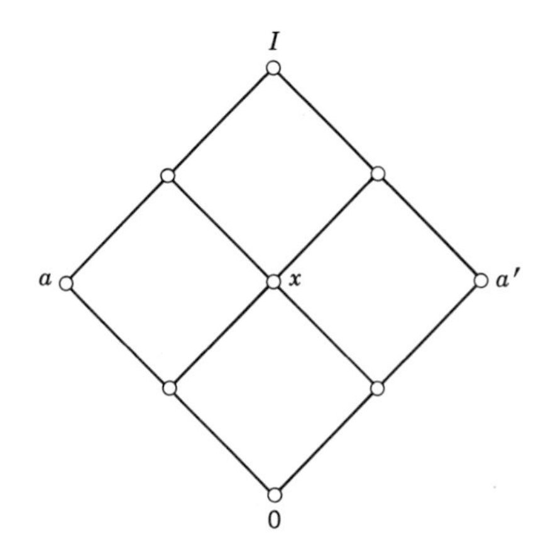

Fig. 9-6.

Lemma 3. *In any Boolean lattice L, for any $a \in L$ the assignment $\beta\colon x \to (x \wedge a, x \vee a)$ is an isomorphism.*

Proof. Since we know from Lemma 1 that the assignment is a monomorphism, it suffices to prove that β is onto, i.e., that every (y,z) with $y \leqq a$ and $z \geqq a$ is the image under β of some $x \in L$. We shall show that this is the case for $x = y \vee (z \wedge a')$, in other words, that $\gamma\colon (y,z) \mapsto y \vee (z \wedge a')$ has β for its right inverse[1] under right composition. The steps are as follows:

[1] Actually, since β is a bijection (as we will show), γ is a two-sided inverse of β.

Specifically, we have the identity

$$\begin{aligned}[y \vee (z \wedge a')] \wedge a &= (y \vee z) \wedge (y \vee a') \wedge a = (y \vee a') \wedge a \\ &= (y \vee a') \wedge (y \vee z) \wedge a = (y \vee a') \wedge a \\ &= (y \wedge a) \vee (a' \wedge a) = y \vee O = y\end{aligned} \tag{i}$$

Proof of (i). The first equality is by distributivity, the second is by L2–L3, and the third follows since $y \vee z \geqq z \geqq a$, which implies $(y \vee z) \wedge a = a$. Now using distributivity and $y \leqq a$, we easily get the last two equalities.

It remains to prove the identity

$$\begin{aligned}[y \vee (z \wedge a')] \vee a &= (a \vee y) \vee (z \wedge a') = a \vee (z \wedge a') \\ &= (a \vee z) \wedge (a \vee a') = z \wedge I = z\end{aligned} \tag{ii}$$

Proof of (ii). The first equality is evident from L2 and L3, dropping parentheses and permuting; the second follows since $y \leqq a$. The remaining equalities are dual to the final equalities in the proof of (*i*).

Theorem 11. *In a Boolean algebra L, for any element a with complement a′, the function* $\alpha: x \mapsto [x \wedge a,\ x \wedge a']$ *is an isomorphism with inverse* $(y,z) \mapsto y \vee z$:

$$\alpha: L \cong [O,a] \times [O,a'] = A \times A' \tag{24}$$

Proof of (24). From Lemma 3, we know that the assignment $\beta: x \mapsto (x \wedge a,\ x \vee a)$ is an isomorphism of Boolean algebras (see Sec. 9-5, Lemma 1, Corollary 2); that is, $\beta: L \cong [O,a] \times [a,I]$. On the other hand, $\alpha': x \mapsto x \wedge a'$ is also an isomorphism $\alpha': [a,I] \cong [O,a']$ with inverse $\alpha: x \mapsto x \vee a$ since it is isotone and

$$\begin{aligned}(x \wedge a') \vee a &= (x \vee a) \wedge (a' \vee a) = x \vee a = x \qquad \text{if } x \in [a,I] \\ (y \vee a) \wedge a' &= (y \wedge a') \vee (a \wedge a') = y \wedge a' = y \qquad \text{if } y \in [O,a']\end{aligned}$$

Theorem 12. *Every finite Boolean algebra L is isomorphic to* $\mathbf{2}^n$ *for some* $n \in \mathbf{N}$.

The proof is by induction. If $L \cong \mathbf{1}$ or $\mathbf{2}$, the result is trivial with $n = 0$ and 1, respectively.

Otherwise L contains some a with $0 < a < 1$. By Theorem 11, $L \cong A \times A'$, where A and A' have fewer elements than L. Hence by the (weak) induction hypothesis, $A \cong \mathbf{2}^r$, and $A' \cong \mathbf{2}^s$ for some r, s, whence $L \cong A \times A' = \mathbf{2}^{r+s}$.

As a final result, we extend the construction used to prove Theorem 11.

Theorem 13. *In a Boolean algebra L, for any element a with complement a', let*

$$\alpha\colon x \mapsto x \wedge a,\ \alpha'\colon x \mapsto x \wedge a',\ \alpha\dagger\colon x \mapsto x \vee a,\ \alpha^*\colon x \mapsto x \vee a' \qquad (25)$$

Then Im $\alpha = [O,a]$, Im $\alpha' = [O,a']$, Im $\alpha = [a,I]$, and Im $\alpha^* = [a',I]$. Under right composition

$$\alpha\alpha\dagger = \alpha\dagger\alpha = p_a,\ \alpha'a^* = \alpha^*\alpha' = p_{a'} \qquad (26)$$

$$\alpha\alpha' = \alpha'\alpha = p_0,\ \alpha\dagger\alpha^* = \alpha^*\alpha\dagger = p_1 \qquad (26')$$

$$\alpha\alpha^* = \alpha^*,\ \alpha^*\alpha = \alpha,\ \alpha'\alpha = \alpha\dagger,\ \alpha\dagger\alpha' = \alpha' \qquad (26'')$$

Sketch of proof. The verifications of the identities (26), (26') and (26'') are trivial. To verify that $\alpha\alpha^* = \alpha^*$, it suffices to compute

$$(\alpha)\alpha = (x \wedge a) \vee a' = (x \vee a') \wedge (a \vee a') = (x \vee a') \wedge I = x \vee a'$$

The other four equations are proved similarly. The proof of the final statement is similar to that of Lemma 2.

Corollary. *The set* $A = \{I_L,\alpha,\alpha',\alpha,A^*,p_ap_{a'},p_0,p_1\}$ *forms a monoid under right composition.*

REFERENCES

ABBOTT, J. C.: "Sets, Lattices and Boolean Algebra," Allyn and Bacon, 1969.
BIRKHOFF, G.: "Lattice Theory," 3d ed., Am. Math. Soc., 1967.
LIEBER, L. R.: "Lattice Theory" (illus. by H. G. Lieber), Galois Institute, 1959.
RUTHERFORD, D. E.: "Introduction to Lattice Theory," Hafner, 1965.
SZASZ, G.: "Introduction to Lattice Theory," 2d ed., Academic, 1963.

Rings and Ideals

10-1. INTRODUCTION

In the next four chapters, we shall analyze a class of algebraic systems called *rings*. Rings have two basic binary operations of *addition* and *multiplication;* in them, sums and products are assumed to satisfy most of the familiar laws of elementary algebra. However, the commutativity of multiplication is often not assumed, and the possibility of division is rarely assumed; thus **Z** is a fairly typical (commutative) ring.

Definition. A ring is an algebraic system $R = [R,+,\cdot\,]$ with two binary operations $+$ and $\cdot$, which is: (*i*) an Abelian group under $+$, (*ii*) a monoid under $\cdot$, and which (*iii*) satisfies the distributive laws

$$a(b + c) = ab + ac \qquad \text{and} \qquad (a + b)c = ac + bc \tag{1}$$

for any elements $a,b,c \in R$.

The integers form a fairly typical ring $\mathbf{Z} = [\mathbf{Z},+,\cdot\,]$ under addition and multiplication. Note that division is *not* usually possible in **Z** (or in rings generally).

Rings are of many kinds. The most familiar rings are *commutative* in the sense that they satisfy the commutative law

$$ab = ba \qquad \text{for all } a,b \in R \tag{2}$$

Familiar commutative rings include **Z**, the finite rings $\mathbf{Z}_n$ of integers mod n (modular numbers) defined in Sec. 2-6, and other familiar number systems.

Other important rings are noncommutative. Indeed, given any nontrivial ring R, commutative or not, one can construct from R a *non*commutative ring as follows.

Example 1. Let R be any ring, and let $M_2(R)$ be the set of all 2×2 matrices $\begin{pmatrix} a & b \\ c & d \end{pmatrix}$ with entries $a,b,c,d \in R$. Define $[M_2(R), +, \cdot\,]$ by the rules

$$\begin{bmatrix} a & b \\ c & d \end{bmatrix} + \begin{bmatrix} a' & b' \\ c' & d' \end{bmatrix} = \begin{bmatrix} a + a' & b + b' \\ c + c' & d + d' \end{bmatrix} \tag{3}$$

$$\begin{bmatrix} a & b \\ c & d \end{bmatrix} \begin{bmatrix} a' & b' \\ c' & d' \end{bmatrix} = \begin{bmatrix} aa' + bc' & ab' + bd' \\ ca' + dc' & cb' + dd' \end{bmatrix} \tag{3'}$$

Then $M_2(R)$ is a noncommutative ring.

Example 2. Let R be any ring, and let n be any positive integer. Let $M_n(R)$ be the set of all $n \times n$ matrices $A = \|a_{ij}\|$, $B = \|b_{ij}\|$, etc., with addition and multiplication defined by the rules

$$\|a_{ij}\| + \|b_{ij}\| = \|a_{ij} + b_{ij}\| \tag{4}$$

$$\|a_{ij}\| \cdot \|b_{ij}\| = \|\sum_{k=1}^{n} a_{ik}b_{kj}\| \tag{4'}$$

Then $M_n(R)$ is a ring with unity given by the identity matrix

$$I = \|\delta_{ij}\| \qquad \text{where } \delta_{ij} = \begin{cases} 1 & \text{if } i = j \\ 0 & \text{if } i \neq j \end{cases} \tag{5}$$

One of our main objects will be to learn how to construct rings having various desirable special properties, especially so-called *fields*. These are commutative rings in which addition, multiplication, subtraction, and division have all their usual properties. The class of all fields includes not only the real field **R**, the complex field **C**, and the rational field **Q**, but also the field $\mathbf{Z}_p$ of integers modulo any prime p and many other finite fields. We shall see that such other finite fields, especially those of "characteristic 2" whose order is a power of 2, are very useful for making codes.

We shall also study various intermediate kinds of systems, notably integral domains and Euclidean domains. In Euclidean domains division can be performed only occasionally, but the concepts of factorization into primes and prime powers are valid. Among the most important Euclidean domains are included the ordinary integers (the Euclidean domain $[\mathbf{Z},+,\cdot]$) and polynomials with real coefficients, with rational numbers as coefficients, or with coefficients in any finite field such as $\mathbf{Z}_p$.

It will be convenient to study rings from an axiomatic standpoint, so as to bring out clearly to just which class of rings any given theorem applies. When we say that a certain result holds in any Euclidean domain, for example, we mean simply that it can be proved from the axioms for a Euclidean domain.

Our first task will be to establish a few elementary results which hold in all rings.

Various elementary properties of rings are just special cases of properties which we have already established more generally in groups and monoids. Thus we have the following property: for given $a,b \in R$, the equation $a + x = b$ has the unique solution

$$x = (-a) + b \qquad \text{in } R \tag{6}$$

As special cases, this implies

(*i*) If $z + a = a$ for some $a \in R$, then $z = 0$.
(*ii*) If $a + x = 0$, then $x = -a$.
(*iii*) If $a + b = a + c$, then $b = c$ (cancellation law for addition).

As in monoids generally, the unity 1 of any ring is unique. More precisely, we can prove

$$\text{If } au = a \text{ for all } a \text{ in } R, \text{ then } u = 1 \tag{7}$$

because setting $a = 1$, we get $u = 1u = 1$.

A slightly less trivial property of rings is the following. Its proof uses the distributive laws (1) above (which provide the only postulate interrelating plus and times).

Lemma 1. *The additive identity* 0 *of any ring* R *also serves as the "zero" for multiplication; thus*

$$0b = 0 = b0 \qquad \textit{for all } b \in R \tag{8}$$

Proof. Since $a = a + 0$ for all $a \in R$, we also have

$$ab = (a + 0)b = ab + 0b$$

But by definition of 0, $ab = ab + 0$; hence, by the transitivity of equality, $ab + 0b = ab + 0$. Using the cancellation law for addition, we infer that $0b = b$. The proof that $0 = b0$ is similar.

Lemma 2. *For all $a,b \in R$, we have*

$$(-a)(-b) = ab \tag{9}$$

Proof. Since $a + (-a) = 0$, we have

$$0 = 0 \cdot (-b) = [a + (-a)](-b) = a(-b) + (-a)(-b)$$

On the other hand,

$$a(-b) + ab = a[(-b) + b] = a0 = 0$$

Hence, by the cancellation law for addition, $(-a)(-b) = ab$ as claimed.

10-2. INTEGRAL DOMAINS AND FIELDS

Rings in general have relatively few interesting properties. Therefore, we shall concentrate our attention on two special kinds of rings about which more can be said. These are, respectively, fields (whose importance was emphasized in the introduction to this chapter) and integral domains, whose most valuable property is their capacity for being extended to a field (see Sec. 10-3). As was stated in the introduction, we shall treat integral domains and fields axiomatically; hence we begin by defining them formally.

Definition. An *integral domain* is a *commutative ring*, which satisfies the following *cancellation law:*

$$ab = ac \quad \text{and} \quad a \neq 0 \quad \text{imply} \quad b = c \tag{10}$$

The most familiar rings [such as the ring $\mathbf{Z}$ of all integers and the ring $\mathbf{Q}[x]$ of all polynomials $p(x) = a_0 + a_1x + \cdots + a_nx^n$ $(n \in \mathbf{N})$ with rational coefficients a_i] are integral domains. The commutative ring $\mathbf{Z}_4$ of Example 3, Sec. 2-4, is not, because $2 \cdot 2 = 0$ in $\mathbf{Z}_4$ though $2 \neq 0$ there.

Theorem 1. *A commutative ring R is an integral domain if and only if in it no product of nonzero factors is zero; that is,*

$$a \neq 0 \quad \textit{and} \quad b \neq 0 \quad \textit{imply} \quad ab \neq 0 \tag{11}$$

Proof. If (11) fails, then $ab = 0 = a0$ for some $b \neq 0$, contradicting (10) with $c = 0$. Conversely, if (10) fails, then $a(b - c) = ab - ac = 0$, where both a and $b - c$ are nonzero.

Condition (11) can also be interpreted as follows: the nonzero elements of any integral domain form a submonoid of the integral domain, considered as a monoid under multiplication.

Note the following contrapositive equivalent of (11):

$$ab = 0 \quad \text{implies} \quad a = 0 \quad \text{or} \quad b = 0 \qquad (11')$$

This has an interesting corollary.

Corollary. *In an integral domain, the only idempotents are* 0 *and* 1.

Proof. By definition, any idempotent satisfies $u^2 = u$, hence

$$0 = u^2 - u = u(u - 1)$$

By (11′), this implies $u = 0$ or $u - 1 = 0$; hence $u = 0$ or $u = 1$.

The Rings $\mathbf{Z}_m$. Condition (11′), when applied to the rings $\mathbf{Z}_m = [\mathbf{Z}_m, +, \cdot]$ of Sec. 2-6, asserts that $\mathbf{Z}_m$ is an integral domain if and only if $m \mid ab$ implies $m \mid a$ or $m \mid b$ in $\mathbf{Z}$, that is, if and only if m cannot divide a product without dividing at least one factor. It is a familiar fact (to be proved formally in Lemma 1, Sec. 10-10) that $m \in \mathbf{P}$ has this property if and only if m is a *prime*. We conclude the following result.

Lemma. *The ring* $\mathbf{Z}_m$ *is an integral domain if and only if* m *is a prime.*

Definition. A *field* is a commutative ring in which the nonzero elements form a group under multiplication.

Since the cancellation law holds in any group, any field is an integral domain.

The most familiar fields are the field $\mathbf{Q}$ of all rational numbers, the field $\mathbf{R}$ of all real numbers, and the field $\mathbf{C}$ of all complex numbers. Somewhat less familiar are the finite fields $\mathbf{Z}_p$ of integers modulo prime numbers p. In these fields, addition and multiplication are performed as with integers (in $\mathbf{Z}$), but all numbers are replaced by their remainders after division by p. Addition and multiplication tables for $\mathbf{Z}_5$ are displayed below. It will follow from Theorem 2 below that $\mathbf{Z}_p$ is a field for any prime p.

+	0	1	2	3	4
0	0	1	2	3	4
1	1	2	3	4	0
2	2	3	4	0	1
3	3	4	0	1	2
4	4	0	1	2	3

·	0	1	2	3	4
0	0	0	0	0	0
1	0	1	2	3	4
2	0	2	4	1	3
3	0	3	1	4	2
4	0	4	3	2	1

Fig. 10-1.

From the definition of a group (Chap. 7), we have the following as a corollary of our definition of a field.

Lemma 3. *Division (except by zero) is possible and is unique in any field.*

The result is true more generally in any division ring (commutativity is irrelevant), as a special case of an elementary property of groups (Chap. 7). Slightly less obvious is the following result.

Theorem 2. *Any finite integral domain D is a field.*

Proof. By Theorem 1, the nonzero elements of D form a (finite) monoid $[D^*, \cdot]$ in which right translations $x \mapsto xb$ and left translations $x \mapsto ax$ are all one-one. But by finiteness, this implies that they are all bijections, so that $ax = xa = 1$ for some $x \in D^*$. Hence each (nonzero) $a \in D^*$ has an inverse in D^*; that is, D is a field.

In any field, the (unique) solution x of $bx = a$ $(b \neq 0)$ is denoted a/b (read "the quotient of a by b"), and we have the following result.

Theorem 3. *In any field, quotients obey the following laws (for $b \neq 0$, $d \neq 0$):*

(*i*) $(a/b) = (c/d)$ *if and only if* $ad = bc$.
(*ii*) $(a/b) \pm (c/d) = (ad \pm bc)/(bd)$.
(*iii*) $(a/b)(c/d) = (ac/bd)$.
(*iv*) $(a/b) + (-a/b) = 0$.
(*v*) $(a/b)(b/a) = 1$ *if* $a,b \neq 0$.

Proof. We prove the five formulas above in succession.

Ad (*i*). The first equation means that $ab^{-1} = cd^{-1}$. This implies the third equation of

$$ad = a(b^{-1}b)d = (ab^{-1})(bd) = (cd^{-1})(db) = c(d^{-1}d)b = cb$$

The other equations displayed are identities in any commutative ring. Conversely, $ad = bc$ implies

$$ab^{-1} = a(dd^{-1})b^{-1} = (ad)(d^{-1}b^{-1}) = (cb)(b^{-1}d^{-1}) = c(bb^{-1})d^{-1} = cd^{-1}$$

Ad (*ii*). The left-hand side of (*ii*) stands for $x \pm y$, where $bx = a$ and $dy = c$. But this implies

$$bd(x \pm y) = bdx \pm bdy = bxd \pm bc = ad \pm bc$$

This says that the right side of (*ii*) also stands for $x \pm y$.

Ad (*iii*). By definition, $(a/b)(c/d) = xy$, where $bx = a$, $dy = c$, $b \neq 0$, and $d \neq 0$. Hence $ac = (bx)(dy) = (bd)(xy)$, where $bd \neq 0$ which shows that $xy = ac/bd$ as claimed.

Ad (*iv*). Similarly, $(a/b) + (-a/b) = x + y$, where $bx = a$ and $by = -a$. Hence $0 = a + (-a) = bx + by = b(x + y)$, making obvious use of the postulates for a ring. By Theorem 1, since $b \neq 0$, this implies $x + y = 0$.

Ad (*v*). Finally, $(a/b)(b/a) = xy$, where $bx = a$ and $ay = b$. Therefore, $ab = (bx)(ay) = b(xa)y = (ab)(xy)$, where we have skipped intermediate steps for brevity. But now $ab = (ab)1$ trivially; moreover $ab \neq 0$ by assumption and (11). Hence $xy = 1$ by the cancellation law (10) applied to the equation $(ab)(xy) = (ab)1$ in which $ab \neq 0$. This completes the proof.

EXERCISES A

1. Prove that in any commutative ring

$$(a + b)^n = a^n + \binom{n}{1} a^{n-1}b + \binom{n}{2} a^{n-2}b^2 + \cdots + b^n$$

2. Write down the expansion of $(a + b)^3$ for the case of a noncommutative ring.
3. (a) Show that if A is a 2×2 matrix, then A commutes with $\begin{pmatrix} 2 & 0 \\ 0 & 1 \end{pmatrix}$ if and only if A is diagonal.
   (b) Is the preceding result true over any field? Over any ring? Justify.
4. Prove that over any commutative ring, for all $m,n \in \mathbf{N}$,

$$a^m a^n = a^{m+n} \qquad (ab)^n = a^n b^n \qquad (a^m)^n = a^{mn}$$

5. Prove that for $a \neq 0$, $b \neq 0$ in any field,

$$a^r a^s = a^{r+s} \qquad (ab)^r = a^r b^r \qquad (a^r)^s = a^{rs}$$

for all $r,s \in \mathbf{Z}$.

A RING IN WHICH $x^2 = x$ FOR ALL x IS CALLED A BOOLEAN RING.

6. Show that any Boolean ring is commutative.
7. Show that $x + x = 0$ for all x in any Boolean ring.
8. Show that a Boolean ring cannot be an integral domain if it contains more than two elements.
   (*Hint:* Consider the equation $x(x - 1) = 0$.)
9. Is the one-element system $\{0\}$, with $0 + 0 = 0 \cdot 0 = 0$, a ring? an integral domain? a field?

★10. Show that in the definition of a ring, it is sufficient to assume that $[R, +]$ is a group and one need not assume that $a + b = b + a$.
(*Hint:* Expand $(1 + 1)(a + b)$ in two different ways.)

★10-3. FIELDS OF QUOTIENTS

Formulas (*i*)–(*v*) of Theorem 3 point the way for constructing from any given integral domain D a related field $Q = Q(D)$ called the *field of quotients* of D, which is the closure of D under division in any field containing it. In this sense, $Q(D)$ is the *minimal* extension of D to a field.

We begin by stating the main theorem to be proved. The proof itself, though included for completeness, is noteworthy only as a rather elaborate illustration of the use of the technique of equivalence relations with the substitution property to construct new algebraic systems out of known ones. The reader can omit it without essential loss of understanding.

Theorem 4. *Let D be any integral domain, and let Q be the set of formal quotients a/b, c/b, . . . of its elements, with nonzero denominators. Then condition (i) in Theorem 3 defines an equivalence relation E on Q. Moreover, formulas (ii)–(iii) define single-valued operations of addition, subtraction, and multiplication on the quotient set Q/E defined by this equivalence relation. Finally, under these operations, Q/E is a field called the field of quotients of the given domain D and designated $Q(D)$.*

Motivation. In the special case $D = \mathbf{Z}$ of the (integral) domain of all integers, this construction yields the familiar field $\mathbf{Q} = Q(\mathbf{Z})$ of all rational numbers; moreover, the construction is most easily visualized by thinking of this familiar special case. Applied to the domain $\mathbf{R}[x]$ of all polynomials with real coefficients, the same construction yields the field $\mathbf{R}(x) = Q(\mathbf{R}[x])$ of all rational functions[1] with real coefficients. The construction has many other important applications.

Proof. The mapping $a/b \mapsto (a,b)$ is a bijection from the set of formal fractions to the Cartesian product $D \times D^*$, where $D^* = D - \{0\}$ is the set of all nonzero elements of D. Formulas (*ii*) and (*iii*) of Theorem 3 define a system $[D \times D^*, +, \cdot\,]$ with two binary operations of "addition" and "multiplication." [Note that we need the property (1) of integral domains to be sure that $ac/bd = (a/b)(c/d)$ has a nonzero denominator bd, that is, that $(ac,bd) = (a,b) \cdot (c,d) \in D \times D^*$.]

The system $[D \times D^*, +, \cdot\,]$ is not a ring. The claim made is that if we define $(a/b)E(c/d)$ as in (*i*) by the relation $ad = bc$, then the equivalence classes of the resulting system form not only a commutative ring but a field. The proof therefore consists in proving that the postulates

[1] Though commonly called *functions*, they are not functions in the strict sense defined in Chap. 1 because they are undefined when their denominator vanishes. Thus, the function $1/x$ is not defined when $x = 0$.

for a field are fulfilled; this we now do as follows:

(*vi*) E is an equivalence relation.
(*vii*) The relation E has the substitution property for $+$ and $\cdot$.
(*viii*) The quotient algebra $(D \times D^*)/E$ is a commutative ring in which $(0/1)$ is an additive zero and $(1/1)$ is a multiplicative unity.
(*ix*) $Q(D) = (D \times D^*)/E$ is a field under the operations defined, and $\mu: a \rightarrow a/1$ is a monomorphism from D to $Q(D)$.

Ad (*vi*). To prove that $(a/b)E(a/b)$ (that E is reflexive), use formula (*i*) of Theorem 3 and verify that $ab = ba$. To prove that $(a/b)E(c/d)$ implies $(c/d)E(a/d)$ (that E is symmetric), verify likewise that $ad = bc$ implies $cb = da$ (by commutativity). Finally, to prove that E is transitive, note that $(a/b)E(c/d)$ and $(c/d)E(e/f)$ imply that $ad = bc$ and $cf = de$, and so that $adf = bcf = bde$. From this relation, canceling $d \neq 0$, we get $af = be$, which proves that $(a/b)E(e/f)$. This shows that E is an equivalence relation on $D \times D^*$. The remainder of the proof is concerned with the quotient set $(D \times D^*)/E$ and makes use of the notion of the substitution property presented in Secs. 2-5 and 2-6.

Ad (*vii*). As in Sec. 2-6, the role of the substitution property is to prove that the operations of addition and multiplication of equivalence classes as defined by formulas (*ii*) and (*iii*) of Theorem 3 are single-valued functions; that is, we must prove the following: if $(c/d)E(e/f)$, then $(a/b) + (c/d)E(a/b) + (e/f)$ and $(a/b) \cdot (c/d)E(a/b) \cdot (e/f)$. We leave the proofs of the two above implications, which follow the pattern of the proofs given already, to the reader.

Ad (*viii*). From formula (*vii*) and Theorem 4 of Sec. 7-4, it follows that $(D \times D^*)/E$ is a group under $+$ and a monoid under $\cdot$. It is also easy to verify that

$$(0/1) + (a/b) = (0b + a)/b = (a/b)$$

and

$$(1/1)(a/b) = (1a/1b) = (a/b)$$

It remains to prove the Distributive Law, the only interrelation between addition and multiplication; we sketch the proof. Manipulations of a familiar sort lead quickly to

$$(a/b)[(c/d) + (e/f)]\, E\, (acf + ade)/(bdf)$$

and

$$(a/b)(c/d) + (a/b)(e/f)\, E\, (acdf + bdae)/(b^2df)$$

But it is easy to verify that if $b \neq 0$ in D, then $(a/j)E[(ab)/(bj)]$ in $D \times D^*$ for all $j \in D^*$, which proves the Distributive Law.

Ad (ix). Finally, since $(a/b)(b/a) = (ab/ba)E(1/1)$, we have a field. Moreover, μ as defined in *(ix)* is a monomorphism since

$$(a/1) + (b/1) = (a1 + b1)/(1 \cdot 1)E(a + b)/1$$

and

$$(a/1)(b/1) = (ab)/(1 \cdot 1)E(ab)/1$$

We again skip over trivial steps.

Throughout the proof sketched above, we have implicitly relied on the commutative law to reduce the number of identities to be proved by a factor of nearly 2. The reader may find it interesting to check through to see just where this has been done.

10-4. SUBRINGS

The notion of a subring of a ring R is an obvious analog of the notions of a subgroup of a group, a submonoid of a monoid, and a subalgebra of a Boolean algebra. Namely, it is a subset of R which is closed under the operations of R, including the zero-ary operations "select 1" and "select 0" and the unary operation $x \mapsto -x$. However, it is unnecessary to assume closure under all these operations, and the following simplified definition is technically most convenient.

Definition. A *subring* S of ring R is a subset of R which contains 1 and with any two elements x and y their difference $x - y$ and product xy.

Evidently, any such subring also contains $1 - 1 = 0$ and with any x and y also $x - (0 - y) = x + y$.

Lemma 3. *Any subring of a ring R is itself a ring under the operations of R.*

Proof. We have proved above closure under all the operations of R. It remains to show that the restriction to S of these operations satisfies the ring postulates. By Theorem 6 of Chap. 7, $[S,+]$ is an additive group (an additive subgroup of $[R,+]$). Likewise, $[S, \cdot\,]$ is a monoid (a submonoid of $[R, \cdot\,]$). Finally, the Distributive Laws (1) hold in the subring S because $x(y + z) = xy + xz$ for all $x,y,z \in R$, and hence for all $x,y,z \in S$ since $S \subset R$. This completes the proof.

Similarly, it is clear that any subring S of a commutative ring R is itself commutative, since the commutative law (2) holds in $S \subset R$ if it holds in R. Likewise, any subring S of an integral domain D is an integral domain because the cancellation law (10) holds in $S \subset D$ if it holds in D. A subring of an integral domain D is therefore called a *subdomain* of D.

A subring S of a field F which is itself a field under the operations of F is called a *subfield* of F. This is precisely the case when $x \in S$ implies $x^{-1} \in S$ for all $x \neq 0$.

The Unital Subring. In any nontrivial ring R, every subring contains the unity 1 of R (the trivial ring is the ring $\{0\}$ in which $1 = 0$). Now consider the additive subgroup U of the group $[R,+]$ generated by 1 in the (commutative) group $[R,+]$. This subgroup consists of all "multiples"

$$n1 = \overbrace{1 + \cdots + 1}^{n \text{ summands}} \ (n \in \mathbf{P}) \qquad (-n)\,1 = -(n1) = n(-1), \quad 01 = 0 \tag{12}$$

of 1; since $1 \neq 0$, its order is at least 2.

Definition. The *characteristic* of a ring R is the order of the *unital subgroup* of all multiples of the unity 1 of R, considered as a group $[R,+]$.

Thus the usual rings $\mathbf{Z}$, $\mathbf{Q}$, $\mathbf{R}$, and $\mathbf{C}$ all have characteristic ∞; the ring $\mathbf{Z}_m$ has characteristic m. If R has characteristic m, then so does any subring or extension of R; we leave the proofs to the reader.

Lemma 4. *The characteristic of a ring R is the least positive integer n (if any) such that*

$$na = a + \cdots + a = 0 \qquad \textit{for all } a \in R \tag{12'}$$

Proof. If m is the characteristic of R, then $m1 = 0$ and so, for all $a \in R$

$$\begin{aligned} ma &= a + \cdots + a = 1a + \cdots + 1a \\ &= (1 + \cdots + 1)a = (m1)a = 0a = 0 \end{aligned}$$

where the dots signify sums of m terms. Hence (12′) holds for $n = m$. On the other hand, if $0 < n < m$, then $n1 \neq 0$, the identity of the group $[R,+]$. This proves Lemma 4.

The characteristic of a ring is its most important numerical invariant. From the definition, we know that $[U,+]$ is group-isomorphic with $[\mathbf{Z}_m,+]$ under the bijection $n1 \leftrightarrow n$ defined by (12). We now show that this bijection is an isomorphism of *rings* (see Sec. 10-5), i.e., that $(n1)(r1) = (nr)1 = (nr)_m 1$ and $1 \cdot 1 = 1$. The second identity (more precisely, $1_R \cdot 1_R = 1_R$) is trivial; the proof of the first breaks down into nine cases according to whether n and r are positive, negative, or zero. If $n > 0$ and $r > 0$ (that is, if $n,r \in P$), then by the general distributive law

$$\begin{aligned}(n1)(r1) &= \overbrace{(1 + \cdots + 1)}^{n \text{ summands}}\overbrace{(1 + \cdots + 1)}^{r \text{ summands}} = \overbrace{(1 + \cdots + 1)}^{nr \text{ summands}} \\ &= (nr)1 \end{aligned} \tag{13}$$

Likewise, if $r = -s < 0$ $(s > 0)$, we have

$$(n1)(r1) = (n1)(s(-1)) = ns[1(-1)] = ns(-1) = (-ns)1 = (nr)1 \tag{13'}$$

The proofs for the other cases are similar.

The subring $U = \{n1\} \subset R$ is called the *unital subring* of R, and we have proved the following theorem.

Theorem 5. *Let R be a ring of characteristic m. Then its unital subring is isomorphic with $\mathbf{Z}_m$ and is contained in every subring of R.*

If R is an integral domain, then its unital subring ($\mathbf{Z} \subset R$ or $\mathbf{Z}_m \subset R$) cannot contain zero divisors. But $\mathbf{Z}_m$ contains zero divisors unless m is a prime. Theorem 6 follows.

Theorem 6. *The characteristic of any integral domain D is either a prime p or ∞.*

Corollary 1. *The unital subring of any finite field is isomorphic to $\mathbf{Z}_p$ for some prime p; it is a subfield.*

Definition. The *center* of a ring R is the set of all elements $a \in R$ such that

$$ax = xa \qquad \text{for all } x \in R \tag{14}$$

The "center" of any commutative ring R is, trivially, R. For F, any field, we shall show in Sec. 10-8 that the center of the noncommutative ring $R = M_n(F)$ consists of the scalar matrices with $a_{ij} = \lambda\delta_{ij}$, $\lambda \in R$. The matrix $\|\delta_{ij}\|$ is also often written as I, as in (5).

Theorem 7. *The center of any ring R is a subring.*

The proof is easy. From $ax = xa$ and $bx = xb$, we infer immediately

$$(a \pm b)x = ax \pm bx = xa \pm xb = x(a \pm b)$$
$$(ab)x = a(bx) = a(xb) = (ax)b = (xa)b = x(ab)$$

Finally, $1x = x1$ for all x, by definition of 1.

EXERCISES B

1. Prove that in the ring $M_n(R)$ of Example 1, the following are subrings:
   (a) The set of diagonal matrices $\|a_{ij}\|$, with $a_{ij} = 0$ whenever $i \neq j$.
   (b) The set of triangular matrices $\|a_{ij}\|$, with $a_{ij} = 0$ whenever $i < j$.
2. Prove that if R is of characteristic r, then $M_n(R)$ is of characteristic r.

3. Is the set of all $n \times n$ relation matrices $\|r_{ij}\|$ a ring under addition (mod 2) and composition?
4. What is the subdomain of **Q** generated by $\frac{1}{3}$? By $\frac{1}{3}$ and $\frac{1}{4}$?
5. (a) What is the subdomain of **R** generated by $\sqrt{2}$?
   (b) Show that $\mathbf{Q}(\sqrt{2}) = a + b\sqrt{2}$ $(a,b \in \mathbf{Q})$ is a subfield of **R**.
6. What is the subdomain of **C** generated by i?
7. What is the subdomain of **C** generated by **Q** and i? Is it a field?
8. Prove in full detail that (v) holds in Theorem 3. [You may assume that (i) through (iv) have been proved.]
9. Show that, in the rational field **Q**, $r = m/n$ $(n \neq 0;\ m,n \in \mathbf{Z})$ is nonnegative if and only if $r = q_1^2 + \cdots + q_k^2$ for some finite set of rational numbers q_i.

★10. Let S be a commutative semigroup in which $a + x = a + y$ implies $x = y$ (cancellation law). Prove that S is monomorphic to a group G.
(*Hint:* Let $G = \{a - b\}/E$, $a,b \in S$, for some suitably defined equivalence relation E.)

10-5. MORPHISMS OF RINGS

We have already studied the morphism concept in groups, monoids, Boolean algebras, and lattices in earlier chapters. In this section, we shall extend it to rings, look at some examples, and derive some elementary properties of morphisms of rings.

In the next section, we shall show that morphisms of rings are related to direct sums of rings in much the same way that morphisms of groups are related to direct products of groups. In Sec. 10-7 we shall show that the epimorphic images of any ring R can be reconstructed (up to isomorphism) from its ideals in much the same way that epimorphic images of any group can be reconstructed from its normal subgroups (see Sec. 7-11).

We begin our discussion by defining a morphism of rings.

Definition. A *morphism* of rings is a function $\theta: R \to S$ (R,S rings) which satisfies the identities

$$\theta(x + y) = \theta(x) + \theta(y) \quad \theta(1) = 1 \quad \theta(xy) = \theta(x)\theta(y) \quad \text{all } x,y \in R \tag{15}$$

Lemma 1. *If $\theta: R \to S$ is a morphism of rings, then $\theta(0) = 0$,*

$$\theta(-x) = -\theta(x)$$

and (if x is invertible) $\theta(1/x) = 1/\theta(x)$.

Proof. By (15), $\theta(0) = \theta(0 + 0) = \theta(0) + \theta(0)$, whence by consequence (i) of (6), $\theta(0)$ is the zero of S. Also, if x is invertible in R, then

$$1 = \theta(1) = \theta(x \cdot (1/x)) = \theta(x)\theta(1/x)$$

using (15) twice, whence $\theta(x)$ is also invertible with $[\theta(x)]^{-1} = \theta(x^{-1})$.

As with groups, a morphism of rings is called a *monomorphism*, an *epimorphism*, or an *isomorphism*, when it is one-one, onto, or bijective, respectively. A morphism $\theta: R \to R$ (same domain and codomain) is called an *endomorphism;* when an endomorphism is bijective (an isomorphism), it is called an *automorphism* of R.

We now generalize the discussion of modular numbers in Sec. 2-6, by proving in detail that $\mathbf{Z}_n$ as defined there was a ring. The proof can be greatly simplified and generalized by first proving the following result.

Theorem 8. *Let R be a ring, let $[S,+,\cdot\,]$ be a system with addition and multiplication, and let $\theta: R \to S$ be epimorphic for $+, \cdot$. Then S is a ring.*

Proof. Under addition, S is an Abelian group by an obvious extension of Theorem 4 of Chap. 7. Under multiplication, S is a monoid with unique[1] identity 1 for the same reason. A similar argument proves, finally, the Distributive Law in S in the following way: For any $a,b,c \in S$, there will exist $x,y,z \in R$ such that $\theta(x) = a$, $\theta(y) = b$, and $\theta(z) = c$. Then

$$a(b + c) = \theta(x)[\theta(y) + \theta(z)] = \theta[x(y + z)] = \theta(xy + xz)$$

will follow since θ is a morphism and R is a ring. Moreover,

$$\theta[xy + xz] = \theta(x)\theta(y) + \theta(x)\theta(z) = ab + ac$$

since θ is a morphism for $+$ and $\cdot$.

Theorem 9. *In any integral domain D of prime characteristic p, the assignment $x \to x^p$ is a monomorphism $\mu: D \to D$.*

Proof. By the binomial theorem, we have

$$(a \pm b)^n = a^n \pm \binom{n}{1} a^{n-1}b + \binom{n}{2} a^{n-2}b^2 \pm \cdots + \binom{n}{n} (\pm b)^n \qquad (16)$$

for any $n \in R$; the proof of this by induction on n is straightforward. In formula (16), multiplication of any $c \in R$ by the (integer-valued) binomial coefficient $\binom{n}{k} = (n!)/(k!)(n - k!)$ means [as in (12)] forming the sum $c + \cdots + c$ with the summand $c \in R$ written $\binom{n}{k}$ times. Hence, if $\binom{n}{k}$ is a multiple of p, the product $\binom{n}{k} c = 0$. But $(p!)$ is divisible by p, whereas neither $(k!)$ nor $(p - k)!$ is divisible by p for $1 < k < p$. Hence $\binom{p}{k} \equiv 0 \pmod{p}$ for all k, $0 < k < p$.

[1] See Chap. 7. By Lemma 1 of Sec. 7-4, $\theta(1_R) = 1_S$.

Therefore, if n is a prime in Eq. (16), all terms except the first and last vanish; thus,

$$(a \pm b)^p = a^p \pm b^p \qquad \text{if char } R = p \tag{17}$$

On the other hand, since $[R, \cdot\,]$ is a commutative monoid,

$$(ab)^p = a^p b^p \qquad \text{in any commutative ring} \tag{17$'$}$$

Formulas (17) and (17′) show that μ is a morphism of rings. We complete the proof of Theorem 9 by showing that it is one-one and hence is a monomorphism of rings.

To show that μ is one-one, suppose that $x^p = y^p$ in an integral domain D. Then, by (17), $(x - y)^p = x^p - y^p = 0$. Since D is an integral domain, this implies $x - y = 0$, or $x = y$, as claimed.

10-6. DIRECT SUMS

In previous chapters, we have defined "direct products" of groups, lattices, Boolean algebras, etc. In this section, we introduce the analogous concept for rings. In the case of rings, however, one usually speaks of "direct sums" and not direct products, even though the construction is basically that of a Cartesian product.

Definition. If R and S are any two rings, their *direct sum* $R \oplus S$ is the set of all ordered pairs (x,y) with $x \in R$ and $y \in S$. Addition, subtraction, and multiplication are defined in $R \oplus S$ by

$$(r,s) \pm (r',s') = (r \pm r', s \pm s') \tag{18}$$
$$(rs)(r',s') = (rr',ss') \tag{18$'$}$$

The unity of $R \oplus S$ is $(1_R,1_S)$ where 1_R is the unity of R and 1_S is the unity of S.

In other words, as an additive group $[R \oplus S, +]$ is the direct product $[R,+] \times [S,+]$ of the additive groups $[R,+]$ and $[S,+]$; whereas as a multiplicative monoid $[R \oplus S, \cdot\,]$ is the direct product of the multiplicative monoids $[R, \cdot\,]$ and $[S, \cdot\,]$. We must verify the Distributive Law

$$\begin{aligned}(r,s)[(r',s') + (r'',s'')] &= (r(r' + r''), s(s' + s''))\\ &= (rr' + rr'', ss' + ss'')\\ &= (r,s)(r',s') + (r,s)(r'',s'')\end{aligned}$$

and its opposite to show that $R \oplus S$ is a ring.

The concept of the direct sum of two rings is related to that of morphism of rings in just the same way as the analogous concept (of direct product) is related to morphism in other kinds of algebraic systems. Thus, we have evident (epi)morphisms $\rho\colon (r,s) \mapsto r$ and $\sigma\colon (r,s) \mapsto s$

from $R \oplus S$ to R and S. Conversely, given any ring T and morphisms $\alpha\colon T \to R$ and $\beta\colon T \to S$, there exists a morphism $\tau\colon T \to R \oplus S$ such that $\rho \circ \tau = \alpha$ and $\sigma \circ \tau = \beta$, namely, $\tau(t) = (\alpha(t),\beta(t))$. This is called the *universality property* for direct sums; a diagram illustrating it is shown in Fig. 10-2.

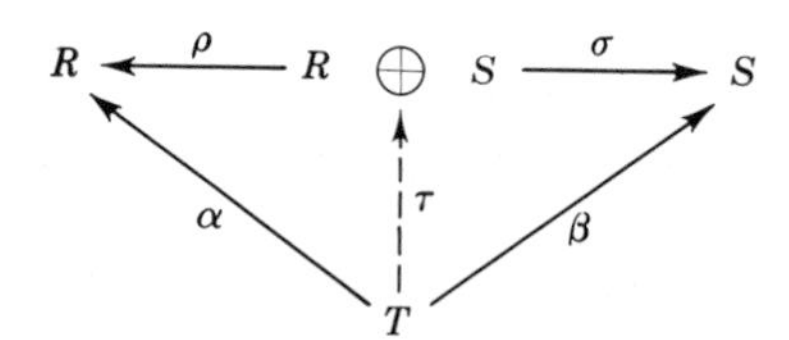

Fig. 10-2. Morphisms and direct sums.

Note that the set of elements of the form $(r,0)$ is *not* a subring of $R \oplus S$, although it is isomorphic to R under addition and multiplication. This is because this set does not contain the unity $(1_R,1_S)$ of R. However, the two elements $e = (1_R,0)$ and $e' = (0,1_S)$ of $R \oplus S$ have some interesting properties.

(*i*) e and e' are *idempotents*; that is, $ee = e$ and $e'e' = e'$.
(*ii*) e and e' are in the *center* of $R \oplus S$.
(*iii*) $e + e' = 1$; that is, $e' = 1 - e$, and $e = 1 - e'$ are *complementary*.
(*iv*) $ee' = 0$.

Conversely, let e be any idempotent in the center of a ring A, and let $e' = 1 - e$. Then

$$e'^2 = (1 - e)^2 = (1 - e)(1 - e) = 1 - e - e + e^2$$
$$= 1 - 2e + e = 1 - e = e'$$

so that e' is an idempotent. Moreover, for any $x \in A$,

$$xe' = x(1 - e) = x - xe = x - ex = (1 - e)x = e'x$$

so e' is in the center of A. Finally, the mapping $x \mapsto (xe,xe')$ defines an isomorphism of rings from A to $R \oplus S$, where $R = [(e),+,\cdot,e]$ and $S = [(e'),+,\cdot,e]$ are the sets (principal ideals) of multiples of e and e', respectively, considered as rings with unities e and $e' = 1 - e$, respectively. Summarizing, we have proved the following theorem.

Theorem 10. *The representations $A = R \oplus S$ of a given ring A as a direct sum are given by the pairs of complementary idempotents e and $e' = 1 - e$ belonging to the center of A.*

Similar results hold for direct sums of any finite number of rings; $A = R_1 \oplus \cdots \oplus R_h$ if and only if the center of A contains h nonzero idempotents $e_1, \ldots, e_h$, such that

$$e_i^2 = e_i \qquad e_ie_j = 0 \qquad \text{if } i \neq j,\ e_1 + \cdots + e_h = 1 \tag{19}$$

Moreover, we have evident isomorphisms

$$R \oplus S \cong S \oplus R \qquad R \oplus (S \oplus T) \cong (R \oplus S) \oplus T \qquad \text{etc.} \tag{20}$$

A particular case of special interest arises when all the R_i are copies (isomorphic replicas) of the *same* basic ring R. The direct sum $R \oplus \cdots \oplus R$ (with h direct summands) is called the hth *power* of R and is denoted R^h. We have, by (20),

$$R^h \oplus R^k = R^{h+k} \qquad \text{for any } h,k \in \mathbf{P} \tag{21}$$

Boolean Rings. The preceding concepts apply with particular force to a special class of commutative rings of characteristic 2 called *Boolean rings,* which are intimately connected with Boolean algebras. This connection is suggested by considering from two standpoints the set of all vectors $\mathbf{x} = (x_1, \ldots, x_h)$ with h binary components, each 0 or 1. This set can be considered either as a Boolean algebra $[\mathbf{2}^h, \wedge, \vee, ', O, I]$ under the operations or as a (Boolean) ring $\mathbf{Z}_2{}^h$, under the operations of product (meet)

$$\mathbf{x} \cdot \mathbf{y} = \mathbf{x} \wedge \mathbf{y} = (x_1y_1, \ldots, x_hy_h) = (x_1 \wedge y_1, \ldots, x_h \wedge y_h)$$

and sum (exclusive-or) $\mathbf{x} + \mathbf{y} = (\mathbf{x} \wedge \mathbf{y}') \vee (\mathbf{x}' \wedge \mathbf{y})$. Note that in the ring interpretation, multiplication is idempotent; since $x_i^2 = x_i$ for all $i = 1, \ldots, h$, we have $\mathbf{x}^2 = \mathbf{x}$ for all $\mathbf{x}$ in the given Boolean ring.

We now use the axiomatic method to generalize the preceding connection to arbitrary Boolean rings and algebras, omitting the assumption of commutativity because it is a necessary consequence of the identity $a^2 = a$.

Definition. A *Boolean ring* is a ring A in which every element is idempotent; thus

$$a^2 = a \qquad \text{for all } a \in A \tag{22}$$

Lemma. *Every Boolean ring is a commutative ring of characteristic* 2.

Proof. Since $(1 + 1) = (1 + 1)^2 = (1 + 1) + (1 + 1)$, we have by cancellation $0 = 1 + 1$, which proves that the characteristic is 2. Likewise, for any a,b,

$$a + b = (a + b)(a + b) = aa + ab + ba + bb = a + b + ab + ba$$

Canceling, $ab + ba = 0$; but $ab + ab = 0$. Hence, canceling again, $ba = ab$.

Corollary. *Under multiplication, every Boolean ring is a semilattice.*

The most interesting property of Boolean rings is their relation to Boolean algebras. This can be stated as follows.

Theorem 11. *Let $\mathcal{A}$ be the class of Boolean rings and $\mathcal{B}$ the class of Boolean algebras, isomorphic members of each being identified. Then the equations*

$$a \wedge b = ab \qquad a \vee b = a + b + ab \tag{23}$$

$$ab = a \wedge b \qquad a + b = (a \wedge b') \vee (a' \wedge b) \tag{23'}$$

define mutually inverse bijections $\alpha: \mathcal{A} \to \mathcal{B}$ and $\beta: \mathcal{B} \to \mathcal{A}$.

Proof. First, let $A = [S, +, \cdot]$ be a Boolean ring with unity 1. By the preceding corollary, the system $B = [S, \wedge, \vee]$ defined from A by formulas (19) is a meet semilattice under $\wedge$. To prove that $[S, \wedge, \vee]$ is a lattice, it therefore suffices to show that $a + b + ab = a$ in A if and only if $ab = b$ (consistency law of Chap. 9).

10-7. IDEALS AND QUOTIENT RINGS

In Chap. 7, we showed that every epimorphic image of any group G was isomorphic with the *quotient group* G/N of G by some *normal subgroup* N, and that every normal subgroup $N \lhd G$ was associated with some epimorphic image in this way. In this section, we shall prove analogous results for rings, referring to *quotient rings* R/H whose elements are the cosets of *ideals* instead of referring to quotient groups and normal subgroups. In rings, the basic definition is the following.

Definition. An *ideal* of a ring R is a nonvoid subset $H \subset R$ such that (*i*) $x \in H$ and $y \in H$ imply $(x \pm y) \in H$, and (*ii*) $x \in H$ and $y \in R$ imply $xy \in H$ and $yx \in H$.

In any ring R, the subset R and the subset $\{0\}$ consisting of 0 alone are, trivially, ideals. Ideals of R other than $\{0\}$ and R are called *proper* ideals of R.

Given any morphism $\theta: R \to S$ of rings, the inverse image of 0 under θ [that is, the set $\theta^{-1}(0)$ of all $x \in R$ with $\theta(x) = 0$ in S] is called the *kernel* of θ. Its basic property is the following.

Theorem 12. *The kernel of any morphism $\theta: R \to S$ of rings is an ideal of the domain R of θ.*

Proof. Considering θ as a morphism of additive groups from $[R,+]$ to $[S,+]$, we see from Theorem 18 of Chap. 7 that the kernel H is an additive subgroup of R. This proves (*i*) in the definition of an ideal. Again, if $x \in H$ and $y \in R$, then (by definition of morphism)

$$\theta(xy) = \theta(x)\theta(y) = 0 \cdot \theta(y) = 0$$

hence $xy \in H$. The proof that $yx \in H$ is similar.

In *commutative rings*, a very important class of ideals can be constructed using the following easily proved result.

Theorem 13. *If R is a commutative ring, the set $(a) = \{xa\}$ of all multiples $xa = ax$ of any fixed $a \in R$ (by a variable element $x \in R$) is an ideal of R.*

Proof. Given $xa \in (a)$ and $ya \in (a)$, trivially, $xa \pm ya = (x \pm y)a \in (a)$ and, for any $r \in R$, $r(xa) = (rx)a \in R$ and $(xa)r = a(xr) \in R$.

Ideals (a) which consist of all the multiples of some fixed $a \in R$ are called *principal ideals.* Clearly, if R is any commutative ring which is not a field, then R will contain some noninvertible $a \neq 0$ whose multiples do not include 1. Hence (a) will be a proper ideal of R.

Conversely, *fields have no proper ideals.* Indeed, let H be any proper ideal in a field F. Then H must contain some nonzero $h \in F$ (being nonvoid). Since F is a field, h^{-1} exists in F; hence H must contain $(xh^{-1})h = x$ for any $x \in F$, which proves that $H = F$. In conclusion, $\{0\}$ and $F = (1)$ are the only ideals of F.

This shows that a commutative ring is without proper ideals if and only if it is a field. We now show by an example that this result fails to be true in general noncommutative rings.

Example 3. Consider the full matrix ring $M_n(F)$ of Example 2, Sec. 10-1, consisting of all $n \times n$ matrices $A = \|a_{ij}\|$ with coefficients in a given field F. In this ring, let E^{hk} signify for $h,k = 1, \ldots, n$, the matrix with (i,j) entry $e_{ij}{}^{hk} = \delta_{hi}\delta_{jk} = 0$ unless $h = i$, $j = k$, and with (h,k) entry 1. If $H \subset M_n(F)$ is any ideal other than $\{0\}$, then it contains some matrix $A \neq 0$ with some nonzero entry $a_{ij} \neq 0$. Hence H must contain all $E^{hi}AE^{jk} = a_{ij}E^{hk}$. Now let $B = \|b_{kh}\|$ be given. Since $a_{ij} \neq 0$, setting $c_{hk} = a_{ij}^{-1}b_{hk}$ and $c_{hk}I$ to be the matrix with entries c_{ij} on the main diagonal and 0 elsewhere, H must contain every $n \times n$ matrix $B = \|b_{hk}\|$, where $b_{hk} = \sum_{h,k} (c_{hk}IE^{hi}AE^{jk})$, and thus $H = M_n(F)$. That is, the full matrix ring over a field F has no proper ideals.

We now come to two fundamental converses of Theorem 12, which constitute uniqueness and existence theorems for epimorphisms, respectively.

Theorem 14. *Let $\theta: R \to S$ and $\theta': R \to S'$ be two epimorphisms of rings with the same domain R and kernel H. Then their images S and S' are isomorphic.*

Proof. Considered as additive groups, by Theorem 18 of Chap. 7, there is an isomorphism $b: S \leftrightarrow S'$ of groups defined by making $x \leftrightarrow x'$ mean that $\theta^{-1}(x)$ and $\theta'^{-1}(x')$ correspond to the same additive coset of H. It remains to prove that this bijection is also an isomorphism for multiplication (i.e., of rings). But, by the definition of a morphism of rings, if $\theta^{-1}(x) = H + a$ and $\theta^{-1}(y) = H + b$, then $\theta^{-1}(xy)$ must contain ab, hence $\theta^{-1}(xy) = H + ab$. Likewise, if $x' = b(x)$ and $y' = b(y)$, then since $\theta'^{-1}(x') = H + a$ and $\theta'^{-1}(y') = H + b$, $\theta'^{-1}(x'y') = H + ab$. Hence $b(xy) = b(x'y')$, since both correspond to the same additive coset of H.

Before proving the second converse of Theorem 14, we prove an easy lemma.

Lemma. *Let H be an additive subgroup of the ring R. Then the partition of R into cosets of H has the substitution property*

$$a \equiv b \pmod{H} \qquad \text{implies} \qquad ar \equiv br \pmod{H} \qquad \text{for all } r \in R \tag{24}$$

if and only if H is a right ideal.

Proof. Condition (24) for H to be an ideal is equivalent by distributivity to the condition that $(a - b) \in H$ implies

$$(ar - br) = (a - b)r \in H$$

for all $r \in R$. And this is the condition that H is a right ideal.

Theorem 15. *Given an ideal H of a ring R, the additive cosets $H + x$ of H form a (quotient) ring R/H under the definitions*

$$(H + x) + (H + y) = H + (x + y) \tag{25}$$

$$(H + x) \cdot (H + y) = H + (xy) \tag{25'}$$

Moreover, the mapping $x \mapsto x + H$ is an epimorphism of R onto R/H, as defined by (25) and (25′).

Proof. Since H is a normal subgroup of the additive Abelian group $[R,+]$ (every subgroup of an Abelian group is normal), R/H is an Abelian group under (25), and the mapping $x \mapsto x + H$ is an epimorphism of additive Abelian groups. It remains to show that (25′) defines a single-valued multiplication of additive cosets of H; that is, we must show that the partition of R into cosets of H has the substitution property for multiplication, or that $H + x = H + x'$ and $H + y = H + y'$ imply $H + xy = H + x'y'$. But this follows since, writing $x' = h + y$ and $y' = h' + y$ $(h,h' \in H)$, we have

$$H + x'y' = H + (h + x)(h' + y)$$
$$H + hh' + hy + xh' + xy = H + xy$$

because $(hh' + hy + xh') \in H$ by the defining properties of ideals.

Remark. In (25′), the product $(H + x)(H + y)$ is intended in the sense of the calculus of complexes[1]: it is the set of all products

$$(h + x)(h' + y) = hh' + hy + xh' + xy \quad (h,h' \in H)$$

It need not itself be a coset of H. Thus the product $((8) + 4)((8) + 4)$ contains only multiples of 16; it is not equal to the complex (8). However, the complex $(H + x) \cdot (H + y)$ is always contained in $H + (xy)$, which uniquely defines $H + (xy)$ as the product $(H + x) \cdot (H + y)$ in the quotient ring R/H.

EXERCISES C

1. Prove in detail the properties of morphisms which are indicated in Fig. 10-2.
2. Show that in any Boolean ring R, for any $a \neq 0,1$, the mapping $x \mapsto (xa, x - xa)$ decomposes R into a direct sum $R = A \oplus B$ of two (proper) epimorphic images of R.
3. Show that any two-element Boolean ring is isomorphic with $[\mathbf{Z}_2, +, \cdot]$.
4. Show that if R is any finite Boolean ring, then $R \cong \overbrace{\mathbf{Z}_2 \cdots \mathbf{Z}_2}^{r \text{ factors}} = \mathbf{Z}_2{}^n$ for some $n \in \mathbf{N}$.
5. Construct an isomorphism of rings $\mathbf{Z}_2 \oplus \mathbf{Z}_3 \cong \mathbf{Z}_6$.
6. (a) Show that cosets of ideals have the following property: If $x \in H + a$ and $y \in H + b$, then $H + xy = H + ab$.
   (b) Interpret this as meaning that "cosets of ideals have the substitution property for multiplication."
7. Show that if A is any Abelian group, then the endomorphisms $\alpha,\beta,\gamma, \ldots$ of A form a ring End A under the definitions $(\alpha + \beta)(x) = \alpha(x) + \beta(x)$ and $(\alpha\beta)(x) = \alpha(\beta(x))$.
8. Show that the monomorphism μ of Theorem 4, has the following universality property: If $\phi: D \to F$ is a monomorphism from D to a field F, then $\phi = \theta \circ \mu$ for some monomorphism of rings $\theta: Q(D) \to F$.
9. Show that in the definition of an *epimorphism* of rings, the condition $\theta(1) = 1$ [that is, $\theta(1_R) = 1_S$] can be dropped.

[1] A *complex* of a group or other algebraic system is a nonvoid subset.

10-8. DIVISIBILITY

The rest of this chapter will be concerned with *commutative* rings. Our main objective will be to establish a unique factorization theorem into primes for an extensive class of Euclidean domains. We will show in Chap. 11 that the class of Euclidean domains includes every ring of polynomials $p(x)$ with coefficients in a field.

But first, we derive some elementary facts about factorization, valid in any integral domain D.

We define the binary relation $a \mid b$ (read "a divides b") in D to mean that $ax = b$ for some $x \in D$. This relation is always reflexive (since $a1 = a$) and transitive since $ax = b$ and $by = c$ imply that

$$a(xy) = (ax)y = by = c$$

Further, $a \mid b$ and $a \mid c$ imply $a \mid (b \pm c)$ because $b = ax$ and $c = ay$ imply $(b \pm c) = ax \pm ay = a(x \pm y)$. [This result simply restates the fact that the set of all multiples of any a is an ideal, the "principal ideal" (a).]

We then define $a \sim b$ (read "a and b are *associates*") to mean that $a \mid b$ and $b \mid a$. The following results, relating divisibility to principal ideals, are then obvious.

Lemma 1. *In a commutative ring R, $a \mid b$ if and only if $(a) \supset (b)$; $a \sim 1$ if and only if a is invertible.*

What is special about integral domains is the fact that in them the nonzero elements form a monoid (Theorem 1). Also, the following lemma holds.

Lemma 2. *In an integral domain, $a \sim b$ if and only if $au = b$ for some invertible (or "unit") u.*

Proof. If $au = b$ for an invertible u, then $bu^{-1} = auu^{-1}$, and so $a \sim b$. Conversely, if $ax = b$ and $by = a$, then $a(xy) = (ax)y = by = a = a1$, hence $xy = 1$ by cancellation, and so x is invertible.

The preceding results are essentially trivial, like many other extremely general results. The rest of our discussion of divisibility will center around generalizations of a familiar but highly nontrivial result of elementary arithmetic, the fact that integers can be factored in one and essentially only one way into prime factors.

The proof of this fact, which goes back to Euclid, depends on the following familiar division algorithm already discussed in Chap. 1.

Division Algorithm. For given integers a and b, with $b \neq 0$, there exist integers q and r such that

$$a = bq + r \qquad 0 \leqq r < b \tag{26}$$

Using this result, one can easily prove another basic result.

Theorem 16. *Any nonvoid set H of integers closed under subtraction either consists of* 0 *alone, or else it consists of the multiples of its least positive member.*

Proof. Unless $H = \{0\}$, it must contain $\pm a$ for some $a \neq 0$; hence it must contain some positive integer. Let b be the *least* positive integer in H; then $(b) \subset H$ trivially. Hence if $a \in H$, then $a - qb = r$ as defined by (26) is in H. Moreover, since $0 \leqq r < b$ and b was *least*, $r = 0$. Hence $(b) = H$, as claimed.

Corollary 1. *In* $\mathbf{Z}$, *all ideals are principal.*

By combining Corollary 1 with Theorem 14 above, we get an important special result.

Corollary 2. *Any epimorphic image of* $\mathbf{Z}$ *is isomorphic to* $\mathbf{Z}_m$ *for some integer* $m > 1$.

10-9. EUCLIDEAN DOMAINS

The preceding properties of the integral domain $\mathbf{Z}$ of ordinary integers have far-reaching generalizations. In order to establish these generalizations, it is easiest to take as assumptions those properties of $\mathbf{Z}$ which are really required in the preceding proofs. Therefore, we begin by stating these assumptions in the form of a definition, as follows:

Definition. A *Euclidean domain* is an integral domain D with a *valuation* $v\colon D^* \to \mathbf{N}$ (D^* is the set of nonzero elements of D, and $\mathbf{N}$ is the set of nonnegative integers), having the following properties:

$$\text{For all } x,y \in D^* \qquad v(xy) \geqq v(x) \tag{27}$$

$$\begin{array}{l}\text{Given } a \in D \text{ and } b \in D^*, \text{ there exists a } q \in D \\ \text{such that } a = bq + r, \text{ where either } r = 0 \text{ or } v(r) < v(b)\end{array} \tag{28}$$

Example 4. The domain $\mathbf{Z}$ is a Euclidean domain with $v(a) = |a|$, the absolute value of a because since $|xy| = |x| \cdot |y|$ and $|y| \geqq 1$ if $y \neq 0$, condition (27) is obvious. Condition (28) just restates the division algorithm of (26).

Example 5. The domain $\mathbf{Z}[\sqrt{-1}]$ of *Gaussian integers* is the subdomain of $\mathbf{C}$ consisting of all complex numbers of the form $m + n\sqrt{-1}$ $(m,n \in \mathbf{Z})$. Under the valuation $v(m + n\sqrt{-1}) = m^2 + n^2$, it is a Euclidean domain.

To show that (27) holds for Gaussian integers, we note that $v(z) = |z|^2$ and use the standard formula $|zz'|^2 = |z|^2 \cdot |z'|^2$ for absolute values of complex numbers. Since $m^2 + n^2 \geqq 1$ unless $m = n = 1$, (27) follows much as in the previous case.

To prove (28) for Gaussian integers, visualize the set of all multiples bq in the complex plane, with $q = m + n\sqrt{-1}$ a variable element of $\mathbf{Z}[\sqrt{-1}]$. These are the vertices of a network of squares of side $|b|$, and thus any $a \in \mathbf{Z}[\sqrt{-1}]$ is at a distance $|a - bq|$, at most $|b|/\sqrt{2}$ from some nearest vertex. This proves (28), with $v(r) = |r|^2 \leqq |b|^2/2 = v(b)/2$.

Example 6. The domain $F[x]$ of all polynomials

$$p(x) = a_0 + a_1x + \cdots + a_nx^n$$

with coefficients a_k in any given field F is a Euclidean domain, with $v(p)$ the *degree* of the polynomial (that is, n if $a_n \neq 0$).

This particular Euclidean domain will be the main subject of Chap. 11. Note that in this example $v(pq) = v(p) + v(q)$.

We now prove a simple general result about Euclidean domains.

Lemma 1. *In any Euclidean domain D*

$$\textit{for } x,y \in D^* \qquad v(xy) = v(x) \tag{29}$$

if y is invertible, whereas $v(xy) > v(x)$ if it is not.

Proof. If y is invertible, then $v(x) = v(xyy^{-1}) \geqq v(xy)$ which, with (27), covers the first case. Moreover, if $(xy) \mid x$, then $xyz = x = x1$ for some $z \in D$, whence $yz = 1$ by the cancellation law since $x \in D^*$; thus y is invertible. It remains to consider the case $(xy) \nmid x$. But in this case, by (28),

$$x = q(xy) + r \qquad \text{where } v(r) < v(xy)$$

On the other hand, $r = x - qxy = x(1 - qy)$, and thus $v(r) \geqq v(x)$ by (27). Hence $v(xy) > v(r) \geqq v(x)$, and the proof is complete.

Corollary. *In a Euclidean domain, an element x is invertible if and only if $v(x) = v(1)$.*

Thus, the invertibles (units) of Z are ± 1; the invertibles of $\mathbf{Z}[\sqrt{-1}]$ are ± 1 and $\pm i$; the invertible polynomials are the nonzero constants (the $p \in D^*$ with $v(p) = 0$).

Theorem 17. *In any Euclidean domain D, every ideal is principal.*

The proof parallels that of Theorem 16. Let H be any ideal of the given Euclidean domain D. If $H = \{0\}$, then the result is trivial. Otherwise, H contains an element $b \neq 0$ with minimum $v(b)$. Clearly, $H \supset (b)$; it remains to prove the reverse inclusion, $H \subset (b)$. But if $a \in H$, the alternative $v(r) < v(b)$ of (28) is impossible since $r \in H$ and b was chosen to minimize v in H. Hence the alternative $a = bq \in (b)$ must hold, proving the theorem.

Euclidean Algorithm. It is a corollary of Theorem 17 that the sum $(a) + (b)$ of any two principal ideals of D is again a principal ideal of D, say (c); that is, there exists some $c = sa + tb$ which is divisor of both a and b. If d is any other common divisor (with $a = qd$ and $b = q'd$), then

$$c = sa + tb = sqd + tq'd = (sq + tq')d$$

Hence c is a multiple of every other common divisor d. Therefore, c is a "greatest" common divisor of a and b in the following sense.

Definition. In an integral domain D, an element c is said to be a *greatest common divisor* of a and b [in symbols, $c = \gcd(a,b)$] when it is a divisor of both a and b and a divisor of every other common divisor.

Note that any associate $cu = c'$ (u invertible) of any greatest common divisor $c = (a,b)$ of a and b is also such a greatest common divisor. For example, in $\mathbf{Z}$, if c is a gcd, then so is $-c$; in this case, the symbol (a,b) commonly refers to the positive gcd (see the ALGOL program for the Euclidean algorithm below).

Not only do we know that such greatest common divisors exist in any Euclidean domain, we can also compute them whenever we know how to compute the q and r of (28). The algorithm for computing $c = \gcd(a,b)$, which goes back to Euclid, is called the *Euclidean algorithm;* it proceeds as follows. As before, we let (a) signify the principal ideal of all multiples ax of a.

If $a = 0$, then trivially $(a) + (b) = (b)$, and we can set $c = b$; likewise, if $b = 0$, we can set $c = a$. Otherwise, we have $a,b \in D^*$, and thus $v(a)$ and $v(b)$ are both defined. Without loss of generality, we can assume that $v(b) = v(a)$.

Now apply (28). If $a = qb$, then $(a) \subset (b)$ and $(a) + (b) = (b)$; hence we can set $c = b$. Otherwise, $a = qb + r_1$, with $v(r_1) < v(b)$; from this $(a) + (b) = (r_1) + (b)$ follows, where

$$\min \{v(r_1), v(b)\} = v(r_1) < \min \{v(a), v(b)\}$$

This process can be iterated until we get a zero remainder r_k, giving

$$(a) + (b) = (b) + (r_1) + (r_1) + (r_2) = \cdots = (r_{k-1})$$

Clearly, $r_{k-1} = \gcd(a,b)$.

The following is an ALGOL program for computing the positive greatest common divisor gcd of two integers a and b, according to the Euclidean algorithm described above:

```
      begin integer a, b, gcd, aa, bb, temp;
      aa := abs (a); bb := abs (b);
      if aa = 0 ∨ bb = 0 then
      begin gcd := aa + bb; go to F end;
      if aa < bb then begin temp := aa; aa := bb:bb := temp end:
D:    r := aa − (aa ÷ bb) × bb:
      if r ≠ 0 then begin aa := bb; bb := r; go to D end:
      gcd := b:
F:    end
```

EXERCISES D

1. Show that in the integral domain $\mathbf{Z}[x]$
   (a) $\gcd(7x^3 + 1, 2x) = 1$.
   (b) There are no polynomials $a(x)$, $b(x)$ such that

$$a(x)(7x^3 + 1) + 2xb(x) = 1$$

2. Show that the relation $a \sim b$ (associate) in any integral domain has the following substitution properties:
   If $a \sim b$, then $ac \sim bc$.
   If $a \sim b$ and $c \sim d$, then $ac \sim bd$.
3. Show that if c and c' are both greatest common divisors of a and b in an integral domain, then $c \sim c'$.
4. (a) Let p be a prime in a Euclidean domain D. Prove that the equation $x^2 = p$ can have no solution in $Q(D)$, the field of quotients of D.
   (b) Prove that $\sqrt[3]{3}$ is an irrational number.
5. (a) Show that $\sqrt{3} \notin \mathbf{Q}(\sqrt{2})$ (see Exercise B5*b*).
   (b) Show that $a + b\sqrt{2} \leftrightarrow a + b\sqrt{3}$ is not an isomorphism of fields from $\mathbf{Q}(\sqrt{2})$ to $\mathbf{Q}(\sqrt{3})$.
   ★(c) Show that there is no isomorphism $\mathbf{Q}(\sqrt{2}) \cong \mathbf{Q}(\sqrt{3})$ of fields.
6. Compute $\gcd(108996, 76219) = d$ in the form $d = 108996s + 76219t$ $(s,t \in \mathbf{Z})$.
7. (a) Make a table of inverses in $\mathbf{Z}_{37}$.
   (*Hint:* Compute 2^{-1}, 3^{-1}, and use $(xy)^{-1} = x^{-1}y^{-1}$.)

★10-10. UNIQUE FACTORIZATION THEOREM

In any integral domain D, every nonzero element x has a trivial factorization $x = uy$ for each invertible u and $y = u^{-1}x$. By a *prime* is meant a (nonzero) element $p \in D^*$ which is not invertible and which cannot be expressed as a product of two noninvertible factors, i.e., which has no nontrivial factorization. Clearly, if p and q are any two primes in D and $px = q$ (that is, $p \mid q$), then x is invertible, and thus $p \sim q$. Hence the primes of D fall into equivalence classes of associate primes, and a prime from one equivalence class cannot divide any prime from any other equivalence class.

In the special domain **Z**, the first few primes are easily listed as ± 2, ± 3, ± 5, ± 7, ± 11, etc. A familiar property of **Z** is the Unique Factorization Theorem, which asserts that any integer can be uniquely factorized into powers of primes [up to the replacement of any prime p by its associate $-p = (-1)p$; in **Z**, -1 is the only invertible besides 1]. We shall now show that a similar result holds in any Euclidean domain, essentially because of the Euclidean algorithm. The proof uses finite induction and the following lemma.

Lemma 1. *If p is a prime in the Euclidean domain D, and $p \mid ab$, then $p \mid a$ or $p \mid b$.*

Proof. The case $p \mid a$ is trivial, hence we suppose $p \nmid a$. Consider gcd (p,a); since $p \nmid a$, this gcd must be a divisor of p *not* associate to p, and hence $c = \gcd(p,a)$ must be invertible. On the other hand, by the Euclidean algorithm (Sec. 10-5), $c = sa + tp$ for some $s,t \in D$. Hence

$$b = c^{-1}cb = c^{-1}(sa + tp)b = c^{-1}sab + c^{-1}tpd$$

By hypothesis, $p \mid ab$, whereas $p \mid c^{-1}tpd$ trivially; hence $p \mid b$, proving Lemma 1.

More generally, call two elements a,d in a Euclidean domain D *relatively prime* [in symbols, $(a,d) = 1$] when $\{sa + td\} = D$. If $(a,d) = 1$ and $d \mid ab$, then a repetition of the above argument gives

$$b = (sa + td)b = sab + tdb$$

whence $d \mid b$ (since d divides both terms sab and tdb in the preceding sum). In summary, if $(a,d) = 1$ and $d \mid ab$, then $d \mid b$. From this generalization of Lemma 1, we obtain as a consequence (for later use) the following lemma.

Lemma 2. *Let $(a,a') = 1$, and let $a \mid c$, $a' \mid c$. Then $aa' \mid c$.*

Proof. Since $a \mid c$, we have $ab = c$. Hence $a' \mid ab$ with $(a,a') = 1$, and $a' \mid b$ by the above generalization of Lemma 1. That is, $b = a'x$ and, so $c = aa'x$ for some $x \in D$, proving that $aa' = c$.

Corollary. *Let $a_1, \ldots, a_n$ be relatively prime divisors of c in a Euclidean domain D. Then* $\left(\prod_{i=1}^{n} a_i\right) \mid c$.

Lemma 3. *In any Euclidean domain D, every nonzero, noninvertible element a can be expressed as a product of prime factors.*

Proof. The result is trivial if a is a prime. Otherwise $a = bc$, where $b,c \in D^*$ are noninvertible. Hence, by Lemma 1 of Sec. 10-9, $v(b) < a$, and $v(c) < v(a)$. It follows, by induction on the numerical value of v, that b and c can be expressed as products of primes; thus $b = \Pi p_i$ and $c = \Pi p_j'$. [In detail, let $P(n)$ be the proposition that any element a with $v(a) - v(1) = n$ can be expressed as a product of prime factors. Then $P(0)$ follows trivially from Lemma 1 of Sec. 10-9; the argument given above shows that the truth of $P(n)$ for b and c implies the truth of $P(n + 1)$ for a.]

Lemma 3 asserts the existence part of the following fundamental existence and uniqueness theorem.

Theorem 18. *In any Euclidean domain, each nonzero, noninvertible element a can be expressed essentially uniquely as an invertible times a product of powers of nonassociate prime factors.*

Proof. Consider any two prime factorizations of a (at least one exists by Lemma 3),

$$a = \prod_{i=1}^{m} p_i = \prod_{j=1}^{n} q_j \tag{30}$$

Since $q_n \mid a$ by Lemma 1 (and induction on m), q_n must divide some $p_{i(1)}$. By the first paragraph of this section, it follows that $q_n = up_{i(n)}$, where u is invertible (a unit). We can therefore write

$$a = p_{i(n)} \prod_{k=1}^{m-1} p_{i(k)} = p_{i(n)}u \prod_{j=1}^{n-1} q_j \tag{31}$$

By the cancellation law, it follows that

$$b = \prod_{k=1}^{m-1} p_{i(k)} = \prod_{j=1}^{n-1} q_j' \qquad q_1' = uq_1 \qquad q_j' = q_j \qquad j > 1 \tag{32}$$

Now let $P(m)$ be the proposition that for given m in (30), $n = m$ and the p_i', q_j are associate in pairs. The case $m = 1$ is trivial. Moreover, assuming $P(m - 1)$ as applied to (32) (induction hypothesis), we can infer that $m - 1 = n - 1$ and that the $p_{i(k)}$ and q_j' in (32) are associate in pairs. Substituting back into (32), we get $P(m)$, which is thus proved. This is one uniqueness theorem.

We now apply Lemma 1 of Sec. 10-8, which asserts that the relation $p_i \sim p_j$ is an equivalence relation, and we choose a representative p_i from each equivalence class E of associate prime factors of a. Given i, write all $p_j \sim p_i$ as $p_j = u_j p_i$, and form the product

$$p_i \prod u_j p_i = \bar{u}_i p_i^{e_i}$$

Now multiplying these grouped factors together, we have

$$a = \prod \bar{u}_i p_i^{e_i} = u \prod p_i^{e_i}$$

This is a prime-power factorization of a. Its uniqueness (up to replacement of each p_i by some associate prime) was proved in the preceding paragraph, completing the proof.

★10-11. PRIME AND MAXIMAL IDEALS

In Sec. 10-7, we showed that the epimorphic images of ring R were (up to isomorphism) precisely the quotient rings R/H of the ring R over its different ideals H. Limiting our attention to *commutative* rings R, we now answer two interesting questions about these quotient-rings.

Which ideals H give rise to quotient rings that are (*i*) integral domains? (*ii*) Fields? This question is completely answered by Theorem 19 below. We first introduce the following definitions:

Definition. An ideal H is *prime* when $ab \in H$ implies $a \in H$ or $b \in H$. An ideal H in a ring R is *maximal* when, for any ideal H', $H \subset H' \subset R$ implies $H' = H$ or $H' = R$; that is, when there is no ideal H' properly including H other than the ring R itself. We then have Theorem 19.

Theorem 19. *If R is a commutative ring, and H an ideal of R, then*

1. *R/H is an integral domain if and only if H is prime.*
2. *R/H is a field if and only if H is maximal.*

Proof. The commutative ring R/H is an integral domain if and only if there are no zero divisors, that is, if and only if $(H + a)(H + b) = H$ implies $H + a = H$ or $H + b = H$. But this is equivalent to the statement that $ab \in H$ implies $a \in H$ or $b \in H$, which is precisely the statement that H is a prime ideal, proving the first conclusion.

Now assume that H is a maximal ideal in R, and consider any nonzero element $H + a \in R/H$. The set $S = \{h + ax \mid h \in H,\ x \in R\}$ is easily seen to be an ideal. Moreover, $H < S$ because $a \notin H$, whence $S = R$ because H is maximal. Thus it follows that $1 = h + ax$ for some $h \in H$, $x \in R$, which in turn means that

$$H + 1 = H + h + ax = H + ax = (H + a)(H + x)$$

This shows the existence of an inverse for any nonzero element $a + H$ of the commutative ring R/H, and so R/H is a field.

Conversely, assume that R/H is a field, and let M be an ideal which properly contains H, so that there is an element a such that $a \in M$ and $a \in H$. Now "R/H is a field" implies that the equation

$$(H + a)(H + x) = (H + b)$$

is solvable for any $b \in R$ [note that $H + a \neq H$ because $a \in H$]; hence $H + aH + Hx + ax = H + b$. But $H + aH + Hx + ax \subset M$ because $H \subset M$ and $a \in M$; therefore $M = M + b$ for any $b \in R$, whence we are forced to conclude that $M = R$, thus proving that H is maximal.

Corollary. *Any maximal ideal is a prime ideal.*

Proof. This follows from Theorem 19 because any field is necessarily an integral domain.

★10-12. GAUSSIAN ELIMINATION

For numerical applications, the most important property of fields is the principle that systems of *simultaneous linear equations* of the form

$$\begin{aligned} a_{11}x_1 + a_{12}x_2 + \cdots + a_{1n}x_n &= b_1 \\ a_{21}x_2 + a_{22}x_2 + \cdots + a_{2n}x_n &= b_2 \\ \cdots\cdots\cdots\cdots\cdots\cdots\cdots& \\ a_{n1}x_1 + a_{n2}x_2 + \cdots + a_{nn}x_n &= b_n \end{aligned} \tag{33}$$

with the number n of unknowns equal to the number of equations, ordinarily have unique, explicitly calculable solutions. We first develop this principle for the case $n = 2$.

Accordingly, let a,b,c,d,e,f be any six constants belonging to the field F. We ask for two numbers x and y which satisfy the simultaneous equations

$$ax + by = e \qquad cx + dy = f \tag{34}$$

If we multiply the first equation by d, the second equation by b, and subtract, we get $(ad - bc)x = de - bf$. Likewise, subtracting c times the first equation from a times the second, we get $(ad - bc)y = af - ce$. Therefore, if the determinant $ad - bc = \Delta$ of the coefficient matrix $\begin{pmatrix} a & b \\ c & d \end{pmatrix}$ of the system (34) is not zero, the system has a unique solution vector (x,y), given by

$$x = \frac{de - bf}{\Delta} \qquad y = \frac{af - ce}{\Delta} \qquad \Delta = ad - bc \tag{35}$$

To prove in detail that formula (35) follows from the postulates for a field is straightforward; conversely, substituting from (35) into (34), we get

$$\frac{ade - abf + baf - bce}{\Delta} = \frac{(ad - bc)e}{\Delta} = e$$

and a similar verification of the second equation.

When the determinant Δ is zero, the system (34) either has no solution or many solutions; this is called the case of a *singular* coefficient matrix.

Similar results hold in the general case. However, rather than develop the general theory of determinants, we shall concentrate our attention on the Gaussian elimination algorithm which enables one to *compute* any particular solution. In considering this algorithm, the student should remember that when real numbers are involved, the operations can only be performed approximately on a computer.

A general system of m simultaneous linear equations in n unknowns can be written compactly in the form

$$\sum_{j=1}^{n} a_{ij}x_j = b_i \qquad i = 1, \ldots, m \tag{36}$$

It is understood that all a_{ij} and b_i belong to a fixed field F, and that one asks for the lists of n elements $x_1, \ldots, x_n$ in F which satisfy all m equations of (36). We now argue by induction on m, as follows.

If all the coefficients a_{1j} of the first equation vanish, then either $b_j \neq 0$ and (36) has no solution, or $b_1 = 0$ and the first equation is fulfilled trivially by any solution of the other $m - 1$ equations, to whose consideration we can pass.

If some $a_{1j} = 0$ (when $F = \mathbf{R}$, to choose j so as to maximize $|a_{1j}|$ is a good choice), then the first equation of (36) is equivalent to

$$x_j = a_{1j}^{-1}(b_1 - \sum_{k \neq j} a_{1k}x_k) \tag{37}$$

and when this holds, the other $m - 1$ equations of (36) are equivalent to

$$\sum_{k \neq j} (a_{ik} - a_{ij}a_{1j}^{-1}a_{1k})x_k = b_i - a_{ij}a_{1j}^{-1}b_1 \qquad i = 2, \ldots, m \qquad (38)$$

This is a system of $m - 1$ simultaneous linear equations in $n - 1$ unknowns, having the same form as (36).

Therefore, we can repeat the process just described, denoting $a_{ik} - a_{ij}a_{1j}^{-1}a_{1k}$ by a'_{ik} (say) and $b_i - a_{ij}a_{1j}^{-1}b_1$ by b'_i, where i ranges from 2 to m and j over $\mathbf{n} - \{j\}$. The exceptional case in which all $a'_{2k} = 0$ is the case that $a_{1j}a_{2k} = a_{2j}a_{1k}$ for all $k \in \mathbf{n}$ other than j. In the nonexceptional case, we can choose k so as to make $a'_{sk} \neq 0$ [when $F = \mathbf{R}$, we can maximize $|a_{2k} - a_{2j}a_{2j}^{-1}a_{1k}|$]. In any case we can get an analog of (37) for x_k, and then replace the remaining $m - 2$ equations by a linear system in $n - 2$ unknowns, from which x_j and x_k have been eliminated.

When $m = n$, after repeating the above process n times, we will normally get an equation giving the value of x_n. By a "back substitution" we can then solve in turn for $x_{n-1}, x_{n-2}, \ldots, x_1$.

ALGOL program is displayed below for computing the solution to a system of simultaneous linear equations by the process of Gaussian elimination just described[1]:

```
procedure gauss (u,a,y);
real array a,y; integer u;
comment This procedure is for solving a system of linear equations by
   successive elimination of the unknowns. The augmented matrix in a
   and u is the number of unknowns. The solution vector is y. If the sys-
   tem hasn't any solution or many solutions, this is indicated by the "go
   to error" where error is a label outside the procedure.
   begin
          integer i,j,k,m,n;
                 n := 0;
                 n := n + 1;
ck0:      for k := n step 1 until u do if a [k,n] ≠ 0 then go to ck1;
          go to error;
ck1:      if k = n then go to ck2;
          for m := n step 1 until n + 1 do
   begin
          temp := a[n,m; a[n,m] := a[k,m]; a[k,m] := temp
   end:
ck2:      for j := u + 1 step - 1 until n do a[n,j] := a[nkj]/a[n,m];
          for i := k + 1 step 1 until u do
          for j := n + 1 step 1 until u + 1 do
```

[1] Jay W. Counts, University of Missouri, Columbia, Mo.

```
a[i,j] := a[i,j] - a[i,n] × a[n,j];
if n ≠ u then go to ck0;
        for i := u step - 1 until 1 do
begin y[i] := a[i,u + 1]/a[i,i];
        for k := i - 1 step - 1 until 1 do
        a[k, u + 1] := a[k, u + 1] - a[k,i] × y[i]
end end;
```

EXERCISES E

1. Prove that $(m,n) = 1$ in a Euclidean domain if and only if $(m,p_i) = 1$ for every prime factor p_i of n.
2. Solve the simultaneous linear equations $x + 2y = 0 \qquad 2x + y = 1$
   (a) over $\mathbf{Z}_5$
   (b) over $\mathbf{Z}_3$
   (c) over $\mathbf{Z}_7$
3. (a) Prove that the following tables define a field:

+	0	1	2	3
0	0	1	2	3
1	1	0	3	2
2	2	3	0	1
3	3	2	1	0

·	0	1	2	3
0	0	0	0	0
1	0	1	2	3
2	0	2	3	1
3	0	3	1	2

   (b) Denoting this field $GF(4)$, establish a monomorphism from $\mathbf{Z}_2$ to $GF(4)$.
   (c) Solve the simultaneous linear equations of Exercise 2 in $GF(4)$.
4. List the nonsingular 3×3 matrices over z_2.

REFERENCES

Rings are treated in the books by Birkhoff and Mac Lane, Mac Lane and Birkhoff, and Herstein listed at the end of Chap. 1, and in N. Jacobson, "Lectures on Abstract Algebra," 3 vols., Van Nostrand, 1953–61.

JANS, J. P: "Rings and Homology," Holt, 1968.
MC COY, N. H.: "The Theory of Rings," Macmillan, 1964.
ZARISKI, O., and P. SAMUEL: "Commutative Algebra," 2 vols., Van Nostrand, 1959–60.

Polynomial Rings and Polynomial Codes

11-1. THE RING $R[x]$

If R is any commutative ring,[1] then expressions like $x^2 + 1$, $2x - x^3 + 1$, and $x^2 - 3x^2$ are ordinarily spoken of as "polynomials" (in x, over R). Moreover, since addition is commutative, $x^2 + 1$ is recognized as being interchangeable with $1 + x^2$. One also commonly omits any term having a factor 0; thus one seldom writes $x^2 + 0 \cdot x + 1$, and one normally simplifies $x^2 + 1 - (1 - x^2) + (2x - 3x^2)$ to $2x - x^2$, without explaining why.

To discuss all such polynomials rigorously, however, would be a very tedious review of principles learned in high school. To avoid the need for this discussion, algebraists commonly define polynomials in a much more restrictive way, specifying as polynomials a small set of standard expressions to which all others can be easily reduced. One such set of polynomials is the following.

[1] In this chapter, only commutative rings will be considered.

Definition. A standard *polynomial* of length $m + 1$ in the *indeterminate* x over a *commutative* ring R is an expression of the form

$$a_0 + a_1x + \cdots + a_mx^m = \sum_{k=0}^{m} a_kx^k, \qquad \text{all } a_k \in R \tag{1}$$

The a_k are called the *coefficients* of the polynomial (1); any or all of them may be 0.

One often writes $a(x)$ as an abbreviation for the expression (1), even though this does not mean that the polynomial $a(x)$ is being considered as a function $a: R \to R$ (see Sec. 11-7). The standard polynomial (1) is also specifiable by the array (vector) of its coefficients, as

$$a = (a_0, a_1, \ldots, a_n)$$

this makes it easy to specify polynomials in ALGOL (see Chap. 4).

The canonical form of the polynomial $a(x)$ of (1) is defined as follows: (*i*) pick the largest k in (1) with $a_k \neq 0$, say $k = n$, and then (*ii*) rewrite (*i*) as

$$a(x) = a_0 + a_1x + \cdots + a_nx^n \qquad a_n \neq 0 \tag{2}$$

In the exceptional case that all $a_k = 0$, the canonical form is 0. The degree of $a(x)$ is the integer n in (2), and it is also symbolized by deg a; if $a(x)$ has the canonical form 0, it is conventionally said to have degree $-\infty$.

Two standard polynomials, the $a(x)$ of length m in (1) and a polynomial of length r,

$$b(x) = b_0 + b_1x + \cdots + b_rx^r = \sum_{k=0}^{r} b_kx^k \qquad \text{all } b_k \in R \tag{2'}$$

are said to be *equal* (as polynomials) when they have the same canonical form. The following result is obvious.[1]

Lemma 1. *The polynomials $a(x)$ and $b(x)$ are equal if and only if (i) $a_k = b_k$ for all k for which both are defined (i.e., all $k \leq m \wedge r$†), and (ii) all other a_k and b_k are zero.*

Equivalently, the $a(x)$ of (1) is said to be equal to the $b(x)$ of (2′) when they have the same degree n (that is, deg a = deg b = n), and $a_i = b_i$ for all $i = 0, 1, \ldots, n$. The set of all polynomials in x over a commutative ring R is denoted by $R[x]$.

[1] From here on, we shall drop the word "standard" since all polynomial expressions with liberal coefficients will be assumed to be in standard form.

† The symbol $m \wedge r$ means (for $m,r \in \mathbf{N}$) "the smaller of m and r," and $m \vee r$ means "the larger of m and r," as in Chap. 9.

To add two polynomials in $R[x]$, add corresponding coefficients. More precisely, if $a(x)$ and $b(x)$ are given by formulas (1) and (2′) above, respectively, then the sum $c(x) = a(x) + b(x)$ is defined as

$$c(x) = c_0 + c_1x + \cdots + c_sx^s \tag{3}$$

where $s = m \vee r$ and

$$c_k = \begin{cases} a_k + b_k & \text{for } k \leqq m \vee r \text{ in any event} \\ a_k & \text{for } m < k \leqq r \text{ if } m < r \\ b_k & \text{for } r < k \leqq m \text{ if } m > r \end{cases} \tag{3′}$$

Likewise the product $p(x) = a(x) \cdot b(x)$ is defined for $a(x)$ and $b(x)$ given by (1) and (2′) as

$$p(x) = p_0 + p_1x + \cdots + p_{m+r}x^{m+r} \tag{4}$$

where

$$p_k = \sum_{i+j=k} a_ib_j \tag{4′}$$

For $k \leqq m \wedge n$, evidently $p_k = a_0b_k + \cdots + a_kb_0$; likewise, clearly $p_{m+r} = a_mb_r$.

Example 1. Let $R = \{0,1\} = \mathbf{Z}_2$, let $f(x) = 1 + 0 \cdot x + 1 \cdot x^2$, and let $g(x) = 1 + 1 \cdot x + 1 \cdot x^2$. Then

$$f(x) + g(x) = (1 + x^2) + (1 + x + x^2) = x$$

and

$$f(x) \cdot g(x) = (1 + x^2)(1 + x + x^2) = 1 + 1 \cdot x + 1 \cdot x^3 + 1 \cdot x^4$$

Trivially, the degree of the sum and product of two polynomials satisfies the inequalities

$$\deg\,(a + b) \leqq \deg a + \deg b \tag{5}$$
$$\deg\,(ab) \leqq \deg a + \deg b \tag{5′}$$

Since, in an integral domain, $a_n \neq 0$ and $b_r \neq 0$ imply $a_nb_r \neq 0$, we have [using (4–4′) and setting $\deg b = r$] the following lemma.

Lemma 2. *If D is an integral domain, then in $D[x]$ we have*

$$\deg\,(ab) = \deg a + \deg b$$

A result which requires more verification is the following theorem (the details of proof are individually easy but numerous):

Theorem 1. *Under the operations* (3), (3′), (4), *and* (4′), *the polynomials over any commutative ring R form a commutative ring $R[x]$, whose constants form a subring isomorphic to R.*

There are many things to prove, each of which is easy to establish in its turn. For example, one must prove that equality, as defined in Sec. 11-2, has the substitution property; that is, if $a(x)$ and $a^*(x)$ have the same canonical form and $b(x)$ and $b^*(x)$ also have the same canonical form, then

$$a(x) + b(x) = a^*(x) + b^*(x) \qquad a(x)b(x) = a^*(x)b^*(x)$$

This follows easily from Lemma 1.

We shall omit proving that the postulates for a commutative ring are satisfied, because a more general result will be proved in Chap. 13 (for formal power series). The fact that the constants (polynomials of length 1) form a subring isomorphic to R, whose 0 and 1 serve as a zero and unity for $R[x]$, is obvious.

Table 11-1. Sums and products in Z[x]
(ALGOL Program for Polynomial Calculations)

```
begin integer array a[-1:100], b[-1:100], s[-1:100], p[1:200]
                    f[-1:100], g[-1:100];
      integer i,k, initial, final;
if a[-1] < b[-1]
   then for i := 1 step 1 until b[-1] do
            begin f[i] := a[i]; g[i] := b[i] end;
   else for i := 1 step 1 until a[-1] do
            begin f[i] := b[i]; g[i] := a[i] end;
comment if deg (a) < deg (b), f specified as a; otherwise f specified as b, and g as a.
   Now compute s = f + g + a + b;
for k := 0 step 1 until f[-1] do s[k] := a[k] + b[k];
for k := f[-1] + 1 step 1 until g[-1] do s[k] := g[k];
s[-1] := g[-1];
bk: if s[s[-1]] = 0
     then begin s[-1] := s[-1] - 1; go to bk end;
comment compute p = fg = ab;
if (f[-1] = 0 ∧ f[0] = 0) ∨ (g[-1] = 0 ∧ g[0] = 0)
     then begin p[-1] = 0; p[0] := 0; go to F end;
p[-1] := f[-1] + g[-1];
for k := 0 step 1 until p[-1] do
begin s[k] := 0;
initial := if k ≦ g[-1] then 0 else k - g[-1];
final := if k ≦ f[-1] then k else f[-1];
for i := initial step 1 until final do
    s[k] := s[k] + f[i] × g[k - i]; end
end
end
```

Table 11-1 presents an ALGOL program for adding and multiplying polynomials in $\mathbf{Z}[x]$ by computer. A polynomial

$$a(x) = a_0 + a_1x \cdots + a_mx^m$$

($m \geq 0$, $a_m \neq 0$ if $m \neq 0$) is represented as an array of integers

$$(m = a_{-1}, a_0, a_1, \ . \ . \ . \ , a_m)$$

To be acceptable, the list must contain $m + 2$ entries with $a_m \neq 0$ if $m \neq 0$. The ALGOL program in Table 11-1 computes the sum $s(x)$ and product $p(x)$ of the polynomials $a(x)$ and $b(x)$ according to the above rules. It is restricted to deg (a), deg (b) ≤ 100.

11-2. POLYNOMIAL RINGS OVER FIELDS

We defined the notion of *valuation* in Sec. 10-9 and called any integral domain which had such a valuation a *Euclidean domain*. We observed that **Z** was a Euclidean domain in which the absolute value function $|a|$ served as a valuation and showed that any Euclidean domain had a division algorithm like that for the integers, implying a unique factorization theorem. We now prove Theorem 2.

Theorem 2. *Over any field F, the polynomials form a Euclidean domain $F[x]$ with the valuation $v(a) = deg\ a$.*

Proof. For any nonzero polynomials $a(x), b(x) \in F[x]$, the product has the leading coefficient $a_m b_n \neq 0$ as in Lemma 2 of Sec. 11-1, and thus

$$\deg ab = \deg a + \deg b \geqq \deg a \tag{6}$$

with equality holding if and only if b is of degree zero, i.e., a constant. Therefore, the degree of a polynomial satisfies condition (27) of Chap. 10.

As a by-product, we obtain the following corollary.

Corollary. *The units (invertibles) of the polynomial ring $F[x]$ are the nonzero constants of F.*

To complete the proof of Theorem 2, it remains to prove condition (28) of Chap. 10. But this is precisely the conclusion of the following lemma.

Lemma 3. (*Division Algorithm.*) *Given polynomials $a(x)$ and $b(x)$ in $F[x]$ with $b(x) \neq 0$, there exist a "quotient" $q(x)F[x]$ and a "remainder" $r(x) \in F[x]$, where either $r(x) = 0$ or $deg\ r(x) < deg\ b(x)$, such that*

$$a(x) = b(x)q(x) + r(x) \qquad deg\ r(x) < deg\ b(x) \tag{7}$$

Proof. The quotient $q(x)$ and remainder $r(x)$ can be computed by the following "polynomial long division" process, commonly taught in high school.

Let $a(x) = \sum_{k=0}^{m} a_k x^k$ and $b(x) = \sum_{k=0}^{n} b_k x^k$, where we suppose $b(x)$ in canonical form (2), with $b_n \neq 0$. There are two cases to consider.

Case 1. If $m < n$, we can set $q(x) = 0$ and $r(x) = a(x)$.

Case 2. If $m \geqq n$, we can define

$$a_1(x) = a(x) - b_n{}^{-1}a_m x^{m-n} b(x) = a(x) - q_1(x)b(x)$$

This will be of degree $m - 1$ at most, since $a(x)$ and $q_1(x)b(x)$ have the same leading coefficient $b_n{}^{-1}a_m b_n = a_m$. We now use induction on m, repeating the process (at most $m - n + 1$ times) until the remainder

$$a_k(x) = a(x) - \left[\sum q_i(x)\right] b(x) = a(x) - q(x)b(x)$$

has degree less than m.

Definition. A *monic* polynomial is a polynomial with leading coefficient 1, that is, a polynomial of the form

$$a(x) = a_0 + a_1 x + \cdots + a_{n-1}x^{n-1} + x^n \tag{8}$$

Clearly, the mapping $b(x) \mapsto b_n$ from nonzero polynomials of $F[x]$ to their leading coefficients is a morphism of monoids; every nonzero polynomial in canonical form can be uniquely factored into the product $cm(x)$ of a nonzero constant $c = b_n$ times a monic polynomial $m(x)$.

Irreducible Polynomials. In Sec. 10-10, we proved that an analog of the unique factorization theorem (into prime factors) held in any Euclidean domain. From Theorem 2 above, therefore, it holds in the Euclidean domain $F[x]$ of all polynomials over any field F. The field F may be $\mathbf{R}$, $\mathbf{C}$, $\mathbf{Q}$, $\mathbf{Z}_p$ (p a prime), or any other field.

We shall develop this consequence of Theorem 2 in more detail in Sec. 11-5. Here, we merely describe what the analog of a prime is like in $F[x]$.

Definition. A polynomial $p(x)$ in $F[x]$ is said to be *reducible* (over F) if $p(x) = a(b)b(x)$ for some nonconstant $a(x),b(x) \in F[x]$. Otherwise, $p(x)$ is said to be *irreducible* (over F).

Note carefully that the reducibility or irreducibility of a given polynomial $p(x)$ depends heavily on the field F under consideration. Thus, $x^2 + 1$ is irreducible over the real field $\mathbf{R}$, whereas over the complex field $\mathbf{C}$, $x^2 + 1 = (x + i)(x - i)$ is reducible. Again, $x^2 - 2$ is irreducible over $\mathbf{Q}$ and over $\mathbf{Z}_3$, whereas over $\mathbf{Z}_{11}$, $x^2 - 2 = (x + 3)(x + 8)$, and over $\mathbf{R}$, $x^2 - 2 = (x + \sqrt{2})(x - \sqrt{2})$ and is reducible in both cases. Similarly, $x^2 + x + 1$ is irreducible over $\mathbf{Z}_2$, whereas over $\mathbf{Z}_3$, $x^2 + x + 1 = (x + 2)^2$ is reducible.

Since $\deg ab = \deg a + \deg b$, linear polynomials are irreducible over any field. The fundamental theorem of algebra asserts that the *only* irreducible polynomials over $\mathbf{C}$ are the linear polynomials $z - c$. Likewise, the only irreducible polynomials over $\mathbf{R}$ are the linear polynomials and the quadratic polynomials $x^2 + px + q$ with negative discriminant $p^2 - 4q$. Over the rational field $\mathbf{Q}$, most polynomials are irreducible.

Our main concern in this book will be with polynomials over the field $\mathbf{Z}_2$.

EXERCISES A

1. Prove the inequalities (5) and (5′).
2. Prove that $R[x]/(x^4 + x^3 + x + 1)$ cannot be a field, no matter what the commutative ring R.
3. Compute $(2x + 1)^{-1}$ in $F[x]/(x^3 - 2)$, when
   (a) $F = \mathbf{Q}$
   (b) $F = \mathbf{Z}_5$
   (c) $F = \mathbf{Z}_7$
4. Show that the field of quotients of $F[x]$ is, for any field F, the field of rational forms $p(x)/q(x)$ with $q \neq 0$.
   (*Hint:* See Theorem 3 of Chap. 10.)
5. Let $F^* = F - \{0\}$, let $(F[x])^* = F[x] - \{0\}$, and let M be the monoid of monic polynomials under multiplication. Establish an isomorphism of monoids $(F[x])^* = F^*M$.
6. (a) For any field F and any positive integers m and n, prove that $(x^m - 1) \mid (x^n - 1)$ if and only if $m \mid n$.
   (b) Conclude that for any positive integer $k > 1$, $(k^m - 1) \mid (k^n - 1)$ if and only if $m \mid n$.
7. Let $C[0,1]$ be the ring of all continuous, real-valued functions on the closed unit interval $0 \leqq x \leqq 1$, with

$$(f + g)(x) = f(x) + g(x) \qquad (fg)(x) = f(x)g(x)$$

   (a) Is $C[0,1]$ an integral domain? Justify your answer.
   (b) Show that the assignment $f \mapsto f(a)$ is an epimorphism $\phi: C[0,1] \twoheadrightarrow \mathbf{R}$ whose kernel is a maximal ideal.

11-3. POLYNOMIAL CODES

In this section, we shall describe a special class of group codes called *polynomial codes*. These are (m,n) codes which encode m-digit message words into n-digit code words by polynomial multiplication; decoding for error detection is straightforward and will be described. As usual with (m,n) block codes, there are $k = n - m$ check digits in each block.

Much as in Chap. 8, we shall let $a_0 \cdots a_{m-1}$ designate the digits of the original message word **a**, where in principle the letters a_i could be from any finite alphabet A. However, in practice, the a_i are generally binary, so that $a_i \in \mathbf{Z}_2$, and there is a natural bijection from m-digit message words **a** to "message polynomials" of length m over $\mathbf{Z}_2$; thus

$$\mathbf{a} \leftrightarrow a_0 + a_1x + \cdots + a_{m-1}x^{m-1} \tag{9}$$

(Note that for polynomial codes, the subscripts $0,1, \ldots ,n-1$ are more convenient than the $1,2, \ldots ,n$ used in Chap. 8.)

Thus, we have the message 01101, where $m = 5$ corresponds to the polynomial $0 \cdot x^0 + 1 \cdot x^1 + 1 \cdot x^2 + 0 \cdot x^3 + 1 \cdot x^4$ so, $a_0 = 0$, $a_1 = 1$, $a_2 = 1$, $a_3 = 0$, $a_4 = 1$, and the message $\mathbf{a} = (a_0,a_1,a_2,a_3,a_4)$ goes into the polynomial $a_0 + a_1x + a_2x^2 + a_3x^3 + a_4x^4$.

In polynomial codes, the letters can be associated with the elements of any finite field (for example, $\mathbf{Z}_3$ or $\mathbf{Z}_5$); we shall now define them in general.

Definition. Let F be any finite field, and let

$$g(x) = g_0 + g_1x + \cdots + g_kx^k \in F[x] \qquad g_0 \neq 0,\ g_k \neq 0 \tag{10}$$

be a given polynomial of degree k. The *polynomial code* with *encoding polynomial* $g(x)$ encodes each message word **a** of the form (9) into the code word $\mathbf{b} = b_0b_1 \cdots b_n$ which corresponds to the code polynomial

$$b(x) = b_0 + b_1x + \cdots + b_{n-1}x^{n-1} = a(x)g(x) \tag{11}$$

We emphasize the requirements $g_0 \neq 0$ and $g_k \neq 0$ in (10). The requirement $g_0 \neq 0$ is necessary in order to avoid wasting the first code word digit; if $g_0 = 0$, all code words will begin with a 0, conveying no information. The requirement $g_k \neq 0$ is needed likewise to avoid wasting the last code word digit.

Example 2. For the binary alphabet 0,1 from the field $\mathbf{Z}_2$, assume we want five-message digits and three check digits. We can construct an eight-digit code word polynomial code using the encoding polynomial, $g(x) = 1 + x^2 + x^3$. A typical message might be $a_0a_1 \cdots a_4 = 01011$, for which the message polynomial would be $m(x) = x + x^3 + x^4$. Then $b(x) = a(x) \cdot g(x) = x + x^5 + x^7$, and $b_0b_1 \cdots b_7 = 01000101$ is the corresponding code word.

Theorem 3. *A polynomial code with encoding polynomial*

$$g(x) = g_0 + g_1x + \cdots + g_kx^k$$

is a matrix code with $m \times (m + k)$ encoding matrix

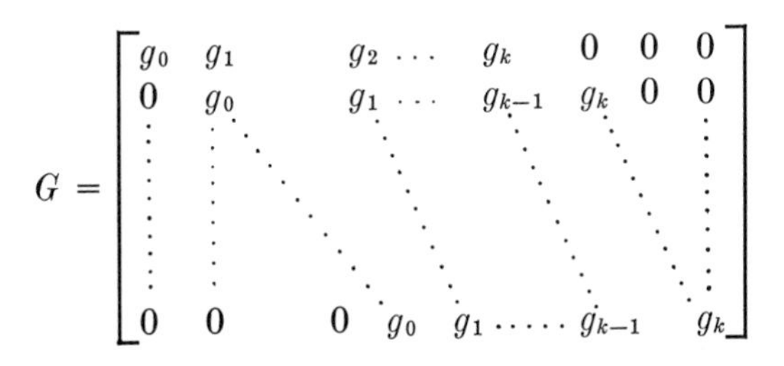

in which the nonzero elements of each jth row lie in a block $g_0 g_1 \cdots g_k$ running from the jth to the $(j + k)$th entry.

The result of Theorem 3 simply applies the definition of the product of the polynomials x^j and $g(x)$ to form the jth row of the encoding matrix, as defined in Chap. 8.

Note that the nonzero entries in each jth row of the encoding matrix G are *shifted* one step to the right beyond where they were in the $(j - 1)$st row immediately above. As we shall see in Sec. 11-5, this property makes it much easier to build (using shift registers) the electronic circuitry necessary to implement polynomial codes physically.

Example 3. For the code polynomial $1 + x + x^3$, and three-digit messages, hence a (3,6) code, we have the encoding matrix

$$G = \begin{bmatrix} 110100 \\ 011010 \\ 001101 \end{bmatrix}$$

The following table lists the appropriate code word **b** for each message word **a** in this polynomial encoding scheme:

$\mathbf{a} \mapsto \mathbf{b} = \mathbf{a}G$	$\mathbf{a} \mapsto \mathbf{b} = \mathbf{a}G$
$000 \mapsto 000000$	$100 \mapsto 110100$
$001 \mapsto 001101$	$101 \mapsto 111001$
$010 \mapsto 011010$	$110 \mapsto 101110$
$011 \mapsto 010111$	$111 \mapsto 100011$

We shall now consider some of the advantageous properties of polynomial encoding. First, we note the following immediately obvious corollary of (11) and the fact, already observed in Chap. 8, that the code words form an additive subgroup of the elementary Abelian group $[\mathbf{Z}_2{}^n,+]$.

Theorem 4. *The minimum separation of two message words in the polynomial code with encoding polynomial $g(x)$ is the minimum weight $w(a(x)g(x))$ of products $a(x)g(x)$.*

We next observe a more special result.

Theorem 5. *A polynomial with coefficients in $\mathbf{Z}_2$ which is divisible by $1 + x$ has an even number of nonzero terms.*

Proof. Let the polynomial be $f(x) = f_0 + f_1x + \cdots + f_nx^n$. Now $f(x) = (x + 1)h(x)$ for some $h(x)$. Give x the value 1. Then

$$f(1) = (1 + 1)h(1) = 0$$

But $f(1) = f_0 + f_1 + \cdots + f_n$, and so the number of nonzero f_i must be even.

In particular, if every code word has an even number of 1's, then we have an *even parity-check code* which will detect any odd number of transmission errors. In the case of one check digit presented in Example 4 below, the hypercube $\mathbf{2}^n$ is a bipartite graph having code words in alternate positions.

Example 4. The encoding polynomial $g(x) = 1 + x$ encodes three-digit message words $\mathbf{a} = a_0a_1a_2$ by the following table:

$\mathbf{a} \mapsto \mathbf{b} = \mathbf{a}G$	$\mathbf{a} \mapsto \mathbf{b} = \mathbf{a}G$
$000 \mapsto 0000$	$100 \mapsto 1100$
$001 \mapsto 0011$	$101 \mapsto 1111$
$010 \mapsto 0110$	$110 \mapsto 1010$
$011 \mapsto 0101$	$111 \mapsto 1001$

$$G = \begin{bmatrix} 1100 \\ 0110 \\ 0011 \end{bmatrix}$$

Notice that the code words each satisfy an even parity check, i.e., contain an even number of 1's.

11-4. ADVANTAGEOUS PROPERTIES

In this section, we shall establish some of the more technical advantageous properties of polynomial codes.

Theorem 6. *If $g(x)$ divides no polynomial of the form $x^k - 1$ for $k < n$, then the code generated by $g(x)$ has minimum distance* 3.

Proof. The code generated by $g(x)$ is the set of polynomials $g(x)a(x)$; deg $a(x) < m$. The fact that polynomial codes form an additive group means that the minimum distance between two code words is equal to the minimum weight of some code word; that is, if $b(x) + e(x) = c(x)$, with $b(x)$ and $c(x)$ being code words, then $e(x)$ is a code word. In our case, if two code words have distance 2, then there is an $e(x)$ with only two nonzero terms $x^i + x^j$, such that $g(x)$ divides $e(x)$; also if there are two code words of distance 1, there then exists x^i such that $g(x)$ divides x^i.

If we postulate that a code word $e(x) = x^i + x^j$ exists, then we can factor this to $e(x) = x^i(1 + x^{j-i})$. Now, if $g(x)$, which has $g_0 = 1$, divides $e(x)$, it must divide $1 + x^{j-i}$. However, $g(x)$ was selected so that it divides no polynomial of the form $1 + x^k$ for $k < n$, so this is not possible, and the nonexistence of the code word of form $x^i + x^j$ is proved.

If $e(x)$ has one nonzero term x^j and if $g(x)h(x) = x^j$, then $g(x)$ must have only one term x^i, $i \leq j$. But we consider only $g(x)$ with nonzero g_0, and if $g(x) = g_0 = 1$, then $g(x)$ divides $x^j - 1$ for any j. Therefore $g(x)$ must have more than one nonzero term, and $e(x)$ must have more than one term if it is divisible by $g(x)$.

Primitive Polynomials. A polynomial $g(x)$ of degree k over $\mathbf{Z}_2$ is called *primitive* when $g(x) \mid (x^m - 1)$ for $m = 2^k - 1$ and for no smaller m.

Example 5. The polynomial $1 + x^2 + x^3 \in \mathbf{Z}_2[x]$ is a primitive polynomial of degree 3; it divides $x^7 - 1$, but not $x^j - 1$ for any $j < 7$.

We shall discuss primitive polynomials again in Sec. 11-8 and again in Chap. 12, where their existence for any positive integer k and various remarkable properties will be explained. For the moment, we shall simply show that the particular primitive polynomial $1 + x^2 + x^3$ generates the (4,7) Hamming code described in Sec. 8-9, showing as a corollary that this is a polynomial code.

The code word assignments for the encoding polynomial $1 + x^2 + x^3$ are listed in the following table:

0000 ↦ 0000000	0100 ↦ 0101100	1000 ↦ 1011000	1100 ↦ 1110100
0001 ↦ 0001011	0101 ↦ 0100111	1001 ↦ 1010011	1101 ↦ 1111111
0010 ↦ 0010110	0110 ↦ 0111010	1010 ↦ 1001110	1110 ↦ 1100010
0011 ↦ 0011101	0111 ↦ 0110001	1011 ↦ 1000011	1111 ↦ 1101001

In this example, we can verify directly from the table (the theoretical conclusion of Theorem 6) that the minimum distance between code words is 3. Also, the encoding matrix

$$G = \begin{bmatrix} 1011000 \\ 0101100 \\ 0010110 \\ 0001011 \end{bmatrix}$$

is the same as the encoding matrix for the (4,7) Hamming code of Sec. 8-6, up to a permutation of the rows and columns.

Theorem 7. *The "error polynomial" associated with any undetected error vector* $\mathbf{e} = e_0e_1 \cdots e_{n-1}$ *of a polynomial* (m,n) *code with generator* $g(x)$ *must be a nontrivial multiple of* $g(x)$.

Proof. By definition, if the received message word $\mathbf{c} = c_0c_1 \cdots c_{n-1}$ with polynomial $c(x) = c_0 + c_1x + \cdots + c_{n-1}x^{n-1}$ is erroneous but undetected, then the error $\mathbf{e} = \mathbf{c} - \mathbf{b}$ must correspond to a nonzero code word $g(x)q(x)$; that is, $e(x)$ belongs to the *principal ideal* $(g(x))$ of all multiples of $g(x)$.

To implement *error detection* of messages transmitted by the $(m, m + k)$ polynomial code generated by a given polynomial $g(x)$ of degree k, it is very convenient to use the following *remainder test* (which is moreover straightforward). Taking the received message $\mathbf{c}$ with message polynomial $c(x)$, use the division algorithm to compute $q(x)$ and $r(x)$ in the formula $c(x) = q(x)g(x) + r(x)$, where $\deg r < \deg g = k$. If $r \neq \mathbf{0}$, an *error* has occurred in reception; $\mathbf{c}$ was not a message word.

Some very efficient error-*detecting* polynomial codes can be constructed. To show this, we now define the *exponent* of a polynomial to be the least positive integer e such that $g(x) \mid (x^e - 1)$. Thus a polynomial over $\mathbf{Z}_2$ is primitive if and only if it has degree k and exponent $e = 2^k - 1$.

We also call two adjacent errors a *double error;* this concept is important because errors tend to occur in "bursts." (The idealization of a binary symmetric channel is unrealistic under many practical conditions.)

Theorem 8. *In a polynomial* (m,n) *code, if the encoding polynomial* $g(x) = (1 + x)h(x)$, *where* $h(x)$ *has exponent* $e > n$, *then any combination of two single or double errors will be corrected.*

Proof. If the error pattern is $e(x) = x^i + x^j$, $j \neq i + 1$, then form $x^i(1 + x^{j-i})$; since $h(x)$ will not divide this polynomial, this type of error is corrected. If $e(x) = x^i + x^{i+1} + x^j$ or $e(x) = x^i + x^j + x^{j+1}$ or $e(x) = x^i$, then the number of nonzero terms is odd, and $1 + x$ will not divide any of these polynomials. If the error pattern is of the form

$$e(x) = x^i + x^{i+1} + x^j + x^{j+1}$$

then form $e(x) = (1 + x)(x^i + x^j)$; since $1 + x$ divides $1 + x$, $h(x)$ must divide $x^i + x^j$. But this is not possible since $j < n - 1$, and thus $x^i(1 + x^{j-1})$ cannot be divisible by $h(x)$.

11-5. SHIFT REGISTERS

Each polynomial code encodes input messages by multiplying message word polynomials by some fixed encoding (or generating) polynomial $g(x)$. In this section, we describe a special kind of circuit called a *shift register* which carries out this polynomial multiplication electronically.

Figure 11-1 shows a circuit which multiplies $g(x) = 1 + x^2 + x^3$ by an input polynomial $a(x) = a_0 + a_1x + \cdots + a_mx^m$. The high-order

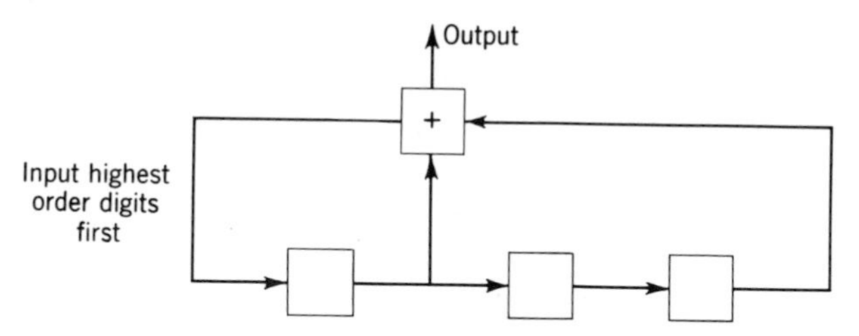

Fig. 11-1. Logical circuit for multiplying $x^3 + x^2 + 11$

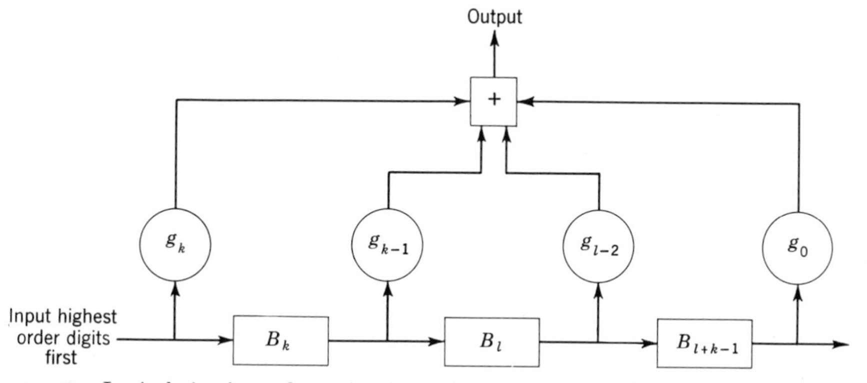

Fig. 11-2. Logical circuit configuration for multiplying by a fixed polynomial $g(x)$.

coefficients of $a(x)$ are assumed to enter first, followed by four 0's. The coefficients of the product $b(x) = a(x)g(x)$ are produced at the output, those of higher order arriving first, b_{m+3} arriving when the first a_m arrives, and b_0 arriving at the end of the $m + 4$ inputs. The memory cells in this circuit are again flip-flops which delay their input one time period. This scheme can be generalized. Figure 11-2 shows a circuit for multiplying an input $a(x)$ by some fixed $g(x)$. In this case, the field of coefficients for $g(x)$ and $a(x)$ need not be binary, the boxes labeled g_i are assumed capable of multiplying g_i whatever is at the output of memory box B_{i+1}, and the memory boxes are not necessarily flip-flops, but are simply circuits capable of storing input symbols. For binary fields the scheme takes a particularly simple form; the box implementing multiplication by g_i is simply a wire (or line) if $g_i = 1$, and does not exist if $g_i = 0$.

Another delicate point should be mentioned. In multiplying by the polynomial $g(x)$, the "identity" of the message digits is lost. It is preferable to maintain these message digits in their original form, and thus a simple change in the encoding procedure is made. To encode, $x^{n-m}a(x)$ is divided by $g(x)$, and the remainder $r(x)$ is subtracted from $x^{n-m}a(x)$ to form the code word polynomial. Now assume the above is performed; then $x^{n-m}a(x) = q(x)g(x) + r(x)$, where $q(x)$ is the quotient of $a(x)x^{n-m}$ divided by $g(x)$. Then, subtracting $r(x)$ from each side, we have $c(x) = x^{n-m}a(x) - r(x) = q(x)g(x)$. Thus a code word polynomial $b(x)$ using this scheme is some polynomial times $g(x)$, and all of the previous encoding properties are maintained, whereas the coefficients $b_{n-k}, \ldots, b_n$ are the same as $a_0, \ldots, a_{m-1}$ and can be used immediately. Figure 11-3

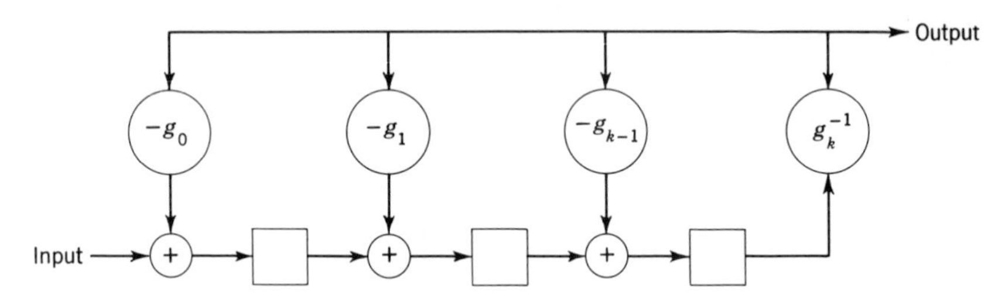

Fig. 11-3. Logical circuit for dividing a polynomial by a fixed polynomial $g(x)$.

shows a circuit for dividing an input polynomial $a(x)x^{n-m}$ by $1 + x^2 + x^3$ over $\mathbf{Z}_2$. The coefficients of the dividend enter in decreasing order (highest degree first). After all the coefficients $b_k, \ldots, b_n$ $(k = n - m)$ of $a(x)x^{n-m}$ have been entered, the coefficients of the remainder polynomial are stored in delay elements (flip-flops), to be read in parallel or shifted out later.

In this scheme, after $b_k, \ldots, b_{n-m}$ have been shifted through the divider, one must wait until $k = n - m$ zeros have been entered at the input before calculating and transmitting the remainder. This delay can be reduced in several ways, which are described in the references at the end of this chapter.

EXERCISES B

EXERCISES CONCERN POLYNOMIALS OVER $\mathbf{Z}_2$.

1. Let the irreducible generator polynomial $g(x)$ of degree k over $\mathbf{Z}_2$ divide $x^n - 1$, but no polynomial $x^r - 1$ with $r < n$.

(a) Show that for $g(x) = 1 + x^3 + x^4 + x^5$, with $k = 6$ and $n = 13$, the matrix

$$G = \begin{bmatrix} 1 & 0 & 0 & 1 & 1 & 1 & 0 & 0 & 0 & 0 & 0 & 0 & 0 \\ 0 & 1 & 0 & 0 & 1 & 1 & 1 & 0 & 0 & 0 & 0 & 0 & 0 \\ 0 & 0 & 1 & 0 & 0 & 1 & 1 & 1 & 0 & 0 & 0 & 0 & 0 \\ \cdot & \cdot & \cdot & \cdot & \cdot & \cdot & \cdot & \cdot & \cdot & \cdot & \cdot & \cdot & \cdot \\ 0 & 0 & 0 & 0 & 0 & 0 & 0 & 1 & 0 & 0 & 1 & 1 & 1 \end{bmatrix}$$

defines a $(k,n) = (7,13)$ polynomial code.

(b) Show that one can form a parity-check matrix by dividing $g(x)$ into $x^{13} - 1$ computing the remainder $r(x)$, and then forming the companion matrix C.

(c) Prove that with this choice of C, $GC^T = 0$.

2. Let $g(x) = 1 + x + x^2 + x^4 = (1 + x)(1 + x^2 + x^3)$ be the reducible generator polynomial for a three-binary-digit message code.
   (a) Show that the minimum weight of nonzero code words is 4.
   (b) Construct a 4×7 matrix G to encode this polynomial code as a group code.
3. (a) In Exercise 1, put G in reduced echelon form $G = [I_3 P]$, where I_3 is the 3×3 identity matrix and P is a 3×4 check-digit matrix.
   (b) Form a 7×3 parity-check matrix H whose columns are linearly independent and such that $GH = 0$.
4. The polynomial $1 \oplus X \oplus X^4$ is primitive over $\mathbf{Z}_2$:
   (a) Draw a shift register and gating network for encoding sets of 11 message digits into 15-digit code words.
   (b) Write the parity-check matrix for this code.
   (c) Derive a generator matrix in reduced echelon form.
5. (a) Show that the code generated by the polynomial $p(x) = x^i + x^j$, $i > j > 0$, *detects* all *odd* numbers of errors.
   (b) Let $g(x)$ be a polynomial of degree k. Show that the set of all code words of length $n > k$ generated by $g(x)$ form an additive group. Thus any polynomial code is a group code.
   (c) For polynomials over $\mathbf{Z}_p$, what is the number of elements in the group of part (b)? What if $g(x) = x^i + x^j$?
6. Given an error n-tuple $e_0, e_1, \ldots, e_n$, an *error burst of length* m is defined as any sequence of m digits whose first and last digits are 1's. For instance, the error n-tuple 010010 has an error burst of length 4, 0110 has an error burst of length 2, 0100 of length 1, 0110010 of length 5, etc. Show that the generator polynomial $g(x) = (1 + x)p(x)$, where $p(x)$ has degree greater than n, the number of digits in the encoded words, yields a polynomial code which will detect any combination of one or two error-burst patterns, each of length 2 or less.

11-6. UNIQUE FACTORIZATION THEOREM FOR POLYNOMIALS

We showed in Theorem 2 of Sec. 11-2 that the polynomials in x over any field F form a *Euclidean domain* $F[x]$. This fact and the results of Chap. 10 specified below have a number of immediate corollaries.

Corollary 1. *$F[x]$ is a principal ideal ring, i.e., a ring in which all ideals are principal.*

This follows immediately from Theorem 10 of Chap. 10.

Corollary 2. *Given any polynomials $a(x), b(x) \in F[x]$, then a greatest common divisor of $a(x)$ and $b(x)$ exists and can be expressed in the form $r(x)a(x) + s(x)b(x)$ for $r(x), s(x) \in F[x]$.*

Moreover, when division can be carried out in F, a greatest common divisor can actually be computed by the Euclidean algorithm described in Sec. 10-9.

Corollary 3. *If $p(x)$ is irreducible in $F[x]$, then*

$$p(x) \mid a(x)b(x) \qquad \textit{implies} \qquad p(x) \mid a(x) \textit{ or } p(x) \mid b(x) \tag{12}$$

Proof. The proof follows immediately from Lemma 2 of Sec. 10-6. However, we shall prove the result directly for $F[x]$ (the reader should compare the proof with that of Lemma 1 of Sec. 10-10).

Suppose that $p(x) \mid a(x)b(x)$ and $p(x) \mid a(x)$. We then have

$$\gcd\,(p(x),a(x)) = r(x)p(x) + s(x)a(x)' = 1 \tag{13}$$

The first equality follows from Corollary 2, and the second from the irreducibility of $p(x)$. Multiplying by $b(x)$ gives

$$r(x)p(x)b(x) + s(x)a(x)b(x) = b(x) \tag{14}$$

Now $p(x) \mid a(x)b(x)$, so that $p(x)$ divides the left-hand side of the above equality. Hence $p(x) \mid b(x)$, which proves Corollary 3.

Corollary 4. *If $p_k(x) \mid f(x)$ for $k = 1, \ldots, r$ and the $p_k(x)$ are non-associate irreducible polynomials, then $\left[\prod_{k=1}^{r} p_k(x)\right] \Big| f(x)$.*

Theorem 9. (*Unique Factorization Theorem.*) *Any nonconstant polynomial $p(x)$ in $F[x]$ can be expressed as a constant c times a product of monic irreducible polynomials. This expression is unique except for the order in which the factors occur.*

This follows directly from Theorem 19 of Chap. 10. We give the following independent proof to make the concept of a Euclidean domain more vivid.

Second proof. If $p(x) = p_0 + p_1x + \cdots + p_nx^n$ is not monic, it is the product of a constant c and some monic polynomial $q(x)$, thus $p(x) = c \cdot q(x)$, where $c = p_n$ and $q_i = p_n^{-1}p_i$. If $q(x)$ is irreducible, we are finished. If $p(x)$ can be factored as $p(x) = r(x) \cdot s(x)$, then we examine

$$r(x) = a \cdot r_1(x) \cdot r_2(x) \cdots r_l(x)$$
$$s(x) = b \cdot s_1(x) \cdot s_2(x) \cdots s_m(x)$$

where the r_i and s_j are monic and irreducible (this is always possible; the proof is by induction).

Then $p(x) = (ab) \cdot r_1(x) \cdots r_l(x) \cdot s_1x \cdots s_m(x)$ where the r_i and s_j are monic and irreducible and $a \cdot b \in F$. This shows that a polynomial $p(x)$ can be written as $p(x) = c \cdot p_1(x) \cdots p_\nu(x)$ where $c \in F$ and the p's are monic and irreducible. We now prove that this factorization is unique. Assume that two such factorizations are possible as follows:

$$p(x) = c \cdot p_1(x) \cdot p_2(x) \cdots p_s(x)$$
$$p(x) = \bar{c} \cdot q_1(x) \cdot q_2(x) \cdots q_t(x)$$

where the p_i and q_j are monic and irreducible.

Now $p_1(x)$ divides $\bar{c} \cdot q_1(x) \cdots q_t(x)$, so it must divide some q_i by Corollary 3. Then

$$\frac{q_i(x)}{p_1(x)} = 1$$

for q_i is monic and irreducible as is p_1. So $p_1 = q_i$, and (canceling)

$$c \cdot p_2(x) \cdots p_m(x) = \bar{c} \cdot q_1(x) \cdots q_{i-1}(x) \cdot q_{i+1}(x) \cdots q_t(x)$$

We then repeat the above step for $p_2(x)$, pairing it with the $q_j(x)$ it divides, and continue this until $p_m(x)$ is reached and found equal to the remaining $q_l(x)$. This pairs each $p_i(x)$ with some $q_j(x)$, showing that the two factorizations are identical up to permutation of the factors.

Actually, factorization into irreducibles is unique for polynomials $p(x_1, \ldots ,x_n)$ in any number of variables over any field. More generally, the same is true for polynomials over any integral domain in which (as in **Z**) factorization into primes is unique (for details see Birkhoff-Mac Lane, Chap. III, Sec. 9).

11-7. COMPLEX ROOTS OF UNITY

In any field F, the *mth roots of unity* are, by definition, the elements $x \in F$ which satisfy the equation $x^m = 1$. It is a familiar fact that in the complex field **C**, the mth roots of unity are the complex numbers $z_k = e^{2\pi ik/m}$; they are the vertices of a regular polygon of radius 1 and m sides with complex z plane. Clearly, they form a *cyclic* group of order m, with generator $z_1 = \zeta = e^{2\pi i/m}$.

As we observed in Chap. 7, the cyclic group $\mathbf{Z}_m$ can be generated by $\phi(m)$ different elements, where $\phi(m)$ is Euler's ϕ-function. Namely, if $(k,m) = 1$, that is, if k is relatively prime to m, then the powers of z_k include every mth root of unity.

The $z_k = \zeta^k$ with $(k,m) = 1$ are called the *primitive* mth roots of unity. For $m = 8$, they are ζ, ζ^3, ζ^5, and ζ^7; for $m = 15$, they are ζ^k with $k = 1, 2, 4, 7, 8, 11, 13$, and 14. Thus $\phi(8) = 4$, and $\phi(15) = 8$.

In general, over any field F, the roots x_k of any polynomial equation $p(x) = 0$ correspond to the linear factors $(x - x_k)$ of $p(x)$. In particular, for the complex polynomial $z^m - 1$, we have the factorization

$$\prod_{k=0}^{m-1} (z - \zeta^k) = \prod_{k=0}^{m-1} (z - z_k) = z^m - 1 \tag{15}$$

This can be proved either by complex analysis or by the following purely algebraic argument.

To make the algebraic proof rigorous, one must be careful to distinguish the *polynomial* $p(x)$, which is an algebraic expression defined in Sec. 11-1 above, from the *polynomial function* $\tilde{p}\colon F \to F$ which assigns to each element $x \in F$ a value $\tilde{p}(x) \in F$. As we shall see in Sec. 11-8, the two are interchangeable over $\mathbf{Z}$ or any other infinite integral domain. However, they are *not* interchangeable notions over $\mathbf{Z}_2$ or any other finite integral domain (field).

Theorem 10. *Given $a \in R$, a commutative ring, $(x - a) \mid p(x)$ in $R[x]$ if and only if $\tilde{p}(a) = 0$ in R.*

Proof. If $(x - a) \mid p(x)$ in $R[x]$, then $p(x) = (x - a)q(x)$ for some $q(x)$, and thus $\tilde{p}(a) = (a - a)\tilde{q}(a) = 0$ in R.

Conversely, for any $p(x)$

$$\begin{aligned} p(x) - p(a) &= \sum_{k=0}^{n} a_k x^j - \sum_{k=0}^{n} p_k a^k = \sum_{k=1}^{n} p_k(x^k - a^k) \\ &= \sum_{k=1}^{n} p_k[(x - a)(x^{k-1} + x^{k-2}a + \cdots + a^{k-1})] \\ &= (x - a) \sum_{k=1}^{n} p_k \left(\sum_{i=0}^{k-1} x^{k-i} a_i \right) \end{aligned}$$

The expression in square brackets is a polynomial in x over R; hence we have shown that $(x - a) \mid p(x) - p(a)$ in $R[x]$, for any $a \in R$. Consequently, if $\tilde{p}(a) = 0$, then $(x - a) \mid p(x)$, completing the proof.

Over **C**, it follows that for $z_k = e^{2\pi ik/n}$ as before,

$$(z - z_k) \mid (z^m - 1) \qquad \text{for } k = 0,1, \ldots ,m - 1 \tag{16}$$

Hence, by Corollary 4 of Sec. 11-7, $\prod_{k=0}^{m-1} (z - z_k) \mid (z^m - 1)$. Since the two polynomials are both of degree m and monic, this equation can only hold if they are the same. This proves (15) over **C**.

Cyclotomic Polynomials. It is interesting to factor $z^m - 1$ over the integers[1] into irreducible factors; we shall now consider in detail the case $m = 15$.

Example 6. Over any field, we have the factorization

$$x^{15} - 1 = (x - 1)(x^2 + x + 1)(x^4 + x^3 + x^2 + x + 1)c_8(x) \tag{17}$$

where $c_8(x)$ is called the *cyclotomic* polynomial of degree 8; thus

$$c_8(x) = x^8 - x^7 + x^5 - 2x^4 + x^3 - x + 1 \tag{17$'$}$$

Over **C**, $c_8(x) = \prod_P (z - z_k)$, where P is the set of all primitive 15th roots of unity. It is an important theorem of classical algebra that $c_8(x)$ and every other cyclotomic polynomial is irreducible over **Q**.

However, over $\mathbf{Z}_2$, the octic polynomial $c_8(x)$ is *reducible;* it has the factorization

$$c_8(x) = (x^4 + x^3 + 1)(x^4 + x + 1) = x^8 + x^7 + x^5 + x^3 + x + 1 \tag{18}$$

Note that the last expression equals that of (17′) over $\mathbf{Z}_2$. The irreducible factors in (18) are primitive polynomials of degree 4 over $\mathbf{Z}_2$ and are associated with primitive 15th roots of 1 in a finite field $GF(16)$ to be described in Chap. 12.

For convenience, we list here primitive polynomials of degrees 5, 6, and 7 over $\mathbf{Z}_2$, associated similarly with the equations $z^{31} = 1$, $z^{63} = 1$, and $z^{127} = 1$, respectively:

$$x^5 + x^4 + 1 \qquad x^6 + x + 1 \qquad x^7 + x^6 + 1$$

EXERCISES C

1. Using the Euclidean algorithm, compute in $\mathbf{Q}[x]$ the following gcd:

$$(x^5 + 3x^4 + 2x^3 - 2x^2 - 3x - 1,\ 3x^7 + 6x^6 + 5x^5 + 7x^4 + 8x^3 + 5x^2 + 4x + 2)$$

[1] It is classic (G. Birkhoff and S. Mac Lane, "A Survey of Modern Algebra," 3d ed., p. 75, Macmillan, 1965) that a polynomial with integral coefficients has just the same factorization over **Z** as it does over the rational field **Q**.

2. (a) Show that if $p(x)$ and $q(x)$ are polynomials of $F[x]$, then $(p(x)) \subset (q(x))$ if and only if $p(x) \mid q(x)$.
   (b) Show that if $p(x)$ is reducible, then $(p(x))$ is not maximal.
3. Show that a quadratic or cubic polynomial $p(x) \in F[x]$ (F a field) is irreducible if and only if $p(c) \neq 0$ for all $c \in F$.
4. Show that, $x^3 + x + 1$ is irreducible over $\mathbf{Z}_2$.
   (*Hint:* Use Exercise 3.)
5. Write down addition and multiplication tables for $\mathbf{Z}_2[x]/(x^3 + x + 1)$.
6. (a) Show that $x^2 + x + 1$ is the only irreducible quadratic polynomial over $\mathbf{Z}_2$.
   (b) In $GF(4) = \mathbf{Z}_2[x]/(x^2 + x + 1)$, compute x^{-1} and $(1 + x)^{-1}$.
7. (a) Show that $x^3 + x + 1$ and $x^3 + x^2 + 1$ are irreducible over $\mathbf{Z}_2$.
   (b) Show that there are no other irreducible cubic polynomials over $\mathbf{Z}_2$.
8. Show that $x^4 + x + 1$ is irreducible over $\mathbf{Z}_2$.
   (*Hint:* Consider all possible linear and irreducible quadratic factors.)

★11-8. POLYNOMIAL FUNCTIONS

In this section, we shall discuss the interpretation of symbolic polynomial expressions (or polynomial forms) as *functions*. It is easy to verify that the set R^S of *all* functions from any set S to any commutative ring R is itself a commutative ring under addition and multiplication defined as follows: For any $f,g \in R^S$ and $s \in S$

$$(f + g)(s) = f(s) + g(s) \tag{19}$$
$$(fg)(s) = f(s)g(s) \tag{20}$$

This result applies in particular when $S = R$. In the ring R^R, the *identity function* 1^R: $x \mapsto x$ for all $x \in R$ clearly plays a special role. By a *polynomial function* is meant an element $\tilde{p} \in R^R$ in the *subring* $R\langle x \rangle$ of functions f: $R \to R$ generated by the constant functions a [with $\tilde{a}(x) = a$ for all $x \in R$] and the identity function x (or 1_R). Thus, a given polynomial

$$p(x) = p_0 + p_1x + \cdots + p_nx^n$$

determines a function $\tilde{p} \in R^R$ by the assignment $\tilde{p}$: $a \mapsto \tilde{p}(a)$, $a \in R$.

Example 7. Let $R = \mathbf{Z}_3$. Then $R[x]$ is the set of all polynomials $f(x) = a_0 + a_1x + \cdots + a_nx^n$, $a_i \in R$. The polynomial $f(x) = 1 + x$ determines the function $\tilde{f}$: $R_3 \to R_3$; thus

$$\tilde{f}(0) = 1 \qquad \tilde{f}(1) = 2 \qquad \tilde{f}(2) = 0$$

Thus, this $\tilde{f} \in \mathbf{Z}_3{}^{\mathbf{Z}_3}$ makes the assignments $0 \mapsto 1$, $1 \mapsto 2$, and $2 \mapsto 3$.

For each such polynomial $g(x) \in R[x]$ there is a function $\tilde{g}(x) \in R^R$. The constant functions are $\tilde{g}(x) = 0$, $\tilde{g}(x) = 1$, and $\tilde{g}(x) = 2$. The identity function is $\tilde{g}(x) = x$; this is sometimes written 1_R and is, of course, $0 \mapsto 0$, $1 \mapsto 1$, and $2 \mapsto 2$.

The correspondence $p(x) \mapsto \tilde{p}$ just defined can be described in another way. Each element a of a commutative ring R defines an *evaluation map* $f_a: R[x] \to R$ through the assignment $f_a(p(x)) = \tilde{p}(a)$. The following result is obvious.

Lemma. *For each $a \in R$, the evaluation map $f_a: p \mapsto \tilde{p}(a)$ is a morphism of rings. Moreover, the map $\mu: p \mapsto \tilde{p}$ from $R[x]$ to R^R defined by the formula $\tilde{p}(a) = f_a(p)$ is a morphism of rings.*

In other words, polynomial functions can be added and multiplied either by adding or multiplying the polynomials and then substituting into the resulting polynomial, or by adding or multiplying the values of the individual functions themselves. In particular, μ assigns to each $a \in R$ the constant function $\phi(x) \equiv a$, and it assigns to the polynomial x (or $0 + x$) the function $1_R: x \mapsto x$ for all $x \in R$. Finally, further consideration makes it clear that the range of the morphism μ of the lemma is the ring $R\langle x \rangle$ of polynomial functions on R to R. The preceding statements can be summarized as follows.

Corollary. *The map μ of the lemma is an epimorphism from $R[x]$ to the subring $R\langle x \rangle$ of polynomial function $f: R \to R$.*

In high-school algebra, polynomial expressions, or "forms," as defined by (1) are not usually distinguished from the corresponding functions. This is legitimate there because the epimorphism described above is an isomorphism in the only polynomial rings considered there, $\mathbf{Q}[x]$ and $\mathbf{R}[x]$. This is not true in general, as will be shown later in this section.

In high-school algebra, polynomial expressions or forms f as defined by (1) are not usually distinguished from the corresponding functions. This is legitimate there because the epimorphism described above is an isomorphism in the only polynomial rings considered there, $\mathbf{Q}[x]$ and $R[x]$. This is not true in general, as will be shown later in this section.

We now derive a simple sufficient condition for the epimorphism $\bar{\theta}: R[x] \mapsto R\langle x \rangle$ from the ring of polynomial expressions (or forms) (1) to the ring of polynomial functions[1] to be an isomorphism. Trivially, a necessary *and* sufficient condition is that

$$p(x_i) = q(x_i) \qquad \text{for all } x_i \in R \text{ implies} \qquad p(x) = q(x) \qquad \text{in } R[x] \quad (21)$$

Equivalently, considering the kernel of $\bar{\theta}$, it is apparent that such a condition is

$$p(x_i) = 0 \qquad \text{for all } x_i \in R \text{ implies} \qquad p(x) = 0 \qquad \text{in } R[x] \quad (22)$$

This condition is not fulfilled in all commutative rings R, not even in all fields.

[1] $\bar{\theta}(p(x)) = \tilde{p}(x)$ defines $\bar{\theta}$, of course.

Indeed, the polynomial $x^3 - x$ assumes the value 0 for all

$$x \in \mathbf{Z}_3 = \{0,1,2\}$$

More generally, if R is any *finite* commutative ring with n elements $a_1, \ldots, a_n$, let $c(x)$ denote the polynomial $(x - a_1)(x - a_2) \cdots (x - a_n)$. Then $\tilde{c}(a_i) = 0$ for all a_i in R; hence $\tilde{c}(x) = 0$ in $R\langle x\rangle$. We shall now use Theorem 10 to show that this cannot happen if R is an infinite integral domain.

Definition. A *zero* of a polynomial $p(x) \in R[x]$ is a value $a \in R$ for which $\tilde{p}(a) = 0$.

Theorem 11. *A polynomial $p(x)$ of degree n over an integral domain D has at most n zeros in D.*

Proof. Let a be a zero of $p(x)$. Then $p(x) = (x - a)q(x)$ by Theorem 10, where $q(x)$ has degree $n - 1$. Since $(b - a)q(b) = 0$ if and only if $b = a$ or $\tilde{q}(b) = 0$ and $q(x)$ has no more than $n - 1$ factors, by induction $p(x)$ has no more than n zeros.

Corollary. *Over an infinite integral domain D, two polynomials which define the same function have identical coefficients; i.e., they are formally equal.*

Proof. Let $p(x)$ and $q(x)$ be two polynomials in $D[x]$ which define the same functions; i.e., let $\tilde{p}(a) = \tilde{q}(a)$ for all $a \in D$. Consider

$$r(x) = p(x) - q(x) = r_0 + r_1x + \cdots + r_nx^n$$

then $\tilde{r}(a) \equiv \tilde{p}(a) - \tilde{q}(a) \equiv 0$ for all $a \in D$. By Theorem 11, setting $\tilde{r} = \tilde{p}$, we see that this is possible only if D has n or fewer elements or if $p(x) = q(x)$; hence the proof is complete.

Lagrange Interpolation Formula. We have shown that there is at most one polynomial function of degree n which assumes given values $y_0,y_1, \ldots, y_n$ at $n + 1$ given points $x_0,x_1, \ldots, x_n$ of a given integral domain D. If D is a *field*, however, one can compute *at least* one such *interpolating polynomial* by an ingenious algorithm due to Lagrange.

Theorem 12. *Given a field F and sets of $n + 1$ elements $x_1, \ldots, x_n \in F$ and $y_0, \ldots, y_n \in F$, there exists a polynomial $p(x)$ of degree at most n such that $p(x_0) = y_0, \ldots, p(x_n) = y_n$.*

Proof. Consider the polynomial

$$q_i(x) = (x - a_0) \cdots (x - a_{i-1})(x - a_{i+1}) \cdots (x - a_n) \qquad (23)$$

If $i \neq j$, then $q_i(a_j) = 0$, but for $i = j$, $q_i(a_j) \neq 0$. Let $L_i = q_i(a_i)$. Since $L_i \neq 0$, we can form

$$p(x) = \sum_{i=0}^{n} \frac{y_i}{L_i} q_i(x) = \sum_{i=0}^{n} y_i \left(\prod_{j \neq i} \frac{x - a_j}{a_i - a_j} \right) \tag{24}$$

Since $q_i(a_j) = 0$ for $i \neq j$, $p(a_i) = y_i$ for $i = 0, \ldots, n$. This is called the *Lagrange interpolation formula.*

★11-9. FORMAL DERIVATIVES

By analogy with the calculus, one can define the *formal derivative* of the formal power series $f(x) = \sum_{k=0}^{\infty} f_k x^k$, with coefficients f_k in the field F, as follows:

$$f'(x) = \sum_{k=0}^{\infty} k f_k x^{k-1} = \sum_{l=0}^{\infty} (l+1) f_{l+1} x \qquad l = k - 1 \tag{25}$$

In (25), the factors k and $l + 1$ are natural numbers; thus kf_k signifies the sum $f_k + \cdots + f_k$ (k summands).

It is easily verified that formal differentiation obeys the following familiar rules for differentiation of functions derived in the calculus:

$$(f + g)' = f' + g' \qquad \text{and} \qquad (cf)' = cf' \qquad \text{(linearity)} \tag{22}$$

and

$$(fg)' = fg' + gf' \qquad (f^m)' = mf^{m-1}f' \tag{23}$$

More generally, the above hold if $f = \sum_{k=-\infty}^{\infty} f_k x^k$ and $g = \sum_{k=-\infty}^{\infty} g_k x^k$ are extended power series.

The interesting result is Theorem 13.

Theorem 13. *Over a field F of characteristic ∞, a (nonconstant) polynomial $f(x)$ is without multiple factors if and only if* gcd $(f,f') = 1$.

Proof. Write f as a product of powers of irreducible factors; thus

$$f = cp_1(x)^{e_1} \cdots p_r(x)^{e_r} \qquad c \neq 0 \text{ in } F$$

Case 1. If some $e_i > 0$, then $p_i(x) \mid f'$ as well as f, since

$$f' = c \sum_{k=1}^{r} \left[e_k p_k^{e_k - 1} p_k' \left(\prod_{j \neq k} p_j^{e_j} \right) \right]$$

Hence $p_i(x) \mid (f,f')$ and $(f,f') \neq 1$.

Case 2. If every $e_i = 1$ above, then for all i

$$p_i'(x) \equiv cp_i'(x) \prod_{j \neq i} p_j(x) \not\equiv 0 \qquad \mod p_i(x)$$

where we have used the fact that the irreducible $p_i(x)$ cannot divide $p_i'(x)$ [or any $p_j(x)$ or c] over a field of characteristic ∞. This shows that no irreducible polynomial which divides f can divide f', whence $(f,f') = 1$.

Remark. A polynomial without multiple factors is called *square-free*, and it is an important fact that the property of being separable is invariant under field extension over fields of characteristic ∞. This is because gcd (f,f') can be calculated directly by the Euclidean algorithm in a way which is unaffected by field extension, and f is without multiple factors if and only if $(f,f') = 1$. For much the same reason, for each degree n there exists a special polynomial $D(f_0, \ldots, f_n)$ called the *discriminant*, which is nonzero if and only if f is a separable polynomial.[1] For $n = 2$, $D = f_1^2 - 4f_0f_2$, for example.

Over a field of characteristic ∞, one can say more. In this case, if f is of degree n, then f' is of degree $n - 1$; hence gcd (f,f') is of degree $n - 1$, or less, and in particular $f' \not\equiv 0 \pmod f$. (Over $\mathbf{Z}_p$, on the other hand, $f(x) = x_p - c$ has derivative $f' = 0$; hence gcd $(f,f') = f$ is of degree n.)

EXERCISES D

1. Show that one can evaluate any quartic polynomial function $q(x)$ with only three multiplications and five additions.[2]
   (*Hint:* Set $q(x) = a_0\{[g(x) + \lambda_2][g(x) + x + \lambda_3] + \lambda_4\}$ for some suitable $g(x) = x(x + \lambda_1)$.)
2. Factor $x^6 - 1$ into irreducible factors
   (a) Over $\mathbf{R}$
   (b) Over $\mathbf{Q}$
   (c) Over $\mathbf{C}$
3. Show that in any field the roots of the equation $x^n = 1$ form a cyclic group whose order divides n.
4. Let $R[c] \subset D$ be any commutative ring generated by a subring R of an integral domain D and an element $c \in D - R$. Extend the mapping $1_R: R \to R$ and the assignment $x \mapsto c$ to an epimorphism of rings $f: R[x] \twoheadrightarrow R[c]$.

[1] Provided we have a field whose characteristic exceeds n.

[2] See L. Lyusternik, "Handbook for Computing Elementary Functions," Pergamon, 1965.

Finite Fields

12-1. EXTENSIONS OF FIELDS

If F is a subfield of a field G, then G is called an *extension* of F. Thus the extensions of a given field F correspond to the monomorphisms from F into larger fields. The complex field **C** is an extension of the real field **R**, and **R** is an extension of the rational field **Q**. Every field F is an extension of its unital subfield (prime subfield), hence it is an extension of **Q** if its characteristic is ∞ or of some $\mathbf{Z}_p$ if its characteristic is finite.

Many aspects of the theory of field extension are illustrated by complex numbers; hence we shall recall their properties in some detail.

Example 1. The complex field **C** consists of all complex numbers of the form $z = x + iy$ ($i = \sqrt{-1}$; $x,y \in \mathbf{R}$). Addition and multiplication in **C** are defined by the rules

$$(x_1 + iy_1) + (x_2 + iy_2) = (x_1 + x_2) + i(y_1 + y_2) \tag{1}$$
$$(x_1 + iy_1)(x_2 + iy_2) = (x_1x_2 - y_1y_2) + i(x_1y_2 + y_1x_2) \tag{2}$$

The assignment $x \mapsto x + i0$ is a monomorphism from $\mathbf{R}$ into $\mathbf{C}$, which embeds $\mathbf{R}$ in $\mathbf{C}$ as a subfield. One can consider $\mathbf{C}$ as a two-dimensional vector space over $\mathbf{R}$, spanned by the basis vectors (unit vectors) $1 = 1 + i0$ and $i = 0 + i1$. Since $i^2 = (-1) + i0$, we write $i = \sqrt{-1}$; $\mathbf{C}$ is "generated" as a field (and even as a commutative ring) by i and $\mathbf{R}$. This fact is expressed by writing $\mathbf{C} = \mathbf{R}[i] = \mathbf{R}[\sqrt{-1}]$.

More generally, *any* extension G of a field F can be considered as a *vector space* over F. This means[1] that $[G,+]$ is an Abelian group, and further that [letting Greek letters denote elements of G (vectors) and Roman letters elements of F (scalars)]

$$a(b\xi) = (ab)\xi \qquad (a + b)\xi = a\xi + b\xi \qquad a(\xi + \eta) = a\xi + a\eta$$

for all $a,b \in F$ and $\xi,\eta \in G$, and finally, that $1\xi = \xi$ for the unity 1 of F (and G) and any $\xi \in G$.

In any vector space V, a finite *basis* is a set of n vectors $\beta_1, \ldots, \beta_n$ such that every $\xi \in V$ has a unique representation as a linear combination $\xi = x_1\beta_1 + \cdots + x_n\beta_n$ with scalars $x_i \in F$. A fundamental theorem about vector spaces asserts that if *one* basis of V (over F) has n elements, then *every* basis of V has n elements. The number n is then called the *dimension* of V (over F), and it can be shown that any $n + 1$ vectors $\xi_0, \ldots, \xi_n$ of V are linearly dependent if V is n-dimensional; that is, there exist scalars $f_0, \ldots, f_n \in F$, not all zero, such that

$$f_0\xi_0 + \cdots + f_n\xi n = 0$$

Also, if V is *spanned* by any finite subset of vectors $\xi_1, \ldots, \xi_r$, in the sense that every $\xi \in V$ has some expression

$$\xi = c_1\xi_1 + \cdots + c_r\xi_r (c_i \in F)$$

then V has a *basis*.

Example 2. Over any field F, the polynomials of length $n + 1$ (degree n or less) form an $(n + 1)$-dimensional vector space over F, with basis $1,x,x^2, \ldots, x^n$ (this basis is, of course, not unique; thus 1, $1 + x$, $1 + x^2, \ldots, 1 + x^n$ also form a basis). The commutative ring $F[x]$ of *all* polynomials over F is an infinite-dimensional vector space over F.

We shall not use the theory of infinite-dimensional vector spaces in this book. Clearly, any finite field F is a finite-dimensional vector space over its unital subfield $\mathbf{Z}_p$.

In general, if a field G has finite dimension n, considered as a vector space over the subfield F, then G is said to be an extension of F of *degree n* (in symbols, $[G:F] = n$).

[1] See Birkhoff-Mac Lane, Chap. 7, where all the facts about vector spaces used below are proved from the axiomatic definition.

Thus, the complex field $\mathbf{C}$ is an extension of degree 2 (a quadratic extension) of the real field $\mathbf{R}$, which is generated by its subfield $\mathbf{R}$ and one other element $i = \sqrt{-1}$.

In general, extensions of fields which are generated in this way are called *simple* field extensions. That is, a field G is a *simple* extension of its subfield F when, for some $c \in G$, $G = \{F,c\} = F(c)$ is *generated* by F and c. As in Sec. 10-3, this means that every element $x \in G$ can be expressed as a quotient; thus

$$x = \frac{p(c)}{q(c)} = \frac{\sum_{k=1}^{n} p_k c^k}{\sum_{k=1}^{m} q_k c^k} \tag{3}$$

of polynomials in c with coefficients $p_k, q_k \in F$ and $q(c) \neq 0$.

In this chapter, we shall be mainly concerned with *finite fields.* Any finite field G has some finite (prime) characteristic p; hence it is an extension of $\mathbf{Z}_p$ of some finite degree n. Since the general element of G can be considered as an n-vector $\mathbf{x} = (x_1, \ldots, x_n)$ with arbitrary components $x_i \in \mathbf{Z}_p$, it follows that *every finite field has prime-power order* $p^n = q$. We shall now exhibit as simple extensions of $\mathbf{Z}_2$ two finite fields $GF(4)$ and $GF(8)$, having four and eight elements, respectively.

Example 3. In $\mathbf{Z}_2$, $x(x + 1) = 0$ for all x (that is, for $x = 0$ and $x = 1$). Therefore we adjoin an imaginary root j of the equation $x(x + 1) = 1$. Equivalently, we form the quotient ring $GF(4) = \mathbf{Z}_2[x]/(x^2 + x + 1)$ (see Sec. 12-3). (Note that over $\mathbf{Z}_2$, $x^2 + x + 1 = x^2 + x - 1$.) To compute addition and multiplication tables for $GF(4)$, simply replace j^2 by $j + 1 = j - 1$ wherever it occurs. This is legitimate since, by hypothesis, $j^2 + j + 1 = 0$; hence $j^2 = -j - 1 = j + 1$ (over $\mathbf{Z}_2$). The result is as follows:

Table 12-1. Addition and multiplication in $GF(4)$

$+$	0	1	j	$j+1$
0	0	1	j	$j+1$
1	1	0	$j+1$	j
j	j	$j+1$	0	1
$j+1$	$j+1$	j	1	0

$\times$	0	1	j	$j+1$
0	0	0	0	0
1	0	1	j	$j+1$
j	0	j	$j+1$	1
$j+1$	0	$j+1$	1	j

The additive group of $GF(4)$ is easily verified to be the 4-group. Under multiplication the nonzero elements form a cyclic group of order 3, with generator j (or $j + 1 = j^2$); and $j^3 = 1$. Hence the ring $GF(4)$ is a *field.*

Example 4. Likewise, the cubic polynomial $x^3 + x^2 + 1$ is irreducible over $\mathbf{Z}_2$ since it is cubic and defines a function which does not vanish there. We define $GF(8)$ (read "the Galois field of order eight") as the quotient ring

$$GF(8) = \mathbf{Z}_2[x]/(x^3 + x^2 + 1)$$

Its general element can be written as $a_2x^2 + a_1x + a_0$, or simply as a string of three binary digits $a_2a_1a_0 = \mathbf{a}$, as in Table 12-2. Moreover, since two such quadratic polynomials can be multiplied as in $\mathbf{Z}_2[x]$ and the remainder computed after division by $x^3 + x^2 + 1$ from the Division Algorithm, $GF(8)$ has the following multiplication table (addition is performed just as the group-theoretic direct product $[\mathbf{Z}_2, +]^3$):

Table 12-2. Multiplication table in $GF(8)$

×	000	001	010	011	100	101	110	111
000	000	000	000	000	000	000	000	000
001	000	001	010	011	100	101	110	111
010	000	010	100	110	101	111	010	011
011	000	011	110	101	001	010	111	100
100	000	100	101	001	111	011	010	110
101	000	101	111	010	011	110	100	001
110	000	110	001	111	010	100	011	101
111	000	111	011	100	110	001	101	010

As in $GF(4)$, the set G^* of nonzero elements of $GF(8)$ forms under multiplication a *cyclic group* (of order 7). The semigroup $[G^*, \cdot]$ has an identity 001; hence it is a monoid in which every element is invertible since, inspecting the multiplication table, the identity 001 appears once in every row and column after the first. Hence $[G^*, \cdot]$ is a group of order 7, which (Chap. 7, Theorem 15, Corollary 2) is necessarily cyclic since 7 is a prime.

12-2. SIMPLE EXTENSIONS

Now let G be any extension of finite degree $n = [G:F]$ of any field F. Let $c \in G$, and consider the $n + 1$ elements $1, c, c^2, \ldots, c^n$ in G. Since $[G:F] = n$, these elements c^k must be linearly dependent over F; that is, there must exist $f_0, f_1, \ldots, f_n \in F$, not all zero, such that

$$f(x) = \sum_{k=0}^{n} f_k c^k = 0$$

This proves the following result.

Theorem 1. *If the field G is an extension of F of finite degree n, then every $c \in G$ is a root of a polynomial equation of degree at most n with coefficients in F.*

Now suppose G is a *simple* extension of F of degree n, so that $G = F[c]$ is generated by F and some one element $c \in G - F$. Then, dividing through the equation $\sum_{k=0}^{n} f_k c^k = 0$ by its leading coefficient, we will have

$$m(c) = c^s + \alpha_1 c^{s-1} + \cdots + \alpha_s = 0 \qquad \text{all } \alpha_i \in F \tag{4}$$

for some *monic* polynomial $m(x)$ of degree $s \leqq n$. Choose $m(x)$ so as to *minimize* s. Since $G = F[c]$ is simple, every element $g \in G$ can be written in the form $p(c) = \sum_{k=0}^{t} f_k c^k$ of a polynomial in c with coefficients $f_k \in F$. By the Division Algorithm, on the other hand, we will have

$$p(c) = q(c)m(c) + r(c) = r(c) \qquad \text{since} \qquad m(c) = 0 \tag{5}$$

for some polynomial of degree at most $s - 1$. Hence, considered as a vector space over F, G will be spanned by $1, c, \ldots, c^{s-1}$, and thus $[G:F] \leqq s$, proving that $n \leqq s$. Using (4), we see that this proves

Theorem 2. *If G is a simple extension $G = F[c]$ of F of finite degree n, then*

$$m(c) = c^n + \alpha_1 c^{n-1} + \cdots + \alpha_n = 0 \qquad \textit{all } \alpha_i \in F \tag{6}$$

for some monic polynomial $m(x)$ of degree $n = [G:F]$.

We shall now show how to reconstruct G from the monic polynomial $m(x)$ of (6).

Theorem 3. *In Theorem 2, $m(x)$ is irreducible, and $G \cong F[x]/(m(x))$ is isomorphic to the quotient ring of the polynomial ring $F[x]$ over the principal ideal $(m(x))$ of all multiples of $m(x)$.*

Proof. First, we show that $m(x)$ is irreducible. If $m(x) = p(x)q(x)$, then (setting $x = c$) we get $p(c)q(c) = m(c) = 0$, whence $p(c) = 0$ or $q(c) = 0$ since G is a field. Since $m(x)$ has minimum degree among polynomials $p(x)$ with $p(c) = 0$, it follows that either $p(x)$ or $q(x)$ must have degree n. Hence $m(x)$ cannot be factored into polynomials of lower degree, and so, by definition, it is irreducible.

The rest of the proof is a simple application of the theory relating epimorphisms of rings onto quotient rings to the *evaluation map* μ: $p(x) \mapsto p(c)$ from $F[x]$ to the set of all sums $\sum_{k=0}^{\nu} a_k x^k (a_k \in F)$. As in Sec. 11-9, this is a morphism, which is an *epimorphism* μ: $F[x] \twoheadrightarrow G$ since $G = F[c]$. The kernel K of this epimorphism is the ideal of all polynomials $p(x)$ with $p(c) = 0$. Since $F[x]$ is a principal ideal ring (Sec. 11-7), the kernel K

therefore consists of the multiples of any polynomial $p(x)$ of least degree with $p(c) = 0$, hence of the multiples of $m(x)$. This proves that $K = (m(x))$. Finally, by Theorem 14 of Chap. 10, this implies that

$$G \cong F[x]/K = F[x]/(m(x))$$

Simple Transcendental Extensions. If F is any field and $F[x]$ the integral domain whose elements are the polynomials in x with coefficients in F, then the field of quotients $Q(F[x])$ of polynomials is called the *simple transcendental extension* of F by the indeterminate x and is denoted by $F(x)$. If F is infinite, then this extension is the same as (isomorphic to) the field of rational functions $p(x)/q(x)$ (quotients of polynomial functions), two such functions being identified if they have the same value at all but a finite number of points.

★12-3. COMPUTATION IN $\mathbf{R[x]/(m(x))}$

We now discuss in more detail how to compute sums, differences, products, and quotients in simple field extensions. The formulas apply more generally to quotient rings $R[x]/(m(x))$ of the ring of polynomials over any commutative ring R, over the principal ideal $(m(x))$ of multiples of any monic polynomial

$$m(x) = x^n + \sum_{k=1}^{n} c_k x^{n-k} \qquad \text{all } c_k \in R \tag{7}$$

Namely, using (7) to write

$$x^n \equiv \sum_{k=1}^{n} (-c_k) x^{n-k} \qquad \text{mod } (m(x)) \tag{8}$$

we can systematically replace any power x^{n+h} of x $(h \geqq 0)$ larger than the $(n - 1)$st power, in $R[x]/(m(x))$, by a linear combination of lower powers of x, simply by subtracting suitable multiples $x^h m(x) \in (m(x))$ of $m(x)$. As a result, each additive coset of the principal ideal $(m(x))$ in the ring $R[x]$ has just one representative (a coset leader) of length n:

$$a_0 + a_1 x + \cdots + a_{n-1} x^{n-1} \qquad \text{all } a_i \in R,\ i = 0, \ldots, n-1 \tag{9}$$

Under addition and multiplication by elements $c \in R$ (scalars), such polynomials of length n form an n-dimensional vector space (module) R^n over R (note that any or all of the a_k can be zero). Thus

$$\sum_{k=0}^{n-1} a_k x^k + \sum_{k=0}^{n-1} b_k x^k = \sum_{k=0}^{n-1} (a_k + b_k) x^k \tag{10}$$

and

$$r \sum_{k=0}^{n-1} a_k x^k = \sum_{k=0}^{n-1} (ra_k) x^k \qquad \text{for any } r \in R \tag{10'}$$

In this vector space, the particular elements

$$1 = 1 + \sum_{k=1}^{n-1} 0 \cdot x^k \qquad x = 0 + 1 \cdot x + \sum_{k=2}^{n-1} 0 \cdot x^k, \; \ldots, \; x^{n-1}$$

act like unit basis vectors.

The computation of products is only a little more difficult; we illustrate the process by the example of a *cubic* polynomial $m(x) = x^3 + x + 1$. We first compute a list by simple recursion, using the equation $x^3 = -x - 1$ repeatedly; thus

$$\begin{aligned}
1 \cdot (b_0 + b_1 x + b_2 x^2) &= b_0 + b_1 x + b_2 x^2 \\
x \cdot (b_0 + b_1 x + b_2 x^2) &= b_0 x + b_1 x^2 + b_2(-x - 1) \\
&= (-b_2) + (b_0 - b_2)x + b_1 x^2 \\
x^2 \cdot (b_0 + b_1 x + b_2 x^2) &= x[(-b_2) + (b_0 - b_2)x + b_1 x^2] \\
&= (-b_2)x + (b_0 - b_2)x^2 + b_1(-x - 1)
\end{aligned}$$

Hence, by bilinearity, we have

$$\begin{aligned}
\left(\sum_{k=0}^{2} a_k x^k\right)\left(\sum_{k=0}^{2} b_k x^k\right) &= a_0[b_0 + b_1 x + b_2 x^2] \\
&\quad + a_1[(-b_2) + (b_0 - b_2)x + b_1 x^2] \\
&\quad + a_2[(-b_1) + (-b_2 - b_1)x + (b_0 - b_2)x^2]
\end{aligned} \tag{11}$$

An alternative computation scheme goes as follows. Since $x^3 = -x - 1$, $x^4 = -x^2 - x$,

$$\begin{aligned}
\left(\sum_{k=0}^{2} a_k x^k\right)\left(\sum_{k=0}^{2} b_k x^k\right) &= (a_0 b_0) + (a_0 b_1 + a_1 b_0)x + (a_0 b_2 + a_1 b_1 + a_2 b_0)x^2 \\
&\quad - (a_1 b_2 + a_2 b_1)(x + 1) - (a_2 b_2)(x^2 + x)
\end{aligned} \tag{11'}$$

The quadratic polynomials of (11) and (11′) are both equal to

$$\begin{aligned}
&(a_0 b_0 - a_1 b_2) + (a_0 b_1 + a_1 b_0 - a_1 b_2 - a_2 b_1 - a_2 b_2)x \\
&\qquad + (a_0 b_2 + a_1 b_1 + a_2 b_0 - a_2 b_2)x^2
\end{aligned} \tag{12}$$

In summary, sums and products can be computed in the commutative ring $R[x]/(x^3 + x + 1)$ by using formulas (10) and (12).

We now generalize these formulas.

Theorem 4. *Let* $m(x) = x^n + \sum_{k=1}^{n} c_k x^{n-k}$ *be any monic polynomial with coefficients in the commutative ring* R. *Then the quotient ring* $R[x]/(m(x))$ *consists of all polynomials* $\sum_{k=0}^{n-1} a_k x^k$ *of length* n *with coefficients* $a_k \in R$. *Addition is defined by* (10), *and multiplication by*

$$\left(\sum_{k=0}^{n-1} a_k x^k\right)\left(\sum_{k=0}^{n-1} b_k x^k\right) = \sum_{k=0}^{n-1}\left(\sum_{i=0}^{k} a_i b_{k-i}\right)x^k + \sum_{k=n}^{2n-2}\left(\sum_{i=k-n+1}^{n-1} a_i b_{k-i}\right)x^k \tag{13}$$

where

$$x^n = -\sum_{k=1}^{n} c_k x^{n-k} = p_1(x)$$

$$x^{n+1} = \sum_{k=2}^{n} (-c_k)x^{n-k+1} + c_1 \sum_{k=1}^{n} (c_k x^{n-k}) = p_2(x) \tag{13'}$$

and, recursively, for $j = 1, \ldots, n-3$,

$$p^{n+j+1}(x) = xp^{n+j}(x)$$

where

$$x(a_0 + a_1 x + \cdots + a_{n-1}x^{n-1}) = \sum_{i=1}^{n-1} a_{i-1}x^i - a_{n-1}\left(\sum_{k=1}^{n} c_k x^{n-k}\right) \tag{14}$$

The proof is left to the reader; it proceeds, for fixed $m(x)$, by induction on j.

EXERCISES A

1. Compute $(1 + \sqrt[3]{3})^{-1}$ in $\mathbf{Q}(\sqrt[3]{3})$.
2. (a) For any element γ of a finite extension G of a field F, show that there is a unique monic polynomial $m_\gamma(x)$ of least degree such that $m_\gamma(\gamma) = 0$.
   (b) Show that if $p(x) \in F[x]$ and $p(\gamma) = 0$, then $m_\gamma(x) \mid p(x)$.
3. (a) If α is primitive in $GF(q)$ $(q = p^n)$, prove that α^j is primitive if and only if $(j, q-1) = 1$.
   (b) Infer that $GF(q)$ has exactly $\phi(q-1)$ primitive elements, where ϕ is Euler's ϕ-function.
4. (a) Find a primitive element α of $\mathbf{Z}_3[x]/(x^2 - 2)$.
   (b) Represent each power $\alpha^2, \ldots, \alpha^8$ of α in the form $a + b\sqrt{2}$ with $a,b \in \mathbf{Z}_3$.
   (c) Is this representation unique?
5. (a) Is $GF(4)$ a subfield of $GF(8)$?
   (b) For which $m,n \in \mathbf{P}$ is $GF(p^m)$ as subfield of $GF(p^n)$? Justify your answer.
6. Show that the mapping $x \mapsto x^3$ is never an automorphism of the additive group of the Galois field $GF(2^n)$ for $n > 1$.
7. For which r is the mapping $x \mapsto x^r$ an automorphism of the multiplicative monoid of the Galois field $GF(2^n)$?

12-4. EXISTENCE THEOREM

Having at hand the construction described in Sec. 12-2, it is relatively easy to give a rigorous general justification for the procedure used to construct $GF(4)$ and $GF(8)$. The essential properties needed are that R be a *field* and $m(x)$ be *irreducible*, as will now be shown. To this end, we first give a fresh proof of a result already obtained in Sec. 10-1.

Lemma 1. *Every maximal ideal M of a commutative ring R is prime.*

Proof. Suppose $ab \in M$, yet $a \notin M$. Consider the set $M + (a)$ which consists of all sums $x + ra$ of an element $x \in M$ and a multiple ra $(r \in R)$ of a. This is an ideal, because for any $x,x' \in M$ and $r,r' \in R$, we have

$$(x + ra) + (x' + r'a) = (x + x') + (r + r')a$$

where $(x + x') \in M$ and $(r + r') \in R$, and for any $v \in R$,

$$v(x + ra) = vx + (vr)a$$

where $vx \in M$ and $vr \in R$. This ideal $M + (a)$ properly contains M since $a \notin M$. Since M is a maximal ideal, it follows that $M + (a) = R$, and so $1 = x + ra$ for some $x \in M$, $e \in R$. Finally, since $ab \in M$ by hypothesis, this implies

$$b = 1b = (x + ra)b = xb + rab \in M + M = M$$

This shows that the ideal M is prime, as claimed.

From Lemma 1, the next result follows easily.

Theorem 5. *In the integral domain $F[x]$, F any field, the following three conditions on a nonconstant polynomial $p(x)$ are equivalent: (i) $p(x)$ is irreducible, (ii) the principal ideal $(p(x))$ is maximal, and (iii) the principal ideal $(p(x))$ is prime.*

Proof. Suppose that $p(x)$ is irreducible and that the ideal

$$Q > P = (p(x))$$

Then some $q(x) \in Q$ is not divisible by $p(x)$. Hence gcd $(p(x),q(x)) = 1$ since such a gcd must be a proper divisor of $q(x)$. By the Euclidean algorithm of Sec. 11-7, this implies that $s(x)p(x) + t(x)q(x) = 1$ for some $s,t \in F[x]$. Since $p,q \in Q$, this implies $1 \in Q$, and so $Q = F[x]$. In conclusion, P is maximal, and condition (*i*) implies condition (*ii*).

Likewise, condition (*ii*) implies condition (*iii*) by Lemma 1. Finally, (*iii*) implies (*i*). The proof is by contradiction; if $p(x) = q(x)r(x)$ is the product of polynomials of lower degree, evidently $q,r \in P$ yet $q \notin P$,

$r \notin P$. Since condition (*i*) implies (*ii*) implies (*iii*) implies (*i*), the proof is complete.

We are now in a position to prove our main existence theorem.

Theorem 6. *Let F be any field, and let $m(x)$ be any monic polynomial which is irreducible over F. Then $F[x]/(m(x)) = G$ is a simple extension field of F; F is embedded monomorphically in G as the subfield of constants.*

Proof. From Chap. 10, Theorem 1, we know that $G = F[x]/(m(x))$ as constructed in Sec. 12-2 is a commutative ring which contains F as the subring (and hence subfield) of constants. It remains to show that G is a field. To show this, take for $p(x)$ in Theorem 5 the $m(x)$ of Theorem 6; it follows that $(m(x))$ is a maximal ideal of $F[x]$. From this, the fact that $G = R[x]/(m(x))$ is a field follows from Chap. 10, Theorem 19. Since G is generated by the subfield F of constants and x, the proof of Theorem 6 is complete.

Computation of Inverses. The computation of sums, differences, and products in $F[x]/(m(x))$ was described in Sec. 12-3. However, to compute inverses is more tedious. For very small fields such as those of Examples 1 and 2, it may be simplest to inspect ab for fixed a and all b. There will be precisely one b with $ab = 1$; set $a^{-1} = b$.

For extensions of larger (e.g., infinite) fields, one can use the Euclidean algorithm for polynomials of Sec. 11-7, to compute gcd $(a(x),m(x)) = 1$. This will give polynomials $s(x)$ and $t(x)$ with $s(x)a(x) + t(x)m(x) = 1$. Since $m(x) = 0$ in G, $[a(x)]^{-1} = s(x)$.

12-5. FINITE FIELDS

By definition, a *finite field* is a field having a finite number of elements. Such a field must have some (finite) prime characteristic p. Hence, as we have already observed, its order must be some power $q = p^n$ of the prime p; every finite field has *prime-power* order. Finite fields are also called *Galois fields*. For the rest of this section, we shall suppose that $G = GF(p^n)$ is a given field of prime-power order $q = p^n$.

Theorem 7. *Every element of $G = GF(p^n)$ satisfies the polynomial equation $x^q = x$, where $q = p^n$.*

Proof. Every nonzero element $x \in G$, as a member of the multiplicative group of all $q - 1$ nonzero elements of G, satisfies $x^{q-1} = 1$ by Lagrange's theorem (Chap. 7). Multiplying through by x, we have $x^q = x$ if $x \neq 0$. But for $x = 0$, $x^q = x$ trivially, completing the proof.

Corollary. *For any* $x_i \in G$, $(x - x_i) \mid (x^q - x)$ *in the polynomial ring* $G[x]$.

Proof. Since $x_i^q - x_i = 0$, the result follows by Theorem 4 of Chap. 11. On the other hand, the $(x - x_i)$ are relatively prime: gcd $(x - x_i, x - x_j) = 1$ if $x_i \neq x_j$. Hence, by Sec. 11-6, Corollary 4, the conclusion $(x - x_i) \mid (x^q - x)$ of the preceding corollary implies that $\left[\prod_{x_i \in G} (x - x_i)\right] \Big| (x^q - x)$. Since the divisor $\prod (x - x_i)$ has the same degree q as $x^q - x$ and both polynomials are monic, the two must be equal. This proves the following theorem.

Theorem 8. *In* $G = GF(q)$, $q = p^n$, $x^q - x = \prod_{x_i \in G} (x - x_i)$.

We shall now prove a sharper result.

Theorem 9. *In any finite field* $GF(p^n)$, *the multiplicative group* G^* *of all nonzero elements is cyclic.*

Proof. The multiplicative group G^* in question is Abelian of order $q - 1$. If it were not cyclic, then by Theorem 16 of Chap. 7 there would be some proper divisor r of $q - 1$ such that $x_i^r = 1$ for all $x_i \neq 0$ in G^*. The argument used to prove that $\prod_{x_i \in G} (x - x_i)(x^q - x)$ could then be applied to prove that

$$\prod_{x_i \in G^*} (x - x_i)(x^r - x) \qquad r < q - 1$$

But this is impossible since a polynomial of degree $q - 1$ cannot divide a polynomial of degree $r < q - 1$.

Corollary 1. *Any finite field of characteristic* p *is a simple algebraic extension of* $\mathbf{Z}_p$.

For, let c generate the cyclic group G^* under multiplication. Then, trivially, the subring of $GF(p^n)$ generated by c includes $G^* \sqcup 0 = GF(p^n)$.

Combining Corollary 1 with Theorem 4, we get a further result.

Corollary 2. *Every finite field* $GF(p^n)$ *is isomorphic to* $\mathbf{Z}_p[x]/(m(x))$, *where* $m(x)$ *is a suitable polynomial of degree* n *irreducible over* $\mathbf{Z}_p$.

Note also that $m(x) \mid (x^{q-1} - 1)$.

Corollary 3. *If* $m \mid (p^n - 1)$, *then* $GF(p^n)$ *contains* $\phi(m)$ *elements of multiplicative order* m, *where* ϕ *is Euler's* ϕ*-function.*

Proof. Let $[G^*, \cdot\,]$ be the cyclic group of nonzero elements of $GF(p^n)$ under multiplication, with generator c of order $q - 1 = p^n - 1$. Then, by Lagrange's theorem, the order of each $c^r \in G^*$ is a divisor $m = m(r)$ of $q - 1$. For each such divisor m of $q - 1$, moreover, the powers c^r of order m have as exponents $r = (q - 1)/m$ and those of its integral multiples $k(q - 1)/m$ $(k = 1,2, \ldots ,m)$ with $(k,m) = 1$. The number of such relatively prime k is, by definition, Euler's ϕ-function $\phi(m)$.

Finally, we note for future reference the following obvious corollary.

Corollary 4. *In* G^*, $c^r c^s = c^{r+s}$.

Example 5. Consider $GF(16) = GF(2^4)$. As in Example 6, Chap. 10, the irreducible factors of $x^{15} - 1$ over $\mathbf{Z}_2$ are given by

$$x^{15} - 1 = (x - 1)(x^2 + x + 1)(x^4 + x^3 + x^2 + x + 1)c_8(x) \qquad (15)$$

where $c_8(x)$ is the following cyclotomic polynomial (irreducible over $\mathbf{Z}$):

$$\begin{aligned} c_8(x) &= \frac{x^{12} + x^9 + x^6 + x^3 + 1}{x^4 + x^3 + x^2 + x + 1} \\ &= \frac{x^{10} + x^5 + 1}{x^2 + x + 1} = (x^4 + x^3 + 1)(x^4 + x + 1) \end{aligned} \qquad (16)$$

Hence, referring to Theorem 4, we have

$$\begin{aligned} GF(16) &\cong \frac{\mathbf{Z}_2[x]}{(x^4 + x^3 + 1)} \cong \frac{\mathbf{Z}_2[x]}{(x^4 + x + 1)} \\ &\cong \frac{\mathbf{Z}_2[x]}{(x^4 + x^3 + x^2 + x + 1)} \end{aligned} \qquad (17)$$

As in Example 1, the elements of $GF(4) \subset GF(16)$ are 0,1, and the two roots of $x^2 + x + 1 = 0$. These have multiplicative order 3 in $[G^*, \cdot\,]$. But $GF(8) \not\subset GF(16)$ because any proper extension of $GF(8)$, considered as a vector space over $GF(8)$, would have to have order 8^r, $r > 1$.

Under the isomorphism $\mathbf{Z}_2[x]/(x^4 + x^3 + 1) = GF(16)$, the element x has multiplicative order 15 in $[G^*, \cdot\,]$, the multiplicative group of nonzero elements of $GF(16)$; such elements are called *primitive*. Whereas under the *natural* isomorphism

$$\frac{\mathbf{Z}_2[x]}{(x^4 + x^3 + x^2 + x + 1)} \cong GF(16)$$

x goes into an element of multiplicative order 5 since

$$x^5 - 1 = (x - 1)(x^4 + x^3 + x^2 + x + 1) = 0$$

which is not primitive in $GF(16)$.

In general, one can show that x is primitive in $\mathbf{Z}_2[x]/(m(x)) \cong GF(2^n)$ if and only if $m(x)$ is a *primitive polynomial* in the sense of page 325.

EXERCISES B

1. For which values of $k = 1,2,3,4,5,6$ is $\mathbf{Z}_7[x]/(x^2 + k)$ a field?
2. Describe the Galois field $\mathbf{Z}_5[x]/(x^2 + 3)$, giving tables of reciprocals for elements in the form $(a + b\sqrt{-3})^{-1}$ and α^r (α primitive).
3. If x_1, x_2, and x_3 are the roots of the cubic polynomial $x^3 + x + 1 = 0$, find a polynomial whose roots are x_1^2, x_2^2, and x_3^2. (*Hint:* Find an equation of the form $x^6 + c_1x^4 + c_2x^2 + c_3 = 0$ implied by $x^3 = -x - 1$.)
4. (a) Show that if $p(x) = a_0x^n + a_1x^{n-1} + \cdots + a_n$ is irreducible over a field F, then $a_n \neq 0$.
   (b) Show that the roots of $a_nx^n + a_{n-1}x^{n-1} + \cdots + a_0 = 0$ are the reciprocals $1/x_i$ of the roots of $p(x)$.
5. (a) Prove that if α is a root of the equation $x^n - 1 = 0$, then so is every power of α.
   (b) Prove that if ω is a primitive root of $x^n = 1$, then every other root is a power of ω.
   (c) Show that the root field of $x^n - 1$ over $\mathbf{Q}$ is a simple extension of $\mathbf{Q}$, with a commutative Galois group. Is this group cyclic?

THE *quaternion extension* OF ANY COMMUTATIVE RING R IS THE SET OF NUMBERS $q = a_0 + a_1i + a_2j + a_3k$ WITH DISTRIBUTIVE MULTIPLICATION DEFINED BY $i^2 = j^2 = k^2 = -1$, $ij = -ji = k$, $jk = -kj = i$, AND $ki = -ik = j$.

★6. Show that the quaternion extension of $\mathbf{R}$ is a division ring.
★7. Show that for no prime p is the quaternion extension of $\mathbf{Z}_p$ a division ring.

12-6. COMPUTATIONS IN $GF(2^n)$

We now consider $G = GF(2^n)$, any binary Galois field, as a vector space over $\mathbf{Z}_2 = \{0,1\}$, relative to the specific basis $1, x, \ldots, x^{n-1}$ in any (isomorphic) presentation of $GF(2^n)$ as $\mathbf{Z}_2[x]/(m(x))$, where $m(x)$ is irreducible and of degree n over $\mathbf{Z}_2$. Thus, in Example 5, using (17), we can set $m(x) = x^4 + x^3 + 1$.

We can encode each polynomial $\sum_{k=1}^{n} b_kx^{n-k}$ of length n, considered as an element (coset leader) of $GF(2^n) \cong \mathbf{Z}_2[x]/(m(x))$, by the binary word $\mathbf{b} = b_1b_2 \cdots b_n$. Thus, in $GF(16) \cong \mathbf{Z}_2[x]/(x^4 + x^3 + 1)$, encode 0 as 0000, 1 as 0001, x as 0010, and the higher powers of x as in the following table, easily computed by the method of Sec. 12-3:

$c = 0010$	$c^6 = 1111$	$c^{11} = 1101$
$c^2 = 0100$	$c^7 = 0111$	$c^{12} = 0011$
$c^3 = 1000$	$c^8 = 1110$	$c^{13} = 0110$
$c^4 = 1001$	$c^9 = 0101$	$c^{14} = 1100$
$c^5 = 1011$	$c^{10} = 1010$	$c^{15} = 0001$

To add two elements of $GF(16)$ in this binary notation, add them as vectors over $\mathbf{Z}_2$. To multiply two elements of $GF(16)$, so represented, first decode using the inverse function table

$$\begin{array}{lll} 0001 = c^{15} & 0110 = c^{13} & 1011 = c^5 \\ 0010 = c & 0111 = c^7 & 1100 = c^{14} \\ 0011 = c^{12} & 1000 = c^3 & 1101 = c^{11} \\ 0100 = c^2 & 1001 = c^4 & 1110 = c^8 \\ 0101 = c^9 & 1010 = c^{10} & 1111 = c^6 \end{array}$$

then multiply using the relation $c^r c^s = c^{r+s}$ (exponents mod 15), and finally encode back using the original table.

12-7. BOSE–CHAUDHURI–HOCQUENGHEM CODES

The most powerful multiple error-correcting codes for random independent errors which have been discovered to date are the *Bose-Chaudhuri-Hocquenghem* (BCH) polynomial codes. These codes were discovered independently around 1960 by Bose and Chaudhuri and by Hocquenghem. For moderate code word lengths, these codes are very good indeed. Moreover, there is a systematic way to construct BCH codes of any length; the number of check digits required is then a function of the number of errors to be detected or corrected. Finally, there is a reasonably efficient procedure for determining which digits are in error when error correction is desired.

The symbols used in BCH codes are from a finite field $GF(q)$, where in practice, for implementation by binary computing devices, $q = 2^s$ is usually some power of 2. An encoding polynomial $g(x)$ is formed with coefficients in $GF(q)$, and message polynomials $a(x)$ are multiplied by $g(x)$ to form $b(x) = g(x)a(x)$, just as in Chap. 11.

As in Chap. 8, two n-symbol words are said to have *distance d* if they differ in d places. Thus, two four-digit words with symbols in $\mathbf{Z}_2$ which have distance 3 are 0111 and 0000. [In $GF(3^6) = GF(729)$, two words with distance 4 are 012021 and 021012.] Now suppose that we wish to construct a code with minimum distance at least d between code words.

The encoding polynomial $g(x)$ is formed in a systematic way, for a given minimum distance d, as follows. First, let $GF(q^r)$ be an extension of degree r of the field $GF(q)$, where $q^r = p^{rs} \geqq d + 1$. Then let $\alpha \in GF(q^r)$ be a primitive element, with multiplicative order $q^m - 1$.

Let $m_1(x)$ be the minimal polynomial for α, let $m_2(x)$ be the minimal polynomial for α^2. The encoding polynomial

$$g(x) = \operatorname{lcm}\,[m_1(x), m_2(x), \ldots, m_{d-1}(x)] \tag{18}$$

will be shown to yield a code with code word length $q^r - 1$ and minimum word distance at least d.

Example 6. We wish to construct a binary ($q = 2$) BCH code with code word length $n = 15$, which will have minimum distance $d = 5$. First a primitive polynomial is selected to form $GF(2^4)$. We choose $x^4 + x^3 + 1$ as this primitive polynomial. Let α be a root of $x^4 + x^3 + 1$; we then form successive powers of α as on page 351, thus:

$$\begin{aligned}
&\alpha \\
&\alpha^2 \\
&\alpha^3 \\
&\alpha^4 = \alpha^3 + 1
\end{aligned}$$

For, substituting α in $x^4 + x^3 + 1$ gives $\alpha^4 + \alpha^3 + 1 = 0$ and so $\alpha^4 = \alpha^3 + 1$. Multiplying by α, we get further

$$\begin{aligned}
\alpha^5 &= \alpha^4 + \alpha = \alpha^3 + \alpha + 1 \\
\alpha^6 &= \alpha^4 + \alpha^2 + \alpha = \alpha^3 + \alpha^2 + \alpha + 1 \\
\alpha^7 &= \alpha^4 + \alpha^3 + \alpha^2 + \alpha = \alpha^2 + \alpha + 1 \\
\alpha^8 &= \alpha^3 + \alpha^2 + \alpha \\
\alpha^9 &= \alpha^4 + \alpha^3 + \alpha^2 = \alpha^2 + 1 \\
\alpha^{10} &= \alpha^3 + \alpha \\
\alpha^{11} &= \alpha^4 + \alpha^3 = \alpha^3 + \alpha^2 + 1 \\
\alpha^{12} &= \alpha^4 + \alpha^3 + \alpha = \alpha + 1 \\
\alpha^{13} &= \alpha^2 + \alpha \\
\alpha^{14} &= \alpha^3 + \alpha^2 \\
\alpha^{15} &= \alpha^4 + \alpha^3 = 1
\end{aligned}$$

We next let $m_i(x)$ signify the minimal polynomial of α^i above and compute the minimal polynomials $m_1(x)$, $m_2(x)$, $m_3(x)$, and $m_4(x)$ for the above field. Since α is a root of $x^4 + x^3 + 1$, α^2 and α^4 are also roots of $m_1(x) = 0$, by the identity $f(\alpha)^{p^j} = f(\alpha^{p^j})$, valid for polynomials f in any field of characteristic p. Since $p = 2$, it follows that α, α^2, and α^4 have the same minimal polynomial, that is,

$$m_1(x) = m_2(x) = m_4(x) = x^4 + x^3 + 1$$

Likewise [since $x \mapsto x^p$ is an automorphism of $GF(q)$ and $GF(q^m)$ for $q = p^n$] α^3, α^6, α^{12}, and $\alpha^{24} = \alpha^9$ have the same minimal polynomial. This is the quartic polynomial

$$m_3(x) = m_6(x) = m_{12}(x) = m_9(x) = (x - \alpha^3)(x - \alpha^6)(x - \alpha^{12})(x - \alpha^9)$$

which can be calculated over $GF(16)$ as follows[1]:

$$\begin{aligned}
&(x - \alpha^3)(x - \alpha^6)(x - \alpha^{12})(x - \alpha^9) \\
&\qquad = (x - \alpha^3)(x - \alpha^3 - \alpha^2 - \alpha - 1)(x - \alpha^2 - 1)(x - \alpha - 1) \\
&\qquad = x^4 + x^3 + x^2 + x + 1
\end{aligned}$$

[1] The point here is that α^3, α^6, α^9, and α^{12} are the four conjugate roots of $m_3(x)$.

Therefore, $x^4 + x^3 + x^2 + x + 1$ is the monic polynomial over $\mathbf{Z}_2$ whose roots (zeros) in the extension field $GF(16)$ are $\alpha^3, \alpha^6, \alpha^9, \alpha^{12}$. Hence, in this case, the least common multiple of $m_1(x), m_2(x), m_3(x), m_4(x)$ will be the product $m_1(x)m_3(x)$

$$\begin{aligned} g(x) &= m_1(x)m_3(x) = (x^4 + x^3 + 1)(x^4 + x^3 + x^2 + x + 1) \\ &= x^8 + x^4 + x^2 + x + 1 \end{aligned}$$

Since this polynomial $g(x)$ is of degree 8, message words in this BCH code will have $15 - 8 = 7$ information digits $a_0 \cdots a_6$. Thus a typical message might be 1000100, for which $a(x) = 1 + x^4$ and

$$a(x)g(x) = x^{12} + x^6 + x^5 + x^2 + x + 1 = b(x)$$

In binary form, this product would be 111001100000100, and this would be the code word transmitted. We will now show that the minimum distance for this code is (at least) 5.

Theorem 10. *Let α be a primitive element of $GF(q^r)$, and let the code word length be at most $q^r - 1$. Then the BCH code with symbols in $GF(q)$ and encoding polynomial $g(x) = \text{lcm}\ [m_1(x), m_2(x), \ldots, m_{d-1}(x)]$ has minimum distance at least d.*

Proof. First observe that $g(x)$ is the polynomial of least degree over $GF(q)$ with roots $\alpha, \alpha^2, \alpha^3, \ldots, \alpha^{d-1}$. Since $c(x) = a(x)g(x)$, it follows that each code word polynomial $b(x)$ has roots $\alpha, \alpha^2, \alpha^3, \ldots, \alpha^{d-1}$.

Since polynomial codes are group codes, the minimum distance must be at least d if no code word $c_0 \cdots c_{n-1}$ has fewer than d nonzero c_i or, equivalently, if no polynomial $c(x)$ of degree less than $q^r - 1$ with roots $\alpha, \alpha^2, \ldots, \alpha^{d-1}$ has fewer than d nonzero terms. We first suppose that this is possible and then show that this supposition leads to a contradiction (proof by contradiction).

If a polynomial $c(x)$ has fewer than d nonzero terms, we can write it as

$$c(x) = c_{n_1}x^{n_1} + c_{n_2}x^{n_2} + \cdots + c_{n_{d-1}}x^{n_{d-1}}$$

where each c_{n_i} is in $GF(q)$. Now suppose that the above equation had $\alpha, \alpha^2, \ldots, \alpha^{d-1}$ as roots; the following system of equations would then hold for the c_{n_i} (not all of which are zero):

$$\begin{aligned} c_{n_1}\alpha^{n_1} + c_{n_2}\alpha^{n_2} + \cdots + c_{n_{d-1}}\alpha^{n_{d-1}} &= 0 \\ c_{n_1}\alpha^{2n_1} + c_{n_2}\alpha^{2n_2} + \cdots + c_{n_{d-1}}\alpha^{2n_{d-1}} &= 0 \\ c_{n_1}\alpha^{(d-1)n_1} + c_{n_2}\alpha^{(d-1)n_2} + \cdots + c_{n_{d-1}}\alpha^{(d-1)n_{d-1}} &= 0 \end{aligned}$$

In other words, suppose that $\alpha, \alpha^2, \ldots, \alpha^{d-1}$ were all zeros of the polynomial

$$c(x) = c_{n(1)}x^{n(1)} + c_{n(2)}x^{n(2)} + \cdots + c_{n(d-1)}x^{n(d-1)}$$

where $q^r > n(1) > n(2) > \cdots > n(d-1)$, each $c_{n(i)}$ was in $GF(q)$, and the multiplicative order of α in $GF(q)$ was greater than $n(1)$. Rewriting $c_{n(i)}$ as b_i to simplify the notation, we would then have $d-1$ simultaneous equations

$$\sum_{i=1}^{n} b_i \alpha^{kn(i)} = \sum_{i=1}^{n} b_i \alpha^{n(i)k} = 0 \qquad k = 1,2, \ldots ,d-1$$

Now consider these as simultaneous *homogeneous* linear equations which must be satisfied by the $d-1$ unknowns b_i $(i = 1, \ldots ,d-1)$; the coefficient matrix A is that of the Vandermonde determinant

$$A = \|a_{ik}\| = \|\alpha^{n(i)k}\|$$

It is *nonsingular* with determinant[1]

$$|A| = \prod_{i>j} (\alpha^{n(i)} - \alpha^{n(j)}) \neq 0 \qquad \text{since } n(j) < n(i) < q^r$$

Hence all $b_i = c_{n(i)} = 0$, the desired contradiction.

12-8. MINIMUM DISTANCE PROPERTIES

We shall now restrict attention to BCH codes with binary-valued symbols and establish their desirable minimum distance properties.

Theorem 11. *A binary BCH code with code word length* $2^m - 1$ *and with minimum distance d can always be constructed with* $[(d-1)/2]m$ *or fewer check symbols.*

Comment. There is a good reason why d is odd in the above; it will be shown that if one attempts to construct a binary code with a minimum distance which is even, say $2m$, then the code will automatically have minimum distance $2m + 1$.

Proof. The basic Theorem 10 shows that at most $m \cdot (d-1)$ check symbols would be required for a $(2^m - 1)$-digit code word binary code with minimum distance d. Now consider that if a polynomial function $p(\alpha) = 0$ with α in $GF(2^m)$, then $p(\alpha^2) = 0$, $p(\alpha^4) = 0, \ldots ,p(\alpha^{2i}) = 0$. Therefore, if we find a minimal polynomial for α, we automatically have a minimal polynomial for α^2, and if we find a minimal polynomial for α^3, we automatically have the minimal polynomial for α^6, etc.

Each minimal polynomial $m_i(x)$ has degree m or less, since $\alpha^i \in GF(2^m)$. Therefore the product of the $(d-1)/2$ minimal polynomials has degree $[(d-1)/2]m$ at most. [Sometimes minimal polynomials have

[1] For this classical result of determinant theory, see Birkhoff-Mac Lane, p. 284, Exercise 6.

degree less than m, and sometimes the least common multiple of the polynomials would have degree less than the product of the polynomials, so $[(d-1)/2]m$ is an upper bound on the number of check digits.]

The multiplicative order of the α selected must be $2^m - 1$, but this is always possible since, by Theorem 9, any Galois field $GF(2^m)$ has at least one primitive element; we choose this for α.

Example 6 had $2^m - 1 = 15$, $m = 4$, and $k = 8$. Since our minimum distance was 5, $[(d-1)/2]m = (5-1)/2 \cdot 4 = 8$ as was the case.

Hamming codes are a special case of BCH codes. They are simply those codes over $\mathbf{Z}_2$ with minimum distance 3 and are therefore single-error correcting. To see this, use the *primitive* polynomial $p(x)$ of degree m to generate a code with m check digits and $2^m - 1 - m$ information digits. Let α be a root of $p(x)$. Now $1,\alpha,\alpha^2, \ldots ,\alpha^{2^m-2}$ comprise the nonzero elements of $GF(2^m)$ and can be represented as the set of all nonzero binary m-tuples. If we write these m-tuples as columns, we can form a parity-check matrix H for a Hamming single-error-correcting code as in Sec. 8-9. Correspondingly, if we take the symbols of a code word $v = (v_0,v_1, \ldots ,v_{n-1})$ to be the coefficients of a polynomial

$$v(x) = v_0 + v_1x + v_2x^2 + \cdots + v_{n-1}x^{n-1}$$

then code words correspond to those polynomials for which α is a root of $v(x)$. Now $p(x)$ is the minimal polynomial for α over $\mathbf{Z}_2$, and the polynomial $v(x)$ corresponding to any code word must satisfy the relation $v(x) = m(x) \cdot p(x)$ for some $m(x)$. For encoding we consider $(a_0,a_1, \ldots ,a_{k-1})$ to be the coefficients of $m_0 + m_0x + \cdots + m_{k-1}x$ and multiply this by $p(x)$ to form a code word polynomial $v(x)$.

As was noted in Chap. 11, it is more convenient, however, to divide $x^m a(x)$ by $p(x)$ and add the remainder $r(x)$ to $x^m a(x)$.

Figure 12-1 shows a circuit for dividing by $x^4 \oplus x^3 \oplus 1$. This is a primitive polynomial of degree 4, where the coefficients of $a(x) \cdot x^m$, the dividend, enter in descending order (a_{k-1} enters first). C_2 is a 1 for $k = 11$ clock pulses, and then C_1 is a 1 for $m = n - k = 4$ clock pulses. $C_1 \neq C_2$ also holds.

The above material can be correlated with results in Chap. 11 as follows. We have shown that a code generated by a primitive polynomial $p(x)$ can detect all double errors and therefore has distance 3. In Chap. 11, a polynomial $q(x)$ was defined to have exponent e when $q(x)$ divides $x^e + 1$ and no polynomial $x^k + 1$ with $k < e$. If an error polynomial $e(x)$ has two terms, it must be of the form $x^i(1 + x^j)$, however, if $j - i < n \leq e$ for $n = 2^m - 1$, and a primitive polynomial $p(x)$ can divide no polynomial $x^k + 1$, where k is less than $2^m - 1$, by the definition of primitive polynomial. Also, $n = 2^m - 1$ is the number of digits in the code words

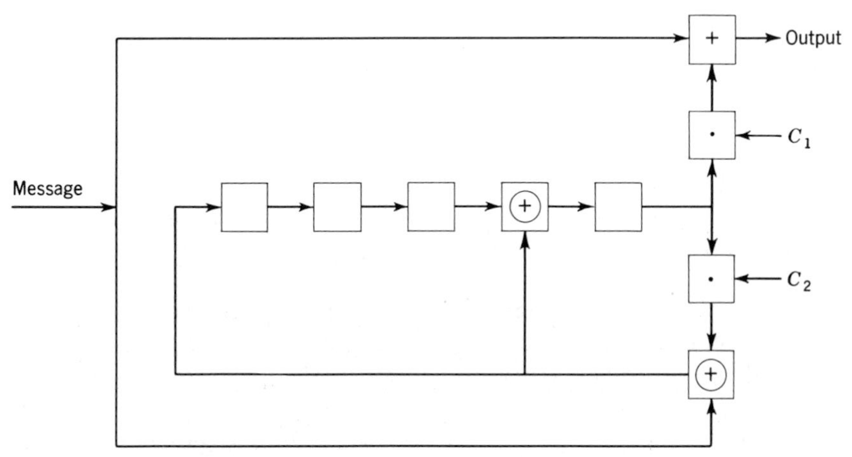

Fig. 12-1. Polynomial division circuit.

[or the maximum degree of our code polynomials $b(x)$]. Further, $p(x)$ cannot divide x^i, for then it would not be irreducible since x would be a factor.

The power of these BCH codes for error detection can be seen by several examples. A widely used code in European data communication systems is a (231,255) code with symbols in $\mathbf{Z}_2$; the code is formed over $GF(2^8)$ which has, of course, a primitive element α of multiplicative order $2^8 - 1 = 255$. The encoding polynomial $g(x)$ has degree 24, and α, α^2, α^3, α^4, α^5, and α^6 are all roots of $g(x)$ giving a minimum distance of 7. The code is primarily used for error detection, so error detection of up to six errors is guaranteed. Of course, most other errors will also be detected. In fact, since the proportion of code words is $2^{-24} \approx 1/16 \times 10^6$ of the total number of 255-bit words, if random words were introduced into the error-detection circuitry, only about 1 in 16 million would be accepted. Extensive tests of the code on actual European communications over a period of years did not turn up any cases where errors caused by the channels were not detected by the code.

A number of shorter codes which correct fewer errors are presently in use, and coders and decoders can be purchased from several commercial concerns.

Such BCH codes are reasonably effective for code words of moderate length,[1] in the sense that they are not too far from being "perfect" or "quasi-perfect," the optimality conditions for which we defined in Chap. 8.

[1] See W. W. Peterson, "Error-Correcting Codes," M.I.T. Press, 1961, for further details.

For a detailed description of the first multiple-error-correcting BCH code which was actually operated, we refer the reader to the literature.[1] This was a 127 code-digit, 92 message-digit, five-error correcting code defined by the generator polynomial

$$\begin{aligned} g(x) &= m_1(x)m_3(x)m_5(x)m_7(x)m_9(x) \\ &= (1 + x + x^7)(1 + x + x^2 + x^3 + x^7)(1 + x^2 + x^3 + x^4 + x^5 + x^7) \\ &\quad (1 + x + x^2 + x^4 + x^5 + x^6 + x^7)(1 + x + x^2 + x^3 + x^4 + x^5 + x^7) \end{aligned}$$

If a 1,000-digit message is transmitted using this code, through a binary symmetric channel whose probability of error is $q = 0.01$ per digit, then the probability that an error will occur after decoding is only 1.2×10^{-6}. For a probability of 0.006 error per digit, the code will have a probability of 7.7×10^{-9} of an error in a 1,000-digit message, and for a probability of error per digit of 0.002, the code provides a probability of 5.8×10^{-14} that an error will escape uncorrected in a 1,000-digit message.

Although BCH codes are quite good for most code words of moderate length, their "goodness" decreases as the length of code words grows larger. As a result, there is continuing effort toward the discovery of a code with even better properties for very long words.

EXERCISES C

1. Consider the BCH code generated over $\mathbf{Z}_2$ by

$$p(x) = 1 + x + x^2 + x^4 + x^5 + x^8 + x^{10}$$

   (a) Show that for some primitive element $\alpha \in GF(2^4)$, $m_1(x) = 1 + x + x^2 + x^3 + x^4$ is the monic polynomial of least degree such that $m_1(\alpha) = 0$.
   (b) Show that $m_3(x) = 1 + x + x^4$ and $m_5(x) = 1 + x + x^2$ are the corresponding polynomials for α^3 and α^5.
   (c) Show that $p(x)$ generates (5,15) group code.
   (d) If this code were arranged in a Standard Array (Chap. 8), choosing coset leaders having minimum weight to head each row, sketch the general appearance of the table.
2. Design a two-error-correcting BCH code with code words of length greater than 15. Exhibit the generator polynomial $g(x)$, and explain how you found it.
3. Let $g(x) = \text{lcm } [m_1(x), m_2(x), \ldots, m_{d-1}(x)]$ generate a BCH code with minimum distance d, and let $m_i(x)$ be the minimal polynomial of α^i $(i = 1, \ldots, d-1)$ for some primitive element α of $GF(2^r)$, so that the binary code words have length $2^r - 1$.
   (a) Show that one can make $\deg g \leqq (d-1)r/2$.
   (b) When can one make $\deg g < (d-1)r/2$? Explain.

[1] T. C. Bartee and D. Schneider, "An electronic decoder for BCH error-correcting codes," *IRE Trans. Information Theory*, vol. IT-8, no. 5, September, 1962.

★12-9. ROOT FIELDS OF POLYNOMIALS

The rest of this chapter will provide an introduction to one of the most fascinating topics in pure mathematics: the Galois theory of fields. It is an interesting fact that Galois not only invented this theory, which he used to prove the insolvability by radicals of the general quintic equation, but he also created the general theory of finite fields (which are often called *Galois fields* for this reason).

As a corollary of our discussion, we will establish the existence of a unique Galois field $GF(p^n)$ of each prime-power order, which can be characterized as the root field of the equation $x^{p^n} = x$. For computations of any *individual*, specific finite field $GF(p^n)$, of course, it is more efficient simply to produce the right primitive irreducible polynomial with coefficients in $\mathbf{Z}_p$. A table of such polynomials for $p = 2$ follows:

Table of primitive polynomials[1] over $GF(2)$

Degree 2: $1 + x + x^2$	Degree 9: $1 + x^4 + x^9$
Degree 3: $1 + x + x^3$	Degree 10: $1 + x^4 + x^{10}$
Degree 4: $1 + x + x^4$	Degree 11: $1 + x^2 + x^{11}$
Degree 5: $1 + x^2 + x^5$	Degree 12: $1 + x + x^4 + x^6 + x^{12}$
Degree 6: $1 + x + x^6$	Degree 13: $1 + x + x^3 + x^4 + x^{13}$
Degree 7: $1 + x^3 + x^7$	Degree 14: $1 + x + x^6 + x^{10} + x^{14}$
Degree 8: $1 + x^2 + x^3 + x^4 + x^8$	Degree 15: $1 + x + x^{15}$

[1] More extensive tables may be found in R. W. Marsh, "Table of Irreducible Polynomials over $GF(2)$ through degree 19," NASA, 1957, and in W. W. Peterson, "Error-Correcting Codes," M.I.T. Press, 1961.

Definition. An extension N of a field is a *root field* of the polynomial $f(x)$ with coefficients in F when (*i*) $f(x)$ can be factored into linear factors in N, so that $f(x) = (x - x_1)(x - x_2) \cdots (x - x_n)$, where all $x_i \in N$, and (*ii*) N is generated by F and the x_i (as a subfield).

When condition (*i*) holds, $f(x)$ is said to "split" (into linear factors) in N; hence any root field of $f(x)$ is a "splitting field" for $f(x)$ over F; moreover, $f(x)$ also splits in any extension of its root field N.

Example 8. Let $f(x) = x^3 - 2$, and let $F = \mathbf{Q}$ be the rational field. Then N is a six-dimensional vector space over $\mathbf{Q}$ with basis elements $1, \frac{1}{3}\sqrt{2}, \frac{2}{3}\sqrt{2}\,\omega, \frac{1}{3}\sqrt{2}\,\omega, \frac{2}{3}\sqrt{2}$, where $\omega = (1 + \sqrt{3}\,i)/2$ is an imaginary cube root of unity. The subfield $\mathbf{Q}(\omega) = \mathbf{Q}[x]/(x^2 + x + 1)$ since $\omega^2 + \omega + 1 = 0$ and $x^2 + x + 1$ is irreducible over $\mathbf{Q}$. Moreover, $\mathbf{Q}(\omega)$ is the root field of

$$x^2 + x + 1 = (x - \omega)(x - \omega^2)$$

it is also the root field of $x^3 - 1 = (x - 1)(x^2 + x + 1)$ over $\mathbf{Q}$.

Theorem 12. *Let the monic polynomial* $m(x) = x^r + \sum_{k=0}^{r} f_k x^k$ $(f_k \in F)$ *be irreducible over the field* F. *Over the extension field* $G = F[x]/(m(x))$, *the corresponding polynomial* $m(t) \in G[t]$ *is reducible, with the linear factor* $t - x$.

Proof. Since $(t - x) \mid c_k(t^k - x^k)$ for all $k \in \mathbf{N}$, clearly $(t - x) \mid [m(t) - m(x)]$ in the polynomial ring $F[x,t]$. Consequently, since $m(x) = 0$ in $G = F[x]/(m(x))$, then $t - x$ is a divisor of $m(t)$ in the (commutative) quotient ring $F[x,t]/m(x) \cong G[t]$ as claimed.

Example 9. Let $F = \mathbf{Z}_p(u)$ be the simple transcendental extension of $\mathbf{Z}_p$ consisting of all rational forms $f(x)/g(x)$, $g \neq 0$. Over $\mathbf{Z}_p(u)$, $x^p - u$ is irreducible since u cannot be factored. Now let $G = F[t]/(t^p - u)$; in G, $x^p - u = x^p - t^p = (x - t)^p$ (by the binomial theorem, as in Chap. 10). Hence $x^p - u$ factors into linear factors over $G = \{F,t\}$, which is therefore the root field of $x^p - u$ over F.

Example 10. Let $GF(p^n) = G$ be a finite field of order $p^n = q$, hence characteristic p and unital subfield $\mathbf{Z}_p$. Then by Theorem 6,

$$x^q - z = \prod_{x_i \in G} (x - x_i) \tag{19}$$

can be factored into linear factors over G, and the subfield

$$\{\mathbf{Z}_p, x_1, \ldots, x_q\} = G$$

(as a set).

Corollary. *Any finite field of order* $q = p^n$ *is the root field of* $x^q - x$.

Caution: We have referred to *the* root field, but its uniqueness (up to isomorphism) won't be proved until the next section, where we shall prove that any two root fields of a given polynomial over F are isomorphic (uniqueness). Here, we now complete the proof of existence.

Theorem 13. *Let* F *be any field, and let* $f(x)$ *be any polynomial of degree* n *over* F. *Then a root field of* $f(x)$ *over* F *exists, of degree of most* $n!$.

Proof. Without loss of generality, we can assume that $f(x)$ is monic (divide out its leading coefficient). Factoring $f(x) = p_1(x) \cdots p_r(x)$ into irreducible factors, we then construct $G_k = F[x]/(p_k(x))$ for any $k = 1, \ldots, r$ as in Theorem 11. In G_k, the coset $x + (p_k(x)) = y_1$ will moreover satisfy $p_k(y_1) = 0$, since $p_k(x) = 0$ in $F[x]$ and $x \mapsto y_1$ under the natural morphism $\theta = F[x] \rightarrow F[x]/(p_k(x))$. But $p_k(y_1) = 0$ in G_k implies $(x - y_1)p_i(x)$ in G_k. Hence, unless $p_k(x)$ is linear, $f(x)$ has at least one more linear factor over G_k than it does over F; moreover, G_k is generated by F and y_1.

After repeating the preceding construction at most n times, where n is the degree of $f(x)$, we will obtain an extension N of F (of degree at most $n!$) with the following properties: (*i*) over N, $f(x) = (x - y_1) \cdots (x - y_n)$ is a product of linear factors, and (*ii*) every element $t \in N$ is a polynomial in $y_1, \ldots, y_n$ with coefficients in F. This shows that N is a root field of F, and the proof is complete.

Repeated Factors. Now let $G = \{F, x_1, \ldots, x_n\}$ be the root field of the monic polynomial f over the field F. Over G, f can be factored into linear factors $f(x) = (x - x_1)(x - x_2) \cdots (x - x_n)$. We ask the following question: When can $f(x)$ have repeated factors? That is, when can $x_i = x_j$ for some $j \neq i$?

When this is the case, clearly $f(x) = (x - x_i)^2 g(x)$ for some monic polynomial $g(x)$. Hence, taking the formal derivative of both sides of this equation, we have $f'(x) = (x - x_i)^2 g(x) + 2(x - x_i)$. Therefore, $x - x_i$ is a common divisor of f and f', proving the following lemma (see Theorem 13 of Chap. 11).

Lemma 1. *If the monic polynomial $f(x)$ has a repeated factor $x - x_i$ over its root field, then $x - x_i$ is a common divisor of f and f'.*

Next, observe that over any field of characteristic p, the formal derivative of $x^q - x (q = p^n)$ is -1 since $qx^{q-1} = p^n x^{q-1} = 0$. Taken with Lemma 1, this implies

Lemma 2. *For $q = p^n$ (p a prime), the polynomial $x^q - x$ is without multiple factors over any field of characteristic P.*

Corollary. *In the (any) root field G of $x^q - x$ over $\mathbf{Z}_p$, with $q = p^n$, the factors $x - x_i$ of $x^q - x = \prod_{i=1}^{q} (x - x_i)$ are distinct; hence so are the roots x_i of the equation $x^q - x = 0$.*

Theorem 14. *If $q = p^n$ is a power of p, any root field G of $x^q - x$ over $\mathbf{Z}_p$ is a field of order q.*

Proof. By part (*ii*) of the definition of a root field,

$$G = \{F, y_1, \ldots, y_r\}$$

where every y_i satisfies $y_i^q = y_i$. On the other hand, since $\theta: x \to x^p$ is a one-one endomorphism of G, the same is true of $\theta^n: x \to x^q$. Therefore, the set of all $y \in G$ which satisfy $y^q = y$ is a *subfield* of G which contains F and all y_i; hence it contains G. Since a polynomial equation of degree q can have at most q distinct roots, the result now follows from the corollary of Lemma 2.

In view of Theorem 13, Theorem 14 implies the following consequence.

Corollary. *For any prime power p^n, there exists a Galois field $GF(p^n)$ of order p^n.*

Primitive Polynomials. We now consider the cyclic multiplicative group of nonzero elements of $GF(p^n)$; its order is $p^n - 1$. Hence, if $\phi(p^n - 1)$ denotes Euler's ϕ-function, which signifies the number of positive integers $k < p^n - 1$ which are relatively prime to $p^n - 1$, then the equation $x^{p^n-1} = 1$ will have $\phi(p^n - 1)$ *primitive roots* in $GF(p^n)$, all of which are distinct by the corollary of Lemma 2. Each of them generates $GF(p^n)$ over $\mathbf{Z}_p$ and hence is one of n (distinct) roots of an irreducible monic polynomial of degree n. This proves

Theorem 15. *Over $\mathbf{Z}_p$, there are $\phi(p^n - 1)/n$ monic polynomials of degree n whose roots have exponent $p^n - 1$.*

★12-10. ISOMORPHISMS OF ROOT FIELDS

It is much harder to prove that any two root fields G and H of the same polynomial $f(x)$ over a given F are isomorphic. The basic idea of the proof is that of "extending" isomorphisms. We now explain this idea in a simple case.

Factor $f(x)$ into irreducible factors $p_i(x)$; let $u_1, \ldots, u_n$ and $v_1, \ldots, v_n$ be the roots of f in G and H, respectively, listed so that u_1 and v_1 are both roots of the same irreducible factor $p_1(x)$. The assignment

$$a_0 + a_1u_1 + \cdots + a_{n-1}u_1^{n-1} \mapsto a_0 + a_1v_1 + \cdots + a_{n-1}v_1^{n-1} \tag{20}$$

defines an isomorphism $F(u_1) \to F(v_1)$, which can also be regarded as the composite of the two isomorphisms $\cong$ in the mapping diagram (draw it) associated with

$$G_1 = F(u_1) \cong \frac{F[x]}{(p_1(x))} \cong F(v_1) = H_1$$

Moreover, this isomorphism maps u_1 into v_1 and leaves every element of F fixed; it is an *extension* of the isomorphism 1_F (the trivial automorphism of F).

To deal with *iterated* algebraic extensions, one must refine the preceding construction as follows.

Lemma. *If $\alpha: F \cong F'$ is any isomorphism of fields, while u and v are any two roots of irreducible polynomials*

$$p(x) = \sum_{k=1}^{m} b_k x^k \qquad q(x) = \sum_{k=1}^{m} b'_k x^k \qquad b'_k = \alpha(b_k) \tag{21}$$

whose coefficients correspond under α, then the assignment

$$a_0 + a_1 u + \cdots + a_{n-1}u^{n-1} \mapsto a'_0 + a'_1 v + \cdots + a'_{n-1}v^{n-1}$$

with $a'_k = \alpha(a_u)$ gives an extension of q to an isomorphism $\bar{\alpha}: F(u) = F(v)$.

In this refined form, one can iterate the preceding construction to prove the following result.

Theorem 16. *Let $\alpha: F \cong F'$ be any isomorphism of fields, and let $p(x)$ and $q(x)$ correspond under α. Then the isomorphism α can be extended to an isomorphism $\bar{\alpha}$ from the root field of $p(x)$ over F to the root field of $q(x)$ over F'.*

The uniqueness of root fields (up to isomorphism), and hence of the finite field of order p^n, follows as the special case $\alpha = 1_F$, $p(x) = q(x)$.

Galois Groups. Now suppose that $p(x)$ is irreducible over F; let $N = F(u_1, \ldots, u_n)$ be the root field of p over F; and let ϕ be any permutation of the u_i. Then the first step in the preceding construction gives an extension $\alpha_1: F(u_1) \to F(u_{1\phi}) \cong F[x]/(p(x))$ of 1_F, which maps u_1 into $u_{1\phi}$. The rest of the construction gives an automorphism $\alpha: N \to N$ which extends α_1.

Hence if we define the *Galois group* $G(N/F)$ as the group of all automorphisms of N which leave every $x \in F$ fixed (i.e., are extensions of 1_F), we can state the following theorem.

Theorem 17. *The Galois group of the root field of any polynomial irreducible over F is transitive on the roots of the polynomial.*

In particular, the order of the Galois group is a multiple of n and a divisor of $n!$, where n is the degree of the given polynomial.

It is easy to compute the Galois group of any finite field $GF(p^n)$ over its unital subfield $\mathbf{Z}_p$. Let $c \in GF(p^n)$ generate the cyclic multiplicative group G^* of all nonzero elements of $GF(p^n)$. Any automorphism

α of $GF(p^n)$ must map c into some power c^r of itself since it cannot map c into 0. Hence, for any nonzero element $x = c^k$ of G^*, we have

$$\alpha(x) = \alpha(c^k) = [\alpha(c)]^k = c^{rk} = x^r$$

whereas trivially $\alpha(0) = 0 = 0^r$, and $\alpha(1) = 1 = 1^e$. This shows that every automorphism of a finite field is of the form $x \mapsto x^r$ for some integer r. Furthermore, since the right composite of $\alpha: x \mapsto x^r$ and $\beta: x \mapsto x^s$ is of the form $x \mapsto (x^r)^s = x^{rs}$, the Galois group of $GF(p^n)$ is monomorphic to the cyclic group $[G^*, \cdot]$. Since any subgroup of a cyclic group is cyclic, it follows that the Galois group of $GF(p^n)$ is cyclic. Finally, if x_1 and x_2 are roots of the same irreducible factor $f(x)$ of $x^q - x$ (over $\mathbf{Z}_p$), then $\mathbf{Z}_p(x_1) \cong \mathbf{Z}_p[x]/(f(x)) \cong \mathbf{Z}_p(x_2)$, by Theorem 3, and the composite of these two isomorphisms is an automorphism of $GF(p^n)$ which maps x_1 into x_2. Therefore the Galois group of $GF(p^n)$ acts *transitively* on the roots of $f(x)$. Being Abelian, it is, however, at most *simply* transitive.[1] We conclude that the Galois group of $GF(p^n)$ is cyclic of order n over $\mathbf{Z}_p$.

Example 11. Let $F(x_1, \ldots, x_n)$ be the field of all rational functions (or forms) with coefficients in a field F of characteristic ∞. Let

$$S = F(s_1, \ldots, s_n)$$

be the subfield generated by F and the elementary symmetric polynomials $s_j = x_1{}^j + \cdots + x_n{}^j$ $(j = 1, \ldots, n)$ in the indeterminates x_i. A basic theorem[2] asserts that S contains all the symmetric rational functions over F, that is, all those which are invariant under the symmetric group of all permutations of the x_i. It follows that the Galois group of $F(x_1, \ldots, x_n)$ over S is the symmetric group of degree n and order $n!$.

EXERCISES D

1. Let A be any set of automorphisms of a field F. Show that the elements $x \in F$, such that $\alpha(x) = x$ for all $\alpha \in A$, form a subfield $S(A) \subset F$.
2. Let T be any of a field F. Show that the automorphisms α of F, such that $\alpha(x) = x$ for all $x \in T$, form a subgroup $G(T)$ of Aut F.
3. In the notation of Exercises 1 and 2, prove that

$$G(S(A)) \supset A \qquad S(G(T)) \supset T$$
$$S(G(S(A))) = S(A) \qquad G(S(G(T))) = G(T)$$

[*Hint:* $T \supset T_1$ implies $G(T) \subset G(T_1)$, and $A \supset A_1$ implies $S(A) \subset S(A_1)$.]
4. Let α be algebraic over $\mathbf{Z}_p$, and let $m(x)$ be its (irreducible) minimal polynomial. Show that the Galois group of $\mathbf{Z}_p[\alpha]$ over $\mathbf{Z}_p$ is Aut $\mathbf{Z}_p[\alpha]$. (You may assume that $\mathbf{Z}_p[\alpha]$ is normal over $\mathbf{Z}_p$.)
5. Compute the Galois group of $\mathbf{Z}_5[x]/(x^2 + 3)$ over $\mathbf{Z}_5$.

[1] See M. Hall, "Theory of Groups," p. 58, Cor. 5.3.
[2] Birkhoff-Mac Lane, Chap. XV, Theorem 10, Cor.

6. Describe the Galois group of $GF(16)$ over $GF(4)$, giving an explicit formula for its proper automorphism.
7. What is the root field of $x^6 - 1$ over $\mathbf{Q}$? Over $\mathbf{Z}_7$?

★8. Repeat Exercise 6 for $GF(81)$ over $GF(9)$.

9. Describe the Galois group of $\mathbf{Q}[\omega]/\mathbf{Q}$
   (a) For $\omega = (1 + \sqrt{-3})/2$.
   (b) For ω, a primitive fifth root of unity.

★10. Describe the Galois group of $\mathbf{Q}[\sqrt[3]{2}]/\mathbf{Q}$.

★11. Describe the Galois group of $\mathbf{Q}[\sqrt[5]{3}]/\mathbf{Q}$.

REFERENCES

ARTIN, E.: "Galois Theory," Notre Dame Mathematical Lectures, No. 2, 1944, 1959.
BERLEKAMP, E. R.: "Algebraic Coding Theory," McGraw-Hill, 1968.
PETERSON, W. W.: "Error-Correcting Codes," M.I.T. Press, 1961.

Recurrent Sequences

13-1. RADAR AND COMMUNICATIONS SYSTEMS

An important development in electronics during the early 1940s was radar, which refers to an electronic system which locates objects in the atmosphere or in space. "Radar" is an acronym for *radio detection and ranging,* and its operation depends upon the fact that electromagnetic energy transmitted at a radio frequency is reflected by objects in the path of the wave. The basic scheme used in a tracking radar is shown in Fig. 13-1. Similar schemes are used to track satellites or other space vehicles and in airport tracking systems for use with landing aircraft.

Basically, the radar system transmits electromagnetic energy which has been shaped into a beam narrow in both height and width by a suitable antenna. By moving the antenna, this beam can be pointed in any desired direction. The frequency f of the electromagnetic energy transmitted generally lies in the range from 200 million to 10 billion cycles per second (2×10^8 to 10^{10} cps). The wavelength $\lambda = c/f$ is correspondingly between 3 centimeters and 1.5 meters, since the speed of propagation of electromagnetic waves is about 300,000 kilometers per second.

When such a beam of radiated electromagnetic energy is concentrated on a target (object), the target reflects a certain amount of the energy back to the antenna. Since the transmitted energy can be directed in space, whether or not energy is reflected enables the radar system to determine the presence or absence of a target in the direction in which the transmitting antenna is pointed. Generally the receiving antenna (which may well be the transmitting antenna also) has a highly selective directional response so that the gain for the system is multiplied by the directivities of both the transmitting and receiving antenna.

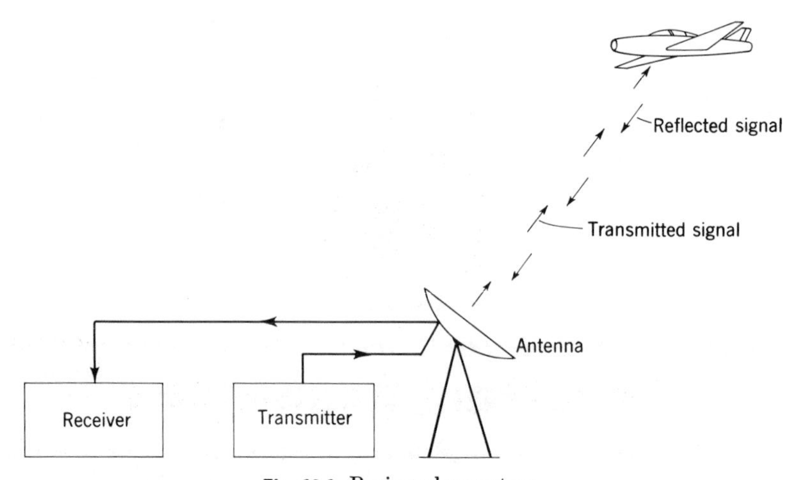

Fig. 13-1. Basic radar system.

The directional response of the system enables the radar to locate an object in two dimensions, giving the angular position of the object relative to the antenna(s). The third required dimension, that of the *range*, or distance of the object from the antenna, was for years determined by simply turning on the transmitted energy for a very short period of time, waiting until the short burst or pulse of energy reflected by the tracked object had returned to the receiver, and then measuring the time elapsed between the transmission and receival of the pulse. By dividing this time interval by 2 and multiplying by the velocity of electromagnetic propagation ($C \simeq 186{,}000$ miles per second, or 300,000 kilometers per second), the distance from the antenna to the object could be determined.

There are some difficulties with this system. The detectability of a target is a function of the amount of energy which can be concentrated on a target (and hence reflected) because there is natural background noise with which the radar must contend and therefore a threshold which the energy in the received signal must exceed. As the range to the target increases, less and less energy is received back at the antenna. In order to overcome this, one would like either to integrate the signal received over long periods of time or to transmit extremely large bursts of power during short intervals.

The radar designer is therefore faced with a dilemma. The shorter the pulses of transmitted energy, the greater the range resolution which the system can achieve; and the longer the pulses of energy which are transmitted, the greater the range at which a target of given size can be detected. Considerations of this sort lead one to wish to muster a great amount of energy during a very short time. But this is quite difficult from the viewpoint of the transmitter design because of the magnitude of the electrical voltages and currents required, as well as other physical considerations.

Several solutions to this problem of ranging have been offered. One of them is to shift the frequency of the transmitted signal during the pulse and then attempt to match the transmitted and returned wave forms; this gives higher range resolution and makes a long pulse possible; it is called *pulse compression.*

Another technique has been widely used, particularly for long-range radars, such as satellite tracking radars and radars which make maps of the Moon or measure ranges of Venus. This technique consists of transmitting a long sequence of pulses or signals of electromagnetic energy, coded in such a way that one can match the transmitted and received energy unambiguously in only one way.

This chapter will deal with coding techniques for generating sequences which are *periodic* with a very *long period.* These sequences are then used to generate the transmitted radar signals, as will be explained in Sec. 13-6; our first concern will be with techniques for encoding them. Section 13-9 will discuss "matching," or "correlating," returned signals which have been attenuated and to which noise has been added with the transmitted sequence of signals.

13-2. DIFFERENCE CODES

A *difference code* is a code whose encoding function converts a given m-digit starting or initial block $\mathbf{a} = a_0, a_1, \ldots, a_{m-1}$ into an *infinitely long sequence* $\mathbf{s} = s_0, s_1, s_2, \ldots$, defined recursively by a linear *difference equation* of the form

$$c_0 s_i + c_1 s_{i-1} + \cdots + c_m s_{i-m} = 0; \; i = m, m+1, \ldots; \; c_m, c_0 \neq 0 \quad (1)$$

The alphabet A from which the s_i (and c_i) are taken is understood to be some *finite field* $GF(q)$ of the type studied in Chap. 12; the code sequence† **s** is understood to begin with the block

$$s_0 = a_0,\ s_1 = a_1,\ \ldots,\ s_{m-1} = a_{m-1} \tag{1'}$$

which is thus its initial segment.

Example 1. Consider the difference equation

$$s_i + 2s_{i-1} + s_{i-2} = 0, \qquad \text{over } GF(3) = \mathbf{Z}_3 \tag{2}$$

If we let $s_0 = 1$ and $s_1 = 1$ (and thus $a_0 = 1$ and $a_1 = 1$), the code sequence generated will be $\mathbf{s} = 1,1,0,2,2,0,1,1,\ \ldots$. If we let $s_0 = 1$ and $s_1 = 2$, the sequence generated will be $\mathbf{s} = 1,2,1,2,1,2,\ \ldots$. If $s_0 = 2$ and $s_1 = 1$, we have $\mathbf{s} = 2,1,2,1,2,\ \ldots$. If $s_0 = 0$ and $s_1 = 2$, we have $0,2,2,0,1,1,0,2,\ \ldots$, and so on. Since there are $3^2 = 9$ different sets of initial values for the s_i in this difference equation, nine different sequences can be generated by the difference equation.

Encoding functions like that of Example 1 can be implemented electronically using flip-flops arranged in shift registers. For values (message digits or letters) from $\mathbf{Z}_2$, the implementation is particularly simple.

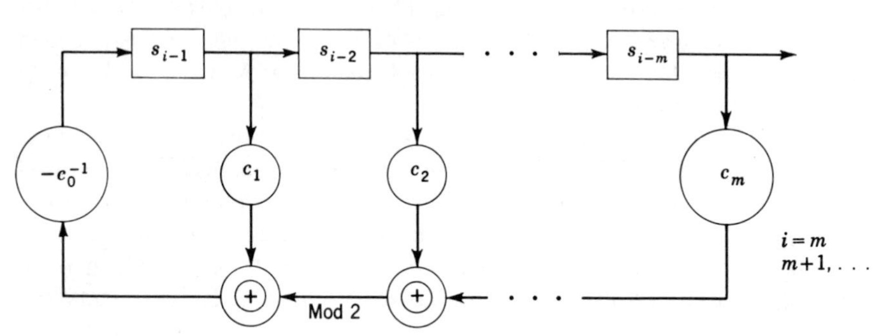

Fig. 13-2.

Thus, Fig. 13-2 indicates how to obtain solutions (code words) for the ΔE (difference equation‡) (1) when the coefficients c_i and the values s_i are in $GF(q)$. There are m flip-flops or delay elements required; these are designated $S_{i-1}, \ldots, S_{i-m}$ in the figure. Also, m multipliers are required, each of which multiplies by the appropriate coefficient c_j. These

† Code sequences defined by (homogeneous) linear difference equations (with constant coefficients) such as (1) or (4) are often called *linear recurring sequences* or, more simply, *recurrent sequences*.

‡ Here and below, we shall often use the convenient abbreviation ΔE for *difference equation*.

values are added and multiplied by $-c_m^{-1}$ to give each new value for s_i. The register of S_i is shifted at each time interval $t = 0,1,2,3, \ldots$ which we associate with the index $i = 0,1,2, \ldots$. At each time $t = i$, the register is shifted to the right so that the value in S_j is shifted into S_{j-1}, except that $-c_0^{-1}(c_{-1}S_{i-1} + c_2S_{i-2} + \cdots + c_mS_{i-m})$ is read into S_{i-1}.

The shift register for Example 1 is shown in Fig. 13-3.

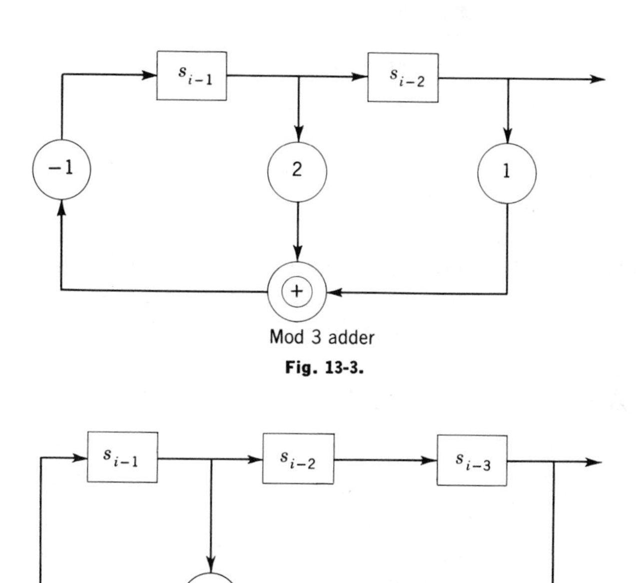

Fig. 13-3.

Fig. 13-4.

Example 2. If the field is $GF(2) = \mathbf{Z}_2$, then the scheme takes a particularly simple form, since each c_i has value either 0 or 1. Figure 13-4 shows a shift register with feedback for the difference equation $s_i + s_{i-1} + s_{i-3} = 0$. If we start this register with $s_0 = 1$, $s_1 = 0$, and $s_2 = 0$, we will generate as output the code sequence

$$\mathbf{s} = \{s_i\}_{i=0}^{\infty} = 100111101 \cdots$$

Notice that for the field $\mathbf{Z}_2$, the multipliers c_i are always 0 or 1 and thus can be implemented by a connection or no connection at the output of each flip-flop S_i.

Typically, the code sequences transmitted using difference codes are periodic with a very long period; indeed, this is one of their most useful properties. Postponing until Sec. 13-6 the technique for computing the period of a given difference code, we here simply establish their all-important periodic character.

Definition. If for a sequence $\mathbf{s} = s_0, s_1, \ldots$ there exists an r such that $s_0 = s_r$, $s_1 = s_{r+1}$, . . . , the sequence is said to be *periodic*, and the least such $r = p(\mathbf{s})$ is called the *period* of the sequence. If for a sequence $\mathbf{s}$ there exist a t and an r such that

$$s_t = s_{t+r},\ s_{t+1} = s_{t+r+1},\ \ldots \tag{3}$$

the sequence is said to be *eventually periodic.*

Theorem 1. *Every sequence* $\mathbf{s}$ *associated with a difference code with* ΔE (1) *is periodic if* $c_m \neq 0$.

Comment. Each sequence $\mathbf{s}$ is sure to be eventually periodic, since the shift register is a finite-state machine; the only question is whether s is periodic or only eventually periodic.

Proof. Let α be the encoding function defined by (1); thus

$$\alpha\colon (s_0, \ldots, s_{m-1}) \mapsto (s_1, \ldots, s_{m-1}, \ -c_0^{-1}(c_1 s_{m-1} + \cdots + c_m s_0))$$

Now α is linear and so a morphism of the additive group of all n-tuples over the field $GF(q)$. Further, α takes only $\mathbf{0}$ into the all-zero n-tuple $\mathbf{0}$, for if $\alpha(s_0, \ldots, s_{m-1}) = (0, \ldots, 0)$, then $s_1 = 0, \ldots, s_{m-1} = 0$, and $-c_m s_0 = 0$ implies $s_0 = 0$. Hence α is one-one and is a permutation of the set of all m-tuples over $GF(q)$. Therefore, $\alpha^{r+1} = \alpha$ for some r, and α^r is the identity mapping. Now, since $\alpha^k(s_0, \ldots, s_{m-1})$ is simply $(s_k, \ldots, s_{m+k-1})$ in the sequence initiated by $s_0, \ldots, s_{m-1}$ and continued using the ΔE (1), $s_0 = s_r$, $s_1 = s_{r+1}$, ...

13-3. DIFFERENCE EQUATIONS

More generally, a linear difference equation (ΔE) of order m with constant coefficients, such as in (1), defines from any initial segment (block) of length m a unique infinite sequence of numbers over any field F or even integral domain D. Thus, given any list

$$\mathbf{a} = (a_0, a_1, \ldots, a_{m-1}) = (s_0, s_1, \ldots, s_{m-1})$$

of initial values, one can compute all later values $s_m, s_{m+1}, \ldots$ recursively from (1).

Given elements c_j, a_j from any field F, one can consider generally *constant-coefficient* linear difference equations of the form of (1). The ΔE (1) defines recursively a *solution* of (1) from any m-tuple $\mathbf{a}$ of *initial values*, computable recursively from the ΔE (1), rewritten as

$$s_n = -\sum_{j=1}^{m-1} c_j s_{n-j} \qquad \text{all } n \geqq m \tag{4}$$

Moreover, the set of all solutions is an m-dimensional vector space over the field F, called the *solution space*, of the ΔE (1), and signified symbolically by $S(\mathbf{c})$, where $\mathbf{c}$ is the list of the $m + 1$ coefficients $c_1, \ldots, c_m$ of (1). In the case of a difference code whose alphabet $F = GF(q)$ has q letters, the solution space contains exactly q^m different code words, each determined by its initial segment.

The following classic illustration from number theory will illustrate the encoding process. (In this illustration, the alphabet can be taken to be $\mathbf{R}$, $\mathbf{Q}$, $\mathbf{Z}$, or even $\mathbf{N}$.)

Example 3. The *Fibonacci equation* is

$$s_{n+1} = s_n + s_{n-1} \qquad s_i \in \mathbf{R},\ n \in \mathbf{P} \tag{5}$$

For the initial values $s_0 = 0$, $s_1 = 1$, the solution gives the Fibonacci sequence for $i \in \mathbf{P}$: 0,1,1,2,3,5,8,13,34,55,89, ...

This ΔE and its solution can also be considered mod p, as a sequence of elements in $\mathbf{Z}_p = GF(p)$. For $p = 11$, we get

$$0,1,1,2,3,5,8,2,10,1,0,1,1,2,3,5,8, \ldots$$

which is *periodic* with period $T = 10$.

In operator notation, let $X: F^\omega \to F^\omega$ and $X^{-1}: F^\omega \to F^\omega$ be the right and left *shift operators*, defined on the space F^ω of all infinite sequences $\mathbf{s} = (s_0, s_1, s_2, \ldots)$ of elements $s \in F$; thus

$$X: (s_0, s_1, s_2, \ldots) \mapsto (0, s_0, s_1, \ldots) \tag{6}$$

$$X^{-1}: (s_0, s_1, s_2, \ldots) \mapsto (s_1, s_2, s_3, \ldots) \tag{6'}$$

Clearly, $X^{-1} \circ X = X \diamondsuit X^{-1} = 1$ is the identity; moreover, the ΔE (1) can be written in the form

$$[c_0 I + c_1 X^{-1} + \cdots + c_m X^{-m}]\mathbf{s} = \mathbf{0} = (0,0,0, \ldots) \qquad c_0 \neq 0,\ c_m = 1 \tag{7}$$

The polynomial $c(x) = c_0 + c_1 x + \cdots + c_m x^m$ is called the *left* characteristic polynomial of the ΔE (1); the reciprocal polynomial (with reciprocal roots)

$$\bar{c}(x) = c_0 x^m + c_1 x^{m-1} + \cdots + c_m$$

is called the *right* characteristic polynomial of (1); both are useful.

Characteristic Solutions. Over the complex field, the (left) characteristic polynomial in (7) can be used to compute elementary solutions of the ΔE (1) as follows.

Theorem 2. *The sequence* $(1,\alpha,\alpha^2, \ldots)$ *defined by* $s_k = \alpha^k$ *satisfies the* ΔE (1) *if and only if* $c(\alpha) = 0$.

Proof. Substituting into (1), we have

$$\sum_{j=0}^{m} c_j s_{i-j} = \sum_{j=0}^{m} c_i \alpha^{i-j} = \alpha^{i-m} c(\alpha) \qquad \text{for all } i$$

from which the result is obvious. Note that the condition $c_m \neq 0$ in (1) excludes $\alpha = 0$ as a root.

Thus, for the Fibonacci equation (3), the characteristic polynomial is $x^2 - x - 1$, whose characteristic roots α_i are $(1 \pm \sqrt{5})/2$. The general complex solution of (3) is thus

$$s_k = a\left(\frac{1+\sqrt{5}}{2}\right)^k + b\left(\frac{1-\sqrt{5}}{2}\right)^k \tag{8}$$

where a and b are arbitrary complex constants. Since the first term grows more rapidly, it is clear that any solution with $a \neq 0$ will grow as, asymptotically, $s_{k+1}/s_k \to (1+\sqrt{5})/2 \simeq 1.618034 \cdots$. For instance, $\frac{89}{55} = 1.6181818 \cdots$.

Likewise, for the ΔE $u_{n+1} = 2u_n + u_{n-1}$, the roots of the characteristic equation are $1 \pm \sqrt{2}$. For the initial values 0 and 1, the resulting sequence continues as 2, 5, 12, 29, 70, 169, ... The ratio

$$\frac{169}{70} = 2.4142857142857 \cdots$$

is a good approximation to $1 + \sqrt{2} = 2.414214 \cdots$.

EXERCISES A

1. List the first 10 digits of the eight sequences of binary digits which satisfy the ΔE $s_i = s_{i-2} + s_{i-3}$ over $\mathbf{Z}_2$.
2. For the ΔE of Exercise 1, considered over $\mathbf{Z}_3$, list the first 10 digits of the three basic sequences of *ternary* digits beginning with 100,010,001. From these compute the sequence 111 $\cdots$.
3. Determine the period of each of the sequences of Exercise 1.
4. Determine the period of each of the four sequences of Exercise 2.
5. (a) What are the characteristic roots of the ΔE $s_i = 2s_{i-1} + 2s_{i-2}$ over the *real* field?

(b) Compute a rational approximation to $\sqrt{3}$, having an error less than 0.001, from the terms of the sequence beginning with $s_0 = 1$, $s_1 = 3$.

6. (a) What are the characteristic roots of the ΔE $s_i = s_{i-1} + 3s_{i-2}$ over the real field?

   (b) Setting $s_0 = 1$, $s_2 = 2$, compute a rational approximation to $\sqrt{13}$ which has an error of less than 0.001. (*Hint:* Let $s_0 = 1$, $s_1 = 3$.)

7. (a) Show that the characteristic polynomial of the ΔE (recurrence relation) $s_n = s_{n-3}$ is reducible over any field F.

   (b) Show how to utilize this fact to obtain a convenient basis of solutions of the ΔE.

   (c) Discuss the periods of the solutions of the ΔE, for digits in $\mathbf{Z}_p$ (p any prime).

8. Given any second-order ΔE of the form $s_n = a_1 s_{n-1} + a_2 s_{n-2}$, obtain a recurrence relation for the ratio of successive terms, $r_i = s_i/s_{i-1}$, of the form $r_i = F(r_{i-1})$. [*Hint:* For the Fibonacci equation, $F(r) = 1 + (1/r)$.]

★9. Show that Newton's method for computing the dominant root of $x^2 - a_1 x - a_2 = \phi(x) = 0$, $X_i = F(x_{i-1})$, where $F(x) = x - [\phi(x)/\phi'(x)]$, has a higher order of convergence than the method of Exercise 8.

13-4. FORMAL POWER SERIES

Given a commutative ring R, one can consider not only polynomials with coefficients in R but, more generally, *formal power series* of the form

$$a_0 + a_1 x + a_2 x^2 + a_n x^n + \cdots = \sum_{n=0}^{\infty} a_n x^n \qquad a_n \in R \tag{9}$$

Here the word "formal" is intended to emphasize the idea that the convergence or divergence of the series is not relevant to the discussion and is not meaningful in most fields (e.g., in $\mathbf{Z}_2$). Such formal power series are very helpful for analyzing the solutions of difference equations, the basic connection being provided by the following construction.

Definition. Given an infinite sequence $\mathbf{s} = s_0, s_1, s_2, \ldots$, the *generating function* of $\mathbf{s}$ is the formal power series

$$s(x) = s_0 + s_1 x + s_2 x^2 + \cdots = \sum_{n=0}^{\infty} s_n x^n \tag{10}$$

Note that the right shift operator of (6) for $\mathbf{s}$ has the effect $s(x) \mapsto xs(x)$ on the generating function $s(x)$ of $\mathbf{s}$.

We shall show in the next section that the solutions of the ΔE (1) are the quotients $a(x)/c(x)$ whose numerators are the polynomials of length m (degree $\leq m - 1$) and whose denominator is the characteristic polynomial $c(x)$ defined below (7). First we must show that the set $R[[x]]$ of all formal power series with coefficients in any commutative ring R is itself a commutative ring.

In $R[[x]]$, addition and multiplication are defined by the rules

$$\sum_{k=0}^{\infty} a_k x^k + \sum_{k=0}^{\infty} b_k x^k = \sum_{k=0}^{\infty} (a_k + b_k) x^k \tag{11}$$

$$\sum_{k=0}^{\infty} a_k x^k - \sum_{k=0}^{\infty} b_k x^k = \sum_{k=0}^{\infty} \left(\sum_{i=0}^{k} a_i b_{k-i} \right) x^k \tag{12}$$

The unity of $R[[x]]$ is

$$1 = 1 + 0x + 0x^2 + \cdots + 0x^n + \cdots \tag{13}$$

with $a_0 = 1$ and all other $a_k = 0$.

Example 4. Let $R = \{0,1\} = \mathbf{Z}_2$, let $f(x) = 1 + x^2$, and let

$$g(x) = \sum_{k=0}^{\infty} a_k x^k$$

where for all k, $a_k = 1$; that is, $g(x) = 1 + x + x^2 + \cdots + x^n + \cdots$. Then$f(x) + g(x) = x + x^3 + x^4 + \cdots + x^n + \cdots$, and

$$f(x)g(x) = 1 + x + 0 \cdot x^2 + 0 \cdot x^2 + \cdots + 0 \cdot x^2 + \cdots$$

Hence $g(x) = (1 + x)/(1 + x^2)$. [Note that $1 + x^2 = (1 + x)^2$ over $\mathbf{Z}_2$; $f(x)$ is a reducible polynomial.]

Theorem 3. *If R is any commutative ring, then formulas* (11) *and* (12) *define* $R[[x]]$ *as a commutative ring.*

Proof. First, $R[[x]]$ is an Abelian group under the addition (11). To prove this, we observe that the formal power series $0 = \sum_{k=0}^{\infty} 0 \cdot x^k$ acts as an identity for addition. Again, $\sum_{k=0}^{\infty} (-a_k) x^k$ is an additive inverse of $\sum_{k=0}^{\infty} a_k x^k$ since $a_k + (-a_k) = 0$ in R for all $k \in N$. Finally, addition is associative; the three formal power series $\sum_{k=0}^{\infty} a_k x^k$, $\sum_{k=0}^{\infty} b_k x^k$, and $\sum_{k=0}^{\infty} c_k x^k$ have the sum

$$\sum_{k=0}^{\infty} (a_k + b_k + c_k) x^k = \sum_{k=0}^{\infty} [a_k + (b_k + c_k)] x^k = \sum_{k=0}^{\infty} [(a_k + b_k) + c_k] x^k$$

Next, $R[[x]]$ is a commutative monoid under the multiplication (12) because $1 + \sum_{k=1}^{\infty} 0 \cdot x^k$ acts as a left identity for multiplication; if $a_0 = 1$ and $a_i = 0$ for $i \neq 0$, then $\sum_{i=0}^{k} a_i b_{k-i} = b_k$ in (12). Multiplication is commutative since $\sum_{i=0}^{k} a_i b_{k-i} = \sum_{j=0}^{k} b_j a_{k-j}$ for $j = k - i$, R being commutative. Multiplication is also associative since

$$\left(\sum_{k=0}^{\infty} a_k x^k\right)\left(\sum_{k=0}^{\infty} b_k x^k\right)\left(\sum_{k=0}^{\infty} c_k x^k\right) = \sum_{k=0}^{\infty}\left(\sum_{S(k)} a_h b_i c_j\right) x^k$$

where $S(k)$ is the set of all lattice points[1] $(h,i,j) \in \mathbf{N}^3$ in the first octant with $h + i + j = k$. This set $S(k)$ may be visualized as the triangle of all lattice points (with integral coordinates $h,i,j \in \mathbf{N}$) in the first octant on the plane $x + y + z = k$ in (x,y,z) space.

Finally, the distributive law holds

$$\begin{aligned}\sum_{k=0}^{\infty} a_k x^k \left(\sum_{k=0}^{\infty} b_k x^k + \sum_{k=0}^{\infty} c_k x^k\right) &= \sum_{k=0}^{\infty}\left[\sum_{i=0}^{k} a_i(b_{k-i} + c_{k-i})\right] x^k \\ &= \sum_{k=0}^{\infty}\left(\sum_{i=0}^{k} a_i b_{k-i} + \sum_{i=0}^{k} a_i c_{k-i}\right) x^k \\ &= \sum_{k=0}^{\infty}\left(\sum_{i=0}^{k} a_i b_{k-i}\right) x^k + \sum_{k=0}^{\infty}\left(\sum_{i=0}^{k} a_i c_{k-i}\right) x^k\end{aligned}$$

completing the proof of Theorem 3.

Invertibles. We next ask: what are the invertibles of the ring $R[[x]]$ of formal power series? Clearly, if $p(x) = \sum_{k=0}^{\infty} p_k x^k \in R[[x]]$ is invertible, then there exists a $q(x) = \sum_{k=0}^{\infty} q_k x^k \in R[[x]]$ such that $p(x)q(x) = 1$, that is, such that the equations

$$\begin{aligned} p_0 q_0 &= 1 \\ p_0 p_1 + p_1 q_0 &= 0 \\ p_0 q_2 + p_1 q_2 + p_2 q_0 &= 0 \\ p_0 q_n + p_1 q_{n-1} + \cdots + p_n q_0 &= 0 \end{aligned} \tag{14}$$

[1] In n-dimensional space $\mathbf{R}^n$ (the Cartesian product of n infinite lines), a lattice point is a point $\mathbf{x} = (x_1, \ldots, x_n)$ with integral coordinates (all $x_i \in \mathbf{Z}$).

are satisfied. From (14), we see that a necessary and sufficient condition that $p(x) = \sum_{k=0}^{\infty} p_k x^k$ be an invertible in $R[[x]]$ is that p_0 be an invertible in R. The q_i of (14) are determined by setting $q_0 = p_0^{-1}$ and computing recursively $q_n = -p_0^{-1} \sum_{i=1}^{n} p_i q_{n-i}$, for $n = 1,2,3, \ldots$. We have proved the following result.

Theorem 4. *A necessary and sufficient condition that* $\sum_{k=0}^{\infty} p_k x^k$ *be invertible in* $R[[x]]$ *is that* p_0 *be invertible in* R.

Corollary 1. *For any coefficient field* F, *the invertibles of* $F[[x]]$ *are the* $\sum_{k=0}^{\infty} p_k x^k$ *with* $p_0 \neq 0$.

Corollary 2. *In* $F[[x]]$, *every nonzero ideal* $H > \{0\}$ *is a principal ideal of the form* (x^n), *where* n *is the least integer for which some* $\sum_{k=n}^{\infty} p_k x^k$ *with* $p_n \neq 0$ *belongs to* H.

EXERCISES B

1. Write the nine sequences which satisfy the difference equation

$$s_i + s_{i-1} + 2s_{i-2} = 0 \qquad \text{over } GF(3)$$

2. Over $GF(2)$, $1 + x^2 + x^3$ is the polynomial associated with the following shift register:

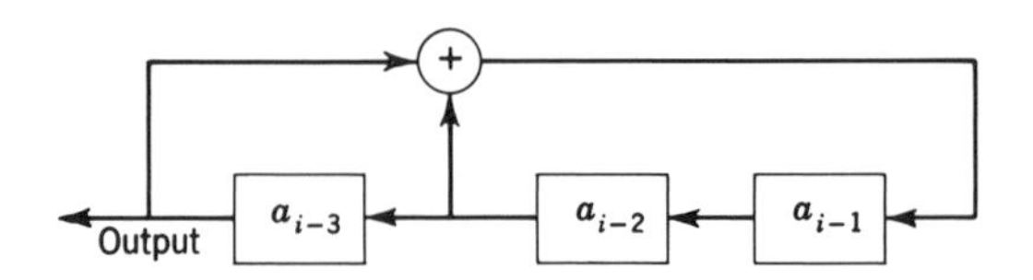

This also corresponds to a difference equation

$$c_0 s_i + c_1 s_{i-1} + c_2 s_{i-2} + c_3 s_{i-3} = 0$$

(a) What are the values of c_0, c_1, c_2, and c_3?
(b) s_i appears in the difference equation. Why is there no box for s_i in the shift register?

3. We initialize the shift register in Exercise 2 by placing binary values in s_{i-3}, s_{i-2}, and s_{i-1}. Suppose we let $s_{i-3} = 1$, $s_{i-2} = 0$, and $s_{i-1} = 0$ (write this state of the shift register as 100).
   (a) What will the output sequence be?
   (b) What is the period of this sequence?
   (c) List the consecutive states the shift register takes on until it reaches the initial state again?
4. Compare the output sequence in Exercise A3 with the output sequence obtained by considering the coefficients of $(1 + x^2)/(1 + x^2 + x^3)$.
5. Draw the shift register corresponding to $1 + x^3 + x^4$.
   (a) What are the consecutive states the shift register takes on (given initial state 0001) until it reads the initial state again?
   (b) What is the output sequence?
6. Show that the relation

$$(1 + x^{-3} + x^{-4})\mathbf{s} = \mathbf{0} \qquad 2(0,0, \ldots ,0)$$

   holds for all sequences generated by the shift register in Exercise 5, where $(I + x^{-3} + x^{-4})$ lists the three shift operators I, x^{-3}, and x^{-4}.
7. Design a *linear* shift register with feedback (using mod 2 or exclusive-or gates only) which generates the following sequence:

   1 1 1 1 0 0 1 1 1 1 0 0 1 1 1 1 0 0 · · ·

   Use as few flip-flops as you can. Also, write the polynomial corresponding to the shift register.
8. When dealing with shift register generators, we generally consider the initial state to be 0 · · · 01. Follow that convention in answering the following:
   (a) Consider the sequences generated by $x^2 + 1$ and by $x^3 + 1$. Add the two sequences. What is the resulting sequence?
   (b) What polynomial (the one of least degree) will generate this sequence [the sequence might have to be shifted to correspond exactly to the sum of the two sequences of part (a)]?
9. What is the inverse of $1 + x^2$ in $\mathbf{Z}_2[[x]]$?
10. What is the inverse of $1 + x^2$ in $\mathbf{Z}_3[[x]]$?
11. Which of the following have inverses in $\mathbf{Z}[x]$? Give the reasons for your answers.
   (a) $1 + x^2$
   (b) $-1 + x^3$
   (c) $2 + x^4$
   (d) $3 + 2x^3$
   (e) $4 + 6x^4$

13-5. APPLICATION TO DIFFERENCE CODES

Given any ΔE (1), with $c_0 = 1$, and $c_m \neq 0$, it follows from Theorem 4 that over any field F, the left and right characteristic polynomials

$$\begin{aligned} c(x) &= c_0 + c_1x + c_2x^2 + \cdots + c_mx^m + 0x^{m+1} + 0x^{m+2} + \cdots \\ c(x) &= c_m + c_{m-1}x + \cdots + c_0x^m + 0x^{m+1} + \cdots \end{aligned}$$

are invertible in $F[[x]]$. From this point we will consider only the *left* characteristic polynomial, calling it simply the characteristic polynomial or characteristic equation.

Theorem 5. *The sequence* **s** *with generating function*

$$s_0 + s_1x + s_2x^2 + \cdots = s(x)$$

where

$$s(x) = \frac{a(x)}{c(x)}$$

and $a(x)$ is of length (or degree) $l < m$, satisfies the ΔE (1), with $c(x)$ the characteristic equation for the ΔE (1).

Proof. First, since

$$c(x) \cdot s(x) = a(x) = a_0 + a_1x + \cdots + a_{m-1}x^{m-1} + 0x^m + \cdots \tag{15}$$

in $F[[x]]$, from the definition of multiplication in $F[[x]]$ the coefficient of x^n in $c(x) \cdot s(x)$ is

$$c_0s_n + c_1s_{n-1} + \cdots + c_ns_0 = 0 \qquad \text{for } n \geqq m \tag{16}$$

For $c_0 = 1$, this is identical to Eq. (1), so the **s** associated with the formal power series $s(x)$ satisfies the ΔE (1).

Corollary. *Each solution* $\mathbf{s} = s_0,s_1,s_2, \ldots$ *of the difference equation* (1) *has in* $R[[x]]$ *a generating function*

$$s(x) = s_0 + s_1x + \cdots + s_kx^k + \cdots = \frac{a(x)}{c(x)} \tag{16'}$$

where $c(x)$ is the characteristic polynomial for (1).

Example 5. Let F be $\mathbf{Z}_2$, and let the characteristic polynomial be $1 + x + x^2$. Then, for the initial block $\mathbf{a} = 10$ with generating function $a(x) = 1 + \sum_{k=0}^{\infty} 0 \cdot x^k$, the generating function

$$s(x) = \frac{1}{1 + x + x^2} = (1 + x + x^2)^{-1} \tag{17}$$

of the encoded sequence is easily computed in $\mathbf{Z}_2[[x]]$; thus

$$s(x) = 1 + x + x^3 + x^4 + x^6 + x^7 + \cdots$$

with *periodic* $\mathbf{s} = 110110110 \cdots$. The computation of

$$\frac{1}{(1 + x + x^2)} = \frac{1}{c(x)}$$

is easily accomplished by polynomial long division, remembering that + and − are the same in $\mathbf{Z}_2$; thus

$$\begin{array}{r|l}
 & 1 + x + x^3 + x^4 + \cdots \\
\hline
1 + x + x^2 & 1 + 0 \cdot x + 0 \cdot x^2 + 0 \cdot x^3 + \cdots \\
 & 1 + x \quad + x^2 \\
\cline{2-2}
 & \qquad x + x^2 \quad + 0 \cdot x^3 \\
 & \qquad x + x^2 \quad + x^3 \\
\cline{2-2}
 & \qquad\qquad x^3 + 0 \cdot x^4 + 0 \cdot x^5 \\
 & \qquad\qquad x^3 + x^4 \quad + x^5 \\
\cline{2-2}
 & \qquad\qquad\quad x^4 \quad + x^5 \quad + 0 \cdot x^6
\end{array}$$

Thus $(1 + x + x^2)^{-1} = 1 + x + x^3 + x^4 + x^6 + x^7 + \cdots$.

Theorem 5 has many corollaries, of which the following is especially noteworthy.

Corollary. *Given the difference equation* (1) *with characteristic polynomial $c(x)$ over a field F, the set $S(c(x))$ of all solutions of* (1) *is the same as the set of all sequences of coefficients of formal power series $s(x) = a(x)/c(x)$ for which deg $a <$ deg c.*

Example 6. The difference equation

$$s_i + s_{i-1} + s_{i-3} = 0 \qquad \text{over } \mathbf{Z}_2 \tag{18}$$

has $2^3 = 8$ different initial conditions s_0, s_1, s_2, each $s_i = a_i$, 0 or 1. Its characteristic polynomial is $1 + x + x^3$. We consider $c(x) = 1 + x + x^3$ as a formal power series in $\mathbf{Z}_2[[x]]$, writing

$$c(x) = 1 + x + 0 \cdot x^2 + x^3 + 0 \cdot x^4 + \cdots$$

Now, picking any $a(x)$ with degree less than 3, $1 + x$ for instance, we have

$$\frac{1 + x + 0 \cdot x^2 + 0 \cdot x^3 + \cdots}{1 + x + 0 \cdot x^2 + x^3 + 0 \cdot x^4 + \cdots}$$

There is no good reason to write out all of the 0 terms, and so the above can be written simply

$$\frac{1 + x}{1 + x + x^3}$$

We now compute the quotient in $\mathbf{Z}_2[[x]]$, by polynomial long division; thus

$$
\begin{array}{r|l}
 & 1 + x^3 + x^4 + x^5 + x^7 + x^{10} + \cdots \\
\hline
1 + x + x^3 & 1 + x \\
 & \underline{1 + x + x^3} \\
 & \qquad x^3 \\
 & \qquad \underline{x^3 + x^4 + x^6} \\
 & \qquad\quad x^4 + x^6 \\
 & \qquad\quad \underline{x^4 + x^5 + x^7} \\
 & \qquad\qquad x^5 + x^6 + x^7 \\
 & \qquad\qquad \underline{x^5 + x^6 + x^8} \\
 & \qquad\qquad\quad x^7 + x^8 \\
 & \qquad\qquad\quad \underline{x^7 + x^8 + x^{10}} \\
 & \qquad\qquad\qquad\qquad x^{10}
\end{array}
$$

This gives the sequence $10011101001 \cdots$. All eight code sequences can be computed similarly; they are:

$$
\begin{array}{l}
00000000000 \cdots \\
10011101001 \cdots \\
01001110100 \cdots \\
11010011101 \cdots \\
00111010011 \cdots \\
10100111010 \cdots \\
01110100111 \cdots \\
11101001110 \cdots
\end{array}
$$

Note that each of the seven nonzero code sequences is a translate of the one we have computed; each is also *periodic* of period 7. In the next section, we shall discover the reason for this.

13-6. RECURRENT SEQUENCES

It is of interest to consider recurrent sequences $\mathbf{s} = s_0 s_1 s_2 s_3 \cdots$ of numbers from a field F (or ring R) without fixing in advance any difference equation or difference equations which $\mathbf{s}$ is assumed to satisfy. This is especially true of sequences which are periodic or eventually periodic, in the sense defined in Sec. 13-2.

By definition, $\mathbf{s}$ is *periodic* if, for some positive integer n, we have $s_0 = s_n$, $s_1 = s_{n+1}$, . . . , $s_i = s_{n+i}$, . . . , all $i \in \mathbf{N}$. The least n for which this is true is called the (least) *period* of $\mathbf{s}$. We remark that the set of all n such that $s_i = s_{i+n}$ for all $i \in \mathbf{N}$ consists of the multiples of its least member: every period is a multiple of the least period.

This is true because the greatest common divisor of the set of all periods is itself a period; to prove this, we show that the Euclidean algorithm can be applied to any two periods n,n' to get $d = (n,n') = \gcd(n,n')$ as a period. By the Euclidean algorithm, $d = kn - ln'$ for suitably chosen positive integers k,l. Hence

$$s_i = s_{i+kn} = s_{i+kn-ln'} = s_{i+d} \qquad \text{for all } i \in \mathbf{N}$$

We now generalize this result.

To see the connection, observe that $s_i = s_{i+n}$ for all $i \in \mathbf{N}$ if and only if $\mathbf{s}$ satisfies the difference equation whose characteristic polynomial is $x^n - 1$. Again, $\gcd(x^n - 1, x^{n'} - 1) = x^d - 1$, if $d = \gcd(n,n')$; we omit the proof.

We now define a *recurrent sequence* in general (over a field F) as any sequence $\mathbf{s} = (s_0,s_1,s_2, \ldots)$ which satisfies one or more linear recurrence relations of the form (1); thus

$$s_i + \sum_{j=1}^{m} c_j s_{i-j} = 0 \qquad i = m, m+1, m+2, \ldots \tag{19}$$

In the notation of formal power series, we let $s(x)$ stand for the generating function of $\mathbf{s}$ and let $c(x)$ stand for the left characteristic polynomial of the ΔE (1),

$$c(x) = c_0 + c_1x + c_2x^2 + \cdots + c_{m-1}x^{m-1} + x^m$$

considered as an element

$$c(x) = c_0 + c_1x + \cdots + c_{m-1}x^{m-1} + x^m + 0 \cdot x^{m+1} + 0 \cdot x^{m+2} + \cdots$$

of $F[[x]]$.

Now, consider the set of *all* polynomials $\bar{p}(x) = \sum_{k=0}^{r} p_kx^k$ such that applying the *left* shift operator corresponding to (7) we have

$$(p_0I + p_1X^{-1} + p_2X^{-2} + \cdots + p_rX^{-r})\mathbf{s} = \mathbf{0} \tag{20}$$

The following result is easily verified.

Lemma. *The set of all polynomials* $\bar{p}(x)$ *for which* (20) *holds is, for any given recurrent sequence* $\mathbf{s}$, *an ideal* $J(\mathbf{s})$ *of the polynomial ring* $F[x]$.

Proof. If $\bar{p}(X^{-1})\mathbf{s} = \mathbf{0}$ and $\mathbf{q}(X^{-1})\mathbf{s} = \mathbf{0}$, then trivially

$$[\bar{p}(X^{-1}) \pm \bar{q}(X^{-1})]\mathbf{s} = \mathbf{0}$$

and for any $\bar{r}(x)$,

$$[\bar{r}(X^{-1})\bar{p}(X^{-1})]\mathbf{s} = \bar{r}(X^{-1})[\bar{p}(X^{-1})\mathbf{s}] = \bar{r}(X^{-1})\mathbf{0} = \mathbf{0}$$

(The assumption that **s** is recurrent is needed to make the set under consideration nonvoid.)

Since $F[x]$ is a Euclidean ring and hence a principal ideal ring (Chap. 11), it follows that for a given recurrent sequence **s** there is a unique (left) monic characteristic polynomial $\bar{c}_{\mathbf{s}}(x)$, such that

$$\bar{p}(X^{-1}) = 0 \qquad \text{if and only if } \bar{c}_{\mathbf{s}}(x) \mid \bar{p}(x) \tag{21}$$

Definition. The polynomial $\bar{c}_{\mathbf{s}}(x)$ for which (20) holds is called the *minimal* polynomial for the recurrent sequence **s**.

Since an irreducible polynomial has no proper factors, this proves in particular the following result.

Theorem 6. *If the characteristic polynomial $c(x)$ of the ΔE (1) is irreducible, then each nontrivial solution* **s** *of* (2) *satisfies precisely those linear ΔE's with constant coefficients whose characteristic polynomial is a multiple of $c(x)$.*

Translates. Having in mind difference codes over finite fields whose code sequences are all periodic, we now define the *left translate* of the periodic code sequence **s** through t units as the code sequence $\mathbf{s}^*$ with $s_i^* = s_{i+t}$. In the operator notation of (6′), $\mathbf{s}^* = X^{-t}\mathbf{s}$. Note that it is *not* true that $X^t\mathbf{s}^* = \mathbf{s}$ because $X^t\mathbf{s}^*$ is not periodic in general (only eventually periodic since $s_i = 0$ for $i < t$).

Example 7. Over $\mathbf{Z}_2$, the difference equation $S_{i+4} = S_i$ with characteristic polynomial $x^4 + 1 = x^4 - 1$ has two solutions $0000 \cdots$ and $1111 \cdots$ of period 1, two solutions $0101 \cdots$ and $1010 \cdots$ of period 2, and twelve solutions of period 4:

$$00010001 \cdots \qquad 00110011 \cdots \qquad 01110111 \cdots \tag{22}$$

and their translates.

Since $x^4 + 1 = (x + 1)^4$ over $\mathbf{Z}_2$, it is clear that the minimal polynomial of each of the above sequences must be $x + 1$, $x^2 + 1 = (x + 1)^2$, $x^3 + x^2 + x + 1 = (x + 1)^3$, or $x^4 + 1$. The minimal polynomial of $00110011 \cdots$ is $x^3 + x^2 + x + 1 = (x + 1)^3$; the minimal polynomial of the other two sequences of (22) and their translates is $x^4 + 1$.

It is easy to show that all the translates of a given periodic sequence have the same minimal polynomial. This is not true, however, for *non*-periodic but ultimately periodic sequences such as $0111111 \cdots$, whose minimal polynomial is $x^2 + x$, associated with the ΔE $s_{i+2} + s_{i+1} = 0$ (valid for all $i \in \mathbf{N}$).

13-7. THE PERIODS OF SEQUENCES ASSOCIATED WITH RELATIVELY PRIME POLYNOMIALS

The periods of sequences generated by difference equations are of particular interest. When the characteristic polynomial for a given sequence is primitive or when the irreducible factors of the characteristic polynomial are known, then much can be determined about the period of the sequences generated by the difference equation. This section will outline some of the properties of these sequences.[1]

Theorem 7. *Let $f(x)$ and $g(x)$ be two relatively prime nonzero polynomials over a field F, and let $h(x) = f(x)g(x)$ be their product. Then $S(h(x)) = S(f(x)) + S(g(x))$, where $S(f(x)) + S(g(x))$ is the set of all sequences of the form $\mathbf{s} + \mathbf{s}^*$ with $\mathbf{s} \in S(f(x))$ and $\mathbf{s}^* \in S(g(x))$.*

Remark. Since $f(x)$ and $g(x)$ are relatively prime, their product is also their least common multiple.

Example 8. The two polynomials $1 + x + x^2$ and $1 + x + x^3$ are relatively prime. Their product is $1 + x^4 + x^5$. The set $S(1 + x + x^2)$ contains the all-zero sequence plus three sequences of period 3, and the set $S(1 + x + x^3)$ contains the all-zero sequence and seven sequences of period 7. The intersection $S(1 + x + x^2) \cap S(1 + x + x^3)$ contains only the all-zero sequence. The set $S(1 + x^4 + x^5)$ contains $S(1 + x + x^2)$ and $S(1 + x + x^3)$ and also 21 sequences of period 21, each of which may be obtained by adding an element of $S(1 + x + x^2)$ to an element of $S(1 + x + x^3)$. For example,

$$
\begin{array}{rl}
011011011011011011011011011 \cdots & \in S(1 + x + x^2) \\
+\underline{111010011101001110100111010} \cdots & \in S(1 + x + x^3) \\
100001000110010101111100001 \cdots & \in S(1 + x^4 + x^5)
\end{array}
$$

The sequences of period 21 in $S(1 + x^4 + x^5)$ are the 21 different translates of the sequence given above.

Proof. To prove Theorem 7, we first show that $S(f(x)) \cap S(g(x)) = 0$ because if $\mathbf{s}$ is in both $S(f(x))$ and $S(g(x))$, then the minimal polynomial $c_s(x)$ must be a common divisor of $f(x)$ and $g(x)$; hence $c_s(x) = 1$, and $\mathbf{s}$ has $s_i = 0$. Hence $S(h(x))$, which contains both $S(f(x))$ and $S(g(x))$ and therefore all sums $\mathbf{s} + \mathbf{s}^*$ with $s \in S(f(x))$ and $s^* \in S(g(x))$, must contain the *direct* sum of $S(f(x)) + S(g(x))$, whose dimension as a

[1] An excellent reference on this subject is N. Zierler, "Linear Recurring Sequences," *J. Soc. Industrial Applied Math.*, vol. 7, no. 1, pp. 31–48, March, 1959.

vector space over F is $\deg f + \deg g = \deg h$. Consequently the set of such sums $s + s^*$ exhausts $S(h(x))$, completing the proof.

Corollary. *If* $f_1(x), \ldots, f_r(x)$ *are relatively prime in* $F[x]$, *then* $S(f_1 \cdots f_r) = S(f_1) + \cdots + S(f_r)$.

Period and Order. We now recall that the *period* $p(\mathbf{s})$ of a periodic sequence $\mathbf{s}$ is the least positive integer r such that $s_i = s_{i+r}$ for all $i \in \mathbf{N}$. We define the *order* of a polynomial $f(x) = \sum_{k=0}^{m} f_k x^k$ with $f_0 f_m \neq 0$ to be the least $e \in \mathbf{P}$ such that $f(x) \mid (x^e - 1)$. We then write $\text{ord } f(x) = e$.

Theorem 8. *Let* $\mathbf{s} \in S(f(x))$ *and* $\mathbf{s}^* \in S(g(x))$, *and let* $f(x)$ *and* $g(x)$ *be relatively prime. Then the period* $p(\mathbf{s} + \mathbf{s}^*)$ *is equal to the least common multiple of* $p(\mathbf{s})$ *and* $p(\mathbf{s}^*)$.

Proof. Certainly the least common multiple r of the periods of the sequences $\mathbf{s}$ and $\mathbf{s}^*$ is divisible by $t = p(\mathbf{s} + \mathbf{s}^*)$. By definition of $p(\mathbf{s} + \mathbf{s}^*) = t$, $s_0 + \bar{s}_0^* = s_t + s_t^*$, and $s_1 + s_1^* = s_{t+1}$ so the sequence $s_i - s_{i+t}^* = s_{i+t} - s_1^*$ belongs to both $S(f(x))$ and $S(g(x))$; however, gcd $(f(x), g(x))$ is 1, so this sequence must be in $S(f(x)) \cap Sg(x) = \mathbf{0}$, and $p(\mathbf{s})$ and $p(\mathbf{s}^*)$ both divide t. Therefore $s_{i+t} = s_i$ and $s_{i+t}^* = s_i^*$ whence $s_{i+t} + s_{i+t}^* = s_i + s_i^*$.

Lemma 1. *For any* $\mathbf{s} \in S(f(x))$, $p(\mathbf{s})$ *divides ord* $f(x)$.

Proof. Since $f(x)$ divides x ord $f(x) - 1$, $S(f(x)) \subset S(x \text{ ord } f - 1)$, and ord $f(x)$ is divisible by each $p(\mathbf{s})$, $\mathbf{s} \in S(f(x))$.

Lemma 2. *If* $f(x)$ *is the minimal polynomial for a sequence* $\mathbf{s}$, *then*

$$\textit{ord } f(x) = p(\mathbf{s})$$

Since $\mathbf{s} \in S(x^{p(\mathbf{s})} - 1)$ and $f(x)$ divides $x^{p(\mathbf{s})} - 1$, ord $f(x) \leq p(\mathbf{s})$. But Lemma 1 asserts that $p(\mathbf{s})$ divides ord $f(x)$, so ord $f(x) = p(\mathbf{s})$.

Corollary. *If* $f(x)$ *is irreducible, then every nonzero* $\mathbf{s} \in S(f(x))$ *has period* $p(\mathbf{s}) = $ *ord* $(f(x))$.

If $\mathbf{s} \in S(f(x))$, with $f(x)$ irreducible, then $\mathbf{s} \in S(g(x))$ implies either $g(x) = f(x)h(x)$ or $f(x) = g(x)h(x)$; but $f(x) = g(x)h(x)$ is impossible if $f(x)$ is irreducible and deg g, deg $f > 0$. Hence an irreducible polynomial $f(x)$

is minimal for all nonzero sequences in $S(f(x))$. Consequently, by Lemma 2, $p(\mathbf{s}) = \operatorname{ord} f(x)$.

13-8. MAXIMAL PERIOD SEQUENCES

When building a shift register generator for coding, we would usually like to get the largest possible period from the number of delay or memory elements we use. Hence our optimization problem consists in finding the largest period which can be generated by a polynomial $f(x)$ of degree n.

Theorem 9. *Let $f(x)$ divide $x^{q^n-1} - 1$, $f(x)$ over $GF(q)$ and $d(f(x)) = n$, and let $f(x)$ divide no polynomial of the form $x^t - 1$ with $t < q^n - 1$. Then $p(\mathbf{s}) = q^n - 1$ for all nonzero $\mathbf{s}$ in $S(f(x))$.*

Proof. If $f(x)$ divides x^{q^n-1}, then $p(\mathbf{s}) \mid q^n - 1$ for all $\mathbf{s}$ in $S(f(x))$. Now suppose $p(\mathbf{s}) \mid t$ for some $t < q^n - 1$. Then $s \in S(x^t - 1)$ so $f(x) \mid x^t - 1$ contrary to supposition. Thus $p(\mathbf{s}) = q^n - 1$ for all nonzero $\mathbf{s}$. A sequence of this type is called a *sequence of maximal period.*

Example 9. The polynomial $x^4 + x^3 + 1$ over $\mathbf{Z}_2$ divides $x^{15} + 1$ and no polynomial of the form $x^k + 1$ where $k < 15$. The associated difference equation is $s_i + s_{i-3} + s_{i-4} = 0$. If we let $s_0 = 1$, $s_1 = 0$, $s_2 = 0$, and $s_3 = 0$, the sequence generated will be 1000100110101111 0001 $\cdots$. This sequence has period 15 as predicted. All 15 translates of this sequence can be generated by the difference equation, and these 15 sequences plus the sequence $\mathbf{0}$ comprise $S(x^4 + x^3 + 1)$.

Consider again, the function $\alpha: F^n \to F^n$ associated with our $f(x)$, where $(s_0, \ldots, s_{n-1}) = [s_1, \ldots, s_{n-1}, \ -f_0^{-1}(f_1 s_{n-1} + \cdots + f_n s_0)]$. As was shown, for $f_0 f_n \neq 0$, α is a permutation on $q^n - 1$ elements, and the greatest possible t for $\alpha^t = 1$ and $\alpha^w \neq 1$ for all $w < t$ would be $t = q^n - 1$; that is, the period $q^n - 1$ is maximal for the sequences associated with a polynomial of degree n over a field of q elements. Another way of looking at this is the following. Our function α must take each n-tuple of elements of q values into another unique n-tuple, and if $\alpha(\mathbf{b}) \neq \alpha^m(\mathbf{b})$ for $m < q^n - 1$, $\mathbf{b}$ an n-tuple, then all nonzero n-tuples are exhausted in the sequence $\mathbf{b}, \alpha(\mathbf{b}), \ldots, \alpha^{q^n-1}(\mathbf{b})$. Since $\alpha(\mathbf{0}) = \mathbf{0}$, this is optimal.

The above also shows that if we examine successive n-tuples in a period of a sequence $\mathbf{s} \in S(f(x))$ with $p(f(x)) = q^n - 1$, then each possible nonzero n-tuple will occur exactly once. Also, each nonzero $(n - 1)$-tuple will occur twice, and the $(n - 1)$-tuple $(0, \ldots, 0)$ will occur only once, etc. The implications of this have caused linear sequences of maximal period to be widely used to generate pseudo-random number sequences.

Examination of Example 6 shows that all seven nonzero binary 3-tuples occur cyclically in the nonzero sequences generated by $1 + x + x^3$. Likewise, in Example 9, all 15 nonzero binary 4-tuples appear cyclically.

Since each set of sequences $S(f(x))$ associated with a polynomial $f(x)$ and its difference equation is closed under translation, addition, and scalar multiplication, if we add an $\mathbf{s}$ and $\mathbf{s}^*$ in $S(f(x))$, the sum is a sequence also in $S(f(x))$.

EXERCISES C

1. (a) Design a shift register that corresponds to a polynomial of degree 3, such that output sequences of different periods are generated given different initial states (000 is not considered as an acceptable initial state).
   (b) What polynomial are you using? Exhibit the initial states, the sequences of states, the output sequences, and the shift register.
   (c) Can the above be done for a primitive polynomial? Why or why not?
2. (a) The polynomial $1 \oplus X \oplus X^4$ is primitive over $\mathbf{Z}_2$. Using this polynomial, draw the block diagram for a shift register which will generate an output sequence with period 15.
   (b) Write a representation for $GF(2^4)$ using the internal states of the shift register and also identifying the elements of $GF(2^4)$ as $0,1,\gamma,\gamma^2, \ldots ,\gamma^{14}$ along with the corresponding internal states $0000,1000,0100, \ldots$. Thus $0000 = 0$, $1000 = 1$, $0100 = \gamma, \ldots$.
3. (a) Display a matrix M_1 over $\mathbf{Z}_2$ which will satisfy the relation $\gamma^i M_1 = \gamma^{i+4}$ (mod 15) for $0 \leq i \leq 15$, considering the internal states γ^i generated above as row vectors.
   (b) Draw the block diagram for a circuit which will realize this matrix, using only "exclusive or" gates.
4. (a) Form a matrix M_2 which will satisfy $\gamma^i M_2 = \gamma^{2i}$ (mod 15) for $0 \leq i \leq 15$.
   (b) Form a matrix M_3 which will satisfy $(\gamma^i M_3)^2$ (mod 15) $= \gamma^i$ for $0 \leq i \leq 15$, that is, a matrix which will take the "square root" of an element γ^i in the representation of $GF(16)$.

★13-9. THE AUTOCORRELATION FUNCTION

When a radar or communication system is designed, certain techniques are used to analyze and optimize the system's performance. In order to detect a reflected signal in a radar system, or to "lock on" and track transmitted signals in a communications system, the type of signal transmitted is analyzed using a mathematical function called an *autocorrelation function*.

Definition. The *autocorrelation function*[1] of a real periodic function (signal) $f(t)$ of period T is defined as follows:

$$A(\tau) = \frac{1}{T}\int_0^T f(t)f(t+\tau)\,dt \tag{23}$$

Figure 13-5*a* shows a periodically pulsed system for a radar where $f(t)$ consists of a single pulse occurring at time mT with $m = 0,1, \ldots, k, \ldots$ and T the period of the function $f(t)$ (see page 390).

Figure 13-5*b* shows the autocorrelation function for $f(t)$.

The type of periodic functions $f(t)$ which have been discussed in this chapter are finite-valued, i.e., the value $f(t)$ at any time t is one of a finite set of values. Also, time is assumed to be discrete instead of continuous, so we replace $f(t)$ with $s(i)$, $i = 0,1, \ldots, m, \ldots$, and replace the autocorrelation function $A(\tau)$ in (23) with the discrete auto-

[1] More generally, the limit

$$\Gamma(\tau) = \lim_{n\to\infty}\frac{1}{nT}\int_0^{nT} f(t)f(t+\tau)\,d\tau \tag{24}$$

when it exists, is called the *covariance* of a real function $f(t)$ defined for positive t, and $\Gamma(\tau)/\Gamma(Q)$ is called its autocorrelation function. The consistency of this more general definition with (23), for periodic functions of period T, is easily verified.

Because of noise, no actual returned signal is perfect. Hence an actual signal received by a radar must be regarded as the sum of the intended periodic signal $P(t)$ and a superimposed random noise $g(t)$; thus

$$f(t) = P(t) + g(t) \qquad P(t+T) = P(t) \tag{25}$$

Substituting from (25) into (24), we see that the measured covariance over n periods is the sum of the following three terms; thus

$$\Gamma_n(\tau) = \int_0^T P(t)P(t+\tau)\,d\tau + \frac{1}{nT}\int_0^{nT}[P(t)g(t+\tau) + P(t+\tau)g(t)]\,d\tau + \frac{1}{nT}\int_0^{nT} g(t)g(t+\tau)\,d\tau \tag{26}$$

In general, its random nature tends to make noise uncorrelated over successive periods. Hence we have

$$\Gamma_n(\tau) = \int_0^T P(t)P(t+\tau)\,d\tau + O\left(\frac{1}{\sqrt{n}}\right) \tag{26'}$$

where we have invoked the deep Central Limit Theorem of probability to conclude that the uncorrelated integrals in (26) grow like $0(1/\sqrt{n})$. To detect a sequence of radar signals with random noise, therefore, it suffices to make the duration nT of the periodic signal long enough so that $\Gamma_n(\tau)$ dominates the $O(1/n)$ contribution from the noise.

correlation function

$$A(j) = \frac{1}{r} \sum_{i=0}^{r-1} s(i) \cdot s(i + j)$$

where the $s(i)$ are the values in a recurrent sequence and r is the period of the sequences.

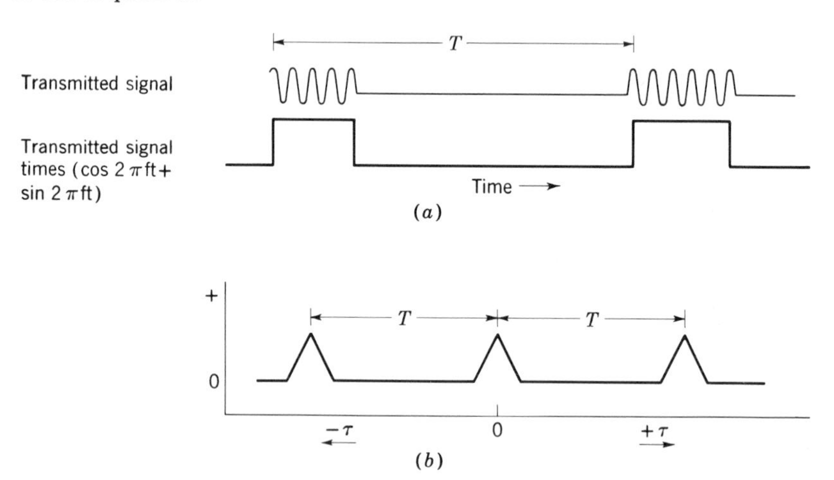

Fig. 13-5. Autocorrelation of transmitted pulse. (*a*) Detection process; (*b*) autocorrelation function for periodic pulse with period *I*.

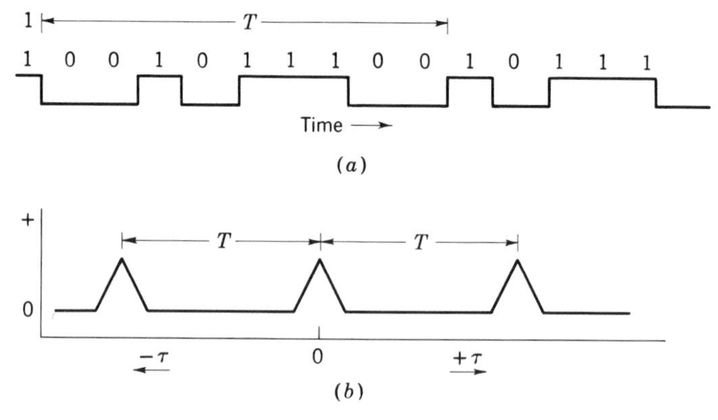

Fig. 13-6. (*a*) Basic radar system with sequences; (*b*) autocorrelation for sequence of binary signals.

A simple example of a recurrent or periodic sequence with period r consists of the sequence with $s(1),s(2), \ldots ,s(j) = 0$ where $j \neq mr$, and with $s(0) = s(r - 1) = s(mr - 1) = 1$ for all $m = 1,2, \ldots ,k$. This sequence has value $s(i) = 0$ at all points, except for one interval at the beginning of each period, and is commonly called a *periodically pulsed system*. This, of course, corresponds to the system in Fig. 13-5. Notice that the autocorrelation function has value $1/r$ for $A(j)$ when $j = 0$, $r - 1$, $2r - 1$, . . . , and 0 for all other j.

A radar system which is periodically pulsed operates as follows. A pulse is transmitted, the pulse is then reflected by the target (object being tracked), the time interval from transmission of pulse to reception of the reflected pulse is measured, and the range to the target is calculated from this. Another pulse may than be transmitted and received, and this process can be continued. Notice that a second pulse must not be sent until the previous reflected pulse has been received since this pulse would cause an ambiguity as to which pulse was received, and a range ambiguity would result. For the periodically pulsed sequence $f(t)$, this means the period T must be greater than the transit time from radar to target and back. For a discrete system with period r, if time passes in increments of λ seconds, then $r\lambda$ must be greater than the transit time for electromagnetic energy to the target and back.

In terms of sequences, this means that the period r for a sequence s must be long if the ranges to be measured are great.

Now consider a radar system which transmits a periodically pulsed signal and receives a perfect duplicate[1] of the transmitted signal. Call the transmitted signal (sequence) s and the returned signal s^*; then s^* is simply s delayed; that is, $s(j) = s^*(j - d)$, $j = 0,1,2,$ . . . , with d being the amount of delay. If we correlate s with s^*, then the value of j for which $A(j) = 1/r$ will equal $d, 2d, 3d,$ . . . , which will then determine the range to the target.

Now let us consider that in a realistic case the returned signal will not only be attenuated by varying amounts, but that noise will be added to the signal. The receiver must then "guess" the value of the delayed attenuated transmitted signal plus noise during each time interval j. Call the delayed transmitted signal $\mathbf{s}^*$ as before, and call the sequence which is guessed $\mathbf{s}$. Then $\mathbf{s}^*$ will disagree with the delayed $\mathbf{s}$ when the receiver guesses wrong; that is, $s^*(j) \neq s(j)$ when the receiver errs. A new sequence can then be constructed called the *error sequence* with

$$e(j) = s^*(j) + s(j)$$

For the simple periodically pulsed sequence, if $e(j)$ contains many 1's, or errors, then many periods must be examined before erroneous ranges are eliminated in the presence of random noise.

[1] This would be the case if we consider an amplified returned signal in a channel with no noise and fixed attenuation.

It is natural to ask whether a better class of sequences can be found to eliminate this problem. Such sequences would have to have long period and a good autocorrelation function. The next section deals with the use of recurrent sequences of maximal period for that purpose.

★13-10. AUTOCORRELATION THEOREM

We shall now find that recurrent (periodic) sequences with maximal period have an autocorrelation function with the same shape as single-pulse systems. In this case, however, the returned signal energy can be integrated throughout the entire period of the sequence, which increases the radar's ability to detect and track targets.

Note that sequences of maximal period are produced by difference codes whose characteristic polynomials are *primitive polynomials* (of degree n and period $p^n - 1$) over $\mathbf{Z}_p$.†

Let us now restrict our discussion to binary-valued sequences of maximal period, i.e., sequences with values in $\mathbf{Z}_2$. Consider a polynomial $f(x)$ of degree n and the associated difference equation. Let $f(x)$ divide $x^{q-1} - 1$ ($q = 2^n$) but not $x^t - 1$ for any $t < 2^n - 1$. There are $2^n - 1$ possible initial nonzero n-tuples of $\mathbf{s}_0, \ldots, \mathbf{s}_{n-1}$ for our difference equation, or, conversely, $2^n - 1$ nonzero polynomials of degree less than n. Since $\deg f = p(s) = 2^n - 1$, each nonzero $\mathbf{s}^* \in S(f(x))$ is a left translate of every other nonzero $\mathbf{s} \in S(f(x))$. Further, since $S(f(x))$ is closed under addition, if we add $\mathbf{s}$ and $\mathbf{s}^*$, we obtain another translate of $\mathbf{s}$ (and $\mathbf{s}^*$, for that matter) or $\mathbf{0}$ in case $\mathbf{s} = \mathbf{s}^*$.

Each $\mathbf{s} \in S(f(x))$ contains exactly $2^{n-1} - 1$ zeros and 2^{n-1} 1's, in a given period (segment of length $2^n - 1$).

This leads to the following *autocorrelation theorem.*

Theorem 10. *Let $S(f(x))$ be a set of linear sequences of maximal period, $d(f(x)) = n$. Define the discrete autocorrelation function*

$$A(r) = \sum_{i=0}^{2^n-1} B(s_i + s_{i+r}) \qquad \text{where } B(0) = +1,\ B(1) = -1 \tag{27}$$

and where $+1$ and -1 are to be considered as real numbers. Then

$$A(r) = \begin{cases} 2^{n-1} & \text{if } (2^n - 1) \mid r \\ -1 & \text{otherwise} \end{cases}$$

† For a further treatment, see D. A. Huffman, "The Synthesis of Linear Sequential Coding Networks," Proc. Third Symposium on Information Theory, pp. 77–95, Academic Press, New York, 1956.

Notice that B is a morphism from $[\mathbf{Z},+]$ considered as an additive group into the multiplicative group consisting of the values $\{-1,+1\}$, for $B(a + b) = B(a) \times B(b)$, with binary addition on the left and real-number multiplication on the right. Here 0 is the additive 0 in our $\{0,1\}$ system, and $+1$ is the multiplicative identity in the group $\{-1,+1\}$.

Now, if we add a sequence $\{s_i\}_{i=0}^{2^n-1}$ element by element to a translate $\{s_{i+r}\}_{i=0}^{2^n-1}$, then we obtain a sequence $\{s_i + s_{i+r}\}_{i=0}^{2^n-1}$ which is simply another translate of s_i (or all 0). If a translate $\{s_i + s_{i+r}\}_{i=0}^{2^n-1}$ contains $(2^n/2) - 1$ zeros and $2^n/2$ 1's, our sum $\sum_{i=0}^{1} B(s_i + s_{i+r})$ becomes $+[(2^n/2) - 1] - (2^n/2) = -1$. For all zeros we get $(2^n - 1)$ zeros and a value $+(2^n - 1)$ for $\sum_{i=0}^{2^n-1} B(s_i + s_{i+r})$.

Figure 13-7 is a graph of the autocorrelation function for *a* sequence of maximal period.

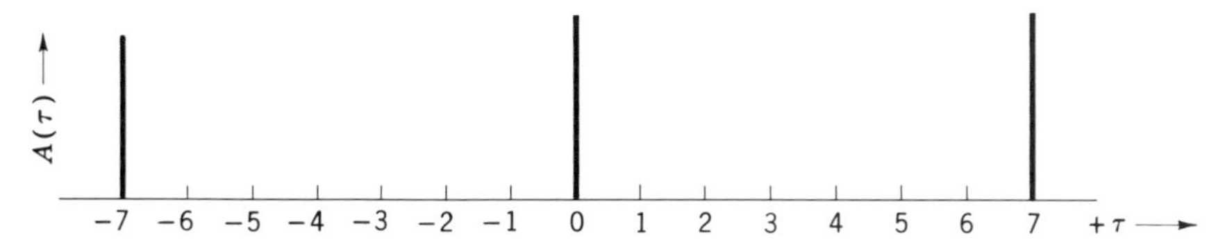

Fig. 13-7. Autocorrelation function for sequences of maximal period.

Consider again the polynomial $x^3 + x + 1$ of Example 6, which generates 101001110100111 · · · and all seven translates of this sequence. Each seven consecutive digits in the sequence contain four 1's and three 0's. If we add digit-by-digit one of these translates to another, we have, for instance,

$$\begin{array}{r} 101001110100111\ \cdots \\ +111010011101001\ \cdots \\ \hline 010011101001110\ \cdots \end{array}$$

which is simply another translate, and thus each seven consecutive digits contain four 1's and three 0's. The function B, when evaluated for four 1's and three 0's, has value -1. Thus each value of the function

$$A(r) = \sum_{i=m}^{m+7} B(s_i + s_{i+r})$$

will be -1 unless $r = 0,7,14$, in which case the sum $s_i + s_{i+r}$ is 0 for i and the function $A(r)$ has value 7.

The above can be generalized. For instance, given a primitive polynomial of degree 20 over $\mathbf{Z}_2$ such as $x^{20} + x^3 + 1$, a sequence of period $2^{20} - 1 \simeq 1{,}000{,}000$ can be generated with an autocorrelation function having values of -1 for all shift or sequence translates except for $r = 0$ and multiples of $2^{20} - 1$ at which values of $2^{20} - 1$ are obtained.

This is about right for radar observations of the moon. For polynomials of degree 30, periods of $2^{30} - 1$, or about 10^9, are obtained. These are adequate for observations of Venus. Satellite communication systems can be obtained which generate sequences transmitted at intervals of 10^{-6} sec which repeat only once a year, using polynomials of degree 50. Notice it would be impossible to examine manually these sequences to determine these characteristics; they can be used only because of the considerable knowledge obtained through the algebraic properties of the polynomials used.

★13-11. EXTENDED FORMAL POWER SERIES

Even in the ring $F[[x]]$ of all formal power series over a field F, only the elements with $p_0 \neq 0$ are invertible; $F[[x]]$ is never a field. We now extend $F[[x]]$ to a *field* $F((x))$, the field of *extended formal power series* over F. Here, by an extended formal power series is meant a series

$$p((x)) = \sum_{k=-\infty}^{\infty} p_k x^k \qquad p_k \in F \tag{28}$$

where all but a finite number of p_k with $k < 0$ vanish. When all $p_k = 0$, we write $p((x)) = 0$; for any other extended formal power series, there will be a *least* $k \in \mathbf{Z}$ such that $p_k \neq 0$; this least k is called the *order* of $p((x))$.

To make $F((x))$ into a field, we define addition and multiplication of extended formal power series by

$$\sum_{k=-\infty}^{\infty} f_k x^k + \sum_{l=-\infty}^{\infty} q_l x^l = \sum_{j=-\infty}^{\infty} (f_j + g_j) x^j \tag{29}$$

$$\sum_{k=-\infty}^{\infty} f_k x^k \cdot \sum_{l=-\infty}^{\infty} g_l x^l = \sum_{j=-\infty}^{\infty} h_j x^j \tag{30}$$

where $h_j = \sum f_i g_{j-i}$ is summed over all i with $f_i g_{j-i} \neq 0$.

Theorem 11. *The set of all extended formal power series over a field F forms a field.*

Proof. We use (29) and (30) for addition and multiplication, so no zero divisors are present, commutativity and associativity are assured, and only inverses are required. Given an extended power series $\sum_{k=-\infty}^{\infty} p_k x^k$, $p_k \in F$, of order s, we write

$$q(x) = \sum_{j=-\infty}^{\infty} q_j x^j = x^{-s} \left(\sum_{k=-\infty}^{\infty} p_k x^k \right)$$

Then $q_0 \neq 0$, all q_l with $l < 0$ are zero, and an inverse of $q(x)$ can be calculated using polynomial long division. Now, call this inverse $\sum_{j=-\infty}^{\infty} \tilde{q}_j x^j$. We have

$$x^{-s} \left(\sum_{k=-\infty}^{\infty} p_k x^k \right) \left(\sum_{j=-\infty}^{\infty} \tilde{q}_j x^j \right) = 1$$

Therefore $x^{-s} \left(\sum_{j=-\infty}^{\infty} \tilde{q}_j x^j \right)$ is the desired inverse of $\sum_{k=-\infty}^{\infty} p_k x^k$.

EXERCISES D

1. Draw a graph of the autocorrelation function for a signal which is 0 for 9 sec, then +10 for 1 sec, then 0 for 9 sec, and +10 for 1 sec, etc.
2. Draw a graph of the autocorrelation function for a shift register with characteristic polynomial $1 + x + x^4$.
3. Draw a graph of the autocorrelation function for a shift register with characteristic polynomial $1 + x + x^2 + x^3 + x^4$. Start the register in 1010. Do different starting values generate different autocorrelation functions? Why?
4. The roots of the following polynomials have multiplicative orders (periods) 3, 4, and 5, respectively:

$$x^2 + x + 1 \qquad x^3 + x^2 + x + 1 \qquad x^4 + x^3 + x^2 + x + 1$$

   Design a shift register with feedback which will have at least one (minimal) period of 30 specifying the polynomial and difference equation associated with it. Use the polynomial of least degree you can find.
5. Prove that taking the square root of an element of a finite field of characteristic 2 [that is, any $GF(2^n)$] is a linear transformation (a transformation L is said to be *linear* when $L(\lambda a + \mu b) = \lambda L(a) + \mu L(b)$ for all a and b in the set).
6. The polynomial $1 + x^3 + x^5$ is primitive over $\mathbf{Z}_2$. The corresponding shift register is

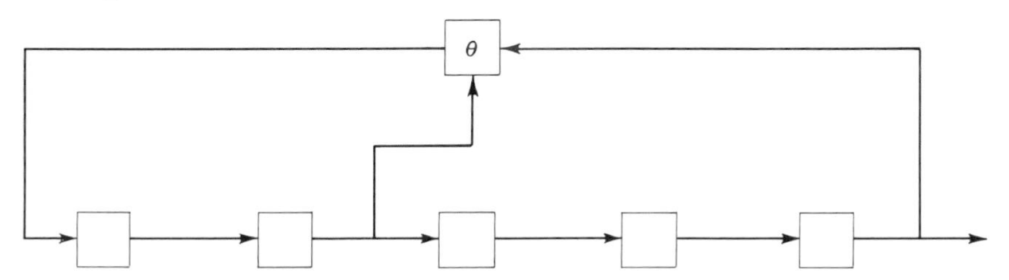

If we call 10000 the element 1 in $GF(2^5)$, 01000 the element α, 00100 the element α^2, etc., the succeeding states of the above register will generate a representation for the multiplicative group of $GF(2^5)$. Using this representation, design a combinational network using only exclusive-OR gates which will have as output the square root of its input when both input and output are in $GF(2^5)$; that is, with input α^i the circuit should have as output $\alpha^j = \alpha^{2i}$ (mod 31).

7. (a) The polynomial $p(x) = 1 + x^2 + x^5$ is irreducible over $\mathbf{Z}_2$. Does the feedback shift register corresponding to $p(x)$ generate maximal period sequences? Explain.
   (b) Draw a feedback shift register that computes $(1 + x^2 + x^5)^{-1}$ in $\mathbf{Z}_2[[x]]$, the ring of formal power series over $\mathbf{Z}_2$. Indicate the initial conditions of your shift register.
8. Show that the polynomial $q(x) = x^4 + x^3 + x^2 + x + 1$ is irreducible over $GF(2) = Z_2$, but that it can be factored as $(x - \alpha_1)(x - \alpha_2)(x - \alpha_3)(x - \alpha_4)$ over $GF(2^4)$ in which the α_i are not primitive.
9. (a) Draw the shift register with feedback which corresponds to $q(x)$.
   (b) The sequence of coefficients of the polynomial $(1 + x)/q(x)$ is among the sequences generated by the shift register of (*a*). In what state would you start the shift register to generate this particular sequence?
   (c) What are the periods of the sequences which can be generated by this shift register?
   (d) For what minimum value of r does $q(x) \mid (x^r + 1)$? Justify. [*Hint:* Use (*c*).]

★10. Show that $F((x)) = Q(F[[x]])$, for any field.

REFERENCES

GILL, A.: "Linear Sequential Circuits," McGraw-Hill, 1966.
GOLOMB, S. W.: "Shift Register Sequences," Holden-Day, 1967.
GOLOMB, S. W. (ed.): "Digital Communications," Prentice-Hall, 1964.
HALL, M., JR.: "Combinatorial Theory," Blaisdell, 1967.
MANN, HENRY (ed.): "Combinatorial Structures in Planetary Reconnaissance," Wiley, 1968.

Computability

14-1. CARDINAL ARITHMETIC

As was stated in Sec. 2-6, one can develop an arithmetic for finite and infinite cardinal numbers very simply from the properties of sets and functions established in Chaps. 1 and 2. As stated there, let Γ be the class of all sets, and let E be the relation between sets defined as follows.

$$S \; E \; T \text{ means that there is a bijection } b\colon S \leftrightarrow T \tag{1}$$

This relation E is clearly reflexive because for each S the bijection 1_S maps S onto itself. It is symmetric because b^{-1} is a bijection from T to S if (1) holds. It is transitive because if $b\colon S \leftrightarrow T$ and $c\colon T \leftrightarrow U$ are bijections, then the left composite $c \circ b\colon S \leftrightarrow U$ is a bijection. Therefore, E is an equivalence relation on Γ.

Since E is an equivalence relation on Γ, we can form the quotient class Γ/E, whose elements are the equivalence classes of Γ under E. The elements of Γ/E form, by definition, the class of all *cardinal numbers.*

Furthermore, E has the Substitution Property, discussed in Chap. 2, for various binary operations on sets.

Theorem 1. *The set-theoretic operations of disjoint sum, Cartesian product, and set power have the Substitution Property for the equivalence relation* (1).

Proof. Let $S \mathrel{E} S^*$ and $T \mathrel{E} T^*$, so that there exist bijections $b: S \leftrightarrow S^*$ and $c: T \leftrightarrow T^*$. Then the composite bijections defined in Sec. 1-6,

$$b \sqcup c: S \sqcup T \leftrightarrow S^* \sqcup T^* \tag{2}$$

$$b \times c: S \times T \leftrightarrow S^* \times T^* \tag{3}$$

$$c^b: \quad T^S \leftrightarrow T^{*S^*} \tag{4}$$

imply the Substitution Property, by the latter's very definition.

These Substitution Properties make it possible to define the binary operations of addition, multiplication, and exponentiation of cardinal numbers in terms of sets and functions, and to derive from the general properties of sets and functions the familiar identities of arithmetic,

$$x + y = y + x \qquad xy = yx \qquad \text{(Commutative)} \tag{5}$$

$$x + (y + z) = (x + y) + z \qquad x(yz) = (xy)z \qquad \text{(Associative)} \tag{6}$$

$$x(y + z) = xy + xz \qquad (x + y)z = xz + yz \qquad \text{(Distributive)} \tag{7}$$

and the laws of exponents,

$$x^{y+z} = x^y x^z \qquad (x^y)^z = x^{yz} \qquad (xy)^z = x^z y^z \tag{8}$$

without going through the tedious verifications by induction sketched in Secs. 1-8 and 1-9.

Theorem 2. *The set-theoretic operations* $X \sqcup Y$, $X \times Y$, *and* Y^X *define single-valued binary operations* $x + y$, xy, *and* y^x *on the class* Γ/E *of all cardinal numbers; moreover, these operations satisfy the nine identities of formulas* (5) *to* (8) *above.*

Sketch of proof. The first statement is a corollary of Theorem 7 of Chap. 2 and the discussion leading to formula (23) there. The truth of the second statement follows from the existence of various obvious bijections, most of which were already mentioned in Chap. 1 (see Exercises A1 to A3 below).

It is also easy to demonstrate, by simple constructions, the existence of *commutative diagrams* such as that of Fig. 14-1 relating some of the bijections used in the preceding proofs. More interesting is the fact that the preceding definitions apply to *infinite sets,* as well as to finite sets.

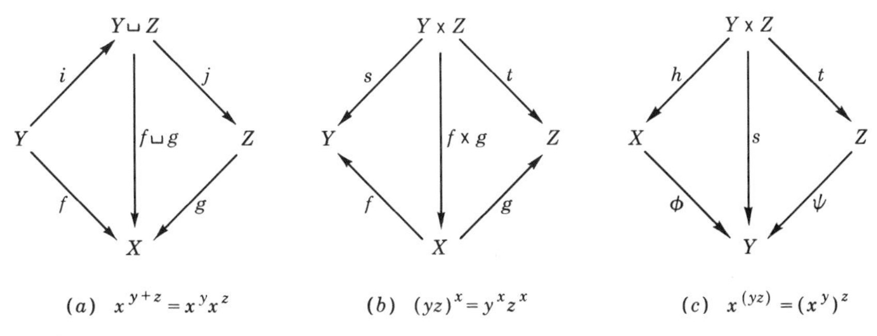

(a) $x^{y+z} = x^y x^z$ (b) $(yz)^x = y^x z^x$ (c) $x^{(yz)} = (x^y)^z$

Fig. 14-1.

This fact makes possible an extension of arithmetic to infinite numbers. A few properties of such infinite (or transfinite) numbers will be described in Secs. 14-2 and 14-3.

14-2. COUNTABLE INFINITY

The two most important infinite cardinal numbers are **d**, the cardinality[1] of **P**, and **c**, the cardinality of **R**. The number **d** is called *countable infinity*, and **c** is called the *power of the continuum*. All of the most familiar infinite sets are bijective to either **P** or **R**. Thus **N**, **Z**, and **Q** are bijective to **P**, whereas **C** and $\mathbf{R}^n$ are bijective to **R**, for example. In more detail, **N** is bijective to **P** under the bijection $\beta: n \mapsto n + 1$ with inverse $\beta^{-1}: m \mapsto m - 1$. Again, the enumeration $0,1,-1,2,-2, \ldots$ of the elements of **Z** establishes an obvious bijection $\nu: n \mapsto 2n + (1 + \operatorname{sgn} n)/2$ from **Z** to **P** (setting $\operatorname{sgn} 0 = 1$).

To construct a bijection between the set of all rational numbers, **Q** and **P** is less simple. Having in mind the representation of rational numbers as quotients of integers (see Chap. 10), we prove here a simpler, closely related result.

Lemma 1. *There is a bijection* $\alpha: \mathbf{N} \mapsto \mathbf{N}^2$.

Explanation. As with finite sets, $\mathbf{N}^2 = \mathbf{N} \times \mathbf{N}$ is the set of all ordered pairs (m,n) with $m,n \in \mathbf{N}$; T^S is the set of all functions $f: S \to T$.

Proof. As in the enumeration scheme by diagonals sketched in Fig. 14-2, we can let α map the diagonal with $k + 1$ entries $(k,0)$, $(k - 1, 1)$, $\ldots$, $(0,k)$ onto the interval

$$\left\{\frac{k(k+1)}{2}, \ldots, \frac{k(k+1)}{2} + k\right\}$$

[1] By the *cardinality* of a set S is meant the cardinal number $n(S)$ of its elements, i.e., the equivalence class of all T with $S\ E\ T$.

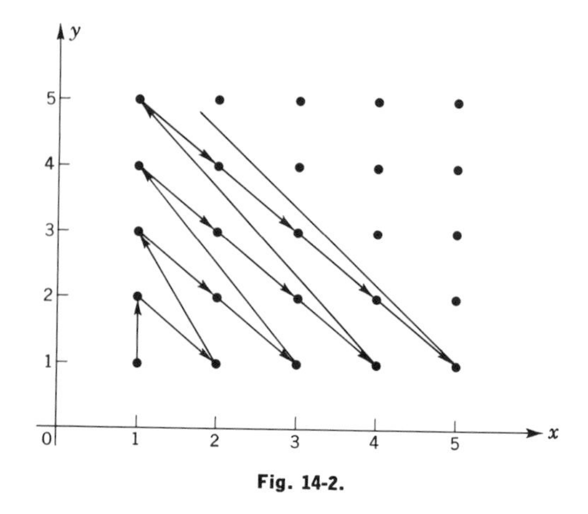

Fig. 14-2.

of the sequence **N**. Since

$$\frac{k(k+1)}{2} + k + 1 = \frac{(k+1)(k+2)}{2}$$

we can set

$$\alpha(i,j) = \left[\frac{(i+j)(i+j+1)}{2}\right] + j$$

in Lemma 1.

We next prove a result which states that no infinite set has smaller cardinality than **P**; **d** is the least infinite cardinal number. To signalize this property, Cantor (who created the theory of infinite cardinal numbers) wrote $\mathbf{d} = \aleph_0$ (pronounced "aleph null").

Theorem 3. *If U is an infinite set, then there is an injection $f\colon \mathbf{P} \to U$.*

Although this result seems obvious, a formal proof requires that one use a very controversial assumption, the so-called *Axiom of Choice*. This assumption seems innocuous at first sight, and, indeed, it seems indisputable when phrased as follows.

Axiom of Choice. Given any set U, there exists a function $\phi\colon \mathcal{P}(U) \to U$ which assigns to each nonvoid subset $S \subset U$ an element $s = \phi(S) \in S$.

Using the Axiom of Choice, we define $f(1) = \phi(U)$, $S_1 = u - f(1)$, and recursively,

$$f(n+1) = \phi(S_n') \qquad S_{n+1} = S_n \sqcup f(n+1)$$

One can also prove recursively that $f(n) \neq f(k)$ for $k < n$; hence, f is one-one from **P** to U.

Corollary 1. *If U is an infinite set, then there is an injection $g: U \to U$ which is not onto and a surjection $j: U \twoheadrightarrow U$ which is not one-one.*

Corollary 2. *For a set S to be finite, each of the following conditions is necessary and sufficient: (1) if $f: S \to S$ is one-one, then f is onto; (2) if $g: S \to S$ is onto, then g is one-one.*

The preceding bijections $\mathbf{P} \leftrightarrow \mathbf{N}$, $\mathbf{P} \leftrightarrow \mathbf{Z}$, and $\mathbf{P} \leftrightarrow \mathbf{Q}$ yield as corollaries of the following formulas of transfinite arithmetic.

Theorem 4. *If $n \in \mathbf{P}$ is a positive integer, then*

$$n + \mathbf{d} = \mathbf{d} + \mathbf{d} = \mathbf{d} \qquad n\mathbf{d} = \mathbf{dd} = \mathbf{d}^n = \mathbf{d} \tag{9}$$

Proof. The disjoint sum $\mathbf{n} \sqcup \mathbf{P}$ is bijective to $\mathbf{P}$ under the mapping $k \mapsto k(k \in \mathbf{n})$ and $l \mapsto l - 1(l \in \mathbf{P} - \mathbf{n})$, hence $n + \mathbf{d} = d$. Again, $\mathbf{P} \sqcup \mathbf{P}$ is bijective to $\mathbf{P}$ under the mapping which injects the first copy of $\mathbf{P}$ onto the odd positive integers $2k - 1$ and the second copy of $\mathbf{P}$ onto the set $\{2k\}$ of all even positive integers. Since $\mathbf{P} = \{2k - 1\} \sqcup \{2k\}$, we have proved that $\mathbf{d} + \mathbf{d} = \mathbf{d}$.

Likewise, $\mathbf{n} \times \mathbf{N} \cong \mathbf{n} \times \mathbf{P}$ is bijective to $\mathbf{P}$ under the mapping $(k,l) \mapsto nl + k(k \in \mathbf{n},\ l \in \mathbf{N})$, which shows that $n\mathbf{d} = \mathbf{d}$. Next, we consider $\mathbf{dd} = \mathbf{d}^2$, the cardinality of the Cartesian product $\mathbf{P} \times \mathbf{P}$. The mapping $R: n \mapsto (g(n),h(n))$ defined by $R(1) = (1,1)$ and

$$R(\sigma n) = (g(\sigma n),h(\sigma n))$$

where

$$g(\sigma n) = g(n) + 1 \quad \text{and} \quad h(\sigma n) = h(n) - 1 \quad \text{if } h(n) > 1$$
$$g(\sigma n) = 1 \quad \text{and} \quad h(\sigma n) = g(n) + 1 \quad \text{if } h(n) = 1$$

is the desired bijection (see Fig. 14-2).

Finally, $\mathbf{d}^1 = \mathbf{d}$, and by induction (assuming that $\mathbf{d}^n = \mathbf{d}$) we have $\mathbf{d}^{\sigma n} = \mathbf{d}^{n+1} = \mathbf{d}^n\mathbf{d}^2 = \mathbf{dd} = \mathbf{d}$, using the result just proved.

14-3. CARDINALITY OF THE CONTINUUM

Now consider the set $\mathbf{R}$ of all real numbers. The function $e^x/(1 + e^x)$ acts as a bijection from $\mathbf{R}$ to the open interval $(0,1) = \{y \mid 0 < y < 1\}$. Hence we have the following lemma.

Lemma 1. *The open interval $0 < x < 1$ has the cardinality of the continuum.*

On the other hand, the infinite series

$$\alpha(\mathbf{a}) = \sum_{k=1}^{\infty} a_k 2^{-k} \qquad \mathbf{a} = a_1 a_2 a_3 \cdots \tag{10}$$

defines a *surjection* onto the closed interval $[0,1] = \{y \mid 0 \leq y \leq 1]$ from the set $\mathbf{2}^{\mathbf{P}}$ of all functions $\mathbf{a}\colon \mathbf{P} \to \mathbf{2}$. Conversely, α is clearly the right inverse under right composition of the *injection* $\tilde{\alpha}\colon [0,1] \to \mathbf{2}^{\mathbf{P}}$ defined by the inequalities

$$\sum_{k=1}^{n} a_k 2^{-k} \leqq x < 2^{-n} + \sum_{k=1}^{n} a_k 2^{-k} \tag{10'}$$

Furthermore, the mappings α and $\tilde{\alpha}$ are *almost* bijections; unless x is a dyadic fraction (and the set of all dyadic fractions is countable), $\tilde{\alpha}(x)$ is the only $\mathbf{a} \in \mathbf{2}^{\mathbf{P}}$ with $\alpha(\mathbf{a}) = x$. Each rational dyadic fraction has two antecedents under α; thus

$$\sum_{j=1}^{m} e_j 2^{-j} + 2^{-m-1} = e_1 e_2 \cdots e_m 1\overbrace{000 \cdots}^{\text{all 0's}} = e_1 e_2 \cdots e_m 0\overbrace{111 \cdots}^{\text{all 1's}}$$

Since the set of dyadic fractions is countable and $2\mathbf{d} = \mathbf{d}$, one can easily alter α to a *bijection* $\beta\colon \mathbf{2}^{\mathbf{P}} \leftrightarrow [0,1]$, by setting

$$\begin{aligned} \beta(e_1 e_2 \cdots e_m 1000 \cdots) &= \sum_{j=0}^{m} e_j 2^{j+1} + 2^{-m-1} \\ \beta(e_1 e_2 \cdots e_m 0111 \cdots) &= \sum_{j=0}^{m} e_j 2^{j+1} + 2^{-m-1} + \tfrac{1}{2} \end{aligned} \tag{11}$$

and defining $\beta(\mathbf{a}) = \alpha(\mathbf{a})$ for all other sequences $\mathbf{a} \in \mathbf{2}^{\mathbf{P}}$.

Composing the standard bijection $\beta\colon \mathbf{2}^{\mathbf{P}} \leftrightarrow 0,1$ so defined with any bijection $[0,1] \leftrightarrow \mathbf{R}$, we obtain a bijection $\mathbf{2}^{\mathbf{P}} \leftrightarrow \mathbf{R}$. Moreover, a similar construction can be made for any integer $n > 1$, using numbers to base n.

This proves the following result.

Theorem 4. *For any positive integer $n > 1$, the set $\mathbf{n}^{\mathbf{P}}$ has the cardinality of the continuum.*

Using the notion of the characteristic function of a set, this has the following consequence (for $n = 2$).

Corollary. *The power set $\mathcal{P}(\mathbf{P})$ has the cardinality of the continuum.*

Theorem 5. (*Cantor.*) *The (infinite) set $\mathbf{2}^{\mathbf{P}}$ is not countable.*

Proof. By definition, $\mathbf{2^P}$ is bijective to the set of all infinite binary sequences

$$\mathbf{x} = x_1x_2x_3 \cdots \qquad x_i = 0 \text{ or } 1$$

Now suppose that one could establish a bijection $\beta: \mathbf{P} \to \mathbf{2^P}$, that is, that one could enumerate all such binary sequences in a sequence of sequences

$$\begin{aligned} \mathbf{x}^1 &= x_1{}^1x_2{}^1x_3{}^1 \cdots \\ \mathbf{x}^2 &= x_1{}^2x_2{}^2x_3{}^2 \cdots \\ \mathbf{x}^3 &= x_1{}^3x_2{}^3x_3{}^3 \cdots \\ & \cdots\cdots\cdots\cdots \end{aligned}$$

Define $\mathbf{y} = y_1y_2y_3 \cdots$ by the condition that each $y_i = 1 - x_i{}^i$ be complementary to the ith term of the *diagonal* binary sequence $x_1{}^1$, $x_2{}^2$, $x_3{}^3$, . . . (Cantor's diagonal argument). Then $\mathbf{y} \in \mathbf{2^P}$; yet trivially $\mathbf{y} \neq \mathbf{x}^i$ for all $i = 1,2,3,$ . . . since $y_i \neq x_i{}^i$. This proves that there is no surjection $f: \mathbf{P} \to \mathbf{2^P}$, which implies Theorem 5 (since any bijection is a surjection).

Theorems 4 and 5 have the following consequence.

Corollary. *The real continuum* $\mathbf{R}$ *is an uncountable set.*

The diagonal argument of Cantor used above can be generalized to arbitrary sets, as follows.

Theorem 6. (*Cantor.*) *Let X be any set. Then there is no surjection* $f: X \twoheadrightarrow \mathbf{2}^X$.

Proof. Let $f: X \to \mathbf{2}^X$ be given. For each $a \in X$, let the function $f_a: X \to \mathbf{2}$ correspond to a under f, where $\mathbf{2}$ is the set $\{0,1\}$ as above. Define the complementary function $c: X \to \mathbf{2}$ by the assignment $c(x) = 1 - f_x(x)$ (for each $x \in X$). Then c cannot coincide with any f_a, because $c(a) = 1 - f_a(a) \neq f_a(a)$. The function $c: X \to \mathbf{2}$ is thus different from every f_a, which shows that the set $\{f_a\}$ of all f_a cannot exhaust the set $\mathbf{2}^X$ of all functions $g: X \to \mathbf{2}$.

Corollary. *Let X be any set. Then there is no bijection* $b: X \leftrightarrow \mathbf{2}^X$.

(Informally, card $X <$ card $\mathbf{2}^X$ for all X.)

Definability and Computability. The attempt of mathematical logicians to formalize theorem proving has uncovered many uncertainties and difficulties. One of the simplest stems from the fact that the set $\mathbf{R}$ of all real numbers is uncountable; it is the following paradox.

Richard Paradox. Let A be any finite alphabet of ν characters (small or capital letters, numerals, punctuation marks, etc.) such as might be printed by a typewriter, including a blank ( ). Then one can form only n^ν sequences or strings of characters of length n. Hence one can list in an infinite sequence all possible finite strings of characters. Omitting some of these as meaningless [not in the "language" (see Sec. 4-8)], we still have an infinite sequence, or exceptionally, a finite set. Consequently, *in any printable language, the set of all statements of finite length is countable* (or finite).

Hence, in particular, the set of possible definitions of real numbers is countable. However, by the corollary to Theorem 5, the set $\mathbf{R}$ of *all* real numbers is uncountable. It follows that *only a minute fraction of all real numbers are definable* (in any printable language); in this technical sense, the continuum is inaccessible to complete linguistic treatment (Richard paradox).

The concept of being "definable" is vague; as we shall see, "computability" is a more precise concept. We shall now define computability for both $\mathbf{R}$ and $\mathbf{2^P} \cong \mathcal{P}(\mathbf{P})$, as follows.

Definition. A real number x is *computable* when an *algorithm* is known which, for any $n \in \mathbf{P}$, enables one to compute in a finite number of steps a dyadic fraction $k/2^r$ ($k \in \mathbf{Z}$, $r \in \mathbf{N}$) such that $x - a < 2^{-n}$. A subset $S \subset \mathbf{P}$ is *computable* when an algorithm is known which, for any $n \in \mathbf{P}$, enables one to decide whether or not $n \in S$.

Clearly, the Richard paradox applies also to the preceding notions of computability; the sets of all computable real numbers and of all computable subsets $S \subset \mathbf{P}$ (or, equivalently, of computable characteristic functions e_S: $\mathbf{P} \to \{0,1\}$) are countable. Since $\mathbf{R}$ and $\mathcal{P}(\mathbf{P})$ are uncountable (Sec. 14-3), *non*computable real numbers and sets $S \subset \mathbf{P}$ exist.

EXERCISES A

1. (a) Given bijections b: $S \leftrightarrow S^*$ and c: $T \leftrightarrow T^*$, construct bijections $b \sqcup c$, $b \times c$, and c^b as described in (6) to (8).
   (b) What substitution properties for E are implied by the existence of these bijections? Explain.
2. Prove (5) to (7) by constructing six bijections, including $A \times (B \times C) \leftrightarrow (A \times B) \times C$ and $(A \sqcup B) \times C \leftrightarrow (A \times C) \sqcup (B \times C)$.
3. Prove (8) by constructing in detail three bijections, such as $C^{A \sqcup B} \leftrightarrow (C^A) \times (C^B)$.
4. Enumerate explicitly the set of all finite sets of integers.

★5. An algebraic number is a number $z \in \mathbf{C}$ such that for some $n \in \mathbf{P}$ and $c_1, \ldots, c_n \in Q$, $z^n + \sum_{k=1}^{n} c_k z^k = 0$. Prove that the set of all algebraic numbers is countable.

FOR ARBITRARY CARDINAL NUMBERS x AND y, $x \leqq y$ MEANS THAT THERE IS AN INJECTION $j: X \to Y$ FOR SOME PAIR OF SETS X AND Y WITH CARD $X = x$, CARD $Y = y$.

6. Show that for any sets A,B with card $A = x$, card $B = y$, and $x \leqq y$, there is an injection $j^*: A \to B$.
7. Show that $x \leqq y$ implies $x + z \leqq y + z$, $xz \leqq yz$, $x^z \leqq y^z$, and $z^x \leqq z^y$, for any cardinal number z.
8. (a) Prove that $\mathbf{c} + n = \mathbf{c} + \mathbf{d} = \mathbf{c}$, for all $n \in \mathbf{N}$.
   (b) Prove that $n\mathbf{c} = \mathbf{d} \cdot \mathbf{c} = \mathbf{c}$, for any $n \in \mathbf{P}$.
9. (a) Prove that $\mathbf{c}^n = \mathbf{c}^{\mathbf{d}} = \mathbf{c}$, for any $n \in \mathbf{P}$.
   (b) Prove that, for $n > 1$, $n^{\mathbf{c}} = 2^{\mathbf{c}} = \mathbf{d}^{\mathbf{c}} = \mathbf{c}^{\mathbf{c}}$.

★10. Construct a bijection $10^{\mathbf{P}} \leftrightarrow \mathbf{R}$. (*Caution:* Beware of ambiguities like $0.999999 \cdots = 1.0000000 \cdots$.)

★11. Let $f: \mathbf{R} \to \mathbf{R}$ have the property that $f(x) > 0$ for all $x \in \mathbf{R}$. Show that for any K, one can find a finite subset F of $x_i \in R$ such that $\sum_F x_i > K$.

14-4. TURING COMPUTABILITY

It is a plausible and widely believed conjecture that the concepts of computability defined in Sec. 14-3 are independent of the language chosen to describe algorithms (e.g., ordinary English or the language of symbolic logic developed by Whitehead and Russell). Further, Turing's thesis (also called Church's thesis) asserts that every effective algorithm may be considered as a (finite) program for some Turing machine of the kind described in Sec. 3-7. We shall call such numbers and sets *Turing-computable.*

Turing supported his view by showing that the class of all computable real numbers is a subfield of $\mathbf{R}$ which includes all rational and all algebraic[1] numbers, and also e,π, the zeros of Bessel functions, and indeed all the real numbers which are customarily used in mathematical analysis.

Likewise, a set $S \subset \mathbf{P}$ may be considered as (Turing) computable when its characteristic function $e_S(n)$ can be obtained as the printout of 0's and 1's in squares (measured from some specified initial square) of a suitably programmed Turing machine or, equivalently, when for each $n \in \mathbf{P}$ precisely one of the two equations $e(n) = 0$ or $e(n) = 1$ appears in this printout.

The following result is not surprising.

Theorem 7. *The standard bijection $\beta: \mathcal{P}(\mathbf{P}) \leftrightarrow [0,1]$ defined by* (9) *and* (10) *carries computable subsets $S \subset \mathbf{P}$ into computable real numbers $x \in [0,1]$.*

[1] We recall that a number is called *algebraic* when it satisfies a polynomial equation of the form

$$x^n + \sum_{k=1}^{n} a_k x^{n-k} = 0 \qquad \text{all } a_k \in \mathbf{Q}$$

For an infinite subset S with infinite complement $S' = \mathbf{P} - S$, this follows because $\left| x - \sum_{k=1}^{m} e_S(k)2^{-k} \right| < 2^{-m}$ and the approximating sum is computable in finite terms. For finite sets S and sets with finite complement, $\beta(S)$ is a dyadic fraction which can be computed exactly in finite terms, completing the proof.

Surprisingly, the converse of Theorem 7 does not seem to be demonstrable, although there is no known specific computable $x \in [0,1]$ such that $\beta^{-1}(S)$ is the characteristic function of a set $S \subset \mathbf{P}$ which is known to be noncomputable.

The source of the difficulty can be seen if one considers the difficulty of proving the plausible (and presumably true) proposition that $J_0(\pi)$ is *not* a dyadic fraction. No matter how many 0's or 1's one calculates in the binary representation of $J_0(\pi)$, how can one ever be sure that their cardinality is infinite?

Similarly, there seems to be no general program for a Turing machine for testing the *equality* of two computable subsets $S \subset \mathbf{P}$ and $T \subset \mathbf{P}$. Although one can show that $S \neq T$ (if this is true) in finite time, no matter how many equations of the form $e_S(n) = e_T(n)$ one verifies, one cannot infer with certainty from this that $S = T$; it only makes $S = T$ highly probable. The relation of equality between computable sets seems not to be computable. A related paradox is the following.

Theorem 8. *The property of being a Turing-computable subset of* $\mathbf{P}$ *is not Turing-computable.*

Proof. As in the Richard paradox, the set of all computable subsets $S \subset \mathbf{P}$ is countable; hence its members can be enumerated as S_1, S_2, S_3, ... Using the diagonal process, however, we can now define a new set T by its characteristic function $e_T(n) = 1 - e_S(n)$. Since this T is none of the S_n, it is not itself computable; this completes the proof.

The preceding paradox illustrates the fact that Cantor's diagonal argument cannot be incorporated into a program for a Turing machine. A similar argument shows that the (countable) set of all real numbers which are computable on a Turing machine is itself *not* Turing-computable; see Minsky, Chap. 9.

14-5. TURING COMPUTABILITY AND PRACTICAL COMPUTABILITY

Of course, as was already shown in Chap. 4, Turing machines can do much more than print out Turing-computable real numbers and Turing-computable subsets of $\mathbf{P}$ (or $\mathbf{Z}$ or $\mathbf{Q}$). They can also be designed to add, subtract, multiply, divide, or in fact to perform any arithmetic operation on integers or on any computable real numbers. Turing machines have also been designed which will locate names in a list, order lists of numbers, and perform many other operations.

This leads to a very basic question: What classes of calculations can be performed by Turing machines, and how do these relate to those things which can be systematically calculated? It was Turing's position (and this is the opinion commonly held today) that any "effective defined procedure" can be carried out by a Turing machine. This is to say that a Turing machine can be designed which can perform (given adequate time) any well-defined procedure. Although not proved (or provable), this viewpoint is used throughout the theory of computability and is also used by logicians whose studies of "effective procedures" often use Turing machines as a vehicle for their proofs.

Turing also showed that it was possible to design a "universal Turing machine" which would perform any calculation which could be performed by any other Turing machine. In effect, an input tape would be coded for this universal Turing machine which would contain first a description of the machine to be simulated and then the data for the problem. Users of modern general-purpose computers will identify the coded description of the Turing machine with a program or description of the algorithm to be performed. It is possible to program a universal Turing machine on a general-purpose computer. Therefore, a general-purpose computer can perform any calculation which can be performed by a Turing machine, except for the limitations of the memory of the general-purpose computer. Moreover, although the tape used in a Turing machine is infinite in length, one can add as much auxiliary storage as needed for any computation. Therefore, real computers can be thought of as having *unlimited* storage.

These considerations lead to the widely held position that any effective procedure or algorithm can be carried out by a general-purpose computer, subject to the limitations of the finite amount of memory in the computer. Thus, a general-purpose computer is like a universal Turing machine with a finite length of tape.

Practical Computability. The logical concept of computability, as defined by Turing and other mathematical logicians, is far more inclusive than that of practical computability by a physically constructable finite-state machine. Thus, as has been observed by Davidon, consider an ultra-fast hypothetical sequential machine, capable of taking a step in the time required for light to traverse a nuclear radius ($\sim 10^{-23}$ sec). If such a machine had been operating since the beginning of the universe (estimated as 4×10^{10} years ago) it could still not have performed $2^{2^8} = 2^{256}$

steps. Therefore, it could not have listed all the Boolean polynomials in the seven letters a, b, c, d, e, f, g, h.

The preceding time limitation is important, especially for scientific computations such as the solution of systems of partial differential equations in which instructions are repeated cyclically many times. In practice, computations requiring $10^{15} \simeq 2^{50}$ steps are prohibitively expensive.

Compared with such practical limitations on the number of steps which can be executed, theoretical limitations imposed by the finiteness of the number of states in finite-state machines are unimportant. Thus, even the "small" PDP-8 digital computer has a core memory of $4{,}096 = 2^{12}$ 35-bit words, giving $2^{35 \cdot 4{,}096} = 2^{143{,}260}$ internal states.[1] In most cases, it is computing time rather than memory capacity which limits what can be done on a computer.

Indeed, except in very frequently used production codes, the most important limiting factor is the difficulty of writing a perfect ("debugged") program of more than 10^3 or possibly 10^4 instructions!

14-6. MATHEMATICAL LINGUISTICS

We shall now resume the study of the grammar of symbolic languages such as ALGOL, which we began at the end of Chap. 4. ALGOL has a phrase-structure grammar in the following sense.

Definition. A *phrase-structure grammar* is a quadruple (V,T,P,A), where V is some finite *vocabulary* or *alphabet* of *characters*, $T \subset V$ is a subset of characters called *terminal symbols* $A \in V - T$ is a special nonterminal signifying only production not consisting entirely of terminal symbols, and P is a finite set of *production rules* of the form $u \to v$, where each u is a (nonempty) string of nonterminals and v is some specified string of characters. The set $L = L(V,T,P,A)$ of all strings of terminal symbols which can be produced by the production rules of a phrase-structure grammar is called a *phrase-structure language*. This is a subset of the set T^* of all strings of characters of T.

In the case of ALGOL, T includes all alphanumeric characters and algebraic symbols, as well as special ALGOL words such as **for,** etc. The set $V - T$ includes descriptors such as ⟨integer⟩, ⟨number⟩, and ⟨Boolean expression⟩, which are used in the ALGOL metalanguage. A few of the numerous production rules of ALGOL will be discussed later in this section.

In applications to ordinary (human or natural languages), the terminal symbols are ordinary words. The characters of $V - T$ are names for parts of speech or other grammatical classes. The production rules include

[1] Auxiliary storage devices currently available have capacities as high as 10^8 bits, giving $2^{100{,}000{,}000}$ different states.

grammatical rules such as A = ⟨sentence⟩ → ⟨subject⟩⟨predicate⟩ and substitutions such as ⟨subject⟩ → ⟨the boy⟩ and ⟨predicate⟩ → ⟨smiles⟩. The terminal productions are grammatical *sentences.*

In various applications to symbolic logic, certain production rules of the form $A \to v$ ($v \in V^*$) are taken as axioms, and the other production rules act as "rules of inference" (i.e., permissible forms of logical deduction). The "sentences" which can be produced by these rules of inference from the axioms are the *theorems* of the formal deductive system under consideration.

We now present four very simple examples of phrase-structure grammars (and hence languages) in which $V - T = \{A\}$, that is, in which there is only one grammatical class.

Example 1. Let $T = \{c\}$ and let P consist of the two production rules

$$A \to c \qquad \text{and} \qquad A \to cA \tag{12}$$

This grammar generates the language $L = \{c,cc,ccc, \ . \ . \ .\}$, that is, the set T^* of all strings of c's having finite length.

Example 2. Consider the grammar with $T = \{0,1\}$, and the following rewriting or production rules:

$$A \to \Lambda \qquad A \to 0A0 \qquad A \to 1A1 \tag{13}$$

where Λ is the null string.[1] This grammar generates the null string and all binary sequences of the form $\mathbf{xx}^R$, where if $\mathbf{x} = x_1x_2 \cdots x_n$, then $\mathbf{x}^R$ is $x_n \cdots x_2x_1$; $\mathbf{x}^R$ stands for $\mathbf{x}$ in reverse order, i.e., the *opposite* of $\mathbf{x}$.

Example 3. Let $T = \{b,c\}$ and let P consist of the production rules

$$A \to bc \qquad A \to bAc \tag{14}$$

The language defined by this grammar consists of all strings b^nc^n ($n \in \mathbf{P}$).

Example 4. Again, let $T = \{b,c\}$, but now let P consist of the production rules

$$A \to b \qquad A \to bA \qquad A \to cA \tag{15}$$

The resulting (phrase-structure) language consists of all strings in T^* which end in b.

Examples from ALGOL. As was explained at the beginning of this section, the production rules for ALGOL arithmetic, as listed in Sec. 4-7,

[1] The null string Λ obeys the rule $x \, \Lambda \, y = xy$; that is, the null string can be absorbed (vanishes) in any sequence.

define it as a phrase-structure language with a less meager vocabulary than the languages in Examples 1 to 4. We shall now develop this remark a little further.

Thus, consider the production rules for unsigned (decimal) integers as a sublanguage, with $T = \{0,1,2, \ldots ,9\} = \langle\text{digit}\rangle$ and

$$V - T = \{\langle\text{digit}\rangle,A\}$$

Then the set ⟨unsigned integers⟩ of all unsigned integers is just T^*; it is generated from T by the production rules

$$A \to \langle\text{digit}\rangle \qquad (\text{or } A \to 0,\ A \to 1,\ \ldots ,\ A \to 9) \tag{16}$$

and

$$A \to A\langle\text{digit}\rangle \qquad (\text{or } A \to A0,\ A \to A1,\ \ldots ,\ A \to A9) \tag{16$'$}$$

To produce the signed integers, set instead $T = \{\langle\text{digit}\rangle,+,-\}$, where $\langle\text{digit}\rangle = \{0,1, \ldots ,9\}$ as above, and give the production rules

$$A \to +\langle\text{digit}\rangle \qquad A \to -\langle\text{digit}\rangle \qquad A \to A\langle\text{digit}\rangle \tag{17}$$

The production rules of ALGOL and those of the simple languages of Examples 10 to 13 are all *context-free* in the following sense.

Definition. A production rule $\xi \to v$ of a phrase-structure language is context-free when, for any production $a\ \xi\ b$, $a\ v\ b$ is also accepted as a production. A phrase-structure language which can be produced by a context-free production (or rewriting) rule is called a *context-free language.*

By contrast, a *context-sensitive production rule* is one which says that $a\ \xi\ b \to a\ v\ b$ for some, but not all a,b. Human languages are very context-sensitive.

An examination of the grammar of ALGOL, as defined in Sec. 4-7, given in Backus normal form, shows that each is, in fact, a rewriting rule for a context-free grammar. This part of ALGOL is therefore a context-free grammar, and, in fact, ALGOL can be considered as a context-free language.[1]

EXERCISES B

1. Show that if the real numbers x and y are computable, then so are $x + y$ and xy. (*Hint:* See Exercise 2, but don't assume it.)
2. Show that if $F(x,y)$ is any continuous function from $\mathbf{R}^2$ to $\mathbf{R}$ and the numbers x and y are computable, then so is $F(x,y)$ computable.
3. Show that the set of all computable real numbers is a subfield of $\mathbf{R}$.

[1] The complete set of Backus normal forms for ALGOL can be found in *Communications ACM*, **6**:(1), January, 1963.

4. Show that if f: $\mathbf{P} \to \mathbf{P}$ and g: $\mathbf{P} \to \mathbf{P}$ are both computable, then so are $f + g$, fg, g^f, and $g \circ f$.
5. Show that for any given $a_0, \ldots, a_n \in \mathbf{P}$, $\sum_{k=0}^{n} a_k x^k$ can be defined as a recursive function.
6. Define $F(m,0) = m + 1$ and $F(m + 1, n + 1) = F(F(m, n + 1),n)$.
   (a) Prove that $F(m,2) = 2m$ and $F(m,3) = 2^m$.
   (b) Prove that $F(1,4) = 2$ and $F(m + 1, 4) = 2^{F(m,4)}$.
   (c) Prove that $F(1,5) = 2$ and $F(m + 1, 5) = F(m,5)^{F(m,5)}$.†
7. Show that if S and T are computable subsets of $\mathbf{P}$, then $S \cup T$ is also computable.
8. Prove (by induction) that the language of Example 1 includes the set of all strings of c's of finite length.
9. Prove that the language defined by (13) includes Λ and all strings $\mathbf{xx}^R$ and nothing else.
10. Prove that the production rules of Exercises 8 and 9 define finite-state grammars.

14-7. FINITE-STATE GRAMMARS

We shall now consider those phrase-structure languages which can be produced (generated) or accepted by a special kind of finite-state machine (automaton) called a *finite-state acceptor.* By definition, such a machine is a finite-state machine which reads an input tape on which are marked terminal symbols from $X = T$ in some language, changing its internal state s each time it reads a new symbol $x \in X$ to a new state $\nu(x,s)$ as directed by its state function ν. When the last symbol has been read, the automaton stops. If the final state is in one of a specified class F of accepting final states, the tape is *accepted;* if the final state is in $S - F$, it is *rejected.*

The preceding idea can be stated formally in the following terms.

Definition. A *finite-state acceptor* is a quintuple $\mathcal{A} = [X,S,\nu,s_0,F]$, where (1) X is a finite set of *input symbols,* (2) S is a finite set of *internal states,* (3) ν is a function from $X \times S$ into S, (4) $s_0 \in S$ is the *starting state,* and (5) $F \subset S$ is a set of *accepting final states.*

In the theory of automata, the inputs are generally thought of as being written on a linear tape, with a single input symbol at each position on the tape. The tape is then moved past a reading head which conveys each input in succession to the automaton.

Example 5. A finite-state acceptor is shown in state-diagram form in Fig. 14-3. Examples of input strings which are accepted by this finite-state acceptor are

001 0001 100 1010 10110

† R. C. Buck, *Am. Math. Monthly,* **70**:128–135 (1963).

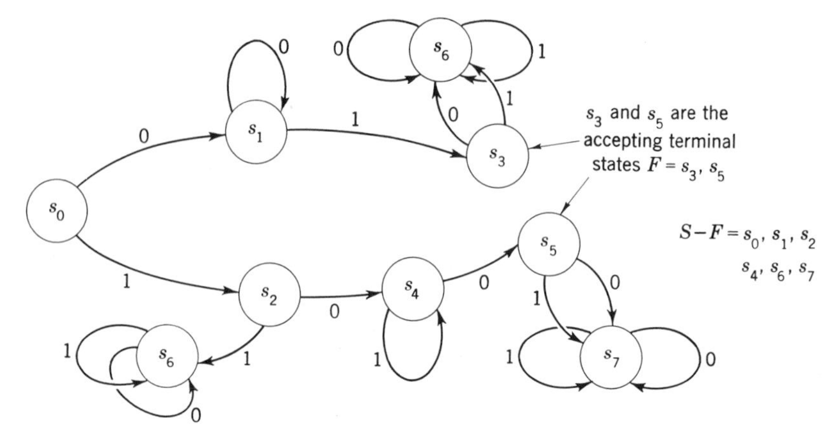

Fig. 14-3. Finite-state acceptor.

We now describe a class of grammars whose production rules will generate just those sets of strings which are accepted by some finite-state acceptor.

Definition. A *finite-state grammar* G is a 4-tuple $[X,T,P,A]$, where

(*i*) X is a finite set of input symbols.
(*ii*) T is a finite set of terminal symbols $T \subset X$.
(*iii*) P is a finite set of productions of the form $Z \rightarrow bY$, $Z \rightarrow b$, where $Z,Y \in X - T$, $b \in T$.
(*iv*) A is a unique *starting symbol* $A \in X - T$.

State diagrams are convenient for representing the operation of small finite-state acceptors. We will now show that a similar technique can be given for constructing a graph representing the rules for a finite-state grammar G. The steps are as follows:

Step 1. Draw a node for each nonterminal (auxiliary) symbol in G.
Step 2. If $X \rightarrow bY$ is a rule of G, draw a link from X to Y, and label the link b.
Step 3. If $X \rightarrow b$ is a rule of G, draw a link from node X to a new node, which we label ACCEPT.

Example 6. Figure 14-4 shows the state diagram (labeled directed graph) of the finite-state grammar G with productions

$$A \rightarrow cA \qquad A \rightarrow bY \qquad Y \rightarrow cA$$
$$Y \rightarrow b \qquad Y \rightarrow bX \qquad X \rightarrow c$$

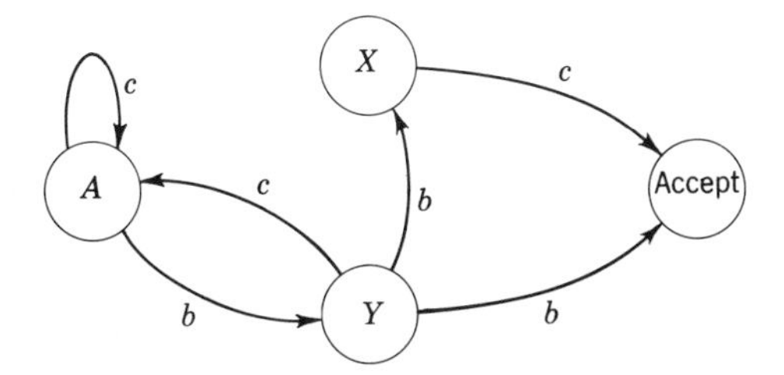

Fig. 14-4. State diagram of a finite-state grammar.

Theorem 9. *The sequence of labels on the arcs of any path of finite length, starting with node A in the labeled directed graph of a grammar G, comprises a string which can be produced (generated) by G, and all strings of G can be obtained in this way.*

Proof. The graph simply represents the rules of G in another form.

Theorem 10. *Given a finite-state grammar G, a finite-state acceptor can be found which accepts just those strings which are generated by the production rules of G.*

Proof. Change A to s_0, and then change each nonterminal symbol used to label a node of the graph for a finite-state language to some (each different) s_j. This gives a graph for a finite-state acceptor.

Theorem 11. *Given a finite-state acceptor $\mathcal{A}$, a finite-state grammar G can be found which will generate just those strings which will be accepted by $\mathcal{A}$.*

Proof. Change each label of a node in a graph for the acceptor from s_j to some nonterminal symbol Y, and then derive the productions for G from the new graph, making the nonterminals from the labels of the nodes of the graph.

The preceding theorems establish a bijection between finite-state grammars and finite-state acceptors. It is natural to ask the following question: Are there context-free languages which are not finite-state languages? (It is obvious that the set of finite-state languages is included in the set of context-free languages since the rules for finite-state grammars include those for context-free grammars.)

Theorem 12. *There are context-free languages which are not finite-state languages.*

Proof. That the context-free language of Example 3 is not finite-state may be seen as follows. After b has been read as many times as there are different states in the machine, we will have $s^{n+p} = s^n$ for some period p. Hence, if the machine accepts $b^n c^n$, it will also accept $b^{n+p} c^n$.

Notice that the language of Example 3 is contained in the language of Example 4, which *is* a finite-state language. (To construct a finite-state acceptor, it suffices to make $\nu(b,s) = f \in F$ for all s.) The opposite of the language of Example 4 is also a finite-state language; to see this, set $\nu(b,s_0) = f \in F$ and $\nu(x,f) = f$ for all input symbols x.

The preceding examples show that, in principle, there is no *fixed* finite-state acceptor for the language of all "well-formed parenthesis sequences" such as ((( )( ))( )), unless some limitation is placed on length.[1]

However, one should not jump to the conclusion that finite-state machines will have trouble in evaluating polynomials of arbitrary length. A simple way out is provided by using *nesting*, as in the polynomial expression

$$a_0 + a_1x + a_2x^2 + \cdots + a_{n-1}x^{n-1} + a_nx^n \\ = ((\cdots((a_n)x + a_{n-1})x + \cdots)x + a_1)x + a_0$$

One can evalute this for arbitrarily large n by having an input tape so marked that, after putting a_n in the accumulator, the instructions

MUL x, ADD a_{n-1}, MUL x, ADD a_{n-2}, . . . ,
MUL x, ADD a_0, HALT

are indicated (in a suitable assembler language).

Such embeddings occur in natural languages as well. Thus Chomsky gives the following:

(the rat (the cat (the dog chased) killed) ate the mole)

Their analysis involves the technique of *parsing* which follows.

14-8. SYNTAX; PUSH-DOWN STORE ACCEPTORS

Notice that the acceptance of a string by a finite-state acceptor includes the parsing or deriving of syntactic information concerning the string because the states of the finite-state machine can be made to correspond to the nonterminal symbols of the grammar. If the semantics associated with these nonterminals is known, then the "meaning" of the string can be determined, and this points to a possible technique for compiling. Programs which first determine the syntax of a string and then use the

[1] In view of the comparisons given at the end of Sec. 14-5, this is not a serious practical limitation.

semantics connected with this syntax to assemble machine-language statements into a compiled program are called *syntax-directed* compilers.

From the standpoint of automaton theory, this compiling process for context-free languages is most easily understood in terms of the concept of a *push-down store,* which we already introduced in Chap. 4. This provides a special structure for the storage of information in a computer, in "layers" of potentially unlimited depth. Each layer can be stored either "in core," where it is immediately accessible (in microseconds), or on magnetic tapes or drums, which may provide a backup store of many millions of words (typically, each containing from 32 to 64 binary digits or bits). As a result, embeddings of reasonable depth can be accommodated with little strain on the memory.

The actual parsing of statements by a compiler can be accomplished in several ways. The languages used are assumed to be unambiguous (that is, if a consistent derivation of a string is found, then no other consistent derivation of the string from the rules of grammar (syntax) can be found). The most direct and obvious way to attempt to parse a given string is by means of a parsing tree. Here the compiler passes down alternate branches of the tree until it finds a consistent derivation. For example, the expression $X \times Y + Z + U$ can be parsed in only one way, and Fig. 14-5 shows the particular parsing which follows the syntactic rules of Sec. 4-7. Notice that the syntactic rules of ALGOL force both the particular precedence of operators and the left-to-right rule for evaluating arithmetic expressions.

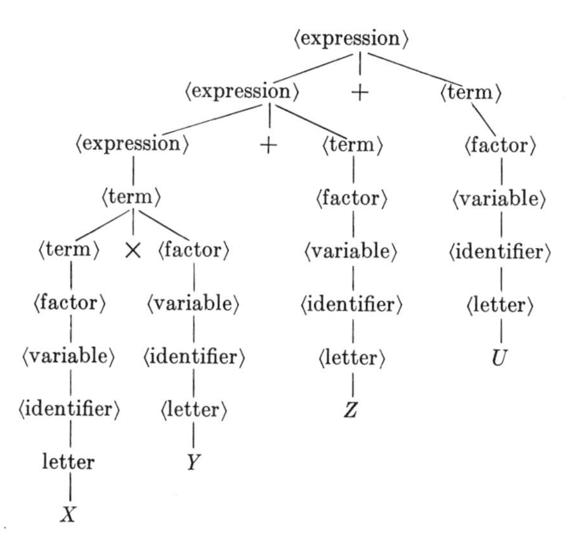

Fig. 14-5. Parsing the expression $X \times Y + Z + U$.

In order to further investigate automata which can translate context-free languages, a useful concept is that of *push-down store,* which we introduced in Chap. 4.

A *deterministic push-down store* automaton consists of a finite-state automaton with an input tape plus a second tape which is used as a memory. The finite-state automaton reads the input tape one symbol at a time, never reversing the direction of travel, just as in previous cases. The push-down store automaton, however, can write or read from its second tape and can move the tape in either direction. This second tape, or memory, is called a *push-down store,* or *stack.* The principal restriction on use of the stack is that it is used on a last-in first-out basis. If we consider the tape to be moved vertically, the tape is moved down when a symbol is written and up when symbols are read. When new symbols are written, these symbols are simply written in order on top of the symbol which was being read when the writing began. When symbols are read, however, they are erased as they pass the reading head. There is also a special symbol σ which is at the bottom of the push-down store. There is no limit to the upper section of the tape, however (the tape is often said to be potentially infinite). As a result, the push-down store automaton is not a finite-state machine.

A *deterministic push-down store acceptor* is simply a push-down store automaton with the rule that if immediately after the last symbol of the input tape is read the automaton also reads the symbol σ at the bottom of the push-down store, then the string on the input tape is said to be accepted, otherwise it is said to be rejected.

A push-down store automaton can be used, in theory, to perform many useful calculations, such as parsing arithmetic expressions for a compiler, and is a very useful concept in both machine design and programming. Because of the utility of a push-down store, several computers have been constructed with memories which are essentially push-down stores.

We define these concepts formally.

Definition. A *push-down store automaton* is a 6-tuple $(X,S,\Delta,\delta,\sigma,s_0)$ where

(*i*) X is a finite set of input symbols.
(*ii*) S is a finite set of input states.
(*iii*) $\Delta \supset X$ is the set of input symbols plus a set of auxiliary symbols.
(*iv*) δ is a function $X \times S \times \Delta \rightarrow S \times \Delta^*$.
(*v*) $\sigma \in \Delta$ (σ marks the end of the store).
(*vi*) $s_0 \in S$ (s_0 is the starting state).

Now consider the grammar $A \rightarrow cAb$, $A \rightarrow cb$, and the finite-state push-down store acceptor, where

$$X = \{c,b\}$$
$$S = \{s_0,s_1\}$$
$$\Delta = \{c,b,S\}$$

$$\begin{array}{ll} (c,s_0,\sigma) \mapsto (s_1,c) & (b,s_2,c) \mapsto (s_2,\Delta) \\ (b,s_0,\sigma) \mapsto (s_3,b) & (b,s_3,c) \mapsto (s_3,b) \\ (c,s_1,c) \mapsto (s_1,c) & (b,s_3,b) \mapsto (s_3,b) \\ (b,s_1,c) \mapsto (s_2,\Lambda) & (c,s_3,b) \mapsto (s_3,c) \\ (c,s_2,c) \mapsto (s_3,b) & (c,s_3,b) \mapsto (s_3,c) \end{array}$$

The grammar given generates the set of strings $\{c^nb^n: n = 1,2, \ldots\}$, and the push-down store acceptor accepts just these strings. Notice that s_3 is a state used as a "trap," for when a sequence of symbols not in the language is given, the acceptor goes to s_3 where it will be locked from then on. The push-down store is used to collect the c's until the first b arrives, then each b causes the erasure of the topmost c in the stack until, if the number of b's equals the number of c's, the acceptor will stop after the last symbol has been read from the stack. At this time the acceptor will be in state s_2, and σ will be at the top of the stack.

This means that push-down store automata can handle nested or embedded sequences, a fact which is widely exploited in program systems. Unfortunately, no deterministic push-down store automaton can be designed for some context-free languages. For instance, the language of Example 2, with strings $\mathbf{x}\mathbf{x}^R$, where $\mathbf{x}$ is any string of symbols and $\mathbf{x}^R$ its opposite,[1] cannot be handled by a stack since the automaton cannot find the center of the string in a single pass. (However, the language of strings $\mathbf{x}\,d\mathbf{x}^R$, where d is a special "marker" symbol, can be handled by a stack.)

In order to find a class of machines which can handle all context-free languages, we must consider either (1) deterministic push-down store automata with two stacks, (2) Turing machines, or (3) nondeterministic push-down store automata.

A push-down store automaton with two push-down stores or stacks is a direct extension of the single push-down store concept. Instead of production rules of the form $(c,s_i,b) \rightarrow (s_j,d)$, we have rules $(c,s_i,b) \rightarrow (s_j,d,e)$ where d is written on one stack and e on the other, for example, or rules such as $(c,s_j,b) \rightarrow (s_k,\Lambda,d)$, which instructs the computer to erase the top symbol from one stack and write d on the other stack.

Notice that two stacks in a push-down store automata offer the same facility for storage and calculation as a tape that can be moved in either

[1] That is, if $\mathbf{x} = abc$, then $\mathbf{x}^R = cba$.

direction, essentially because the two stacks (tapes) can be spliced together into a single tape which can be moved in either direction. Moreover, push-down store tapes (stacks) are imagined as unlimited in length.

Hence deterministic push-down store automata with two tapes are essentially equivalent to Turing machines which, as was stated in Sec. 3-7, are simply deterministic finite-state automata with a single tape, but with the ability to move the tape in either direction. As was also mentioned in Chap. 3, a Turing machine's rules of operation can be expressed using a set of production rules of the form $s_ix_j \to s_kx_lR$ or $s_ix_j \to s_kx_lL$ where the s_r are internal states of the automata and the x_t are the symbols read from or written on the tape. The R moves the tape in one direction (right), and the L moves the tape in the other direction. The input strings are written on the tape as before, the same tape being used to record the output string and for storage during computations.

A final class of more powerful machines consists of nondeterministic automata. These are also often studied in automaton theory. Intuitively, the idea is that the automata do several things at the same time. From a mathematical viewpoint, the functions δ and S for the automaton are multiple-valued. In this way it is possible to define automata, for instance, which place an entry on top of the stack and simultaneously erase the current entry from the top of the stack. Then the rule is given that, for instance, if any possible sequence of choices from those specified by the multiple-valued functions leads to an acceptable final state when the tape has been read, then the tape is said to be *accepted*. It is known that a nondeterministic push-down store automaton can be designed which will accept any given context-free grammar.

★14-9. RECURSIVE FUNCTIONS

In the remainder of this chapter, we shall describe two other technical ideas which are closely related to the concepts of definability and computability discussed earlier. The first of these attempts to extend the concept of a definition by induction from Peano's axioms. Thus it attempts to answer the following question: What functions $f\colon \mathbf{P} \to \mathbf{P}$ are constructively definable from Peano's description of the positive integers as a unary algebra $[\mathbf{P},\sigma]$?

Simple Recursion. Consider the definitions in Sec. 1-7 of the functions $\sigma^m\colon r \mapsto m + r$ (add to m) and $p_m\colon r \mapsto mr$ (multiply by m). These definitions both had the following form:

R1. The value of $f(1) = f_1 \in \mathbf{P}$ was specified.
R2. $f(\sigma n)$ was defined as $h(f(n))$, where $h\colon \mathbf{P} \to \mathbf{P}$ was a specified, computable function.

Thus, in defining σ^m, we set $f_1 = \sigma m$ and $h = \sigma$; in defining p_m, we set $f_1 = m$ and $h = \sigma^m$.

In general, a function defined by any two conditions of the form of R1 and R2 is said to be defined by *simple recursion*, provided h in R2 has already been defined by simple recursion. Any such function is well defined, as the following result shows.

Theorem 13. *Let f: $\mathbf{P} \to \mathbf{P}$ and g: $\mathbf{P} \to \mathbf{P}$ both satisfy the same simple recursive definitions R1 and R2. Then $f = g$.*

Proof. Let S be the set of all $n \in P$ such that $f(n) = g(n)$. Then $1 \in P$ by R1, and $n \in S$ implies

$$f(\sigma n) = h(f(n)) = h(fg(n)) = g(\sigma n)$$

Hence, $n \in S$ implies $\sigma n \in S$. By Peano's induction axiom 3, it follows that $S = \mathbf{P}$; hence, by the definition of equality for functions, $f = g$.

The preceding theorem can be viewed in extended form as asserting the existence of a single axiom which is equivalent to all three of Peano's axioms, as stated in Sec. 1-7. This single axiom is the following.

Peano-Lawvere Axiom. Given any constant $c \in \mathbf{P}$ and any function h: $\mathbf{P} \to \mathbf{P}$, there exists a unique function f: $\mathbf{P} \to \mathbf{P}$ which satisfies R1 and R2.

To indicate the nature of the class of all recursive functions, we now consider the union $\mathcal{B}$ of the following three basic sets of functions:

(*i*) The set of constant functions $\phi(x_i, \ldots ,x_n) = m$ for some fixed $m \in \mathbf{P}$ and all $\mathbf{x} = (x_1, \ldots ,x_n) \in \mathbf{P}^n$.
(*ii*) The set of all (Cartesian) projections $\psi_i(x_1, \ldots ,x_n) = x_i (i \in \mathbf{n})$.
(*iii*) Peano's successor function σ: $\sigma(n) = n + 1$.

We extend the set $\mathcal{B}$ of basic recursive functions ϕ: $\mathbf{P}^n \to \mathbf{P}$, etc., by admitting as recursive also any function h: $\mathbf{P}^n \to \mathbf{P}$ of the form

(*iv*) $h(x_1, \ldots ,x_n) = f(g_1(x_1, \ldots ,x_n) \ldots ,g_k(x_i, \ldots ,x_n))$, where f and all g_i are recursive.

Specifically, we set $\mathcal{B}_0 = \mathcal{B}$, and let $\tilde{\mathcal{B}}_{w+1}$ include all functions of form (*iv*), with f and all g_i in $\bigcup_{i=0}^{w} \mathcal{B}_i$.

Finally, we let $\mathcal{B}_{w+1}$ include all functions which can be defined by a complicated generalization [Rogers, p. 6, (v)] to functions of $n + 2$ variables of conditions R1 and R2 defining simple recursive functions of one variable. The class of primitive recursive functions is then $\bigcup_{w=0}^{\infty} \mathcal{B}_w$.

A very deep theorem[1] asserts the class of all *recursive* functions $\phi\colon \mathbf{P}^n \to \mathbf{P} (n \in \mathbf{P})$ is the same as the class of all *computable* functions $\phi\colon \mathbf{P}^n \to \mathbf{P}$, by which we mean that there exists a *program* $\mathbf{A}_f$ which, for any given $\mathbf{n} \in \mathbf{P}$, will compute $\phi(n)$ on some Turing machine. (On a Turing machine, each integer $n \in \mathbf{P}$ is conventionally supposed to be designated by a string of $n + 1$ ones, which is, of course, very inefficient.)

The set of functions which can be computed in this way is, of course, countable, whereas the set of *all* functions $f\colon \mathbf{P} \to \mathbf{P}$ has the cardinality $\mathbf{c}$ of the continuum and is hence bijective to the sets $\mathbf{2}^{\mathbf{P}}$ and $\mathbf{R}$ discussed earlier in this chapter. There is indeed a very simple bijection between functions $f\colon \mathbf{P} \to \mathbf{P}$ and those sequences $\mathbf{x} \in \mathbf{2}^{\mathbf{P}}$ such that $0 \cdot x_1x_2x_3 \cdots \in \mathbf{R}$ is *not* dyadic. Namely, consider the sequence $x_0x_1x_2x_3 \cdots$ with $x_0 = 0$; its values constitute an *infinite* string of *finite* strings of 0's and 1's, in alteration. We let $f_x(n)$ be the length of the nth such string.

One may hope that the set of recursively definable functions $\phi \in \mathbf{P}^{\mathbf{P}}$ goes into the set of computable nondyadic $\mathbf{x}$ under this bijection. However, this also seems to be very hard to prove.

★14-10. NONCOMPUTABILITY; WORD PROBLEMS

The grammar and syntax rules of fully defined programming languages (see Sec. 14-4) can be regarded as *axioms* for corresponding (abstract) algebraic systems, whose meaningful formulas correspond to acceptable statements or "programs" of the language. The difficulties described in Secs. 14-8 and 14-9, concerning the design of "acceptors" for programming languages, are therefore related to difficulties in the foundations of mathematics, such as those which led to the concept of a recursive function described in Sec. 14-10.

We shall now elaborate further on these difficulties. We shall begin by defining a *recursively enumerable set* $S \subset \mathbf{P}$ of positive integers as the set of values (image or range) of a recursive function; that is, $S \subset \mathbf{P}$ is recursively enumerable if and only if, for some recursive $f\colon \mathbf{P} \to \mathbf{P}$, we have $S = \operatorname{Im} f$:

$$n \in S \qquad \text{means that } \exists\, m \in \mathbf{P} \text{ with } f(m) = n \tag{18}$$

Any computable subset S of $\mathbf{P}$ is recursively enumerable, together with its complement (which is computable). The computable subsets S

[1] See the books by Martin Davis and Hartley Rogers listed at the end of this chapter.

of $\mathbf{P}$ (with computable characteristic functions $e_s\colon \mathbf{P} \to \mathbf{2}$), moreover, form a *field* of sets, as is obvious from the formulas

$$e_{S \cap T} = e_S \wedge e_T \qquad e_{S \cup T} = e_S \vee e_T \qquad e_S = 1 - e_S$$

Unfortunately, recursively enumerable sets do not have such nice properties, though it is easy to show that the set union of two recursively enumerable sets $S = \text{Im } f$ and $T = \text{Im } g$ is $S \cup T = \text{Im } h$, where $h(2n - 1) = f(n)$ and $h(2n) = g(n)$ for all $n \in \mathbf{P}$.

Indeed, the complement of a recursively enumerable set need not be recursively enumerable. The difficulty (which is related to Theorem 8) can be illustrated by considering the specific function $\phi\colon \mathbf{P} \to \mathbf{P}$ defined recursively as follows: $\phi(n)$ is the remainder of $(n!)^{(n!)}$ after division by the $(n!)$th prime. How can one ever be sure that a given $r \in \mathbf{P}$ is *not* in the (presumably, very sparse) set Im ϕ? And, if one cannot be sure, how can one possibly enumerate $(\text{Im } \phi)'$?

The preceding difficulty is typical of the algebraic complexity of the rules for handling the *quantifiers* $\forall x$ (read, "for all x") and $\exists x$ (read "there exists an x"). Although quantifiers satisfy some algebraic rules, it seems to be impossible to determine in general whether or not two expressions involving quantifiers have the same meaning. One knows that $\forall x$, $x \in Q$ ("all x have property Q") is equivalent to $(\exists x \in Q')'$ ("there does not exist x not in Q"); hence one can systematically eliminate $\forall$ from expressions involving quantifiers. But there are no equally simple prescriptions for reducing to canonical form all expressions involving Boolean operations and quantifiers, Cartesian products, powers, etc., that is, for eliminating the redundancy of even the purest part of the logic of mathematics.

Word Problems. In general, problems of this kind (testing two algebraic expressions for equivalence, given certain equations or identities) are called *word problems*. For example, suppose we are given some finite set of equations such as

$$abc = adef \qquad abe = abc \qquad bdf = bde \tag{19}$$

It is natural to ask for a rule which will decide, in a finite number of steps, whether or not a string X in the symbols a, b, c, d, e, f can be derived from a string Y by a sequence of substitutions from the list (19). Presumably, this is equivalent to asking for a Turing machine which will accept $X \equiv Y$ when this is so, and reject it otherwise.

In earlier chapters, we have given solutions to a few such word problems, such as that for Boolean expressions. The word problem is also solvable for commutative groups with a finite number of generators

$a,b,c,\ldots$. However, it is a sad fact that the word problem is *not* solvable in (for example) noncommutative groups.[1] Neither is it solvable, in general, for semigroups.

This being the case, it is hardly to be expected that even if one had a complete set of rules for formal logic, one could find a decision procedure for deciding in a finite length of time whether or not a given symbolic theorem was true, or even whether or not two theorems were asserting the same thing.

The Halting Problem. As a final example of noncomputability, we mention the Halting Problem for Turing machines. This is the problem of finding a decision procedure which would enable one to determine in a finite number of steps, given any Turing machine T and premarked tape $\mathbf{t}$, whether or not the machine will ever make $\delta(s^j,a^j) = \text{HALT}$. Equivalently, the problem is to construct a Turing machine which will accept just those pairs $(T,\mathbf{t})$ which *will* come to HALT, and reject just those pairs $(T,\mathbf{t})$ which will not come to HALT. By an argument similar to Cantor's diagonal process, one can show that this problem has no solution; the Halting Problem for Turing machines is undecidable.[2]

For more thorough discussions of the profound question of what is computable, the reader is referred to the references at the end of this chapter.

EXERCISES C

1. Show that the finite-state grammar of Example 3 produces strings of letters of the form Wbb and $Wbbc$, where W is any string of c's and single b's (bb excluded) and no other string.
2. Show that the finite-state acceptor of Fig. 14-3 will accept input strings if and only if they have the form.
3. Define the following functions recursively, in terms of Peano's successor function $\sigma: n \mapsto n + 1$:
   (a) $f(n) = 3n$ (b) $g(n) = 3n^2$ (c) $h(n) = n^3$
   [*Hint:* $h(0) = 0$ and $h(n + 1) = h(n) + g(n) + f(n) + 1$.]
4. Define the function $n!$ recursively (in terms of σ).
5. Define recursively the function $q(n) = [\sqrt{n}]$ defined as the largest integer q such that $q^2 \leqq n$. Justify your answer. [*Hint:* Consider also $r(n)$, where $q(1) = 1$, $r(1) = 0$, and (1) if $r(n) < 2q + 1$, then $r(n + 1) = r(n) + 1$ and $q(n + 1) = q(n)$, (2) if $r(n) = 2q + 1$, then $r(n + 1) = 0$, and $q(n + 1) = q(n) + 1$.]
6. Show that the set $WFPS$ of all well-formed parentheses strings is defined in Backus normal form by

$$WFPS := \langle (\quad) \rangle \mid (\langle WFPS \rangle \langle WFPS \rangle) \mid (\langle WFPS \rangle) \mid WFPS(\quad)$$

[1] For a discussion of this fact, see W. Magnus, A. Karrass, and D. Solitar, "Combinatorial Group Theory," Interscience, 1966.
[2] See Minsky, Chaps. 8 and 9.

7. Construct a finite-state machine with input alphabet $A = \{``,"\}$ which will accept all well-formed parentheses strings and nothing else.

★8. Design a finite-state acceptor which will accept precisely those pairs $(m,n) \in \mathbf{P} \times \mathbf{P}$ such that $m \mid n$.

★9. Show that if a group G is generated by $a_1, \ldots, a_n$ and if $x^2 = 1$ for all $x \in G$, then G is an Abelian group of order 2^n for some $r \leqq n$.

★10. Show that the semigroup freely generated by a and b, subject to the defining relations $aba = b$ and $bab = a^r$, has $5r + 3$ elements. (R. C. Buck, *Am. Math. Monthly*, **75**:852–856 (1968).)

REFERENCES

CHOMSKY, NOAM: Formal Properties of Grammars in "Handbook of Mathematical Psychology," Wiley, 1963.

DAVIS, MARTIN: "Computability and Unsolvability," McGraw-Hill, 1958.

GINSBURG, S.: "The Mathematical Theory of Context-Free Languages," McGraw-Hill, 1966.

HALMOS, P. R.: "Naive Set Theory," Van Nostrand, 1960.

HAYS, DAVID G.: "Introduction to Computational Linguistics," Elsevier, 1967.

KLEENE, S. C.: "Mathematical Logic," Wiley, 1967.

MINSKY, M.: "Computation-Finite and Infinite Machines," Prentice-Hall, 1967.

ROGERS, HARTLEY, JR.: "Theory of Recursive Functions and Effective Computability," McGraw-Hill, 1968.

TRAKHTENBROT, B. A.: "Algorithms and Automatic Computing Machines," Heath, 1963.

Index

Abelian group, 210, 217ff.
Absorption laws, 6, 130
Accept, 195
Accessible, 59
Accumulator, 100
Action on set, 220
Acyclic, 61
Additive identity, 283
Adjacency matrix, 56
ALGOL, 102ff., 146ff.
Alphanumeric, 231
Alternating group, 223
Analog computers, 63
AND gate, 141ff.
AND-to-OR gate, 143
Antisymmetric, 36
Applicable, 88
Arithmetic expressions, 103, 105
Array, 72, 109, 150
Assembler language, 100, 102
Assignment statement, 107
Associate, 302
Associative laws, 5
  generalized, 25
Asymmetric, 36
Autocorrelation:
  function, 389
  theorem, 392
Automaton, 62ff.
Automorphism, 208
Axiom of Choice, 400
Axiom of Specification, 3

Backus normal form, 119, 140
Basis, 340
Begin, 107, 115
Bijection, 11
Binary algebra, 198ff.
Binary operation, 3, 197
Binary relation, 31
Binary symmetric channel, 232ff.
Bipartite graphs, 57
Bistable, 65, 141
Bit, 65
Block, 107, 115
Block codes, 238
Block diagrams, 141
Boolean algebra, 6, 129ff., 256
Boolean array, 151
Boolean lattice, 274
Boolean morphism, 158, 276
Boolean operation, 5
Boolean polynomial, 136
Boolean ring, 287, 297
Boolean subalgebra, 153
Bose-Chaudhuri-Hocquenghem codes, 352ff.

Cancellation law, 284
Cardinal numbers, 47, 397
Cartesian product, 17, 59
Cayley's theorem, 209
Center, 292, 296
Chain, 39, 57
Characteristic, 291
Characteristic function, 11
Characteristic polynomial, 373
Circuit, 58
Clock algebra, 49
Coding, 99
Codomain, 8
Common multiple, 5
Commutative, 199, 210
Commutative laws, 5
Commutative rings, 282
Commute, 14, 19
Compatible, 89, 90, 93
Compiler, 102, 125
Complement, 5
Complementary, 5
Complemented lattice, 275
Completely specified, 67
Complex, 228
Composite, left or right, 8, 9
Compound statement, 115
Computable, 404ff.
Computers, 63
Conditional statement, 147ff.
Conjugacy class, 226
Conjugation, 226
Connected, strongly, directed graph, 60
Consensus, 187ff.
Consistency, 7
Context-free, 410
Continuum, cardinality of, 401ff.
Coset leader, 245
Countable infinity, 399
Cover, 39ff., 73, 89, 184
Covering, 73, 89
Cycle, 57
Cyclic, 200ff., 215
Cyclotomic polynomial, 333

Decoding, 235, 238, 246
de Morgan's laws, 6, 130, 140
Difference codes, 369ff.
Difference equation, 369ff.
Digital computer, 63ff.
Dihedral group, 211
Dimension, 340
Direct product, 158, 209, 265
Direct sum, 59, 295
Directed graphs, 552
Discriminant, 338
Disjoint, 5, 155
Disjoint sum, 17

Disjoint union, 59
Disjunctive normal form, 155ff.
Distance, 60, 152, 239ff.
Distributive lattice, 154, 268
Distributive laws, 6, 24
Divide, 302
Division Algorithm, 26ff., 303, 319ff.
Domain, 8
Don't-care entries in state tables, 87
Dual, 38
Duality Principle, 38, 131, 258

Edge, 55
Elementary Abelian group, 219
Elementary group, 218
Encoding, 235, 238
Encoding matrix, 243
End, 107, 115
Endomorphism, 208
Epimorphism, 46, 207
Equivalence relations, 42, 44
Equivalent code, 250
Equivalent machines, 73
Equivalent states, 75
Error-correcting codes, 235ff.
Error-detecting codes, 235ff.
Euclidean algorithm, 305ff.
Euclidean domains, 303ff., 319
Euler graph, 57
Euler's ϕ-function, 349ff.
Evaluation map, 335, 343
Even, 223
Executable, 108
Exponentiation, 18
Extended formal power series, 394
Extension, 339, 362
  of fields, 339ff.

False, 147
Fewest literals, criterion of, 179
Fewest terms, criterion of, 179
Fibonacci equation, 373
Field, 282, 285
  of gradients, 288–290
  of quotients, 288
  of sets, 153
Finite fields, 348–351
Finite induction, 22
Finite sets, 15
Finite-state acceptor, 410ff.
Finite-state language, 410ff.
Finite-state machines, 63, 66
Flip-flops, 189ff.
For, 112
Formal derivative, 337, 377–378
Formal power series, 375–378
FORTRAN, 102–103
Free, 203
Free algebra, 159
Free Boolean algebra, 157–159
Free group, 216
Free monoid, 202–203
Free unary algebra, 50
Function, 8

Galois field, 359, 362
Galois group, 363
Gates, 141
Gating networks, 141ff.
Gaussian elimination, 310–313
Gaussian integers, 304
Generated by elements, 154
Generating function, 305, 375
Generating matrix (*see* Encoding matrix)
Generators, 125
Grammar, 117
Graphs, 32, 52ff.
Greatest, 39, 260
Greatest common divisor, 305
Greatest lower bound, 41, 155, 260
Group, 203, 212
Group codes, 234, 244ff.

Halting problem, 422
Hamming code, 252ff.
Hypercube, 249, 254
(*See also* n-cube)

Ideals, 275, 298
Idempotent, 5, 14, 130, 205, 285, 296
Identifier, 106, 119
Identity, 137, 198
right or left, 199
Identity function, 9
If ··· then ··· else, 146
Image, 8
Implicant, 180
Incidence, 56
Inclusion, 2, 133
Incompletely specified, 87
Independent trials, 233
Indeterminate, 316
Induction, 22–24
Induction axiom, 21
Injection, 10
Injective, 10
Inner automorphism, 226
Input, 64, 66, 67, 116
Input devices, 63
Insertion, 44
Integral domain, 284ff.
Internal state, 66
Intersection, 4
Interval, 154, 267
Inverse, 12, 203
Inverter, 141–144
Invertible, 203, 205
Involution, 130
Irreducible, 320
Irredundant, 181
Irreflexive, 36
Isomorphism, 46, 207
Isotonicity, 135

Join, 130, 260

Kernel, 276, 298
of morphism, 227

Labeled directed graphs, 61, 141
Labels, 148ff.
Lagrange interpolation formula, 336–337
Lagrange's theorem, 224
Language, 118
Lattices, 154, 256, 257, 261
Boolean, 274
complemented, 274
morphism of, 275
Least, 39, 260
Least upper bound, 41, 155, 260, 268
Left inverse, 12
Left zero, 199
Length, 61
Links, 55
Literal, 172
Logic design, 171
Logical operator, 147
Loop, 52, 55, 142
Loop free, 52

Machine language, 100, 102
Majority rule code, 236
Mapping diagram, 9
Maximal, 40, 260
Maximal ideals, 309–310, 347
Maximal period sequences, 387–388
Meet, 130, 260
Memory, 63, 64
Merging states, 91
Metalanguage, 117–119
Minimal, 40, 260
Minimal state machine, 73, 80, 95
Minimal states, 73, 80
Minimization, 78, 90
Modular lattice, 270
Modular law, 6

Modular number, 45, 46
Monic polynomial, 320
Monoids, 36, 198, 265
Monomorphism, 46, 207
Morphism, 46, 49, 74, 145, 158, 159, 202, 206ff.
  of lattices, 275
  of machines, 74
  of rings, 293
  of semigroups, 206–207

N, 2
n-cube, 249, 254
NAND gate, 175–178
Negation, 32
Next-state function, 66
Nim, game of, 62
Node, 35, 52
Noncomputability, 420
NOR gate, 176
Normal subgroup, 226–229

Odd, 223
One-one, 10
Onto, bijection, 11
Operations:
  binary, 3
  unary, 4
Optimal path, 162, 168
Optimality Principle, 163–164
Optimization, 162
OR gate, 141ff.
Order, 133, 201
Order of a polynomial, 386
Ordered pair, 3
Output, 64, 66, 67, 116
Output devices, 63
Output function, 66

P, 2
Parity check, 69
  machines, 69
  matrix, 251
Parsing tree, 124
Partial function, 13
Partial order, 133
Partial ordering, 37, 135, 258
Partially ordered set (POSET), 38–41, 257–259
Partition, 5, 42–44, 271
Peano algebra, 48
Peano axioms, 20
Peano-Lawvere axiom, 419–420
Peano's successor function, 8
Perfect code, 249
Periodic, 372, 382
Permutable, 14
Permutation, 221
Permutation matrix, 35
Phrase-structure grammar, 408
Phrase-structure language, 408
Pigeonhole principle, 26, 28
Polynomial codes, 321–327
Polynomial function, 334
Polynomials, 316
Poset, 38–41, 257–259
Postulates, 131
  for groups, 204, 213
Power, 16
  of the continuum, 399
  set, 3
Precedence, 105, 122
Prime ideals, 309–310, 347
Prime implicant tables, 184
Prime implicants, 180, 182
Primitive, 325, 362
Primitive polynomials, 325, 362
Principal ideals, 299
Product, 16
Product-of-sums, 172
Product relation, 37
Product term, 172
Production rule, 118, 408

Programming, 99
Programs, 66, 99–101
Projection, 14, 18
Proper ideals, 298
Proposition, 145, 148
Push-down store, 122ff., 414ff.

Q, 1
Quasi-ordering, 259
Quasi-perfect code, 249
Quaternion, 351
Quine-McCluskey technique, 183
Quoset, 259
Quotient rings, 298

R, 1–2
Radar, 367
Recurrent sequence, 382ff.
Recursive function, 418–420
Recursively enumerable, 420
Reduced echelon form, 250
Reducible, 320
Reducible matrix, 59
Reflexive, 2, 36
Relation, binary, 31ff.
Relation matrix, 33
Relative complement, 155
Restriction, 44
Richard paradox, 404
Right inverse, 12
Right zero, 199
Ring of sets, 265
Rings, 281
Root fields, 359ff.
Root tree, 61
Roots of unity, 331
Rounding, 121

Semidistributive law, 268
Semigroup, 198
Semilattice, 263
Sequential machine, 189ff.
Sequential network, 191
Sets, 1
Shift operators, 373–374
Shift register, 327, 371
Shortest path, 60
Simple circuit, 52, 58
Simple extension, 341ff.
Simple field extension, 341ff.
Simple path, 52, 58
Simple recursion, 419
Simple transcendental extension, 344
Simply ordered, 39
Simultaneous linear equation, 310
Singular matrix, 311
Span, 340
Splitting field, 359
Square-free, 338
Stack, 122ff., 416
State diagram, 68
State table, 68
State vector, 66
States, 63
Step, 112
String, 25, 66, 72
Subalgebra, 265
Subfield, 291
Subgroup, 214
Sublattice, 265
Submonoid, 204, 209
Subring, 290
Subsets, 1
Subsumes, 181
Substitution property, 47
Successor function, 8, 21
Successor relation, 53
Sum, 16, 59
Sum-of-products, 172
Sum term, 172
Surjection, 11
Surjective, 10

Symmetric, 36, 178
Symmetric difference, 133, 140
Symmetric group, 211
Syntax, 118

Table of combinations, 143–144
Tautology, 145
Terminals, 408
Transcendental extension, 344–346
Transitive, 2, 36
Transposition, 221, 223
Trees, 61
True, 147
Turing computability, 405ff.
Turing machines, 83ff.
Type declaration, 106, 107

Unary algebra, 48ff.
Unary operation, 4
Union, 4
Unique Factorization Theorem, 307, 329
Unital subgroup, 291
Unital subring, 292
Units, 319
Universal, 202
Universal bounds, 6, 39, 130
Universal relation, 60
Universality, 18

Valuation, 303, 319
Vector space, 340–341
Venn diagram, 4, 138
Vocabulary, 118

Weight, 239–241
Word problem, 137, 421

Z, 1
Zero, 199, 283
  of a polynomial, 336